The Eel

The Eel

Third edition

F.-W. Tesch

With contributions from
P. Bartsch, R. Berg, O. Gabriel, I.W. Henderson,
A. Kamstra, M. Kloppmann, L.W. Reimer, K. Söffker
and T. Wirth

Translated from the German by R.J. White

Edited by J.E. Thorpe

Blackwell
Science

The Fisheries Society
of the British Isles

Editorial Offices:
Blackwell Publishing Ltd, 9600 Garsington Road, Oxford OX4 2DQ, UK
Tel: +44 (0)1865 776868
Iowa State Press, a Blackwell Publishing Company, 2121 State Avenue, Ames, Iowa 50014-8300, USA
Tel: +1 515 292 0140
Blackwell Publishing Asia, 550 Swanston Street, Carlton, Victoria 3053, Australia
Tel: +61 (0)3 8359 1011

Library of Congress Cataloging-in-Publication Data
Tesch, Friedrich-Wilhelm.
[Aal. English]
The eel / Friedrich-Wilhelm Tesch ; translated from the German by R.J.White ; edited by John E. Thorpe.-- 3rd ed.
p. cm.
Includes bibliographical references and index.
ISBN 0-632-06389-0 (alk. paper)
1. Anguilla (Fish) 2. Eel fisheries. I. Thorpe, J. E. (John E.)
II. Title.

QL638.A55T4713 2003
597′.432--dc21 2003008640

ISBN 0-632-06389-0

A catalogue record for this title is available from the British Library

Set in Times and produced
by Gray Publishing, Tunbridge Wells, Kent
Printed and bound in the UK using acid-free paper
by MPG Books, Bodmin, Cornwall

For further information on Blackwell Publishing, visit our website:
www.blackwellpublishing.com

Contents

Contributors

F.-W. Tesch	Gartenstrasse 1a, 22869 Schenefeld, Germany
P. Bartsch	Naturhistorisches Forschungsinstitut Museum für Naturkunde, Invalidenstrasse 43, 10115 Berlin, Germany (*Chapter 1.2*)
R. Berg	Fischereiforschungsstelle des Landes Bad-Württemberg, Mühlesch 13, 88065 Langenargen, Germany (*Chapter 7.4.4*)
O. Gabriel (deceased)	Bundesforschungsanstalt für Fischerei Institut für Fischeretechnik, Palmaille 9, 22767 Hamburg, Germany (*Chapter 5*)
I.W. Henderson	University of Sheffield, Animal and Plant Sciences, Alfred Denny Building, Western Bank, Sheffield S10 2TN, England (*Chapter 1.8*)
A. Kamstra	Netherlands Institute for Fisheries Research, Haringkade 1, PO Box 68, NL-1970 AB IJmuiden, The Netherlands (*Chapter 6*)
M. Kloppmann	Heidblick 23, 21149 Hamburg, Germany (*Chapter 1*)
L.W. Reimer	Am Bahnhof Mindenstadt 4, 32423 Minden, Germany (*Chapter 7*)
K. Söffker	Heinrich Lehmann Strasse 2, 31542 Bad Nenndorf, Germany (*Chapter 7.4.5*)
T. Wendt	Ernst-Mittelbach-Steig, D-22455 Hamburg, Germany (*Chapter 5*)
T. Wirth	Max-Planck Institut für Infektionsbiologie, Schumannstrasse 21/22, 10117 Berlin, Germany (*Chapter 2.5*)
R.J. White	320 12th Avenue North, Edmonds, Washington 98020, USA
J.E. Thorpe	Institute of Biomedical and Life Sciences, University of Glasgow, Glasgow G12 8QQ, Scotland

Preface

Demand for the German editions and for the English translation of this book on eels has made it necessary to publish a further English edition. It is based on the third German edition (1999) and contains further updatings. The unusually high demand for a book specialising on a single genus is due to the eel's scientific and gastronomic popularity in large areas of Europe and in Japan, and the scientific demand has expanded greatly since first publication of the book in 1973, especially in East Asia and Australia/New Zealand. Subject areas like continental ecology and aquaculture contributed to this. These have drawn in with them other topics like oceanic ecology, genetics, and parasitology: this edition contains a completely new section for genetics.

With regard to the oceans, meteorology and sea currents expanded interest in eels to specialists of those subjects. In the 1970s, the author was even asked whether knowledge of eel occurrence in the various parts of the oceans could also contribute to exploration of ocean currents. In North America, until then interest in eels was limited to the areas of its continental occurrence along the east coast. Besides the economics of the eel, its importance as an oceanographic indicator has expanded interest in this animal among other circles, especially in North America and East Asia

Accordingly, it seemed appropriate to consult with other colleagues in certain subject areas of this book, so specialists in endocrinology, genetics, aquaculture, parasitology and toxicology, among others, were brought in. The descriptions of fishing techniques also benefited from collaboration of a specialist. It became clear that almost no other fish species is caught by such a diverse range of gear, so this chapter represents almost a cross-section of general fish catching technology.

Recent population decline of some eel species created utmost economic and scientific interest. This might be equated with the diminution of other economically important food fishes and blamed on overfishing. But because the critical times of the eel's highest mortality extend over longer periods than those of other fishes, and are primarily at sea, this decline cannot be explained so simply. Therefore, in the present English edition the marine phases receive no less attention than do the continental phases of the life cycle. The data on the marine biology of the eel are increasingly important also because there are considerable deficiencies in knowledge of the Indopacific species relative to the Atlantic species, even though in recent years research activities on eels in the Pacific Ocean have far exceeded those in the Atlantic.

Sincere thanks are due to the Fisheries Society of the British Isles whose generous financial support enabled the translation of this book to go ahead.

Finally, I must express my grateful thanks to all those who have helped in the production of this book. I feel particularly indebted to the translator, the editor and the publisher, who have worked so conscientiously to ensure the appearance of the second English edition.

1
Body structure and functions

Updated by M. Kloppmann

1.1 Introduction

Eels (Anguilliformes) are always relatively elongated fishes. They have no ventral fin and no pelvic girdle. Some groups (e.g. Muraenoidei) exhibit reduced pectoral fins. The number of vertebrae and of myomeres can vary between 105 in some Congridae up to >300 in the meso- and bathy-pelagic Nemichthydae of the deep sea (Nielsen and Smith, 1978). The unpaired fins are confluent, and relicts of the caudal fin are distinguishable at the tail end, at least internally early in ontogenesis (Fig. 1.4). The fin rays (Lepidotrichia) always correspond functionally and numerically exactly with the pterygiophores the elements supporting their endoskeletal epineural. In accordance with its elongated body form, the eel has a rather narrow head that helps hiding in sand, mud and narrow holes. Within the head, the gill apparatus has to be accommodated to the elongated form. From its normal position in fishes below the skull it is displaced backwards almost to behind the skull. The gill construction is, therefore, very long; the pectoral girdle is disconnected from the skull and the post-temporal is reduced. There are no gillrakers and no mesocoracoid in the pectoral girdle. In particular, the skull shows a number of reductions and characteristic peculiarities that are described below. The alimentary tract of the Anguilliformes probably always lacks pyloric appendices (Robins, 1989). The female gonads have no separate outlet and the eggs are expelled through the abdominal pore. As far as we know, the Anguilliformes are monocyclic, which means that the parents die after the first spawning. The strange leptocephalus larvae of the Anguilliformes always have pectoral fins, even in those groups where the adults have none. In early developmental stages they also have a rounded caudal fin that develops in connection with the dorsal and anal fin and which shows no caudal fin peaks. A rostral commissure of the lateral line always exists in young developmental stages and persists usually from larval to adult stages.

From a physiological point of view the eel is a particularly popular experimental animal. This is due not only to its extremely marked resistance to experimental conditions, but also to many distinctive characteristics, namely: its hol-euryhaline osmoregulatory capacities; its phases of differential activity and behaviour patterns; its multistage metamorphosis during ontogeny; its great endurance and ability to navigate during migration; and lastly, even its unusual body shape. The range of publications in these fields of study increased greatly over the past few decades. In the morphological and physiological parts of this chapter, particular reference will be made to those publications that are in some way connected with the ecology of this interesting fish.

1.2 Skeleton

1.2.1 Skull (updated and revised by P. Bartsch)

The structure of the eel's skull will be described using Matsui and Takai's (1959) clear description of the skull of the Japanese eel (*Anguilla japonica*) (Figs 1.1 and 1.2). The skull and other parts of the adult eel's (*A. anguilla*) skeleton, as well as that of other Anguilliformes (*Muraenesox cinereus*), were described adequately, and partly compared, by Törlitz (1922) and Takai (1959). Smith and Castle (1972) made similar studies on *Anguilla*, *Neoconger vermiformes*, *Moringua edwardsi* and *Phytonichthys*, McCosker (1977) on several Ophichthidae, and Smith (1989a) on *A. rostrata*. Detailed monographs on the develop-

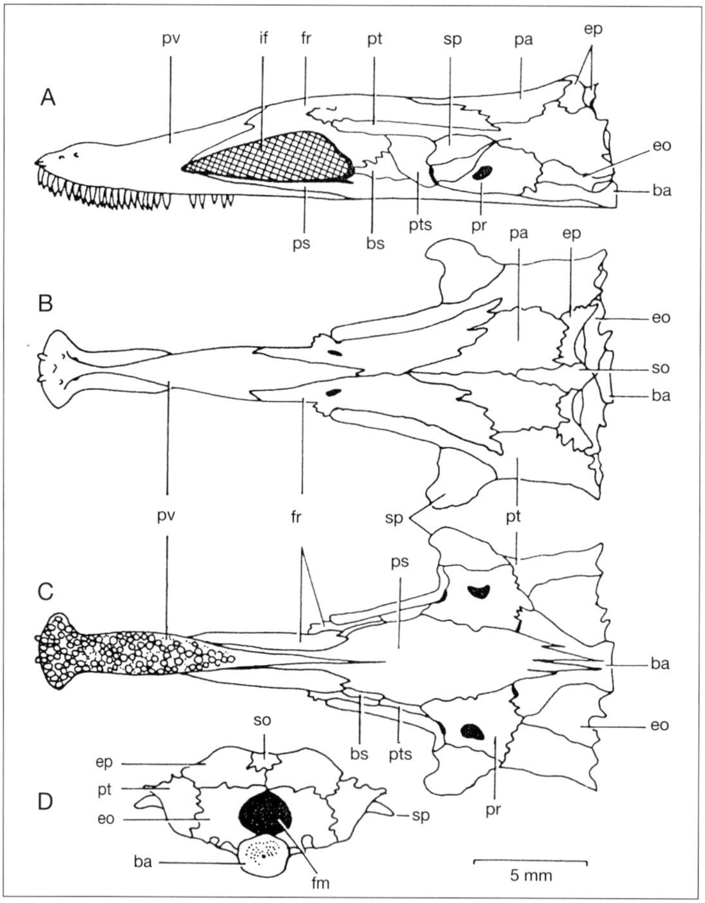

A From the side
B From above
C From below
D From behind

ba Basioccipital
bs Basisphenoid
eo Exoccipital
ep Epiotic
fm Foramen magnum
fr Frontal
if Interorbital window
pa Parietal
pr Prootic
ps Parasphenoid
pt Pterotic
pts Pterosphenoid
pv Premaxillo-ethmovomerine bloc
so Supraoccipital
sp (Auto)Sphenotic

Fig. 1.1 Skull (Neurocranium and upper exocranium) of *A. japonica* (after Matsui and Takai, 1959, modified)

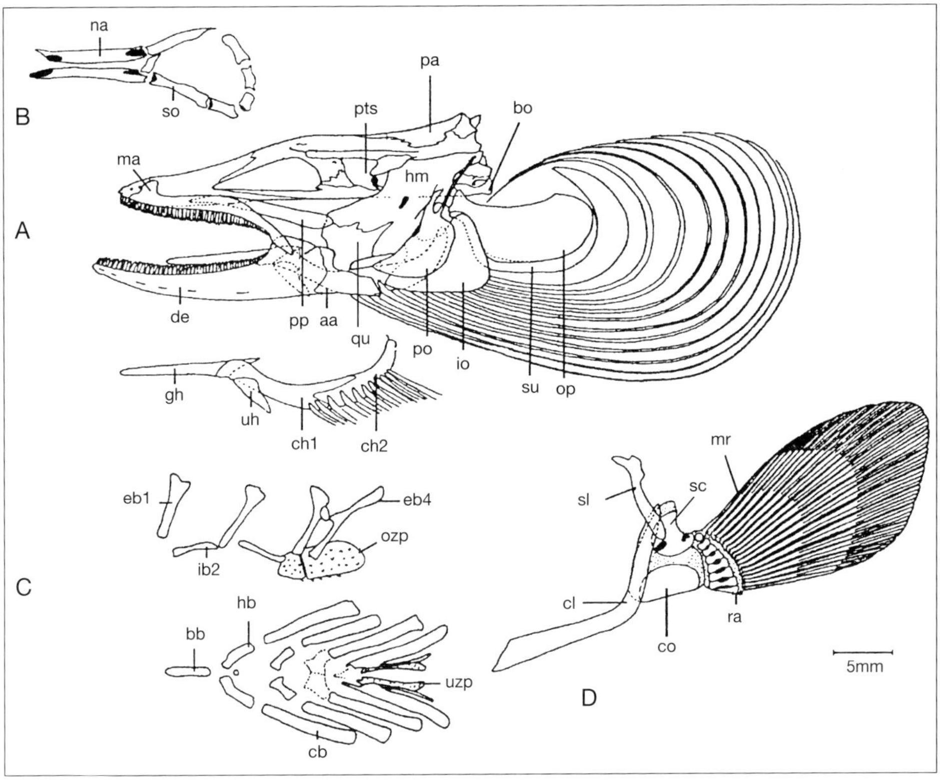

Fig. 1.2 Cranial skeleton and pectoral girdle of the eel (partly after Matsui and Takai, 1959 and Smith, 1989)

A Cranial skeleton with suspensorium and jaws of *A. japonica*
B Ossifications of lateral line organs (*A. anguilla*)
C Inferior hyoid arch of *A. japonica*
D Pectoral girdle

aa Angulo-Retro-Articular
bb Basibranchial
bo Basioccipital
br Branchiostegal rays
cb Ceratobranchials
ch1/ch2 Ceratohyal 1 and 2
cl Cleithrum
co Coracoid
de Dental
eb1 Epibranchial 1
eb4 Epibranchial 4
gh Glossohyal
hb Hyperbranchial
hm Hyomandibular
ma Maxilla
mr Marginal pectoral ray
Na Nasal
op Operculum
ozp Upper dental plate
pa Parietal
po Preoperculum
pp (Ecto-)Pterygoid
pts Pterosphenoid
qu Quadrate
ra Radial
sc Scapula
sl Supracleithrum
so Supraoccipital
su Suboperculum
uh Urohyal
uzp Lower dental plate

ment of the skull of *A. anguilla* larvae were published by Norman (1926) and, on the congrid larva of *Ariosoma balearicum* by Hulet (1987). The skull of the European eel leptocephalus (Fig. 1.3) is quite different in its components and proportions from that of the adults (Fig. 1.1). Also, the proportions of the skulls differ between the two ecological varieties of broad- and narrow-headed European eels (Törlitz, 1922; see Section 3.3.1.4).

Generally, the anguilliform skull differs significantly from that of other groups of genuine bony fish (Teleostei). In the upper jaw a stable, fused bone is formed that is derived from the dentated upper bony elements – the premaxillary, the vomer, and the so-called mesethmoid bone. The latter in eels does not seem, at least in parts, to be a cartilaginously

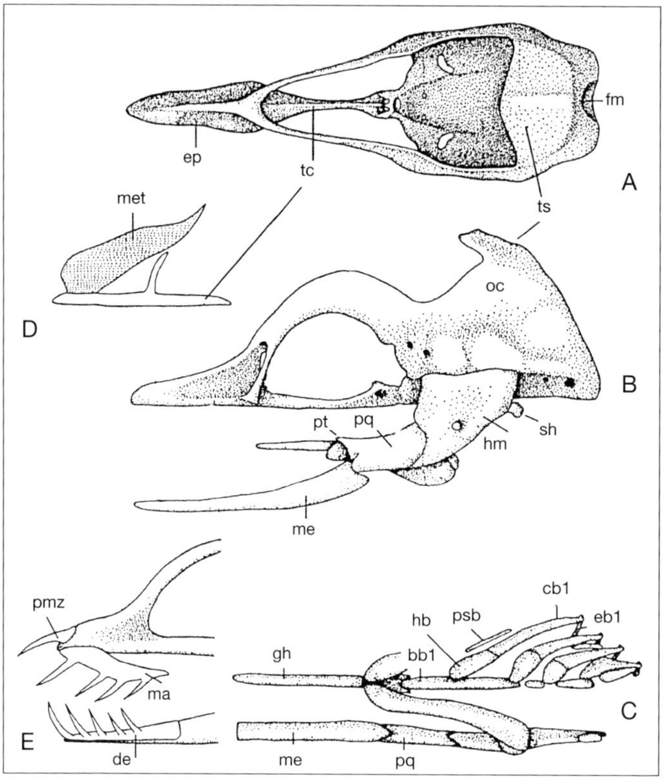

A Chrondrocranium without branchial skeleton (total length of the larva: 31 mm)
B Chrondrocranium from the side
C Chrondrocranium visceral skeleton from below
D Rostral region with mesethmoid bone (total length of the larva: 70 mm)
E Teeth of an early larva (total length of the larva: 11 mm)

bb1 Basibranchial 1
cb1 Ceratobranchial 1
de Dentary
eb1 Epibranchial 1
ep Epiotic
fm Foramen magnum
gh Glossohyal
hb Hyperbranchial
hm Hyomandibular
ma Maxilla
me Meckel's cartiledge
met Mesethmoid
oc Ear capsule
pmz Premaxillar tooth
pq Palatoquadrate
psb Pseudobranchial arch
pt Pterygoid process
sh Stylohyal
tc Trabecula communis
ts Synotic tectum

Fig. 1.3 Cranial skeleton of different stages of leptocephalus larvae (after Norman, 1926)

preformed bone. More probably it originates as an independent, immediate ossification (Norman, 1926: *A. anguilla*) or, it could arise from a ventral excrescence of a covering-bony dermethmoid (Leiby, 1979: *Myrophis punctatus*, Ophichthidae), following the disappearance of the mesethmoid cartilage of the leptocephalus (see also Jollie, 1986; Patterson, 1975, 1977: Teleostei). This 'premaxillo-ethmo-vomerine-block' (Fig. 1.1) limits the mobility of the bordering elements of the upper jaw. It seems to assist an excellent grasping and holding ability (see also Gregory, 1933), which corresponds with a considerable mass of the adductor musculature of the jaw and its extension to the vault of the cranium.

The maxilla in the upper jaw articulates movably with the ethmoidal region. Inwards, it forms a broad flank, leaning on the 'pterygoid' or 'palatopterygoid' bone. However, considering position and genesis of this bone, it seems to represent the ectopterygoid of other groups of the Teleostei only; its cartilaginous processus pterygoideus of the palatoquadratum is reduced or not developed at all, early in ontogeny.

Compared, for example, with the Muraenidae, the mouth opening of the Anguillidae is comparatively short and the suspensorium (movable suspension of the upper jaw on the neurocranium by the hyomandibula, Fig. 1.2) is tilted forward considerably. The hyomandibula articulates with the ear capsule of the neurocranium by means of a rather elongated hinge joint; the joint pit is situated in the pteroticum and is, in the forepart, considerably extended, roundish, into the autosphenoticum. On the ventral side the hyomandibula is connected with the quadratum (Fig. 1.2) by a stable, closely toothed suture. Normally, this connection is mediated by an independent bone, the sympleticum. In the eel this bone cannot be distinguished. But, the eel larva has a distinct cartilaginous processus sympleticus of the hyomandibula. Therefore, it is suggested that it has been ossified continually together with the hyomandibula or, connected with the quadratum (Leiby, 1981). The lower elements of the hyoid arch are suspended on the inner side of the hyomandibula by a ligament only and not by a separate bone that is present in actinopterygians showing a rod-shaped stylo- or interhyal (Fig. 1.2). In the Ophichthidae, a similar element consisting of cartilaginous matter supposedly occurs in its leptocephalus (Leiby, 1981). The ceratohyalia 1 and 2 (the latter sometimes called epihyal) constitute a paired, robust, rounded bone element that bears about 10-11 branchiostegal rays. In the forepart, the ceratohyalia of both sides are connected by an elongated unpaired geniohyal (basihyal). Ventrally, between the hyoid arch branches, there is a urohyale, which is triangular in side view and flat on the lower side. It is embedded in the connective tissue of the paired retractor muscles (Mm. sternohyoidei) of the hyoid arch (Kusaka, 1973, 1975; Arratia and Schultze, 1990).

The articulare, ossifying in the posterior section of the Meckel's cartilage, together with the quadratum, forms the jaw articulation. For the major part, the lower jaw branch is occupied by the extensive dentary that surrounds the Meckel's cartilage and its substituting ossification ('mento- and corono-Meckel's' bone). In the grown eel, neither a separate angular nor a retroarticular is visible. They generally fuse to form a uniform bone element, first the angulare with the retroarticulare (cf. Nelson, 1973 in the Elopiformes; Leiby, 1981 in the Ophichthidae), and fuse later on in ontogeny with the articulare. The mandibular lateral line canal runs enclosed in the dentary, opening outwards by several pores.

As in other teleost fishes, the upper skull has an extraordinarily complicated structure. It is composed of neurocranial elements inserted into each other (endoskeleton) and covering bony (exoskeleton) elements. The narrow base of the skull is supported essentially by the vomer part, which is elongated caudally, and by the parasphenoid (covering bony elements of the palatal cover, Fig. 1.1). In the backward cephalic area, this is indented with the basioccipital, which is the ossification of the hindmost base of the neurocranium. A small basisphenoid sits rostrally on the parasphenoid forming the bony backward limitation of the membranic interorbital septum, by a downward extending appendix of the frontal. A separate opisthoticum or intercalar is absent in adult eels at least. On the other hand, the pteroticum extends far behind and forms, with the exoccipital, the sharp outside edge of the backward part of the head. In *Anguilla*, lacrimal, nasal, suborbitals and postorbitals, are represented merely by thin shell-shaped lateral line canal ossifications, which are situated superficially in the connective tissue; they are omitted in many representations of the skeleton (Fig. 1.2B).

In the Anguillidae, the ossifications of the gill cover are quite large and very completely developed which contrasts with most other families of the Anguilliformes. The preopercular is fixed by connective tissue with the backward outside edge of the hyomandibular and surrounds a great opening (foramen) for the ramus hyoideus of the facialis nerve. The movable operculum articulates with the opercular process of the hyomandibular by a comprehensive socket of a joint. A narrow falciform suboperculum and a large-surface interoperculum complete the ossified operculum to the lower and to the frontal side. In its main area, however, the gill cover membrane, which in eels exhibits a largely expanded branchial space caudally, is supported by the branchiostegal rays (Fig. 1.2A).

The gill arches appear considerably more flexible than in other bony fish and they provide essential assistance in the production of positive and negative pressure during uptake of food (Alexander, 1970). Also, the gill arch elements are rather completely formed in the Anguillidae except the fifth arch that consists of ceratobranchials only supporting the lower pharyngeal tooth plates. The third and fourth epibranchials bear the corresponding upper tooth plates. These form a simple not spectacular device for ingestion at the entrance of the oesophagus (Fig. 1.2C). The fine and pointed conical teeth do not imply a function of food processing as known from the pharyngeal tooth apparatus of cyprinids and of many acanthopterygians (Nelson, 1969; Lauder and Liem, 1983; for a general view). A detailed analysis of the gill arch skeleton and of the appertaining branchial musculature is provided by Nelson (1966, 1967). These studies have also displayed an anagenetic sequence of progressing reduction from *Conger marginatus* (Congridae) through *A. rostrata* (Anguillidae) and *M. javanica* (Ophichthidae), *Kaupichthys diadonotus* (Clopsidae), *Uropterygius knighti* to *Gymnothorax petelli* (Muraenidae).

1.2.2 Vertebral column

The vertebral column in eels (Anguilliformes) is particularly interesting from the morphological, functional and systematic points of view. Hardly any vertebrate order is as polymorphic in this respect as are eels. Often it is used for species determination that is favoured by radiographic determination; even within the family of Anguillidae, the number of vertebrae is one of the most important diagnostic features at the species level (Sections 2.2 and 2.3).

There are only a few comparative studies on the morphology of the vertebral column in the various species of *Anguilla*. But additional groups of Anguilliformes have provided comparative results on the osteology, such as Muraenesocidae (Takai, 1959), Congridae (Asano, 1962; Smith, 1989b) and *A. japonica* (Matsui and Takai, 1959), *A. rostrata*, *N. vermiformis*, *M. edwardsi* and *Phythonichthys* sp. (Smith and Castle, 1972), as well as Ophichthidae (McCosker, 1977).

The vertebral column of *Anguilla japonica* (Fig. 1.4), which has a similar number of vertebrae to the European eel (*A. anguilla*), but more than the American eel (*A. rostrata*) (Table 1.1), is subdivided as follows: total number of vertebrae 116; the 44th is the last abdominal vertebra; the 38th is situated near the anus; pleural ribs occur on the 7th to 38th vertebrae; the haemal arches begin near the region of the 45th vertebra, which then is the first caudal vertebra (Fig. 1.4); dorsal intermuscular bones (epineurals) occur on the 1st to the 86th vertebrae, ventral intermuscular bones (epipleurals) occur on the 38th to the 86th vertebrae. According to Smith and Castle (1972; see also Patterson and Johnson, 1995), the distribution of epipleurals in *A. rostrata* is slightly different, in that they occur on the 34th to the 87th vertebrae. The epineurals of the first vertebrae are always grown together with the neural arches. Moreover, the neural spines of these vertebrae, in the longitudinal axis, are strongly extended and close together, which is a normal phenomenon in the Anguillids. The axial system of the skeleton has to be considered as closely connected

A Lateral view, 1st to 5th vertebrae
B Ventral view, 1st to 5th vertebrae
C 1st vertebra anterior view view
D 5th vertebra anterior view
E last abdominal and first two caudal vertebrae (44th, 45th and 46th vertebrae)
F 44th vertebra anterior view
G 45th vertebra anterior view
H 46th vertebra anterior view
I Caudal skeleton

ar Anal fin ray
ce Centrum
cr Caudal fin rays
dr Dorsal fin ray
el Epipleural inter muscular bone
en Epineural inter muscular bone
ha Haemal arch
hp Haemal canal
hs Haemal spine
hy Hypurals
na Neural arch
np Neural canal
ns Neural spine
pa Parapophysis
rad Distal radial
rap Proximal radial
un Uroneural

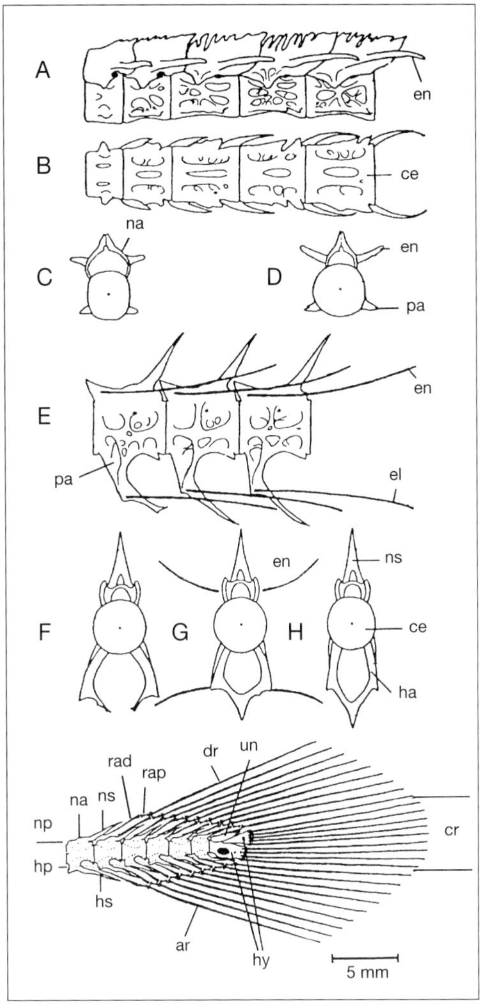

Fig. 1.4 Spinal column of *A. japonica* (after Matsui and Takai, 1959)

functionally with myosepta and the segmental musculature (see Section 1.3.5). The number of muscle segments corresponds approximately with the number of vertebrae, the high number of these segments and the relatively long body expressing themselves in the characteristic high amplitude winding anguilliform movement, which runs uniformly along the whole body (Lindsey, 1978). Also, the eel has S-shaped transverse walls of connective tissue of the body muscle segments showing three-dimensional foldings to anterior and posterior bags; this arrangement is typical for primary aquatic vertebrates with jaws (Gnathostomata). The 'intramuscular' bones, epineuralia and epipleuralia, are embedded in corresponding collagen filaments of the myosepta (Gemballa, 1995).

Finally, it should be mentioned that the skeleton of the vertebrates is never an entirely static system, but alters continuously throughout life and with growth. Lopez *et al.* (1970) have determined the amount of crystalline apatite and amorphous calcium phosphate in the bone of the vertebral column of the European eel. With increasing maturity of the gonads, decalcification takes place resulting in a great decrease in the amount of amorphous calcium phosphate. Lopez (1970) has studied the bone structure. Deformities of the vertebral column are described in Section 7.4.1.

1.2.3 Pectoral girdle and fins

Lack of connection of the pectoral girdle (Fig. 1.2D) with the skull and a reduced posttemporal are specialities that distinguish eels from other fishes; at most, connective tissue provides some association with the vertebral column (Berg, 1958). In comparison with other Anguilliformes, such as *N. vermiformes*, *M. edwardsi* and *Phytonichthys* sp. the eel (*A. rostrata*) has a relatively small supracleithrum. But the cleithrum is large, looks like a boomerang and extends cranially. It provides the base for the endoskeletal pectoral girdle with the bony scapula and coracoid as well as seven bony radialia.

The unpaired fins, the dorsal and anal fin, are confluent with the tail fin, which really exists (Fig. 1.4). The caudal section of the vertebral column (the two consolidated ural vertebrae) clearly shows two hypural elements, which can be considered as the result of fusion of a higher number of hypuralia during larval development. These endoskeletal elements have nine unequivocal caudal finrays that are not discernible because they are included in the dorsal and caudal fin arrangement. Also, there is one uroneural, connected dorsolaterally. Therefore, one may deduce that the strongly declined symmetrical form of the caudal fin of eels may be related to an originally homocercal tail of the teleosts (Whitehouse, 1910; Schmalhausen, 1913; Monod, 1968).

The eel's only paired fins are the pectorals (pectoralis), which do not differ greatly from those of many other species of bony fish. The fin area of the pectoralis is supported by branched and grouped bony fin rays (lepidotrichia); the rays of the unpaired fins are undivided and not branched. However, the pectorals are interesting because of their change in shape during the later phases of development in adults. While the so-called yellow eel has relatively wide, spoon-shaped pectoral fins, these become long and pointed (Fig. 1.5) shortly before the gonads mature. Furthermore, differences exist between the paired fins of male and female Japanese eel, *A. japonica*: the pectoral fin in the female is shorter and more rounded than that of the male (Matsui, 1952).

1.3 Skin and musculature

1.3.1 Structure and function of the skin

Eels survive in many diverse, and often harsh environments. Such an ability stems, at least in part, from the possession of a tough durable integument. Both the epidermis, and especially the corium, are thick. In a 20 cm long eel the total thickness of the skin was 0.15–0.18 mm and in a 56 cm long eel 0.50–0.53 mm (Hebrank, 1980). Jakubowski (1960b) compared the skin of seven teleost species, among them the eel, the flounder (*Pleuronectes flesus*) and the weather fish (*Misgurnus fossilis*). Only *M. fossilis* had an epidermis thicker than that of the eel. The epidermis of the eel is 0.263 mm thick, while that of the flounder

A Eels in advanced stages of maturity
B Yellow eels

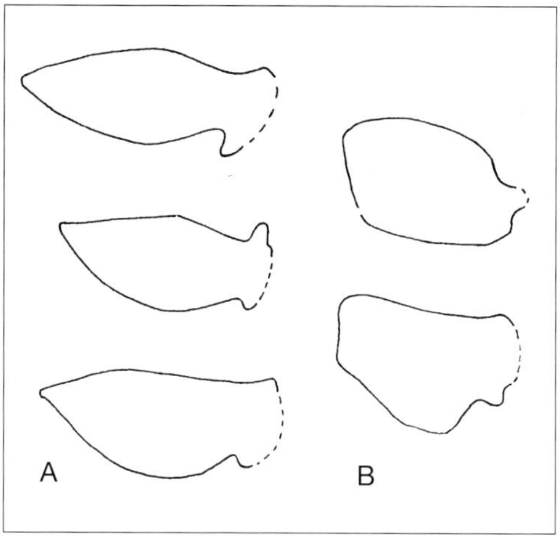

Fig. 1.5 Outline of the pectoral fin (after Wundsch, 1953)

is 0.036 mm (Fig. 1.6). Pfeiffer (1960) studied the skin of 10 species and found that the eel had the thickest epidermis. But in the eel, unlike other teleosts, the corium, particularly that on the head, is thicker than the epidermis. Saglio *et al.* (1988) demonstrated that the skin is thickest in the middle of the eel's body, and thinnest in the caudal area and around the pectoral fin. Also, the skin of female silver eels is thicker than that of the males. Yellow eels have thinner skins than silvery specimens.

The corium of the eel is provided intensively with collagen fibres; they run at an angle of 45° across the longitudinal axis of the fish and have tendinous stability (Hebrank, 1980). Therefore, it is likely that muscular action on the comparatively long body does not damage the skin; the torsion forces of the eel's spiral movements are not dangerous for the animal. In Scandinavia, it is said that eel skin treated with tannin is used for door-hinges. In South Korea and China, clothes and different kinds of bags made of eel skin are available.

Numerous, extremely well-developed club cells occur over the whole epidermis and apparently secrete many substances that have a protective function (Harder, 1964). Such cells may be as large as 0.150 mm by 0.025 mm in diameter in the eel (more than half the total height of the epidermis) (Pfeiffer, 1960), contrasting with species such as the armoured catfish (*Corydoras palaeatus*) in which they are about 0.025–0.035 by 0.013 mm. Henrikson and Maltotsy (1968a–c) have described the ultrastructure of the eel integument; the ontogeny from larval to silver eel stages has been studied by Aust (1936). Major changes take place in the skin during the fourth larval stage of metamorphosis (Table 1.1).

The thick epidermis, a robust protection against mechanical damage, is also relatively impermeable to water and electrolytes. Indeed, Bentley (1962) estimated that 1 ml of water would take 5 years to pass through 1 cm^2 of eel skin at a pressure difference of 1 atm! Transfer of eels from fresh water to brackish water results in an increased epidermal thickness (Thurow, 1957), and fish chronically adapted to fresh water have thinner skins than those adapted to brackish water; moreover, the skin made up to 8.5% of the body weight of freshwater-adapted specimens, compared with 9.4% in brackish water eels. Repeated transfer experiments produced equivocal results, however, and Thurow (1957) suggested that the induced excessive secretion of mucus eventually exhausted the cells producing

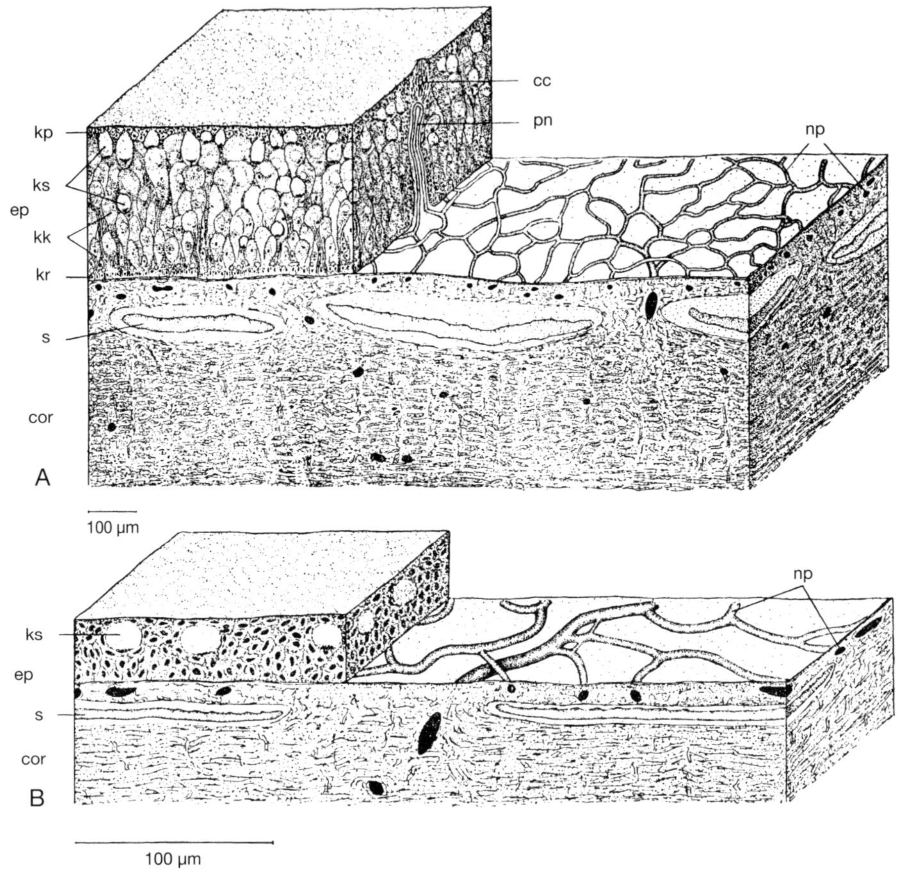

Fig.1.6 Structure and proportions of the epidermis and the corium of the European eel and of the flounder (*Pleuronectes flesus*) (after Jakubowski, 1960a, b)

A	Eel	cor	Corium	ks	Mucous or goblet cells
B	Flounder	kk	Club cell	np	Subepithelial blood vessels
ep	Epidermis	kp	Squamous cell	pn	Vascular loop
cc	Sensory cell	kr	Germinative layer	s	Scale

this substance. It was suggested that the thinner skin of freshwater eels requires regeneration of mucus cells before the eel can successfully enter sea water. Other data are probably more relevant (Portier and Duval, 1922; Portier, 1938 in Remane and Schlieper, 1958): eels transferred from low to high environmental osmolarities adapted less well if the skin was rubbed with a cloth to remove the mucus. In particular, hyperosmolarity of the blood and ion imbalance occurred. It was concluded that the skin – especially its mucus secretion – acted as a barrier against fluxes of water and electrolytes along osmotic and diffusion gradients. The secretion of mucus and its relationship to *N*-acetylneuramine have been examined (Lemoine and Olivereau, 1971) and prolactin from the adenohypophysis has been implicated as a factor controlling the structure and function of the skin (Olivereau and Lemoine, 1971; see Section 1.8.3).

Another physiological aspect of the eel's integument is its possible hindrance in gaseous exchange (see Section 1.4). Jakubowski (1960a), citing Krogh (1924), and Jeuken (1957), reported that fishes such as the eel, and the equally thick skinned *Misgurnus fossilis*, can meet virtually all their oxygen needs cutaneously. Furthermore, Byczkowska-Smyk (1958), in a study of branchial respiration also concluded that a large part of the eel's respiration must be met from cutaneous respiration (see Section 1.4). Indeed in air, species such as the eel, survive far better than do purely branchial breathers (Murygin and Anokhina, 1967). Therefore, the thick epidermis does not seem to prevent gaseous exchange or small exchanges of electrolytes. Jakubowski (1960b) suggested that the secretory cells – both mucus producing and club cells – lying between the blood vessels of the dermis and epidermis, contain sufficient amounts of water to permit ready diffusion of oxygen. Bolognani-Fantin and Bolognani (1964; see also Seutter *et al.*, 1970; Asawaka, 1974; Yamada and Yokote, 1975; Saglio *et al.*, 1988) have discussed the cytological and chemical bases for the production of mucus by eel skin. Mucus is of great adaptive significance, not only when the animals are in water, but it may also prevent dehydration when the animals undertake their brief excursions on land, and may aid survival at low temperatures (Gadeau De Kerville, 1918). Stripped of mucus experimentally, eels survived up to 7 days provided the humidity of the air was normal and the temperature low. Saglio and Fauconneau (1988) considered the free amino acid content of mucus and its significance for osmoregulation as well as for recognition of sexual partners; mature silver eels exhibited more free amino acids than did yellow eels.

1.3.2 Scales

Like other families in the Anguilliformes, the Anguillidae possess scales. However, these scales are rudimentary – at least in comparison with those of other species of fish. The scales are relatively well embedded in the upper layers of the corium below the epidermis (Fig. 1.6), and are not arranged in overlapping rows as they are in other fish, but are irregular, and in some places, distributed like parquet flooring. In general, one row of scales lies at right angles to the next, although the rows of scales immediately above and below the lateral line lie at an angle of approximately 45°.

In *Anguilla* species the first scales do not develop immediately after the larval stage is over – as is normal in other bony fish – but appear much later. Opuszynski (1965), Matsui (1952) and Jellyman (1979b) have shown that, in *A. anguilla, A. japonica* and *A. dieffenbachii* individuals measuring 15 cm or less do not have scales, whereas scales are present in most individuals 17 or 18 cm long. It seems likely that other species of eels also develop scales very late in ontogeny, although Pantulu (1956) reports their earlier appearance in *A. nebulosa*, where specimens of 11 or 12 cm have already developed scales. It seems likely that in *A. anguilla* the formation of scales is not an age-dependent process, and this has been demonstrated in *A. japonica* (Matsui, 1952).

As regards the region of the body where the scales develop first, it seems that there may be further differences between *A. nebulosa* and the so-called 'northern' eels of the temperate regions. In species like *A. rostrata, A. bengalensis, A. dieffenbachi* (Fig. 1.7) and *A. australis* the primary region is in the last third of the body (Smith and Saunders, 1955; Pantulu, 1956; Jellyman, 1979b), in *A. japonica* (Matsui, 1952) and *A. anguilla* (Rahn, 1957c)

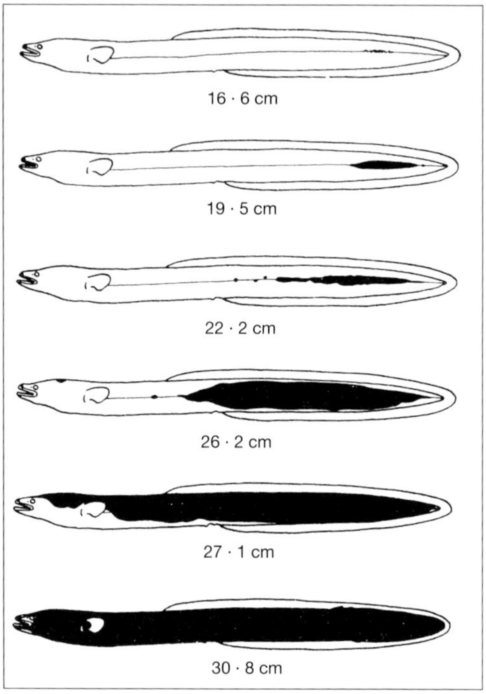

Fig. 1.7 Position and distribution of the first scales (black areas) in *A. dieffenbachii* (after Jellyman, 1979b)

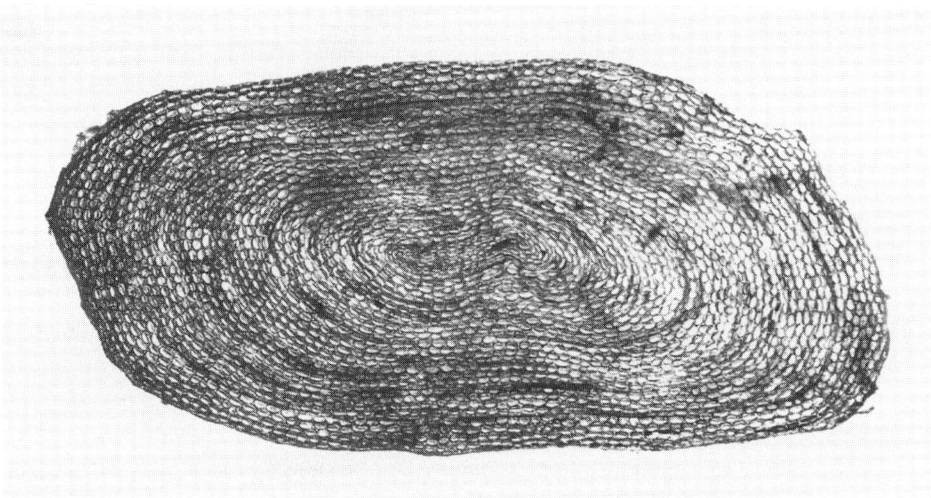

Fig. 1.8 Photograph of an eel scale (photo: Tesch)

perhaps slightly further orally. In *A. anguilla* this primary region has been located only indirectly, by establishing which part of the body had scales with the greatest number of annual rings (Rahn, 1957c). From the anal region, the zones of scale develop and spread forwards and backwards along the lateral line as well as dorsally and ventrally; in normally developing eels 2 or 3 years may elapse between the appearance of the first and the last scales (Gemzøe, 1906; Matsui, 1952). The lips of the upper and lower jaws, the throat, and, it appears the pectoral fin bases too, all remain scaleless.

The morphology of the scales has been described in many papers on the growth of the eel. But, recently they are used rarely for age determinations (see Section 3.3.2.1). The superficial structure of the scale is rather unusual (Fig. 1.8). Its contours suggest it is cycloid, but it has a very elongate-oval shape, although there are many variations. So-called circuli (concentric lines) are also seen in the eel scale. These are not made up of smooth 'ridges' and 'grooves' but from rows of plates, which resemble small medallions.

1.3.3 Pigmentation

The development of pigment provides the most useful means of recognising the different ontogenetic stages in the eel. This not only applies to subepidermal, external pigmentation but also to the internal pigment of the larval phases. During early development, as in the case of many other species of fish, the internal pigment also acts as a means of separating different species. In the leptocephalus, the first internal pigment develops along the notochord, while at the beginning of stage II (Table 1.1) it spreads in a caudo-rostral direction

Table 1.1 Development of pigmentation in *A. anguilla* (abridged from Strubberg, 1913, and adapted from Bertin, 1956).

Stage	Characteristics
I	Larva, fully grown leptocephalus (Fig. 2.3)
II	Semilarva, pigmentation on the posterior end of the spinal chord
III	Semilarva, pigmentation on the nerve chord becomes more extensive, skin pigment also seen at the tip of the caudal fin
IV	Semilarva, pigmentation on the nerve chord reaches the head
VA	Metamorphosis complete, eel-like in form, no external pigment (glass eel) except the caudal spot (Fig. 1.9)
VB	No pigment on the back, body or tail region, except for the skull, caudal spot and some rostral pigment
VIA_I	Development of pigmentation along the whole dorsum, post-anal dorsolateral pigment develops, post-anal, no clear mediolateral pigment (Fig. 1.10a)
VIA_{II}	No pre-anal ventrolateral pigment. Post-anal development mediolateral pigment (Fig. 1.10b)
VIA_{III}	No pre-anal ventrolateral pigment. Clear pre-anal development of mediolateral pigment, post-anally over almost entire dorsum, pigment rows along the myosepta, and in places doubling of the mediolateral melanophores (Fig. 1.10c)
VIA_{IV}	Clear development of pre-anal ventrolateral pigmentation. Initially, in places, a doubling of the mediolateral melanophores in the pre-anal region (Fig. 1.10d), post-anal pigment between the myosepta in the ventral region (Fig. 1.10e), and finally, similar changes in the pre-anal region (Fig. 1.10f)
VIB	Pigment rows along the myosepta becoming indistinct. Lateral line still recognisable, as are the individual melanophores on the head, 'cheek', behind and below the eyes and on the lower jaw (Fig. 1.10g)

(Schmidt, 1906; Gilson, 1908). In comparison with the first, external chromatophores, which are brownish in colour, the internal chromatophores are black and relatively large. Species differences appear to exist in the ontogenetic development of the internal pigment. Pigmentation of the notochord during stage II (Table 1.1) begins posteriorly in *A. anguilla*, whereas in *A. japonica* pigmentation begins anteriorly (Egusa, 1972).

External pigment also develops during the leptocephalus stage and is visible as a dark patch on the fin rays of the tail fin during stage III. This patch remains recognisable during further development and is the only form of external pigmentation until stage VA (Fig. 1.9, Table 1.1). A characteristic of this stage is that developmental changes are largely internal; but if the temperature is raised, large numbers of melanophores appear along the whole length of the body.

The beginning, and to a certain extent the end of stage V (also called VB) is marked by the formation of the so-called 'skull spot', the appearance of which is certainly of great physiological significance; this spot can still be seen in older eels although it does become less distinct. As Gilson (1908) has shown, this pigment is not produced in the corium – as is the caudal spot and most of the pigment that appears after it – but in a sort of fontanelle. Later, this hole is occluded as a result of dermal bone formation. In older eels the skull spot is found in the meninges under the frontal and parietal bones.

Developmental differences in external pigmentation are evident in *A. japonica* and the European eel. In the Japanese eel, the caudal patch does not develop until after the beginning of stage IV, and not in stage III. In contrast, in *A. japonica*, the skull spot and a certain degree of rostral pigmentation are already in evidence in stage III (Egusa, 1972). These differences may also occur in the other Indo-Pacific species, thus making it possible to distinguish between the various metamorphic stages of the different species (Marquet, 1992).

According to studies on other fish species, the melanophores in this region of the head act as regulators controlling the amount of incoming light (Nicol, 1963). The pineal organ, the epiphysis, is photosensitive and must be protected from excess illumination. On the other hand, young fishes, including the leptocephalus, are relatively insensitive to light and receive very little protection from their melanophores (Breder and Rasquin, 1950). As research on the salmon has indicated (Hoar, 1955), the almost total absence of morphological differentiation in the pineal organ in young fish probably is partly responsible for

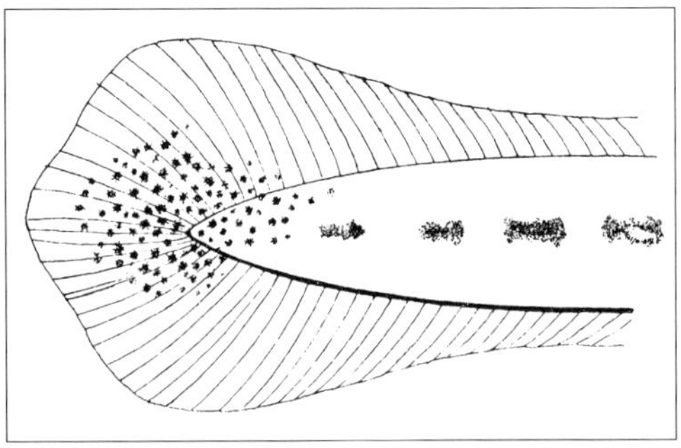

Fig. 1.9 Tip of the tail showing the caudal spot in a glass eel at Stage VA; this spot is even more developed at stage VB (after Gilson, 1908)

this insensitivity. Leptocephali (*A. anguilla*), with increasing age and size, prefer greater depths (Schoth and Tesch, 1984), which may be connected with increasing light sensitivity of their pineal organ.

The first appearance of the skull spot indicates an important step in the life of the eel. It forms at a time when the eels arrive in the coastal waters and abandon their purely pelagic existence (Section 3.1.1). The eel's behaviour during this first phase of pigmentation indicates a marked sensitivity to light (Tesch, 1965). Melanophores then begin to develop along the entire body length and a transient phase of reduced sensitivity to light begins; from time to time glass eels even swim along the banks near the surface.

A list of characteristics can be drawn up that describes the extent of pigmentation during various advanced stages of development (Table 1.1; Fig. 1.10) (Grassi, 1913; Strubberg, 1913; see also Boëtius, 1976). According to this system, stage VI (Schmidt, 1906) has been divided into substages A and B. Essentially, VIB indicates the end of the pigmentation, and stage VII represents the fully pigmented, benthic and migrating young eel (Gilson, 1908). Thus, in stage VIB the eel loses its glass-like transparency.

In addition to the black colouring, provided by the melanophores, other pigments also begin to appear and are already in evidence towards the end of stage VIA. In particular, a green colouration becomes recognisable as a result of the formation of yellow pigment. Heldt and Heldt (1929a) attribute this to the beginning of food intake (see Section 3.3.1), whereby lipoids are formed, thus providing the basis for the yellow colour. Water-soluble flavones are particularly involved in the formation of yellow pigment in the eel – a characteristic that distinguishes the eel from other, less euryhaline fishes (Fontaine and Busnel, 1939).

Stage VIA is further divided into a number of subsections of which only the main divisions, VIA_{I}–VIA_{VI} are given in Table 1.1. According to Strubberg (1913), each of these four subsections is made up of a further one to four subdivisions; this system is based on the fact that pigmentation starts caudally and dorsally and proceeds rostrally and ventrally. In German, regardless of the degree of pigmentation, young eels are referred to as 'Aalbrut', 'Montée' or, if they are not too darkly coloured, as 'Glasaal'. In French, they are called 'civelles'. In English, however, unpigmented young are referred to as 'glass eels', while pigmented ones are called 'elvers'. When pigmentation is complete the yellow eel stage is reached; there are no major external changes after this until the eel returns to the sea. In German, the small, fully pigmented eels, which are not yet suitable for marketing, are called 'Satzaale'. Strictly speaking, the name 'yellow eel' is not correct, because, although many eels do vary from yellow to white on the underside, a large number have almost completely white bellies that change to a light grey on the flanks. However, the term 'yellow eel' has now been adopted universally and is used to distinguish this stage from the silver eel stage. The yellow eel's dorsal surface varies from dark green or brownish-green to black, the former colours giving rise to the term 'Grünaal' or 'green eel'. A similar duality of terms occurs in France; both 'anguille jaune' and 'anguille verte' are frequently used; the use of 'green' or 'vert' might also imply immaturity. In English, the term 'yellow eel' is commonly used, but from time to time one comes across the expression 'golden eel'. However, this use of metallic colour terms can be misleading because a metallic shimmer is the distinguishing characteristic of the 'Blankaal' ('silver eel' or 'bronze eel') or of eels without black pigment (xanthochromatism) (Section 7.3).

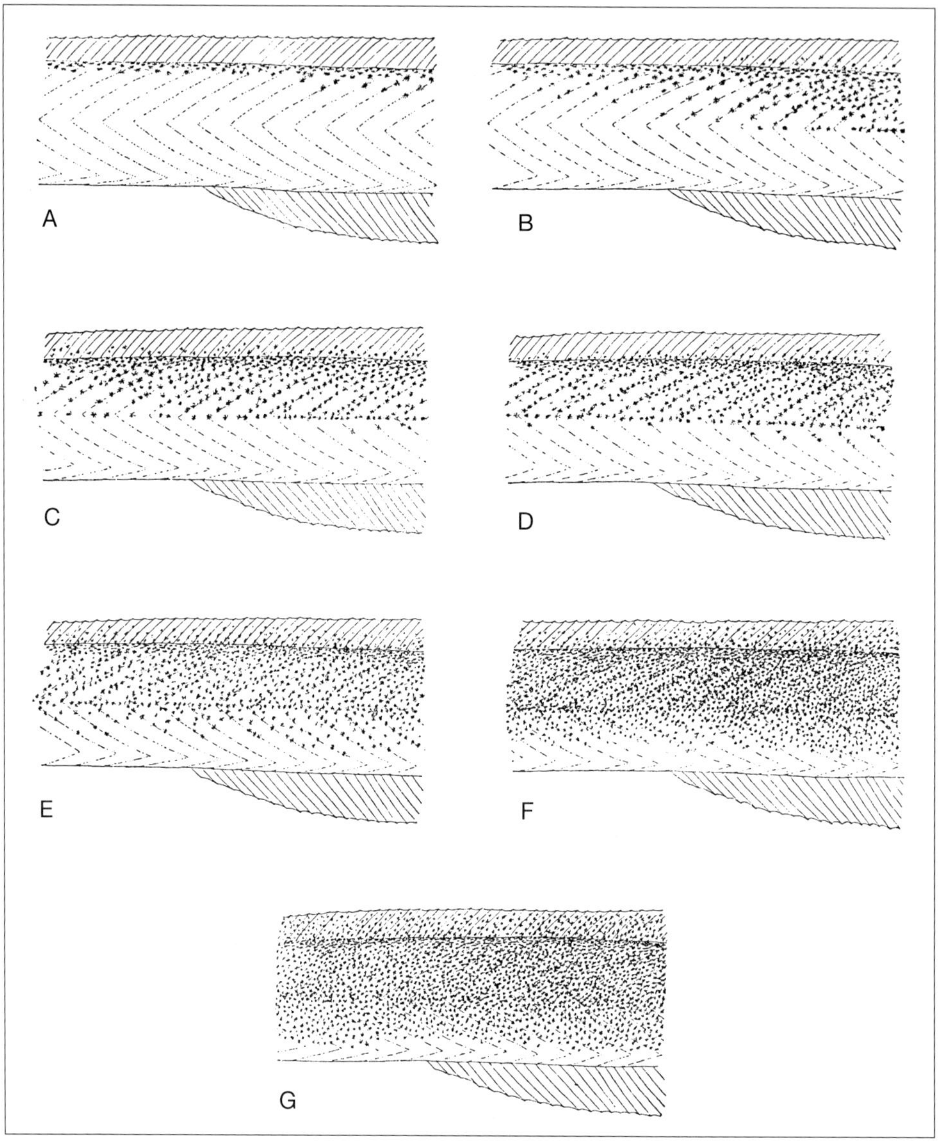

Fig. 1.10 Development of the subepidermal pigment in the glass eel (after Strubberg, 1913)

A Stage VIA_{I} B Stage VIA_{II} C Stage VIA_{III} D,E,F Stage VIA_{IV} G Stage VIB

Naturally, colour in eels from various biotopes can be very different and must depend to a great extent on the background. Fishermen involved in marking and transplanting eels in the North Sea have reported (Tesch, 1967a) that transplanted eels recaptured outside their original habitats were a quite different colour from those that normally lived in the area. Thus, these eels had retained their colour for days, weeks even, without showing adaptation to their new environment. Experimental studies by Neill (1940) and Odiorne (1957) have also shown that eels take a very long time to change colour; as many as 20 days may elapse before one extreme condition of the melanophores will change to another. In the wrasse (*Crenilabrus*) on the other hand, the change begins in a few hours. Hormonal or humoral processes are probably responsible for colour changes in the eel, whereas in *Crenilabrus* and many other species of fish, these changes appear to be under neural control. In the eel, special cells in the middle lobe of the pituitary gland (Fig. 1.31A) exhibited a hormone, which affects the black colour cells (Baker, 1972; Fremberg and Olivereau, 1974). If one believes that the eel's colour tones with the substrate, this can only be true for a particular tonal range. Most eels that originated from underground river areas were an unusual light colour. Thus, lack of light obviously produces contraction of melanin in the melanophores (Section 7.4.2).

With the growth of the body and the gonads (Section 1.7.2), a change in colour takes place. During migration to the spawning grounds in the Atlantic the eel also, as it were, readjusts itself in other, morphological respects to its new environment. Thus, an open water fish develops from a benthic fish. Starting from the side of the body the eel develops a silver tint that gradually spreads ventrally. The eel has then reached the 'silver' or 'bronze eel' stage as it is so aptly called. On the back and dorsolaterally, the eel becomes darker in colour and it looks almost black. The pectoral fin also becomes black. Until the silver tint completely covers the surface area of the belly, the eel is referred to as being 'half silver'. There are, of course, many transitional phases starting with the onset of the silver eel stage and ending with the fully coloured silver eel, and it is often difficult to decide from external features whether a particular individual has already reached this stage physiologically. The best indication of whether the animal is physiologically a fully developed silver eel is the method by which it was caught. Animals which appear 'silver' and that are caught in a downstream current or in a stow net at ebb tide, are sure to be fully developed. However, those eels often caught shortly before their seaward migration should certainly not be considered mature or to have completed their change in colour (Fig. 1.36; see also Fig. 1.27). Experiments on male eels, where maturity of the gonads was produced artificially, have shown that a narrow, whitish-silver band remains on the ventral side, which is itself flecked with red.

The metallic colour varies in silver eels from different places. Some appear a true silver colour while others range from bronze to coppery tones. It seems certain that this is due not only to differences in the degree of maturity of the individual, but also to its habitat and to the original colour adaptation to its former biotope, the effects of which are still visible through the silver sheen.

1.3.4 Teeth

Other groups of anguilliform eels show a great variety of teeth when considering their type, form and size. The teeth of anguillid eels are much more elementary and uniform. The eel has setiform teeth on the maxilla, the dentary and the premaxilla-ethmovomerine bloc (Fig. 1.11). In many species of eel the arrangement of vomerine teeth serves as an important diagnostic feature (Table 2.1, Fig. 2.11).

During ontogenetic development the teeth undergo considerable changes in form. The teeth succeeding the large canines of the larva (largest in the pre-larva: Fig. 2.9) are not as large and become progressively smaller towards the back of the jaw. The larva loses these relatively large teeth during the very first stages of metamorphosis (Fig. 1.12), so that when

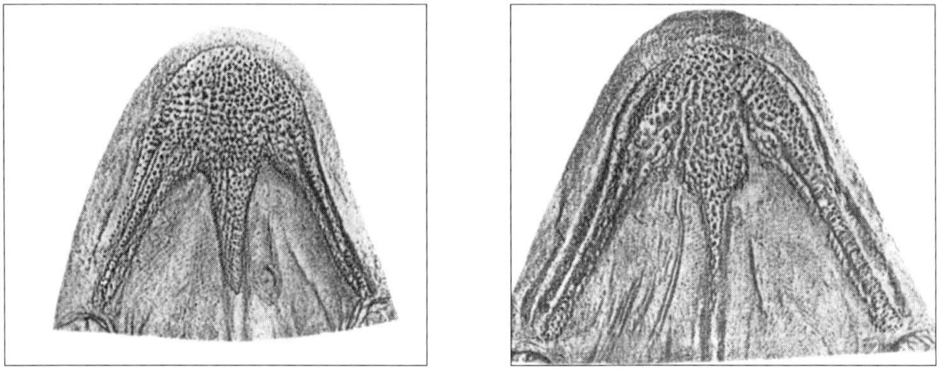

Fig. 1.11 Plasticine imprints of the upper jaw dentition of probably *A. nebulosa labiata*, the left specimen is doubtful and designated as *A. marmorata ? mossambica* type (from Balon, 1975) (see also Fig. 2.11)

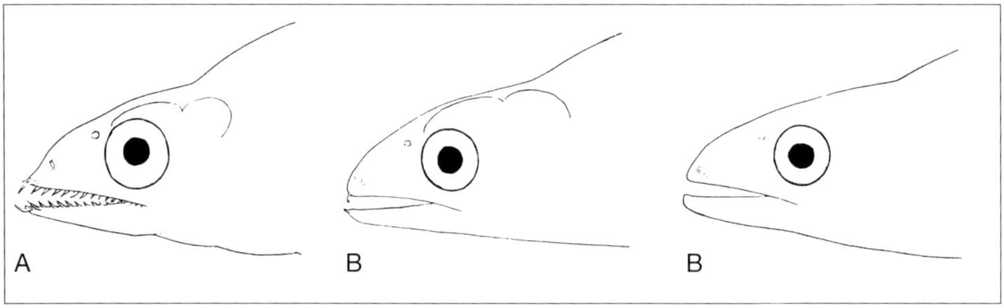

Fig. 1.12 Head of larval *A. anguilla* (after Schmidt, 1908)

A Stage I, with fully developed tooth rows
B Stage II, with almost no teeth
C Stage II, with no teeth

the glass eel stage is reached no larval teeth remain. Schmidt (1916) found no differences between tooth development in *A. rostrata* and *A. anguilla*. Setifom teeth begin to form at stage V but do not break through the epidermis until later (Gilson, 1908; Gandolfi-Hornyold, 1918b); the first teeth appear on the jaws and on the vomerine plate in stage VIA. Initially, the very numerous teeth vary considerably in size, and there are no recognisable rows. In stage VIB too, most of the teeth do not protrude very far out of the epidermis. However, there is considerable individual variation, and so the teeth are less useful than pigmentation in classifying the various developmental stages (Gandolfi-Hornyold, 1918b). In general, completely formed rows of teeth are not seen until the eels are 12–15 cm long, at which time one can use the teeth to distinguish between species (Fig. 2.11); in tropical eel species it may be possible to use this feature at somewhat earlier growth stages.

1.3.5 Musculature

Studies on morphology and functional properties of the musculature of the eel have been published by Willemse (1975) (Fig. 1.13). This musculature shows, on growth, an increase in the number, as well as in the diameter, of muscle fibres (Willemse and van den Berg,

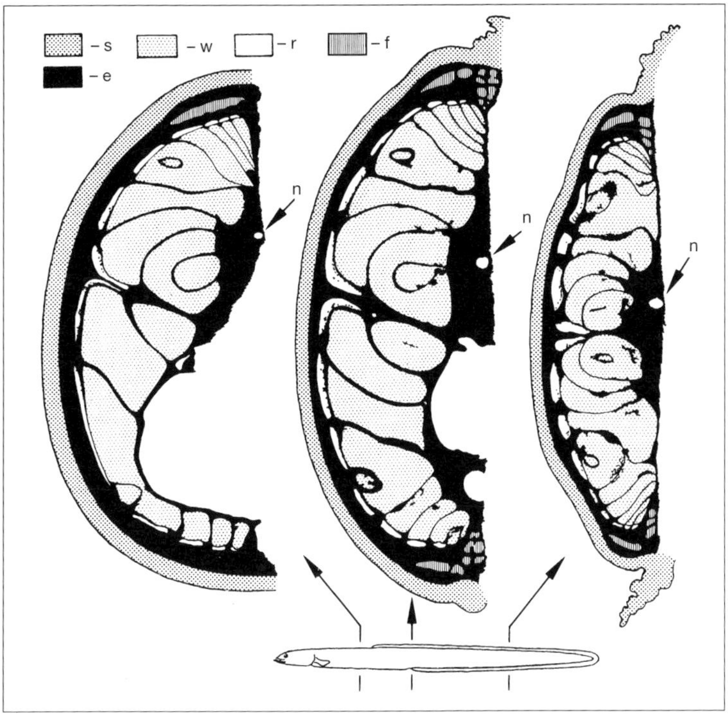

e Connective tissue
f Fin muscles
n Vertebral column
r Red muscle fibres of the lateral muscular system
s Skin
w White muscle fibres of the lateral muscular system

Fig. 1.13 Cross-sections through various regions of the body in a 35 cm long European eel (after Willemse, 1975)

1978) that are differentiated into four different types (Willemse and Ruiter, 1979; Hulbert and Moon, 1978).

The white musculature of the eel develops as follows (Romanello *et al.*, 1989). It corresponds to the three stages: glass, yellow and silver eel. The glass eel shows a uniform structure of the fibres; in the yellow eel the single fibres are differentiated by different diameters. In addition, in contrast to the fibres in glass eels, they are differentiated by immunochemistry and histochemistry; in the silver eel, mainly large fibres are visible. It is concluded that in glass and silver eels the enlargement of the present muscle fibres is characteristic. In the yellow eel the increase in number is essential. Blin and Balea (1955) described the course of the myomeres and myosepta of the conger eel that permits comparison with *Anguilla*. Alexander (1969) has mentioned an unusual feature of the white muscle fibres in *A. anguilla* as compared with those of most other species of fish. Their position relative to the horizontal septum is similar to that in cartilaginous fishes, primitive bony fishes and members of the Salmoniformes. Nelson (1967) has compared the gill arch musculature of *A. rostrata* with that of eels from five other families and genera and found reduction from the condition in *A. rostrata*, through that of *C. marginatus*, *M. javanica*, *K. diodontus*, *U. knighti*, to that of *Gymnothorax petelli*.

Research on the chemical composition of eel muscle has been carried out with regard to concentrations of sodium, potassium, water-containing ninhydrin positive particles, free amino acids, thanolamines, cytathionines and carnosines. After a month in sea water, silver eels show an increase in amino acid content. In particular, there are increases in alanine, glutamine, glutamine acid, glycine, proline, tyrosine and cystathionine. The amount of carnosine, ethanolamine and other amino acids hardly changed at all (Huggins and Colley, 1971). The amount of fat, protein, carbohydrates, ash, water and dry materials is given in Table 8.2.

1.4 Respiratory organs and swimbladder

In the eel, the gill arches, filaments and lamellae conform to the general teleostean pattern. It should be mentioned that the gill lamellae of the eel are in two rows on the filaments that differ from a second group of bony fish supplied with gill lamellae branching from the filaments in one row (Harder, 1964). In eels of between 32 and 37 cm in length, on gill arches one to four of each side, there are 147, 136, 129 and 119 filaments, respectively, giving a total of 1062 filaments for the four arches. The total length of all the filaments is 3422 mm. Each millimetre of gill filament has 30 lamellae, and the total respiratory surface area/kg of body weight has been estimated at 0.987 m^2 (Byczkowska-Smyk, 1958).

Such a surface area for respiratory exchange is of intermediate size compared with that of other fish. For example, the pike perch (*Stizostedion lucioperca*), and the rainbow trout (*Oncorhynchus mykiss*), have some 1.800 m^2/kg, and the viviparous blenny (*Zoarces viviparus*) only 0.475 m^2/kg body weight. This intermediate position of the eel in terms of branchial surface area is reflected in its slow swimming, benthic habits and its concomitant low oxygen consumption. On the other hand, the frequent and wide-ranging movements in search of prey, as well as prolonged migrations at certain phases in the life cycle, must often demand sudden and unpredictable oxygen requirements. Furthermore, Byczkowska-Smyk (1958) has calculated that, in water, the surface area of the gills is insufficient for the respiratory requirements of eels. A second source of gas exchange is thus relevant, namely the skin (see Section 1.3.1). However, oxygen taken in by the skin is barely sufficient to provide the skin's needs (Kirsch and Nonnotte, 1977). In a detailed study, Berg and Steen (1965) showed that in water 90% of oxygen was absorbed across the gills while in air some two-thirds of the oxygen consumption was cutaneous. When the eel is out of water, its branchial chamber is filled with air, from which there is a slow absorption of oxygen. At 20°C, opercular movements replenish the supply of air by pumping at a rate of 60 movements/h; in water, this respiratory rate is 20 times more rapid (Berg and Steen, 1965).

Berg and Steen (1966), after examination of the ventilatory responses to separate exposures of the skin or gills to various gas mixtures (O_2, N_2 and CO_2) suggested reduced environmental oxygen tensions increased the breathing frequency. CO_2 had no effect provided the eel was exposed to air. A number of authors have examined the physiological mechanisms that control the microcirculation of the gills. Gilloteaux (1969) concluded that both neural and endocrine factors moderate the flow of blood through the filaments, and hence oxygen uptake, in eels in which the spinal cord was destroyed (see also Leicht, 1969). It had been suggested earlier (Schultze, 1965) that receptors in the head region were responsible for the regulation of the oxygen uptake over a wide temperature range. In particular, these receptors, sited either within the ventral aorta, or on the afferent branchial arterioles, monitored blood oxygen in the venous (afferent branchial) blood and brought about changes in blood volume and distribution within the branchial vascular bed. The delicate vascular arrangements within the gills, especially the shunting processes along the lamellae, are probably of great significance and certainly respond to various neurohumoral factors (Steen and Kruyssee, 1964; Maetz and Rankin, 1969; Rankin and Maetz, 1971).

The capacity of the adult eel to survive in both air and water is associated with its ability to use both branchial and cutaneous modes of respiratory gas exchange. The eel survives better in air than in poorly oxygenated or polluted water; indeed, its resistance to hypoxic water is apparently no greater than that of other fishes. Given the choice, eels invariably select water of high oxygen tension (Hill, 1969), and it is of interest that when lakes freeze, eels show the highest mortality rate among the fishes present (Rahn, 1963). The benthic mode of life of the eel frequently exposes it to hypoxic zones of water that are often at low temperatures. Low temperatures inhibit the eel's mobility and thus its well-known panic-like escape reflexes. Oxygen supply is, therefore, not the only ecological factor affecting the survival of eels, and it seems that temperature is especially significant in subtropical and temperate climates (Johansen, 1929; Bernhardt, 1963; Scheer, 1964; Tesch, 1964). Hill (1969) studied the eel's tolerance to low oxygen and, at a temperature of 21°C, found a lower limit of 2.5 mg/l O_2 that indicates a not too high sensitivity.

The gills, in addition to their respiratory function, participate in osmoregulation (Section 1.8 and especially Sections 1.8.3, 1.8.7 and 1.8.11). This requires that in sea water the branchial epithelium extrudes excess ions such as sodium and chloride, while in fresh water net uptake of salts takes place. The eel, in particular during its silver eel phase, displays a remarkable capacity to change effectively the polarity of the active transport of the ions, and has been the subject of extensive investigations in many laboratories. Aspects of the biochemistry (Maetz *et al.*, 1969), physico-chemistry (Motais *et al.*, 1969; House and Maetz, 1974) and physiology (Henderson and Chester Jones, 1967; Maetz, 1971) have been studied in detail in eels adapted to fresh and sea water. Many of the fundamental processes responsible for the adult transfers are under intimate endocrine control (Section 1.8). In sea water, the branchial epithelium displays a rich population of chloride cells (Keys–Willmer cells) (Keys and Willmer, 1932). These cells, situated at the base of the gill lamellae, were originally described as the site of chloride transport (Keys, 1931; Schlieper, 1933), and later they have been shown to possess many of the ultrastructural and enzymatic characteristics of electrolyte transporting epithelia (Mizuhira *et al.*, 1964, Shirai and Utida, 1970; Utida *et al.*, 1971; Masoni and Isaia, 1973; see also Motais and Garcia Romeu, 1972). Present evidence thus suggests that the branchial chloride cells extrude sodium and chloride when eels are in sea water. Functioning of the chloride cells in both fresh and sea water, is controlled by activity of Na^+/K^+-ATPases which are influenced by angiotensin (Marsigliante *et al.*, 1996, 1997). There may be more than one type of gill cell and it is possible that the absorptive function of the gills in freshwater-adapted eels reflects the existence of a second type. Among the inconsistencies which may argue against the chloride cells being the unique route of transport is the fact that eels <30 cm do not have readily recognisable chloride cells when they live in sea water (Ogawa, 1962).

The molecular and enzymatic mechanisms have been studied in preparations of eel gills. For example, the extrusion process in sea water is inhibited by materials that block protein synthesis (Maetz *et al.*, 1969) or inhibit the enzyme sodium-potassium-activated adenosine triphosphatase. The latter is present in greater amounts in the gills of seawater-adapted fish (Kamiya and Utida, 1968, 1969; Jampol and Epstein, 1970; Motais, 1970; Milne *et al.*, 1971; Forrest *et al.*, 1973) although Kirschner (1969b) was unable to demonstrate such a difference.

Wherever the precise sites of water and electrolyte movement across the branchial epithelium, the gills play a crucial role in the adaptive processes. As the environmental sodium concentration increases, the gills progressively extrude the sodium gained by drinking (Maetz and Skadhauge, 1968; Mayer and Nibelle, 1970), and their water permeability changes (Keys, 1931). Such adaptive responses are under endocrine control (Section 1.8.5) but may be influenced by other factors such as branchial haemodynamics (Maetz and Rankin, 1969; Kirschner, 1969b) and may be liable to seasonal influences (Utida *et al.*, 1967, 1969).

A final function of the gills – that of nitrogenous waste excretion – is related to acid–base balance as well as to ionic regulation. Ammonium ions pass freely along the gradient from the blood to the surrounding water. Subsequent work has confirmed Keys' (1931) original suggestion of an NH_4^+/Na^+ exchange across freshwater eel gills (Garcia-Romeu and Motais, 1966) and further, that although small amounts of ammonia may be produced in the gills, it emanates largely from the liver (McBean *et al.*, 1966).

Much morphological and physiological information is available on the swimbladder of the eel. To paraphrase Dorn (1961) '... [the swimbladder of the eel] is a physostomous type with a large duct from the oesophagus, although the opening itself is tiny. The duct is expansive and highly flexible and can give the appearance of a primitive lung, indeed it subserves a resorptive function. In other respects the [eel] swimbladder has many physoclistous characteristics. There are two highly developed capillary bundles (retia mirabilia) dorsal and ventral to the opening into the swimbladder (secretory bladder), which are separated from the gas gland, itself a special feature and part of the total epithelial formation.' The gas gland supplies the swimbladder with CO_2 and O_2. The high water pressure at great depths requires strong filling capacities that are provided by glycolytic formation of CO_2 in the epithelium of the swimbladder (Pelster and Scheid, 1991). This process is free of oxygen consumption. Provision with gas is reduced under conditions of hypoxia (Pelster and Scheid, 1992).

A most striking feature of the eel swimbladder (Fig.1.14) is the relatively large anterior section – the pneumatic duct – which has a resorptive function (Steen, 1963b, c), and, when full, clearly represents an oxygen reservoir. Histologically, the pneumatic duct may, in many respects, resemble a primitive vertebrate lung, a fact that adds strength to the theory that this zone could aid aerial respiration in the eel (Dorn, 1961). However, it is only partly true that 'pulmonary' respiration has replaced branchial respiration when the eel is out of the water. When breathing aerially, the eel absorbs only half as much oxygen as it does when breathing in water, and further, the respiratory rate is much reduced (Berg and Steen, 1966). During the first hour in air only one-third of the oxygen consumed is derived from branchial respiration, and after 20 h at moderate temperatures an oxygen deficiency becomes apparent. These facts are very relevant to the husbandry and transport of eels. Eels certainly can be transported for days without water, providing the humidity is high, and perhaps more important, that the temperature is kept low: 7°C is suitable, but 15°C may be fatal.

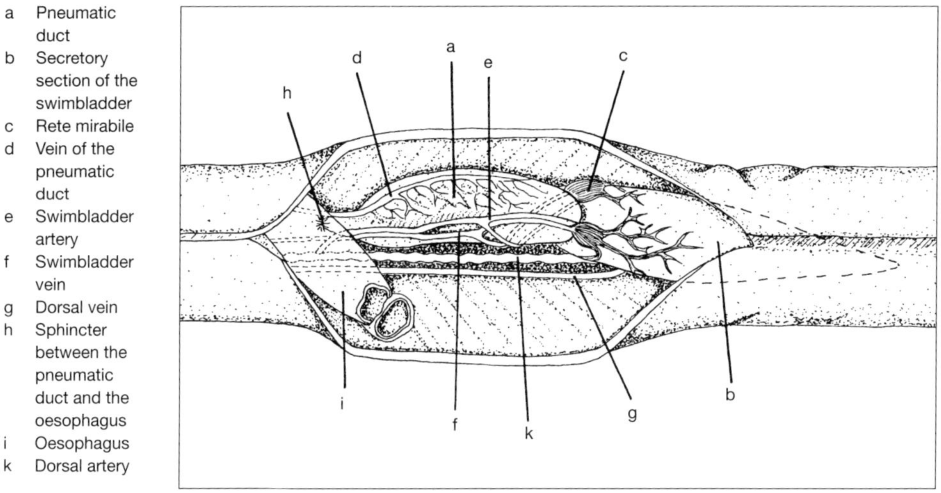

Fig. 1.14 Swimbladder of the European eel in the body cavity (modified after Steen, 1963b)

Light and electron microscope studies on the histology of the eel swimbladder have been published by Dorn (1960, 1961). Krogh (1924) examined the capillary network, especially those within the retia mirabilia and Bandayan *et al.* (1974, 1975) its ultastructure. Steen (1963a–c), as well as Steen and Sund (1977), related the structure to gaseous exchange. They studied the gas secretion of the gas gland by measuring oxygen, carbon dioxide, lactic acid and the pH of the blood in veins and arteries of the retia mirabilia. They also examined the reabsorption of gases from the swimbladder.

A brief summary of the findings to date must include the fact that the swimbladder receives arterial blood which forms a capillary net at a localised area on the inner surface of the swimbladder, and that these capillaries are drained retrogressively by veins running in parallel with them. A key feature is that the capillary system of arterial and venous origin intertwines in such a way that arterial capillaries are surrounded by venous ones and *vice versa* to form a rete. This arrangement ensures an optimal diffusional state between arterial and venous blood; furthermore, the flow in the capillaries is of the counter-current type. In the eel the rete contains about 100,000 capillaries of each type, and the distance between arterial and venous vessels is about 1 μm. Although the organ itself is only 3–5 mm in diameter the aggregate length of capillaries is about 900 m with an area of 210 m^2! Such volume-to-surface area relationships, together with the counter-current exchange situation may readily explain the respiratory, absorptive, secretory and acid-base relationships observed in the eel (Stray-Pedersen and Nicolaysen, 1975).

During development, the morphology and function of the eel's swimbladder change. This is especially the case for the transition between continental yellow and oceanic silver eel stages, when, with diving into greater depths tolerance to greater pressure has to be increased; pelagic trawl catches in the North Atlantic have captured silver eels at >300 m depth (Ernst, 1975) and, during trackings over the continental slope as well as near the Sargasso Sea spawning grounds, silver and nearly mature eels could dive into depths of up to 700 m (Tesch, 1978a, b, 1989; Tesch *et al.*, 1979) (photographs of, perhaps, American eels at a depth of 2000 m (Robins *et al.*, 1979), are unlikely to have observed the right species); moreover, calculations have shown that eels could change their preferred depth by >200 m within as little as half an hour (Tesch, 1995). In order to obtain these abilities the retia mirabilia had to be adapted from the conditions of the continent to those of the deep sea. Their volume increased one-and-a-half-fold, which resulted in a 3-fold length increase as well as in a thickening of the capillaries (Kleckner, 1980a). In the walls of the swimbladder the content of guanine crystals increased one-and-a-half-fold resulting in a diminution of gas losses by 15% (Kleckner, 1980b). These morphological changes enable the swimbladder of the silver eel to produce five times more gas than that of the yellow eel; decreasing volume of the swimbladder, caused by greater depth and increasing pressure, is therefore largely compensated.

1.5 Organs of feeding and digestion

1.5.1 The gastro-intestinal tract

The gastro-intestinal tract and the intestine lie below the swimbladder and gonads (Fig. 1.15). The stomach is narrow and cone shaped (Fig. 1.16) with an obvious diverticulum, giving it a Y-shaped form (Harder, 1964). The gastro-intestinal tract receives the swimbladder duct, which opens into the upper part of the oesophagus (Kukla, 1954), a characteristic of physostome fishes. There are no pyloric appendages such as frequently occur in other bony fishes.

The mucous membranes of the oesophagus and stomach are thrown into long straight folds, which are fully developed only after the glass eel stage (Berndt, 1938). The folds of the mid-gut are more irregular and zig-zagged. The vascularisation, and the ontogeny of the gastro-intestinal tract have been described by Kukla (1954) and Berndt (1938), respectively.

The relative length of the alimentary tract varies with life-cycle stage. It is longest in leptocephali with a total length between 5 and 30 mm (79% of this length) (Schoth, 1982). Fully-grown larvae exhibited a relative length of the intestine of 63% (Berndt, 1938). Until metamorphosis into the glass eel it is gradually reduced to 38% and then increases until the silver eel stage to 40%, after which there is a slight decrease. During the glass-eel stages VIA_1–VIA_2 (after Strubberg, 1913, Table 1.1) there is a slower intestinal growth rate in sea water than in fresh water (Vilter, 1945b). A marked increase occurs at the beginning of voluntary food intake (Vilter, 1945a).

In addition to these gross changes in size there are considerable changes in the morphology of the individual gut segments. Although the stomach is readily recognisable in the larval forms, there is a tendency for the oesophagus, stomach, mid- and hind-guts to be relatively undifferentiated. With the emergence of the glass eel the stomach and mid-gut enlarge concomitantly with changed post-metamorphic feeding patterns. The pelagic leptocephalus feeds on planktonic material, while after metamorphosis and a switch to a benthic habitat a more omnivorous habit is characteristic (Section 3.3.1).

The slightly diminished gut of silver eels is perhaps to be expected as it is then that the prolonged fast begins. However, this occurs alongside many other, more fundamental changes in gastro-intestinal structure and function. Thus, convolution in the gut disap-

R Right side of the body
als Stomach behind the blind stomach
doa Dorsal aorta
els Stomach between oesophagus and blind stomach
fab Fat
gon Gonad
int Intestine
inv Intestinal vein
ltp Lymphoid tissue (pronephros)
pan Pancreas
pcv posterior cardinal vein
pnd Pneumatic duct
spl Spleen

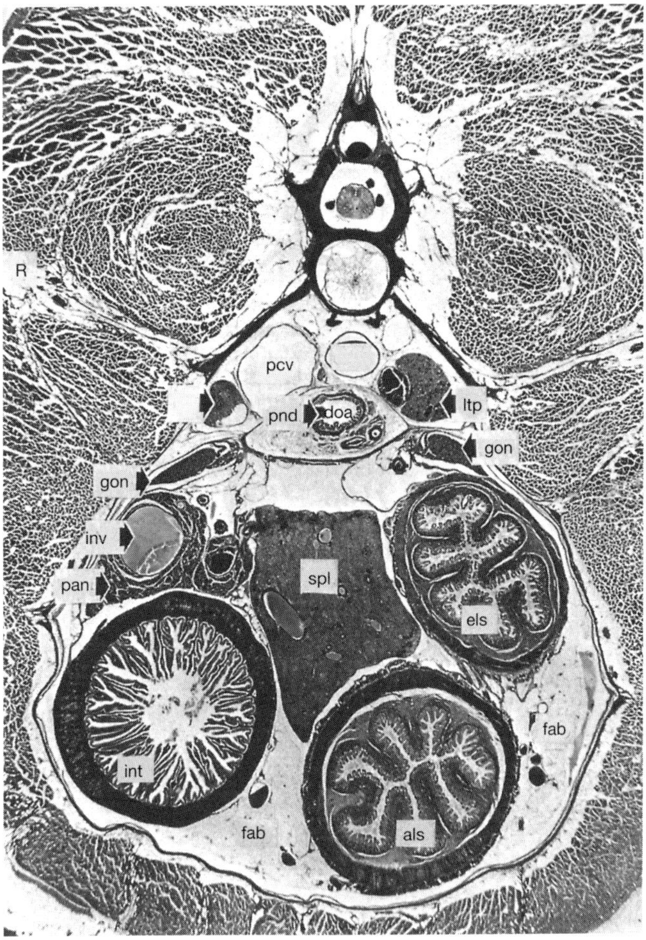

Fig. 1. 15 Cross-section through a female European silver eel close behind the liver (after Willemse,1979)

pears and radical histological changes, especially in the oesophagus and hind-gut, are apparent. For example, in the last mentioned parts the epithelium turns over to histolysis; the number of intestinal appendices and mucus cells diminishes (Pankhurst and Sörensen, 1984). The intestinal musculature begins to degenerate, a process initially obvious at the ends of the muscle fibres and finally in the nuclei. The gut musculature reaches a maximal thickness shortly before the onset of the silver eel stage, after which there is a sharp and rapid decrease (Figs 1.17 and 1.18). One of the more dramatic changes that was thought to take place was the complete closure of the anus and the occlusion of the gut lumen (Schnackenbeck, 1940). However, no closure of the anus and occlusion of the gut lumen is reported by later studies (D'Ancona, 1959a; Boetius and Boetius, 1967a; Villani and Lumare, 1975). Functionally, the gut of the eel is of interest in that the stomach and its diverticulum lie parallel and very close to one another. After feeding, the stomach is distended and greatly compresses the intestine, but when it empties it becomes almost occluded by the full intestine. Food intake, therefore, is possible only at infrequent intervals.

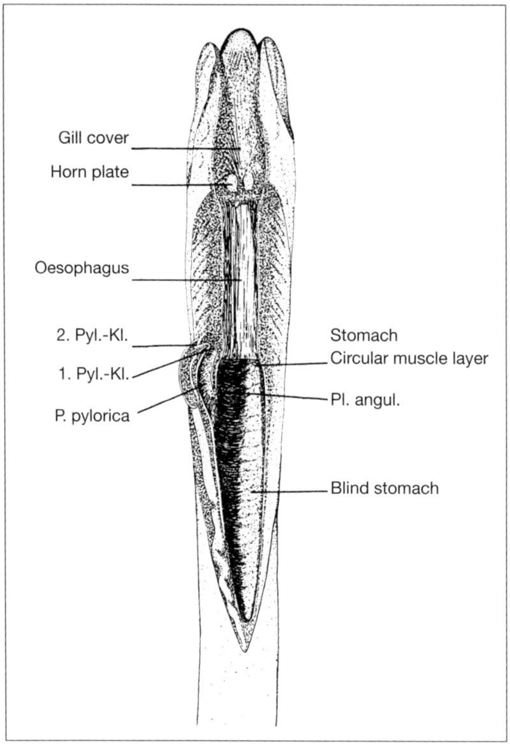

P.pylorica Pyloric part of the stomach
Pyl. Kl. Pyloric valve
Pl. angul. Pyloric angular

Fig. 1.16 Digestive tract of the eel (after Pernkopf, 1930 from Harder, 1964)

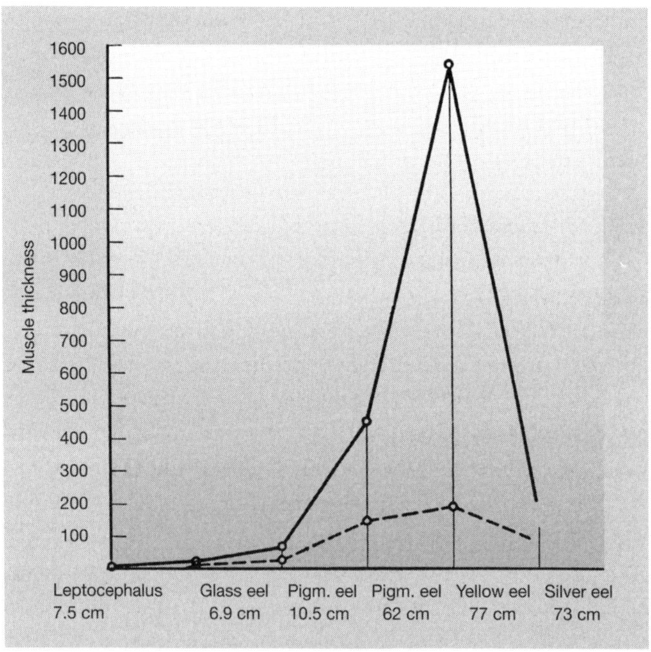

Fig. 1.17 Thickness (μm) of the circular (———) and the longitudinal (-------) muscle layer of the mid gut at particular developmental stages of the eel (after Berndt, 1938)

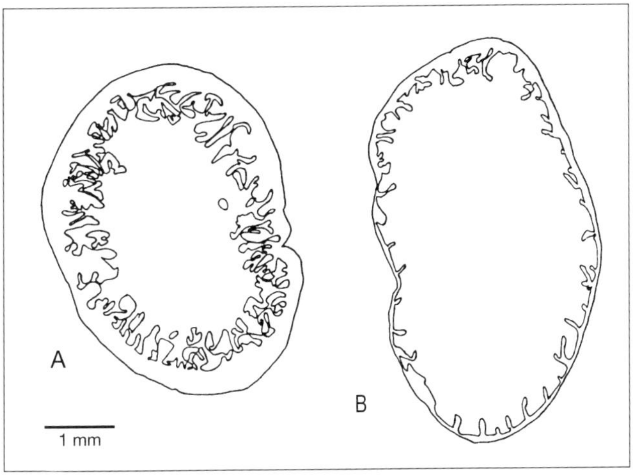

A Sexually immature silver eel (length 55 cm)
B Sexually mature eel (length 73 cm)

Fig. 1.18 Cross-section of intestines of *A. anguilla* at different developmental stages (from Pankhurst and Sorensen, 1984)

In addition to its function in feeding, the gut plays an important role in osmoregulation, especially in sea water. This function was first demonstrated by Smith (1930). He showed that a considerable amount of liquid is absorbed by the gut. Eels in which the anus had been occluded died after 3–4 days (Keys, 1933). The absorbed amount of liquid increased with development to the silver eel stage. But, also in fresh water liquid is imbibed in small amounts (Maetz and Skadhauge, 1968; Hirano, 1967; Sharrat *et al.*, 1964a; Skadhauge, 1969). The foregut with its lengthwise running gut folds is especially responsible for the uptake of water (Simmoneaux *et al.*, 1988). It is also noteworthy that during development from yellow to silver eel the rates of movement of water and ions from the lumen to blood increase and may also show positive changes from spring to autumn (Utida *et al.*, 1969). It has been suggested that these changes in rates of transport and the obligatory absorption in fresh water are adaptive features preparatory to the onset of the prolonged marine migration. The degeneration of the intestine, mentioned above, is not essential for the eel's water-receptive capability (Pankhurst and Sorensen, 1984).

The Japanese eel has been studied extensively with a view to elucidating the mechanisms of water and ion transport across the alimentary walls (Oide and Utida, 1967b, 1968; Utida and Isono, 1967; Utida *et al.*, 1967; Ando, 1975; see also Skadhauge, 1974; Kirsch and Laurent, 1975) as well as to identifying factors that govern rates of, and stimuli for drinking (Hirano *et al.*, 1972; Hirano, 1974). Alkaline phosphatase and especially adenosine triphosphatase activities are greater in seawater-adapted eels than in those adapted to fresh water, and these enzymes have been implicated as basic features in the osmoregulatory role of the gut (Gaitskell and Chester Jones, 1970, 1971; Utida *et al.*, 1972; Utida and Hirano, 1973; Ando, 1974).

Attempts to identify specific cells responsible for ion and water movement (analogous to the branchial chloride cells) on the basis of ultrastructure (Yamagishi *et al.*, 1969) indicate that ultrastructural laminae are correlated with marine environments, and probably osmoregulation. However, further studies are still required, especially with regard to the ontogeny of marine adaptation (Kaneko *et al.*, 2001).

1.5.2 Pancreas

Both endocrine and exocrine functions of the pancreas change with the developmental stage of the eel. The differing amounts and quality of food consumed at the different stages are, of course, primary reasons. Clearly, the exocrine function of producing digestive enzymes will be governed by both quality and quantity of food (herbivorous, omnivorous, etc.) available for digestion. During the various developmental stages metabolic demands also change, for example, the voraciously feeding omnivorous yellow eels must be contrasted with the migratory, fasting silver phase. Such metabolic changes are related to or dependent on, endocrine pancreatic function (Section 1.8.10).

The pancreas of the eel, unlike that in many other fishes is a compact structure surrounding the portal vein on the mid-gut (Figs 1.19 and 1.34). It is readily distinguishable from the gut by its darker colour. In a 35 cm long eel, the pancreas is 6–7 mm wide and 55–60 mm long (Kukla, 1958). Excluding a slightly tapered portion, which extends as far as the bile duct, the pancreas begins at about the point where the stomach joins the intestine

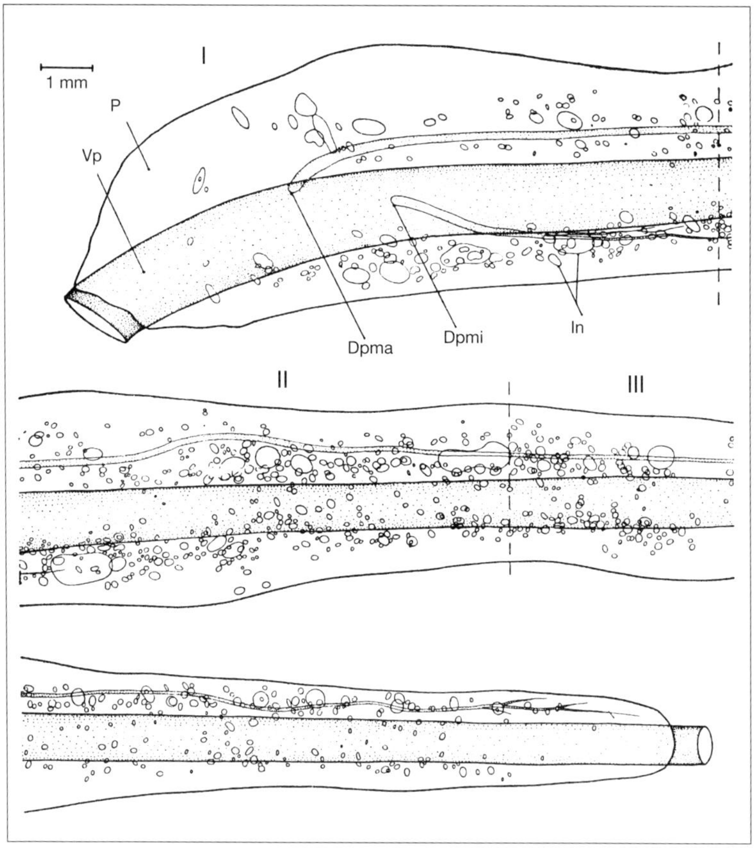

Fig. 1.19 Distribution of the Islets of Langerhans in the European eel (after Kukla, 1958) (see also Figs 1.33 and 1.34)

Dpma Ductus pancreaticus major
Dpmi Ductus pancreaticus minor
In Islets of Langerhans
P Pancreas
Vp Portal vein

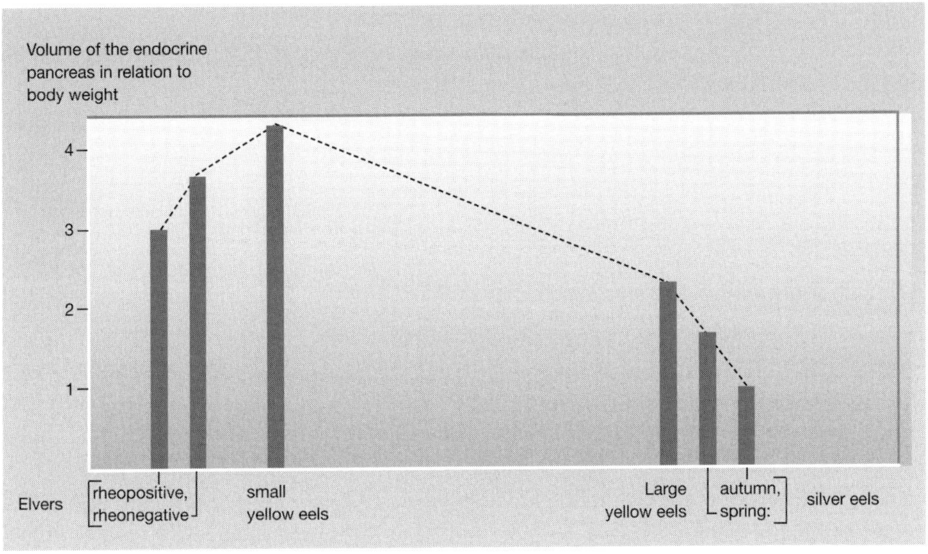

Fig. 1.20 Development of the endocrine pancreas during various developmental stages of the eel (modified after Palayer, 1967)

and it extends as far as the gastric caecum. Kukla (1958) has described the general morphology of the pancreatic vascularisation, and notes that islets of Langerhans (Fig.1.19) are less numerous in the anterior portions of the pancreas and that no extrapancreatic islets are found. Theret and Palayer (1967) have compared the pancreatic cytology of the eel with that of other teleosts, and Epple (1969) compared the pancreatic morphology of the eel with that of other fishes. Considering histophysiological criteria in the final stages of metamorphosing eel larvae, Bremer *et al.* (1975) studied the initial exocrine functions.

The relative volume of the exocrine pancreas of eels fed for 4 weeks was 1.71:1.23 for those fed for 2 weeks. Starvation reduced the volume of the pancreas to one-third (Palayer, 1962). Honma (1966) also reported that feeding changed the sizes of both the exocrine and the endocrine pancreas considerably: injecting thyroxine reduced both components of the pancreas. The endocrine pancreas changes in relation to the stage of development (Fig. 1.20). The volume generally increases up to the time of maximal feeding activity, that is, in the yellow eel. In larger yellow eels, reaching the end of their intensive feeding phase the volume is reduced, an effect especially marked in winter: in spring it resumes greater size. The volume reaches a minimum in silver eels, although activity is still required for gonadal function as well as possibly for osmoregulation in the sea (Epple and Lewis, 1975). As the female gonads are larger, so the islets of Langerhans of female silver eels reduce less strongly than in males. Removal of the pancreas reduced blood cholesterol, but caused no diabetes (Lewis and Epple, 1972; Lewis *et al.*, 1977).

1.5.3 Liver and gall bladder

Unlike other fish the eel possesses a unilobular liver with but a slight indentation at the posterior end. Aberrant types, such as double livers, have been described (Reichenbach-

Klinke, 1952). This anomaly is probably not uncommon; the liver was not exceptionally large and not only slightly divided. The two livers were well developed and each had an outlet canal although there was only one gall bladder.

The liver's role in nitrogen metabolism has been the subject of some controversial studies. The eel has much glutamate dehydrogenase in the liver but only a little in muscle tissue (McBean *et al.*, 1966). Therefore, significant transdeamination is unlikely to take place in muscle, but Janicki and Lingus (1970) have observed *in vitro* transdeaminase activity in the liver of eels. Hepatectomy had no effect on ammonia production (Kenyon, 1967), although some amino acid levels were elevated in the blood. Therefore, the liver does not seem to be essential for amino acid deamination, although excess amino acids cannot be deaminated in the absence of the liver (Watts and Watts, 1974). Hepatectomised eels suffered from hypoglycaemia although they survived for up to 63 days (Inui, 1969). Like other fishes, starved glass eels had small hepatic cells (Inui and Egusa, 1967) developing external cavities in the cells for glycogen storage.

Other aspects of hepatic function including the presence of fatty acids and their esters (Hata and Hata, 1967) and characteristics of bile salts (Hatakeyama and Ukamara, 1928; Hirofuji, 1966) have been reviewed extensively in eels (Haslewood, 1969; Tammar, 1974).

1.6 Circulation of blood

1.6.1 Cardiovascular system

The structure and position of the heart in the pericardial cavity corresponds much with those of true bony fishes (Fig. 1.21), but, the bulbus arteriosus is comparatively small which is a consequence of the much caudally displaced position (Fiedler, 1991). Skramlik (cited by Wunder, 1936) suggests that the eel is unusual with respect to the number and distribution of its many pacemakers. Three (Bielig, 1931) or four (Grodzinski, 1954) have been suggested, the major one being associated with the atrium (Fig. 1.22). The most important pacemaker is probably that in the auricular canal (Bielig, 1931; Grodzinski, 1954). The ventricle pacemaker is completely controlled by this pacemaker. The pacemaker of the sinus venosus at best accelerates the pacemaker of the auricular canal. Grodzinski (1954) suggests that a fourth pacemaker is situated in the dorsal part of the ventricle. Electron micrographic techniques have been applied to the atrial musculature, but have failed to produce unequivocal information on its pacemaker activities (Conteaux and Laurent, 1957). Randall (1968, 1970) reviews these aspects of cardiac function in eels and other fishes generally. Franklin and Davie (1991) described the function of the pericardial sac, which assists the pumping activity of the eel's heart.

Cardiac weight relative to body weight is less in eels than in other fishes. Törlitz (1922) noted that 'broad-headed' eels (Sections 3.3.1.4 and 3.3.2.4) have hearts (including the bulbus arteriosus) weighing some 1.2% of the body weight, while in 'narrow-headed' eels this weight is 0.8%. When the bulbus arteriosus is excluded, the weights are 0.9% and 0.6%, respectively. Wunder (1936) suggested that the 'broad-headed' variety with its marauding behaviour required a large heart, in contrast to the more peaceful 'narrow-headed' eels.

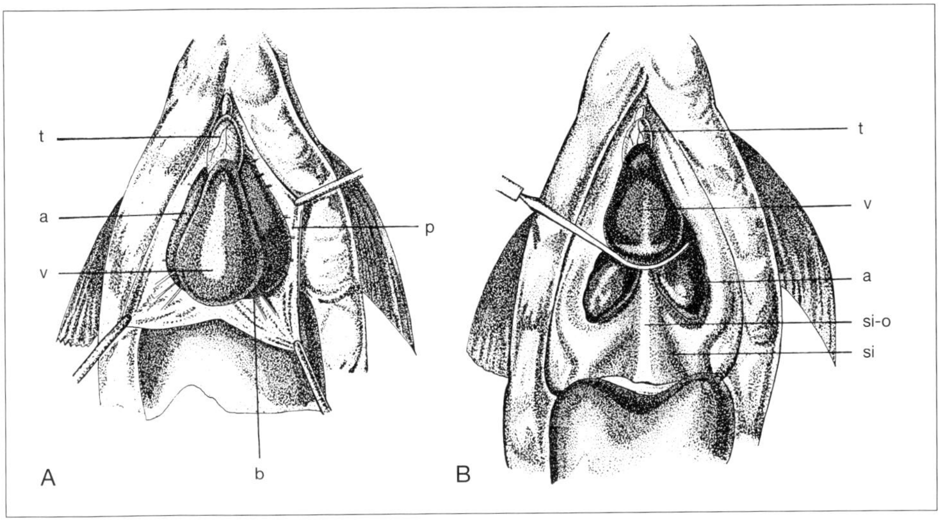

Fig. 1.21 Position of the eel's heart in the pericardial cavity (modified from Wunder, 1936 after Troemer, 1932)

A With the heart in its natural position
B With the ventricle lifted so that the transition of the sinus venosus into the auricular canal can be seen

a Atrium
b Connective tissue threads by which the atrium and the ventricle are attached to the pericardial sac

p Pericardial sac
si Sinus venosus
si-o Transition of the sinus venosus into the auricular canal
t Truncus arteriosus
v Ventricle

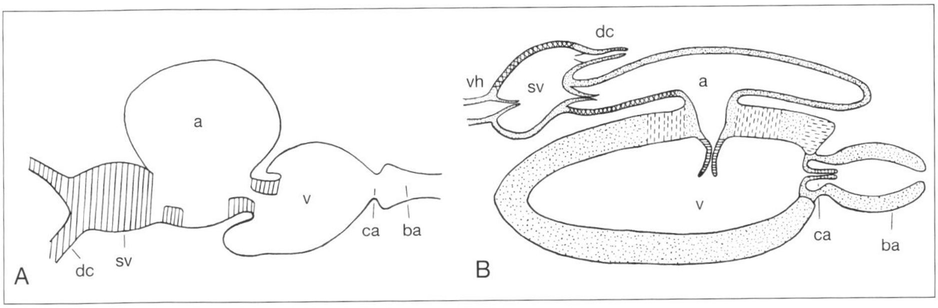

Fig. 1.22 The pacemakers of the eel's heart

A After Bielig (1931)
B After Grodzinski (1954)

a Auricle
ba Bulbus arteriosus

ca Conus arteriosus (also called the bulbus cordis)
dc Ductus cuvieri
sv Sinus venosus
v Ventricle
vh Hepatic vein

Cardiac activity and electrocardiogram (ECG) traces, first described by Gitter (1933) and Wunder (1936), and the different influences on them, have been studied extensively. Cardiac rate both *in vivo* and *in vitro* increases with increasing temperature; thus, intact yellow eels have a cardiac rate of 20/min at 10°C, and 95/min at 30°C (Grodzinski, 1954). Above the latter temperature heart rate decreases sharply, indicating an optimal temperature below this level.

Increases in heart rate also occur in glass eels transferred from fresh water into sea water, although the opposite transfer has little effect (Lhotte, 1945). Chloride ions in sea water activated the parasympathetic nervous system, which dilated the peripheral blood vessels and hence cardiac output. Many drugs have been tested on cardiac function in the eel (Fänge and Östlund, 1954). Adrenalin increased cardiac rate from 20 to 22–33, noradrenalin from 24 to 33 and dopamine from 30 to 36. Drugs seemed to have less influence on the eel than on other fish. Jäger (1975) studied the influence of different drug concentrations on the muscle fibres of the heart and their action potentials; Forster (1976) localised the effect of drugs in the cardiovascular system. Daily rhythmic variation of catecholamine concentration in the brain of the eel was noted by Le Bras (1979). With respect to their influence on the function of the heart a number of other drugs have been tested by Chow and Chan (1975), Chan and Chow (1976) and Mott (1951).

The ECG of the eel, originally considered unusual among fishes (Oets, 1950), resembles that of other anamniote vertebrates. The differences are explicable on the basis of times of depolarisation, and intervals between the QRS and T components (Noseda *et al.*, 1963; Randall, 1970).

The amounts of *N*-acetylhistidine, a substance possibly involved with acetylcholine synthesis, found in the heart of eels adapted to fresh water, are similar to those in the hearts of eels in sea water or brackish water (Hanson, 1966). The levels were similar to those of stenohaline marine teleosts, which characteristically are lower than stenohaline freshwater fishes. As with other physiological comparisons, the eel then belongs to a marine type rather than a freshwater one, and is unlike members of the anadromous Salmonidae and *Osmerus eperlanus*. The latter displays the high cardiac *N*-acetylhistidine of a freshwater fish.

Mott (1950, 1951) has given a detailed angiographic description of the general vascular system of the anaesthetised eel, and describes some action of vasoactive materials on blood pressure and flow. Various regional vascular beds have been described and include the digestive tract (Kukla, 1954), skin (Jakubowski, 1960a) olfactory bulb (Kaluza, 1959) and swimbladder (Steen, 1963a–c).

Eels possess a well-developed lymphatic system, which requires accessory pumping structures or lymph hearts. The lymphatic system (Dunajewski, 1930; Florkowski, 1930) and the caudal heart situated at the tip of the tail (Tright, 1913; Mislin, 1960; Harder, 1964) are morphologically well documented. The caudal heart, originally described as a 'ventrical and valve' system by von Leeuwenhoek (1695: see Mislin, 1960), situated in the tail fin posterior to the last vertebra, has proved suitable for studying the effects of various environmental and pharmacological agents. Temperature (Jankowski, 1968) affected caudal heart rate, which reached 380 beats/min at 34°C, while the cardiac rate was at a maximum (see above). Caudal heart rate always exceeds the cardiac rate at all temperatures. Other agents shown to act on the caudal heart are those emanating from the caudal neurosecretory system (Chan, 1971). Caudal heart rate seems to be governed by both peripheral and central nervous mechanisms.

1.6.2 Blood

The major characteristics of European eel blood include: 1.4×10^6 red cells/mm^3; haematocrit 37%; haemoglobin 63–70 g/100 ml; blood volume 3.2% of body weight; red cell dimensions 23 × 10 μm with surface area of 199 μm^2(or 2.9 dm^2/ml blood (Schucher, 1927 in Wunder, 1936; Steen and Berg, 1966; Zaitseva, 1966, 1967). The values for erythrocyte dimensions and concentrations are considerably higher than those observed in other teleosts (Swarts, 1969), although it should be pointed out that considerable ranges are reported in the publications cited above.

Blood makes up 1.1–1.15% of the body weight of glass eels, and haemoglobin concentrations increase in proportion to body weight: 0.22 g glass eels had a haemoglobin content of 4.7 g/100 ml; up to 1 g they had 10.2 g/100 ml; and young eels >1 g in weight had a haemoglobin concentration of 11.7 g/100 ml (Zaitseva, 1979).

An interesting case of severe anaemia in the eel has been described by Steen and Berg (1966), and a comparison was made with the haemoglobin-free ice fishes *Chamsocephalus aceratus* and *C. esox*. The anaemic eel had a blood volume (1.3% of the body weight) somewhat greater than normal (0.8%), which is probably important for a more frequent passage through the gills. Swimbladder gas content (Section 1.4) and blood pH (normal eel 7.5; anaemic 7.25) were relatively normal. The oxygen tension in both arterial (0.9 *versus* 12 vol.%) and venous blood (0.6 *versus* 6 vol.%) were considerably below normal (see also Itazawa, 1970), but there were no obvious defects in general behaviour. It is possible that the general rate of clearance through the gills is adequate to satisfy the oxygen demands of the fish, although experiments on the ability of such animals to undergo periods of oxygen debt require study. It should be pointed out that the leptocephalus larva of the eel is without blood pigment (Schnakenbeck, 1955). The ontogeny of haemoglobin synthesis in eels requires further study, and Kawamoto (1929) suggested that oxygen partial pressure/saturation/dissociation relationships in the eel were unique. However, the latter was disputed by Riggs (1952), while the presence of the Root effect in eels (Forster and Steen, 1969) could explain earlier findings.

The general haematology, including numbers of red and white cells, is affected by temperature (Durairaj, 1970) and by the general state of the fish's health, including the presence of certain cutaneous and systemic infections (Enomoto, 1969a, b). In the European eel, after infection with *Vibrio anguillarum*, depression of the number of erythrocytes and of haemoglobin and haemotocrit could be observed (Kreutzmann, 1973). Chloramphenicol and, to a lesser extent, oxytetracycline, exhibited negative influence on blood cells and plasma (Kreutzmann, 1977). After punctuation of the heart the loss of blood was compensated within 6 days (Kreutzmann, 1976).

Environmental salinity also affects the numbers of red cells, although considerable variations occur (Fontaine *et al.*, 1945): after 8 days in sea water the red cell count fell from the original value on transfer, 1.7×10^6 cells/mm^3, to about 1 × 10 cells/mm^3. Transfer of eels to sea water, in addition to reducing the number of red cells, also decreased plasma protein concentration; the latter, unlike the former, is a less reversible process (Firly and Fontaine, 1932), and the acid–base balance and pH of the blood must clearly be affected by the changed environment (Fontaine and Boucher-Firly, 1933). Moreover, the longer eels are kept in sea water, the more the blood CO_2 decreased. Thus, the buffering capacity of blood falls in a medium which itself displays a greater buffering capacity.

Artificial maturation of eels changed the composition of their blood (Hilge and Klinger, 1978): the number of erythrocytes decreased although the number of white cells first increased and then decreased.

The electrolyte concentrations in blood plasma of eels adapted to sea water are generally lower than those of truly marine fishes, although when adapted to fresh water, eel plasma has concentrations of electrolytes similar to those of stenohaline freshwater species (Holmes and Donaldson, 1969). The total osmolarities of plasma in both freshwater- and seawater-adapted eels are similar to stenohaline species in these respective environments (neuroendocrine regulation of electrolyte concentration: see Section 1.8).

Studies on the regulation of blood pressure in the eel are known from Chester Jones *et al.* (1966). They show that the corpuscles of Stannius play an important role that follows from experiments removing these corpuscles or injecting extracts of the corpuscles into silver eels (see the corresponding endocrinological sections, Sections 1.8.3 and 1.8.13).

The protein content of eel blood shows seasonal variation (Kvasova, 1966) and probably also varies with sexual maturity (Ochiai *et al.*, 1975; Takashima *et al.*, 1979); furthermore, as many as 12 different antigens have been identified (Fine and Drilhon, 1961; Fine *et al.*, 1963). Of special interest is the occurrence of green chromoprotein, thought to be largely biliverdin (Oide and Utida, 1967a; Yamagushi *et al.*, 1968).

The material shows an annual cycle and is probably associated with nutritional status. Larsson and Fänge (1969) have shown that there are changes in some chemical constituents of the blood (cholesterol, fatty acids, haemoglobin, urea and proteins) during the metamorphosis from the yellow to the silver eel stages. Poluhovich and Park (1972) compared eels from sea water with eels from fresh water: analysis from seawater eels exhibited higher values of Ca, Cl, cholesterin, total protein and osmotic pressure; freshwater specimens showed higher content of glucose, higher pH values and more consumption of oxygen.

The haemoglobins of Japanese (Yoshioka *et al.*, 1968) and of European eels (Itida *et al.*, 1970) have been studied using ion-exchange chromatography. The studies on *A. anguilla* displayed two separable fractions in the haemoglobin; the biochemical reactions of these fractions could be identified (Itida *et al.*, 1970). The general biochemical and physical properties of the eel haemoglobin were reviewed by Christomanos (1968).

Finally, some mention should be made of the relatively high toxicity of the eel's blood as compared with that of other fishes. The so-called 'ichthyotoxin' causes muscular cramp; small doses injected into mammals accelerate the rate of breathing and the heart beat. A 0.1 ml/kg volume of eel blood is sufficient to kill a rabbit; 0.2 ml/kg will kill a dog (Mosso, in Guibé, 1958). Wunder (1936) stated that: 'If placed on mucous membranes, the poison produces inflammation which can last for days; one must be particularly careful that eel blood does not contact the human eye while the eel is being killed. The toxin is destroyed by heating to 58–70°C, and there is no need to fear any effects of the poison after preparation of the fish ...'. The effects have been classified into two components: neurotoxic and cytolytic by Bertin (1956), who also gave further characteristics and a brief history of this substance. Rocca and Chiretti (1964) discovered a toxic protein in the eel serum.

1.7 Urinogenital system

1.7.1 The kidneys

Original descriptions of the kidney in eels include those of Haller (1908) and Jourdain (1859) (see also Harder, 1964; Bridge and Boulanger, 1958). The kidney is primarily a paired organ which shows degrees of caudal fusion, the pronephric portion remaining separate as lymphoid, haematopoetic structures containing few nephrons. The posterior kidney, a mesonephros, performs the usual excretory and endocrine functions (see Section 1.8.12). A large part of the blood from the caudal circulation enters the kidney directly while some can effectively bypass the kidney and enter the hepatic portal system. Histological and biochemical investigations of the kidney in eels include those of Gérard (1958) and Guglielmone (1966, 1967a, b). A detailed study on the vascular supply is provided by Ditrich and Splechtna (1993), and it has been argued that the general form of the nephron in eels resembles that of a stenohaline freshwater fish (Grafflin, 1937). Pequignot and Serfaty (1965) suggested that respiratory responses both of the gills and the kidneys were similar to those of freshwater teleosts. However, the physiologically stenohaline freshwater characteristics of the eel are unusual (Section 1.8.3). Olivereau (1971) and Olivereau and Olivereau (1977) have related the renal histology to water and electrolyte composition as well as to the endocrine status of the eel. The kidney plays a major role, both in sea water and fresh water. Thus, in fresh water, eels have much higher glomerular filtration rates than they do in sea water (Sharrat *et al.*, 1964b). In fresh water, the urine is very dilute with respect to plasma, as a result of intense tubular reabsorbtive activity, especially for sodium and chloride (Chester Jones *et al.*, 1969a). On adaptation to sea water there is an acute reduction in the glomerular filtration rate; at the same time, tubular function, although still carrying reabsorbtive activity, shows divalent cation secretion (Chester Jones *et al.*, 1969a; Oide and Utida, 1967b; Butler, 1969). Renal mechanisms are largely responsible for the volume adjustments as reflected in body weight changes that occur in eels when transferred from fresh to sea water or *vice versa*.

The kidney of eels contains within its vasculature granular epithelioid cells, homologous with the renin-producing juxtaglomerular cells of other vertebrates (Sokabe and Ogawa, 1974; Nishimura and Ogawa, 1973; Krishnamurthy and Bern, 1969). Renal extracts contain a renin-like enzyme, which interacts with a substrate in plasma (Chester Jones *et al.*, 1966; Sokabe *et al.*, 1966). The function of renin in eels is unknown but blood pressure regulation and the regulation of glomerular filtration rate have been suggested (Chester Jones *et al.*, 1966; Sokabe *et al.*, 1973). The changes that occur in plasma renin and renal activity are at present difficult to interpret, although the observations noted in the publications cited argue against an exactly homologous function between the mammalian renin–angiotensin system and that of the eel (Section 1.8).

1.7.2 Gonads and early larvae

The gonads lie freely suspended almost through the entire length of the body cavity (Figs 1.15 and 1.23).

The location of the gonad in the immature eel and the determination as male or female in the differentiated specimen is often difficult; fatty tissue (Fig. 1.23) is some times misidentified as gonad! The gonads are of unequal size, with the right extending further forward (about 1 cm in 30 cm eels), and the left reaching further posteriorly (about 2 cm behind the cloaca in 30 cm eels). The left is about 2–3% longer (Tesch, unpublished), is heavier and contains more eggs than the right gonad (Matsui, 1952). Kokhnenko (1959) related head width to ovarian width; broad-headed specimens had gonads about 5–6 mm

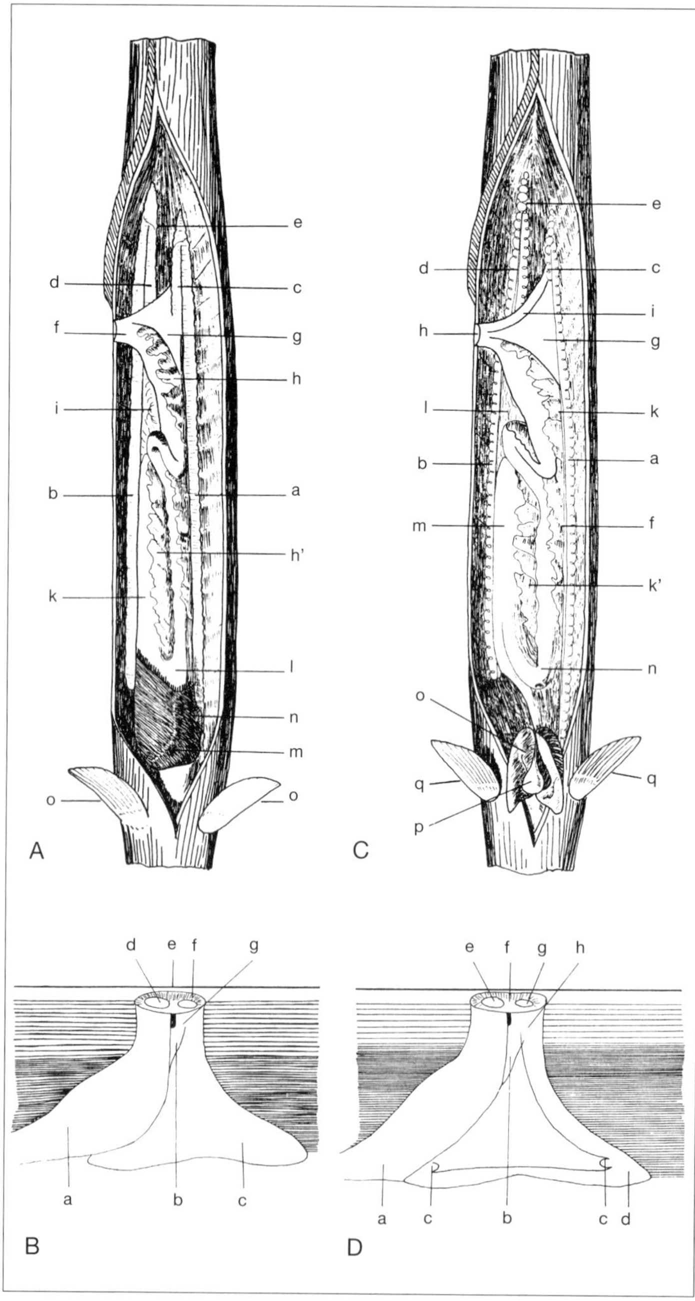

Fig. 1.23 Longitudinal section through the body of the female and the male eel (after Syrski, 1876)

A Female eel
a/b Right/left ovary
c/d Section leading to the right/left ovary
e Dividing wall between the two ovaries
f Anal opening
g Urinary bladder
h/i Fat on the right/left side which is often mistakenly thought to be the testis
h' Fat on the stomach
k Stomach
l Pylorus
m Liver
n Gall bladder
o Pectoral fins

B Anal region of a female eel
a Intestine
b Fissura recto-vesicalis
c Urinary bladder
d Anal opening
e Dividing wall
f Urogenital opening
g Orifice of the genital opening into the urethra

C Male eel
a/b Right/left testis
c/d Section leading to the right/left testis
e Dividing wall between the two testes
f Vas deferens
g Bursa seminalis
h Anal pit
i Urinary bladder (largly covered by the bursa seminalis)
k,l,k' Fat on the right/left side and on the stomach
m Stomach
n Pylorus
o Liver (displaced slightly to reveal the inner side adjacent to the oesophagus and the stomach
p Gall bladder
q Pectoral fin

D Anal region of a male eel
a Intestine
b Fissura recto-vesicalis (covered by the external wall of the bursa seminalis)
c Opening of the anterior and posterior part of the vas deferens into the bursa seminalis
d Urinary bladder
e Anal opening
f Dividing wall
g Urogenital opening
h Orifice of the genital opening into the urinary duct

across, while eels with more pointed heads had ovaries about 9 mm wide. The earliest stage at which it is possible to distinguish male from female gonads macroscopically is when the fish reaches a length of about 20 cm (Walter, 1910; Colombo *et al.*, 1984), although, according to some authors, definite identification cannot be made until a length of 30–35 cm has been reached (Colombo and Grandi, 1996; see also Sinha and Jones, 1966; Kokhnenko and Gorovaya, 1965; Satoh *et al.*, 1962). The development of male and female gonads is similar in European and Japanese eels.

Protrusions appear on the genital ridge or undifferentiated cord of the potential male gonad; these finally come to overlap one another and are referred to as 'lobes' (Fig. 1.24). In the female, the developing folds come to resemble a frill, as ruffles develop along the gonadal cord. Indeed, the female gonads are sometimes referred to as the 'frilled organ' (Fig. 1.24). In the yellow and earlier stages of the silver eel phases, ovaries and testes are insignificant-looking structures, but in the fully mature animal they are impressive structures (Fig. 1.25), preserving their earlier form of 'lobes' and 'frills.'

Fatty degeneration of gonadal tissue (Schnakenbeck, 1934) may give an outward appearance of mature ovaries to the gonads; a microscopic check is necessary in such cases. The gonadal cells of degenerated ovaries are very immature, and an unusually large number of fatty globules surround the oocytes. Such anomalies are possibly more common than was first thought and some instances of eels in supposedly advanced stages of maturity may indeed be examples of fatty degeneration (Rasmussen, 1951; Wundsch, 1953). In the studies cited, the fish were outwardly mature but the oocytes had diameters of 0.1–0.2 mm, a size range no larger than that found in normal immature silver eels. Bezdzenyezhnykh (1973, 1974) found resorptions of the oocytes in silver eels, which had

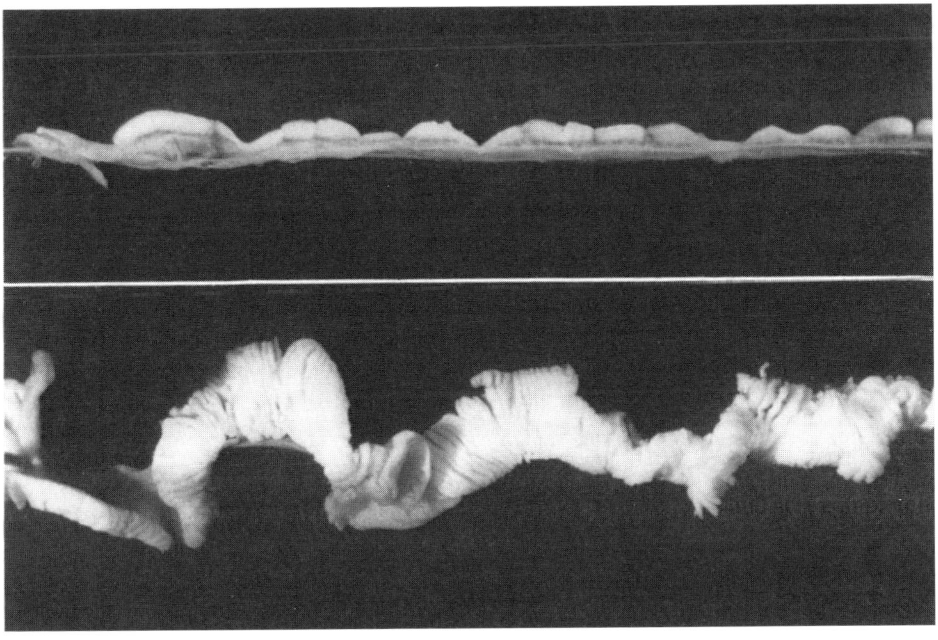

Fig. 1.24 Gonads (enlarged 2×) of the European eel (from Kuhlmann, 1975); above: male gonad from a 42 cm long eel, below: female gonad from a 57.5 cm long eel

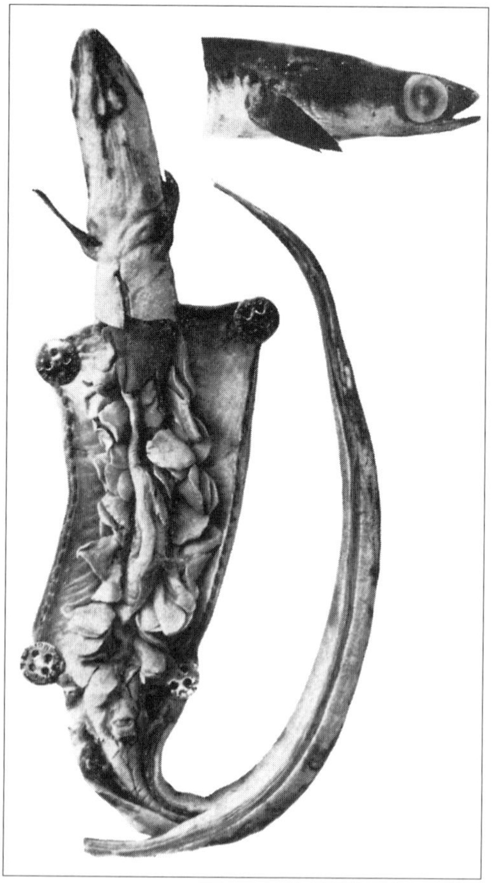

Fig. 1.25 Male eel, 34 cm long, in an advanced stage of maturity, caught in Danish waters (Prästö-Fjord, western Baltic, near the Öresund) (after Schmidt, 1906)

been prevented from migrating for a period of 10 years; he considered this resorption to be phagocytic. The oocytes will occasionally show simultaneous re-growth.

Microscopic examination of a smear or tissue section is of diagnostic value, particularly since the results of microscopic and macroscopic examinations do not always coincide (Sinha and Jones, 1966). The former technique demonstrates oocytes in macroscopic lobes although often only single oocytes occur in the lobes (Peñaz and Tesch, 1970). It is possible that testicular tissue could still develop from this apparently intersexual condition.

The primordial stage (1) of testis development, perhaps the most commonly seen, leads to the more advanced stages of maturity (Fig. 1.26). The sketches shown were made from an eel treated with hormones; the more advanced stages are rarely encountered in nature.

Matsui (1952) and Kuhlmann (1975) (see also Kokhnenko and Bezdzenyezhnykh, 1973b) have described the histology of the developing gonad of the eel. Immature ova are readily distinguishable from immature sperm by virtue of their greater size. A recently published study by Colombo and Grandi (1996) provided a detailed description of histological tissue sections from all development stages of male and female gonads of the

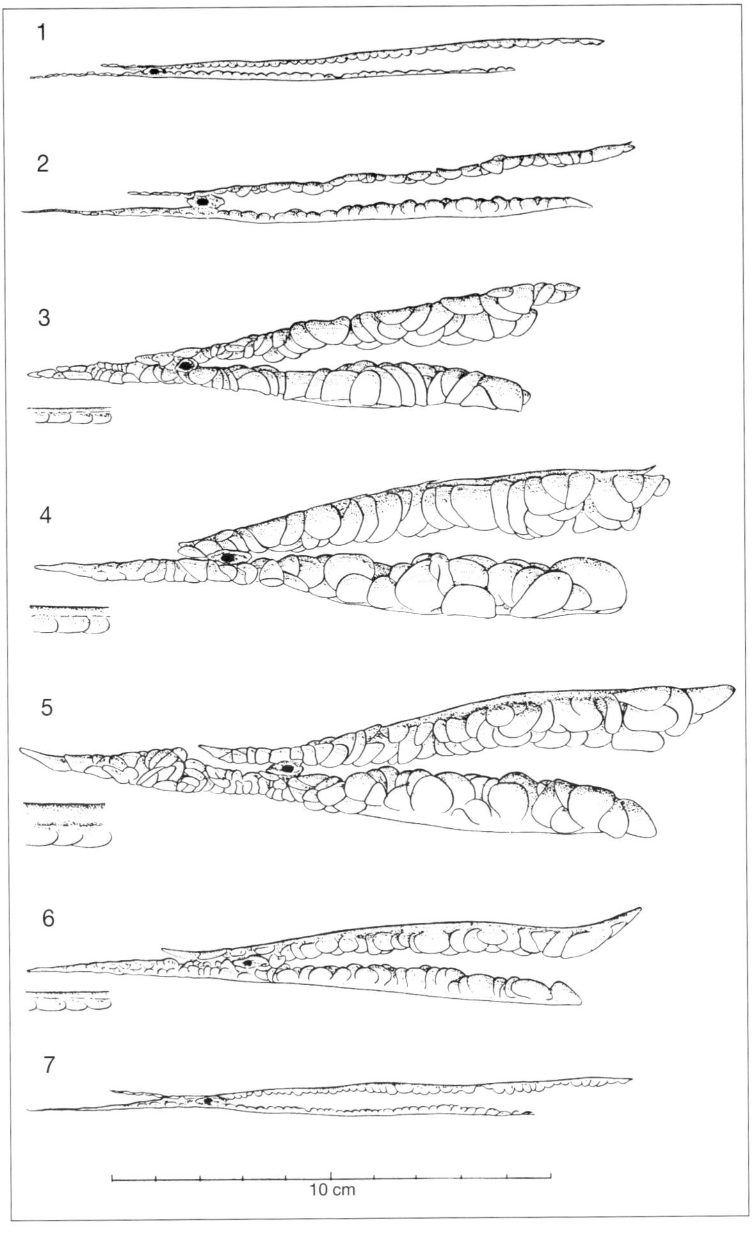

Fig. 1.26 Schematic representation of the male gonads in seven developmental stages. Stage 1 represents a normal, untreated silver eel which was caught at the exit of the Baltic during its spawning migration. The remaining sections show hormonally treated specimens of stages 2–7 (after Boëtius and Boëtius 1967a)

European eel. This study confirmed the earlier results (see above) that a definite differentiation, microscopically, is possible only with animals >30 cm long.

The histological stages of gonadal development have been related to body length, and such information for *A. anguilla* and *A. japonica* is summarised in Table 1.2. Sexual differentiation in these two species of eel occurs at similar body lengths (Colombo and Grandi,

Table 1.2 Histological development of the gonads in the eel after Colombo and Grandi (1996).

6–10 cm long eels	Gonads exhibit pear-shaped ridges attached to the dorsal wall of the abdominal cavity, per cross section they contain 1–4 primary primordial germ cells (PGC1)
10–15 cm long eels	Gonads exhibit blade-like, lamellar shape 5–10 PGC1
15–18 cm long eels	More than 15 PGC per cross-section, mitoses of PGC give rise to PGC2 clusters, they surround some of the PGC1
18–20 cm long eels	40–50 germ cells per cross-section, in addition to PGC1 and clusters of PGC2 oogonia (OOG) and also early meiotic oocytes (OOC1) are visible
20–22 cm long eels	Gonadal lamellae larger than in previous stages, besides germ cells of previous stages: primary spermatogonia (SPG1), some OOC2 and OOC3, but also PGC2
22–25 cm long eels	Macroscopically appearing as the lobed tape typical of a testis, cytologically and histologically composed of three kinds of structure: (1) Structure similar as in 20–22 cm long eels, made of rows of PGC2 and SPG1, of OOG and OOC1 cysts, of OOC2 and few OOC3 intermingled in the same area. (2) Structure composed of female and nearby male areas. Female areas show the presence of OOG and OOC1 cysts and of single OOC2 and OOC3. Male areas consist of rows of PGC2 and SPG1. (3) Macroscopic appearance of a lobed lamella, few PGC, cysts of OOG and OOC1, and a large number of OOC2 and OOC3
25–30 cm long eels	All types of gonads found in 22–25 cm long eels occur also here, some show degenerating and dying oocytes, fully differentiated immature small ovaries with large perivitellinogenic oocytes (OOC4)
Eels >30 cm	Some gonads similar as in smaller eels, many showed early testes, fully differentiated immature testes and fully differentiated immature ovaries. Early testes show rows containing PGC2 and SPG1, some SPGA (terminology of Miura *et al.*, 1991) are present, still few oogonia and oocytes present. Fully differentiated, immature testes exhibit long cords of SPGA and some cysts of SPGB; in the ovaries numbers of OOC4 increased

1996: see also Satoh *et al.*, 1962). Size and gonad development are probably similarly related in other species of eel. Species in which a large adult size is reached, for example *A. dieffenbachia* of the South Pacific, probably show a similar developmental pattern, but with relatively larger proportion. The findings presented in Table 1.2 are to a certain extent contradictory as regards the sexual differentiation of the young eel. Thus, Satoh *et al.* (1962) found no intersexuality in the Japanese eel, although, in a later study, Takahashi and Sugimoto (1978) detected bi-sexual gonads in a Japanese eel. For the European eel, reference should be made to Kuhlmann (1975), who was able to identify an intersexual stage. Furthermore, ecological and experimental studies of the European eel (van Oordt and Bretschneider, 1942; Tesch, 1928; Gandolfi-Hornyold, 1932b; D'Ancona, 1951a; Peñaz and Tesch, 1970) have shown that varying degrees of ambisexuality in young eels can result from environmental factors (see also Sinha and Jones, 1966 and discussion below). It should also be pointed out that gonadal growth is more marked in the autumn than in winter (Satoh *et al.*, 1962).

More information is available on the later development of the gonads of the eel once it has left continental waters. On the continent, a lack of pituitary gonadotropin synthesis and releases results in a blockage of gonadal development. Gonadotropins or pituitary extracts and sex steroids are the main hormones, which have been applied for the induction of artificial maturation in eels. Reviews on this matter are available from Fontaine and Dufour (1991) and Kagawa (2002). At stage 5 in testicular development in such preparations, the testes have maximally distended vasa deferentia in which spematozoa are stored (Fig. 1.26). During stage 6/7 spawning begins, after which the gonad is barely distinguishable from the first development stage.

It seems that male eels show little species specificity to exogenous hormones. The induction of sexual maturation in female eels using exogenous hormones has proved to be more difficult. Fontaine *et al.* (1964) were the first to successfully induce maturity in female *A. anguilla* using extracts from carp hypophysis; three months later, shortly before maturity, desoxycorticosterone acetate was also administered. The most recent results were obtained using the Japanese eel: 21–23 weekly injections of salmon gonadotropin fractions (sGTH) were effective for inducing ovarian maturation (Sato and Aida, 2000). The effects of sGTH I and II were different. Both fractions were required for the induction of ovarian maturation. In the ovaries that complete vitellogenesis, final oocyte maturation and ovulation were obtained by a simultaneous injection of sGTH and 17α, 20β-dihydro-4-pregnen-3-one (DHP). Liao and Chang (2001) applied the same co-injection but for the preceding weekly gonadotropin injections four catfish pituitary extracts each were used. The endocrine mechanisms are described in the endocrinological part, especially in Section 1.8.9.

In sea water, the weight of mature gonads in male *A. anguilla* comprises 3.6% of the body weight, according to Meske and Cellarius (1973), and 9.8% according to Bieniarz and Eppler (1975); in fresh water this percentage varied between 7.3% and 7.5%. Gonads of untreated animals weighed between 0.2% and 0.6% of the body weight. On the basis of electron microscope studies, Ginsburg and Billard (1972) described the morphology of sperms in *A. anguilla*, and Colak and Yamamoto (1974) and Okamura *et al.* (2000) have done the same for *A. japonica*. The latter authors reviewed the earlier studies including other Anguilliformes and detected some discrepancies; further investigations on this spermatozoon using not only TEM but also SEM will be needed to understand its exact three-dimensional structure and evolutionary relationships.

Maturity is accompanied by decalcification of the bones (Lopez *et al.*, 1970), mainly through a reduction in calcium phosphate.

In addition to the yolk granules, the mature eggs also contain droplets of oil (Fig. 1.27) (Fontaine *et al.*, 1964; Yamamoto *et al.*, 1974b; Edel, 1975a) which later converge and fuse gradually (Sugimoto and Takeuchi, 1976; Liao and Chang, 2000). The relationship between egg and yolk granule size possibly differs between species as comparisons between *A. australis* and *A. dieffenbachii* have shown (Todd, 1979). It should be noted here that gonads of silver eels from southern North America have larger and, therefore, more mature eggs than those of animals from Canada. In addition, it is important to point out that, in a comparison with eels of the same age but different sizes, larger eels have more fully developed ovaries (i.e. their oocytes are larger in diameter) than do smaller eels

(Kokhnenko and Bezdzenyezhnykh, 1973a). Body size and fecundity seem to be correlated positively: larger females spawn more eggs than smaller ones (Barbin and McCleave, 1997).

Japanese experiments (Yamamoto *et al.*, 1974a) have shown that hormone injections cause the body weight of female eels to increase by varying amounts. Body weight also showed a marked increase in *A. rostrata* (Edel, 1975a). Body weight of *A. japonica* is reported to increase by more than 100% (Liao and Chang, 2000).

A

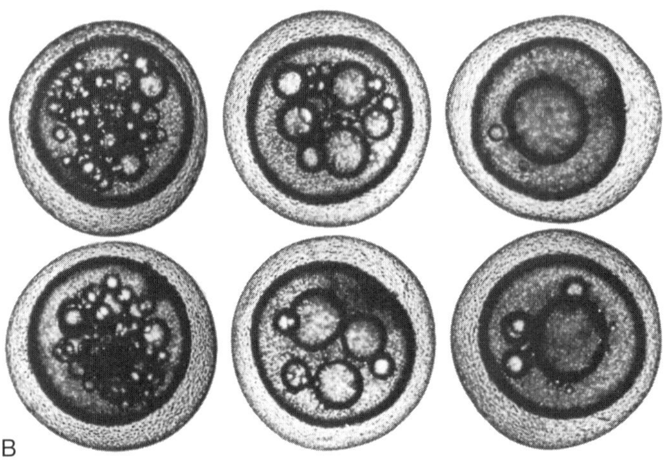

B

A Female and male eel treated with hormones and spawning under experimental conditions showing release of sperms (arrow)

B The eggs from these experiments probably differently developed. Left: early stages with many small oil globules. Central: transitional stage with a small number of large oil globules. Right: late stage with one large oil globule.

Fig. 1.27 Spawning behaviour of the European eel and eggs obtained by these studies (after Boëtius and Boëtius,1980)

Oogenesis in eels does not vary greatly from that in other species (Yamamoto *et al.*, 1974b). The number of eggs in mature *A. rostrata* weighing 560 g totalled 1.3–1.5 million (Edel, 1975a). Similar figures have been published for silver eels (*A. rostrata*) from Chesapeake Bay (Wenner and Musick, 1974); the number of oocytes (Y) can be calculated from the body length (X) as follows:

$$\mathrm{Log}\,Y = -4.2951 + 3.74418 \log X$$

Obtaining male and female *A. japonica* that reached sexual maturity simultaneously, Yamamoto and Yamauchi (1974) and a Chinese research group (Research Group Eel Reproduction Xiamen, 1978) were the first to successfully carry out artificial fertilisation of eel eggs. In addition, reports on unsuccessful experimental fertilisation of the Atlantic eel species came from Europe and North America (Ghittino *et al.*, 1975; Eppler and Bieniarz, 1978). Successful maturation and hatching experiments with European eels are described by Bezdzenyezhnykh *et al.* (1983a, b). They obtained 3 mm long larvae 50–60 h after fertilisation. The Japanese scientists (Yamamoto *et al.*, 1974a) used 20 mature males, treated with synahorin, and five females. The females were injected with extracts of eight salmon pituitary glands/kg eel (dried in acetone). Fertilisation was effected in salt water at a temperature of 23°C. Hatching occurred 38–45 h later. Length increase from the first to the fifth day was 4.8 mm, 5.3 mm, 5.9 mm and 6.2 mm, respectively. Opening of the mouth and anus as well as formation of the pectoral fin occurred on the third day. After 5 days the larvae had developed 53 pre-anal and 48 postanal myomeres which, together, is 15 myomeres less than indicated for *A. japonica* in Table 1.1. Counting myomeres of the smallest eel larvae (<5 mm) captured in the Sargasso Sea did not show such a deficit (Schoth, 1982).

Recent culturing experiments in Japan succeeded in extending the survival of eel larvae to 253 days (Liao and Chang, 2000). The longest body length recorded of a 209-day-old larva, was 31 mm. The main problem of obtaining larger larvae is their nutrition. Extracts of alimentary canal samples of *A. japonica* and conger eel *Gnatophis nystromi* captured in the ocean were analysed by immunological methods; cnidarians and tunicates were used as indicators. From the results and other observations, such as the epithelial structure of the intestine, it is concluded that so-called marine snow, derived from gelatinous plankton, such as cnidarians, is the natural food of eel larvae (Suzuki *et al.*, 2000).

An important and controversial issue from both a scientific and a practical viewpoint is the question of sex determination. From results cited in previous sections it is obvious that microscopically recognisable 'intermediate' (intersexual) stages occur in eels. Thus, the gonadal anlage passes through phases during which either male or female types could develop, and the direction of development may be influenced by environmental factors. This does not exclude genotypic sex determination and the presence of sex chromosomes (Lieder, 1963); *A. japonica*, *A. rostrata* and *A. anguilla* all contain 19 pairs of chromosomes (Chiarelli *et al.*, 1969; Sick *et al.*, 1962) and the 19th obviously differs (Ohno *et al.*, 1973; Passakas and Klekowski, 1973; Passakas, 1976; Park and Kang, 1979). In females of *A. anguilla*, *A. rostrata* and *A. japonica* this pair shows the largest (z) and the smallest (w) metacentric chromosome; in males the pair is equal (zz). This is the same in *Astroconger myriaster* (Park and Kang, 1979) and may be similar in other Anguilliformes. As shown by studies of Passakas and Tesch (1980), the possibility of genetic sex determination may assist interpretation of the phenotypic sex; sex of chromosomes compared with that of gonads can differ considerably when environmental conditions are extreme (see below). Also, later publications indicated that the phenotypic sex differentiation is only indirectly connected with the genetic sex; environmental conditions can have stronger effects than the genetic sex (Wiberg, 1983; Sola *et al.*, 1984; Cau *et al.*, 1992).

It has been impossible to induce intersexuality experimentally or to itemise exactly the environmental effects upon sex determination. However, many experiments and ecological studies are confirming the hypothesis of intersexuality (e.g. Tesch, 1928; Gandolfi-Hornyold, 1932b; Schnackenbeck, 1944, 1953b; D'Ancona, 1950; Matsui, 1952; Müller, 1967; Egusa and Hirose, 1973; Kuhlmann, 1975; Colombo and Grandi, 1996). The studies also indicate that, as density increases the number of male eels increases; decreasing density favours more females.

D'Ancona (1958, 1959b, 1961) suggested that the differential distribution of the sexes in river estuaries, in inland waters, and in the east and west Baltic, can be attributed primarily to the differential levels of migratory drive correlated with the sex genotypes. As a result, males with a low migratory drive, remain for example in the estuary of the Elbe, while the more active females migrate further up the Elbe to produce the well-established preponderance of females in the Upper Elbe, its tributaries and nearby lakes. This hypothesis is supported by many studies on the distribution of the two sexes (e.g. Walter, 1910; Gandolfi-Hornyold, 1925; Ehrenbaum, 1930; Callamand, 1943; Vladykov, 1966; Peñaz and Tesch, 1970; Saint-Paul, 1977; Fernandez-Delgado *et al.*, 1989). A similar opinion has been presented by Svärdson (1976). It has to be mentioned that this opinion is based on eel samples, which originate mainly from the first part of the last century (reproduced by Tesch, 1977, 1983). Today the relationships of stock densities are mostly quite different and, therefore, not comparable.

However, Peñaz and Tesch (1970) as well as Saint-Paul (1977) have been unable to confirm such differential sex distribution in the Elbe area. It is indeed true that in the Elbe the proportion of males is at its greatest just before the river enters the North Sea. It is also true that the proportion of males decreases upriver. In the upper waters there is an extremely high proportion of females, but the proportion of females also increases from the river estuary towards the sea. Around Heligoland, for example, there is an extremely high proportion of females. The proportion of males roughly correlates with population density: high densities with a large proportion of males (as well as hermaphrodites) occur in the estuary. The population density decreases near the sea because few of the invading glass eels remain in that region, most moving inland (Creutzberg, 1961; Tesch, 1971). Further investigations have shown that this situation in the Elbe area had not changed up to the 1980s (Passakas and Tesch, 1980), and it became evident that the different regional sex distribution up and down the river Elbe was a consequence of phenotypic development of genetic males near Heligoland towards females: in the Elbe near Hamburg it was a change of many of the genetic females towards phenotypic males. The proportion of males also drops towards the sea in other cases: the population of the IJsselmeer comprised 94% males but there were only 70% in the Wadden Sea which is nearer the open sea (Tesch, 1928). In small rivers on the east coast of North America males predominate, although the proportion of females increases in the estuaries (Winn *et al.*, 1975). Other studies in lakes

(Rasmussen, 1952; Section 3.3.2.2) also relate male predominance to increased population density as a consequence of young eel stocking. In these studies, large numbers of glass eels were introduced into a lake and the proportion of males increased as the total catch increased from 0% to 62%. In the Northern Irish Lough Neagh, glass eels were stocked from 1932 to 1947 and from 1960 the proportion of males increased from 0% to 86% (Frost 1950; Parson *et al.*, 1977). Research in the Carmargue, the estuary area of the Rhône, revealed vastly different figures for the proportions of males in three nearby waterways in the Rhône plain (Gandolfi-Hornyold, 1930a, 1935). Male to female ratios in the three places were 45 : 55, 12 : 88 and 6 : 94. Thus, a different migratory drive is an inadequate explanation for the sex distribution.

These few examples suggest that the eventual sexual differentiation of eels is not entirely dependent upon genotype. The following factors may play a role:

- The relationship between a high population density and a high proportion of males or hermaphrodites suggests that competition for food and hence relative malnutrition stimulates maleness (Burnet, 1969a). In goldfish *Carassius auratus*, malnutrition, by affecting hypophysial gonadotropic function, inhibits gonadal growth (Bogenschutz and Clemens, 1967). The reason is probably declining gonadotropin in the pituitary gland; similarly, this endocrinological mechanism could prevent development of female gonads. In goldfish, this development is reversible. The same could be true for the labile gonads of the eel. Work by Kuhlmann (1975) suggests that sex determination in *A. anguilla* may be influenced by the quality of its food; feeding elvers on cod ovaries (*Gadus morhua*) resulted in a higher proportion of females. It is also reported that males grow more slowly (Bertin, 1958; Sinha and Jones, 1967c; Peñaz and Tesch, 1970) and are smaller at sexual maturity. The slower growth rate is found largely in those younger age groups in which sexual differentiation takes place (Sinha and Jones, 1966, 1967c; Peñaz and Tesch, 1970). For example, Heligoland eels, with a high proportion of females, grew more rapidly during the first 3 years of their stay in continental waters. On the other hand, the older (already differentiated) eels showed a normal or poor rate of growth, similar to English eels. Laboratory experiments with glass eels, however, do not support these ecological observations (Meske and Cellarius, 1973; Kuhlmann, 1975); they showed that males were growing faster. Egusa and Hirose (1973) found that pond cultured males grew at the same rate until they had reached a length of 40 cm.

  Recent field (Holmgren *et al.*, 1997) as well as laboratory studies (Holmgren and Mosegaard, 1996a, b) displayed a faster weight and length increase in males than in females, but the males stopped growing at a smaller size than females. The last mentioned results do not favour the theory of competition, mentioned above: Colombo and Grandi (1996), therefore, suggest that a social component could influence sex differentiation.
- The eels often live close together, which could provide mutual influence.
- It has been suggested salt or brackish water could assist the development of a higher proportion of males (Callamand, 1943); sodium ion concentration might affect the release or action of gonadotrophic hormones, which differentially influence the ovaries rather than the testes. In addition, the increased salt concentration effectively lowers

the water content of the organism and again, by an action upon the hypophysis, might favour maleness. For example, male *Gambusia holbrooki* have a lower salt content than do females.

- However, the salt concentration can have no influence on the sex determination of the eel; numerous freshwater localities harbour males and saltwater localities females predominantly (Sinha and Jones, 1966; Peñaz and Tesch, 1970). No difference could be detected between eels grown in fresh- and saltwater ponds (Egusa and Hirose, 1973).
- No clear-cut relationship between temperature and sex determination can be established from Kuhlmann's (1975) experiments on the European eel. Exposure to moderate temperatures (20 and 21°C) resulted in a preponderance of males, low and high temperatures (17, 26, 29°C) seemed to favour female production, but extremely high temperatures produced no definite bias towards either sex. It is noteworthy that the same experiments suggested that the geographical origin of the elvers might have some influence on sex determination; elvers from the Tyrrhenian Sea produced more males than did those from the North Sea. The experiments of Holmgren (1996), and Holmgren and Mosegaard (1996a, b) seem to agree with these results. However, in many of the experiments cited above mortality was rather high and identification of the sexes was reliable for the surviving specimens, but not for all those that died.

1.8 Endocrine system (updated and rewritten by I.W. Henderson)

The eel continues to prove to be a most suitable species for investigating endocrine regulatory systems of teleost fishes. It must be recognised of course, that the eel is but one representative of more than 20,000 species of teleost. Hence, it is unwise to discuss eel endocrinology as truly representative of Teleostei generally; rather the eel is a highly specialised fish that in many instances is atypical of even relatively close relatives. A major feature of the eel is its extraordinary ability to adapt to extremes of salinity, both acutely and chronically, and to withstand some remarkable surgical interventions. It is arguable that some of the eel's phenomenal robustness reflects exceptional endocrine regulatory processes. The eel possesses the full endocrine repertoire of bony fish, including their unique corpuscles of Stannius and urophysis, the caudal neurosecretory system seemingly restricted to ganthostomatous fishes. Since the last editions of this book (Tesch 1977, 1983), considerable advances have been made in the understanding of the roles of hormones (some of which had not even been described, including endothelin and the natriuretic peptides) in the context of osmoregulation, reproduction, growth and differentiation, and energy.

1.8.1 The brain

The brain (Section 1.9.1; Fig. 1.28) of the eel, as in other vertebrates, is now recognised as a key neuroendocrine structure, which transduces environmental and internal homeostatic signals *via* neuroendocrine routes towards the hypothalamus and thence to the pituitary gland, or hypophysis. From here, hormones reach out to affect all processes from behavioural changes to adjustments in osmoregulatory capacity, to growth and differentiation. The key sensory and motor pathways link the brain, the hypothalamus, the pituitary gland and central and peripheral end-organs and tissues, one of which may be the brain itself. Although not widely and specifically addressed in the eel, it may be noted that the brain contains many hormones, which classically are present in structures such as the gut and these, probably acting in a paracrine neuromodulatory fashion, influence the pathways noted (Henderson, 1997).

1.8.2 The pineal

Information on the basic structure or function of the eel pineal is surprisingly sparse. However, as in other fishes it is likely to mediate responses to environmental light regimes.

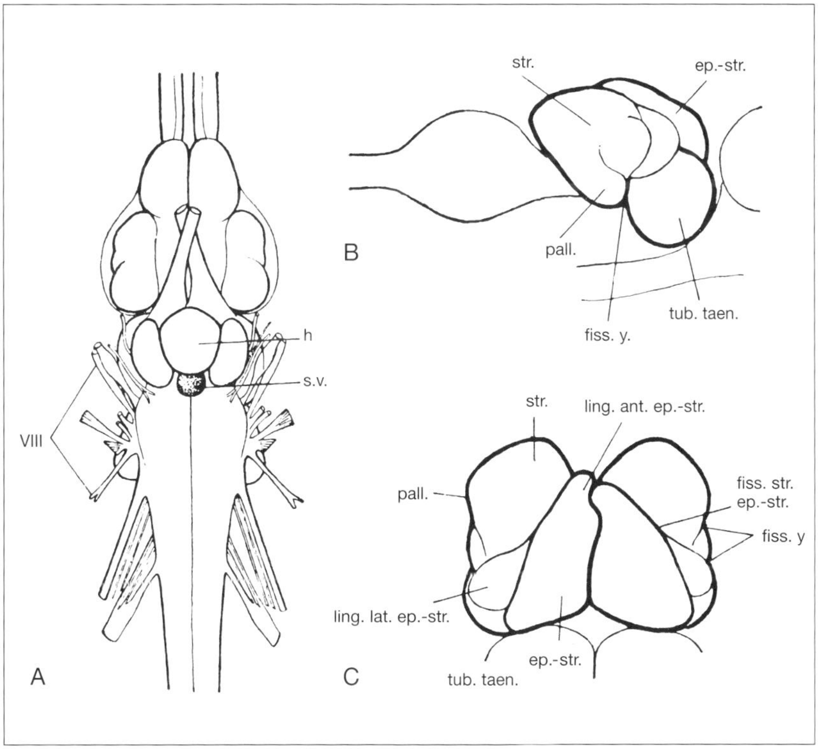

Fig. 1.28 The eel's brain (after Lissner, 1923)

A	from below
B	Corpora striata from the side
C	Corpora striata from above

ep.-str.	Epistriatum
fiss. str. ep.-str.	Fissura striata epistriatica
fiss. y.	Fissura ypsiliformis
h.	Hypophysis
ling. ant. ep.-str.	Lingua anterior epistriatica
ling. lat. ep.-str.	Lingua lateralis epistriatica
pall	Pallium
str.	Striatum
s. v.	Saccus vasculosus
tub.taen.	Tuberculum taeniae
VIII	Nerves

Indeed Vigh-Teichmann *et al.* (1983) provided evidence for the presence of a biochemical indicator of photosensitivity, opsin, in pineal cells of the eel. Neural pathways between the pineal and the hypothalamus and brain stem are present and close associations between retinal and pineal terminal fields have been noted (Ekstrom and van Veen, 1984). Further investigations are required to elucidate both the manner in which the pineal responds and the influence that its key secretory product, melatonin, has on the biology of the eel; of especial interest would be the changing patterns of melatonin secretion during the period leading up to the yellow–silver transformation (Section 1.3.3).

1.8.3 The hypothalamo-hypophyseal system

The pituitary gland (hypophysis) lies just posterior to optic the chiasma and anterior to the saccus vasculosus, on the ventral surface of the brain (Fig. 1.28). At the base of the brain, lying dorsal to the pituitary, is the hypothalamus which together with the pituitary forms the hypothalamo-hypophyseal complex, which has been extensively described in the eel. The hypothalamus contains aggregations of neurosecretory cells (hypothalamic nuclei). These are bilaterally arranged and their axons pass towards the adenohypophysis to communicate with the various cell types therein (Fig. 1.29).

Each of these nuclei produces specific hormones although not all are fully defined for the eel, which regulate the activities of specific cell populations within the hypophysis (see below). Within these hypothalamic nuclei, neuropeptides have been specifically localised, although it is not always a case of one cell one neuropeptide. Among the substances produced are corticotropin-releasing hormone, somatostatin, gonadotropin-releasing hormones, arginine vasotocin, isotocin, melanocyte concentrating hormone, growth hormone-releasing hormone and thyrotropin-releasing hormone. The hypophysis itself consists of two primary divisions; the central neurohypophysis whose cell bodies arise from the pre-optic nuclei of the hypothalamus and whose nerve terminals interdigitate with the second major division the adenohypophysis. The adenohypophysis is further subdivisible into rostral and proximal parts and the posterior pars intermedia. The neurohypophyseal axons come into especially intimate contact with the pars intermedia. Like most other teleosts, the eel possesses a median eminence, which has become incorporated into the hypophysis as a microvascular bed, but there is no distinct hypothalamo-hypophyseal portal system (Fig. 1.30). Based on staining and immunocytochemical and immunohistochemical observations, a number of distinct cell types are present in the pituitary gland (Fig. 1.31).

Gonadotropes are responsible for the production of two gonadotropins, GTH-I and GTH-II; the secretion of these hormones is selectively stimulated by oestrogenic and androgenic steroids to regulate the onset of sexual maturation in the eel (Dufour, 1994; Montero *et al.*, 1995; Huang *et al.*, 1997). The actual release of the gonadotropins is under hypothalamic control by GnRH (s); this is not restricted to the hypothalamus and has a wide distribution in the eel brain including the olfactory bulb, the pre-optic area, neurohypophysis and optic tectum and optic nerve (Nozaki *et al.*, 1985; Dufour *et al.*, 1993). There is also dopaminergic inhibition of GnRH release possibly contributing to prepubertal blockade of gonadotropin release in yellow eels (Dufour *et al.*, 1988).

A Sagittal section, frontal view
B Cross-sections 1–4, frontal view
apol lateral preoptic area
apom median preoptic area
ca anterior commissure
co optic chiasma
lfb lateral forebrain bundle
me median eminence
mfb median forebrain bundle
no optic nerve
npo preoptic nucleus
npp periventricular preoptic nucleus
nrl nucleus recessus lateralis
nta anterior tuberal nucleus
ntl lateral tuberal nucleus
ntp posterior tuberal nucleus;
pva anterior periventricular nucleus of the hypothalamus;
to optic tract

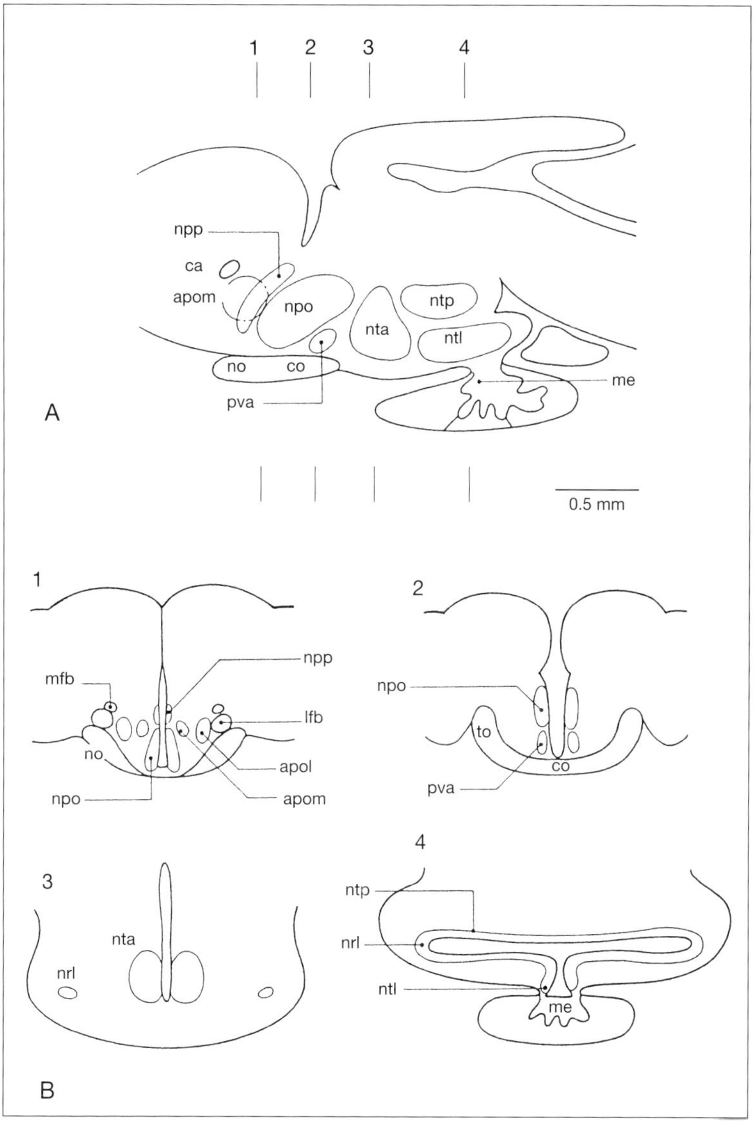

Fig. 1.29 Diagrammatic sagittal section illustrating the relationship between the hypothalamo-hypophyseal system of the eel (from Matsumoto and Arai, 1992)

Thyrotropes produce thyroid-stimulating hormone (TSH), which belongs to the glycoprotein family of hypophyseal hormones, which includes the gonadotropins. These hormones consist of α and β chains which are non-covalently bound. The α subunit is common to the three hormones whilst the β chain donates specificities of action. The synthesis and probable release of TSH reflects feedback of thyroid hormones directly upon the thyrotropes as well as upon hypothalamic release of thyrotropin-releasing hormone (Pradet-Balade *et al.*, 1997).

Somatotropes produce somatotropin or growth hormone (STH or GH). There appear to be two isoforms of GH in the eel (Kishida *et al.*, 1987). Growth hormone is important throughout the life of the eel, but especially during larval growth and during the

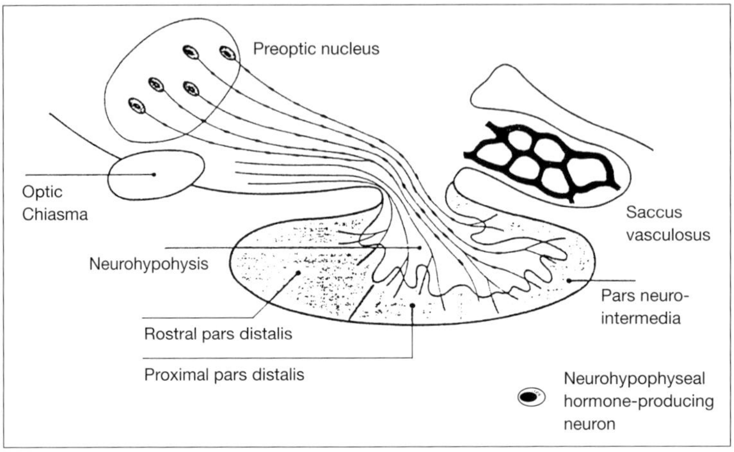

Fig. 1.30 The hypothalamo-neurohypophyseal system of the eel showing the relationships between the neurohypophyseal preoptic nuclei, their axon terminals and the adenohypophysis; the saccus vasculosus and optic chiasma are also shown (from Matsumoto and Arai, 1992)

yellow–silver transformation and osmoregulation in seawater environments (Arakawa *et al.*, 1992; Marchelidon *et al.*, 1996). The mechanisms of action of GH are likely to involve somatomedins (or insulin related growth factors; Inui and Ishioka, 1985; Duan and Inui, 1990a, b; Duan and Hirano, 1992).

Lactotropes secrete prolactin (PRL). Eel prolactin mRNA has been sequenced (Suzuki *et al.*, 1991) and its levels are highest in the rostral pars distalis; of interest, this same study showed pituitary mRNA was markedly depressed when freshwater eel were transferred to sea water (Querat *et al.*, 1994). Neither testosterone nor oestradiol influenced the levels of expression. The functions of prolactin are widespread and include actions on water permeability of, and ionic transfer across the gills, renal function and growth, calcium metabolism and reproduction.

Melanotropes produce melanocyte-stimulating hormone (MSH), which regulates background colour adaptations, probably in concert with melanocyte concentrating hormone (Baker and Rance, 1983; Gilham and Baker, 1984).

Somatolactin-producing cells synthesise and release a hormone that, as it name implies is closely related to both prolactin and growth hormone. This relatively recently discovered hormone, whose cDNA has been cloned for the eel (May *et al.*, 1997) has been held variously to be involved with calcium regulation, pre-migratory adaptation, renal and branchial solute transport and acid–base balance (Rand-Weaver and Kawauchi, 1993); however, definitive investigations of somatolactin in the eel are awaited.

The neurohypophyseal hormones are the classical octapeptides, arginine vasotocin and isotocin. Either or both are almost certainly involved with osmoregulatory processes in both freshwater and seawater environments but there is increasing evidence that they are involved with cardiovascular homeostasis and the regulation of blood to and haemodynamics within individual organs and organ systems (Henderson *et al.*, 1985; Bennett and Rankin, 1986; Oudit and Butler, 1995). Moreover, actions of neurohypophyseal hormones within the pituitary influence, for example, ACTH release, and more distantly affect genital smooth muscle and gonadal steroidogenesis (Henderson, 1997).

A Structure of the pituitary gland (after Vivien, 1958, modified from Kerr 1942)
agr Anterior lobe (pars distalis)
brc Connection with base of the brain
cav Cavity of the anterior lobe
hypf Opening into posterior lobe (with diverticula) of the neurohypophysis
hypr Recessus infundibularis and its diverticulum
mgr Middle lobe (Pars intermedia)
nl Neurohypophysis
pgr Posterior lobe

B Anterior lobe (rostral pars distalis) (modified from Olivereau, 1967)
c Corticotropes
p Lactotropes (prolactin cells)
t Thyrotropes (black)

Proximal pars distalis:
g Gonadotropes (fine stipple)
t Pars intermedia
n Neurohypophysis
s somatotropes (coarse stipple);

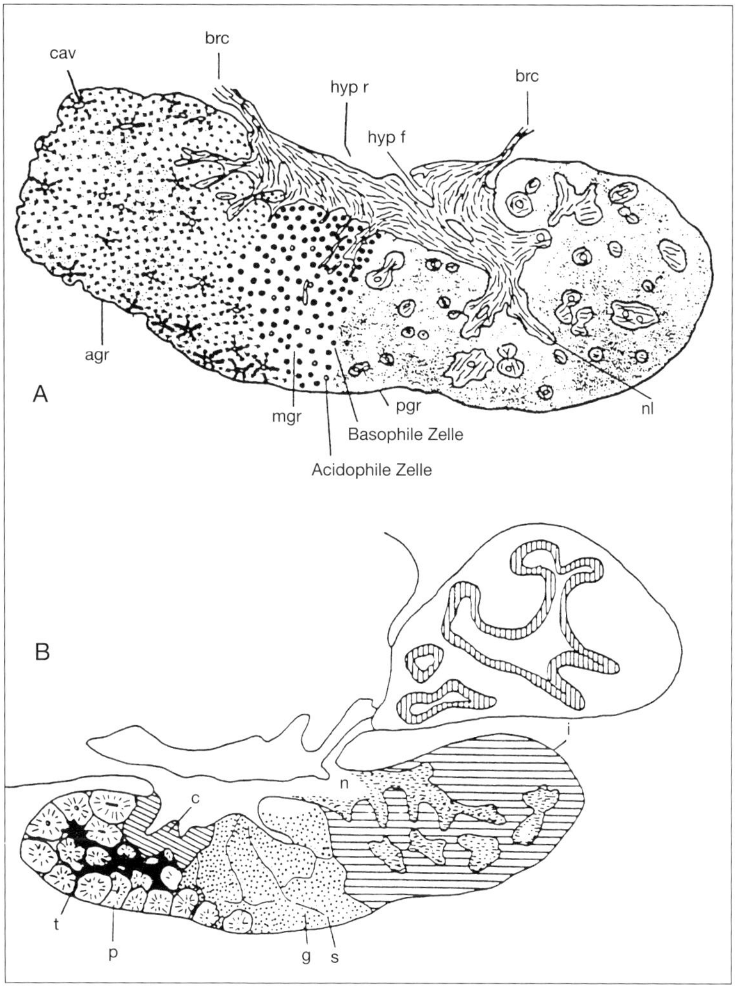

Fig. 1.31 The pituitary gland of *A. anguilla*

1.8.4 The thyroid gland

The eel's thyroid gland is a much less discrete structure compared with that of other vertebrates. The gland itself consists of the typical thyroglobulin-containing follicles somewhat dispersed rostral to the origins of the first branchial artery, lying under the urohyal bone (Fig. 1.32). Thyroid activity is under the predominant control of adenohypophyseal TSH, itself controlled by hypothalamic thyrotropin-releasing hormone although there is evidence that the gonadotropins (glycoprotein products of the related gene family) may, at certain stages of the eel's life history, affect thyroid activity. The principal hormone released by the thyroid gland is thyroxine (tetra-iodothyronine) which undergoes peripheral conversion, mainly in the liver and kidney, to a more active form tri-iodothyronine; however, the plasma concentrations of these two hormones are roughly equal in eels. This

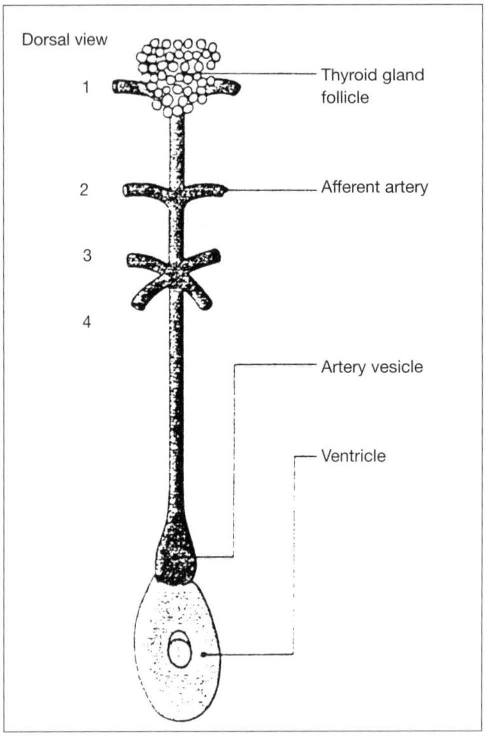

Fig. 1.32 Diagram illustrating the disposition of the thyroid follicles of the eel in relation to the heart and ventral aorta. The follicles are concentrated around the branching of the afferent branchial; follicles occur in the hollow portions of the branchial skeleton (from Suzuki, 1992).

conversion can be increased by prolactin, which increases thyroid hormone activity without necessarily altering thyroidal uptake of iodine for example (de Luze *et al.*, 1984; de Luze, 1987).

1.8.5 Ultimobranchial bodies

The ultimobranchial bodies (the homologues of the calcitonin-producing cells of the mammalian thyroid gland) are distinct structures around the pericardial septum of the heart. The ultimobranchial bodies produce the hormone calcitonin, whose sequence has been determined for the eel (Otani *et al.*, 1976). Its functional roles are uncertain but several studies have indicated a variety of actions on calcium metabolism with the gills, gut, kidney and bone being potential primary or secondary mediators of the hormone's function. Following electrocautery of the ultimobranchial bodies of eels there is a severe hypercalcaemia associated with reduced mineralisation of the bone matrix and cessation of osteoblastic apposition (Lopez *et al.*, 1976). Earlier controversies remain unresolved as to whether, under normal circumstances either endogenous or exogenous calcitonin influence plasma calcium concentrations of the eel or ultimobranchial function (Chan *et al.*, 1968; Pang, 1971; Orimo *et al.*, 1972; Hirano *et al.*, 1981). In other teleosts, calcitonin has been suggested to have its primary functions during reproduction, especially in the vitellogenic female, when total plasma calcium is grossly elevated and during larval growth and differentiating stages of life (Norberger *et al.*, 1989). Nothing is known of the calcitonin status of the eel during these phases.

1.8.6 The thymus gland

Hagen (1936) described the thymus and Hassal corpuscles of eels. This system lies dorsoventrally in the gill chamber abutting on to the opercular cover. There are age-associated changes in this gland with a progressive reduction in size with age with little more than fatty aggregations being present in the adult, but sexually immature eel. The presence of T-lymphocyte cells has been noted in other fish and an association with the immune system is likely in the eel but specific studies have not been published.

1.8.7 The interrenal tissue

The interrenal tissue of the eel occurs in and around the head kidney and walls of the anterior and posterior cardinal veins (Fig. 1.33).

The major corticosteroid produced is cortisol with smaller amounts of cortisone and other 21 carbon metabolites of cholesterol. The blood production rates of cortisol (derived from metabolic clearance rate determination) and the rates of appearance of cortisol into the catheterised cardinal veins are greater in marine than in freshwater-adapted eels. There is also evidence that the activities of the enzyme 11-β-hydroxysteroid dehydrogenase, which balances the interconversions between cortisol to cortisone (and hence the relative biological activities of the more potent, cortisol), also differ in the two environments. The structure and secretory activities of the eel interrenal depends mostly upon hypophyseal ACTH and the renin–angiotensin system (Henderson and Garland, 1980). Secretions from the urophysis, thyroid hormones and natriuretic peptides are among other factors that affect cortisol secretion. A major function of cortisol is the regulation of renal and extrarenal (gill, gut, urinary bladder) osmoregulatory functions in both freshwater and seawater environments and here again there are close interplays between cortisol and prolactin (fresh water) and growth hormone (sea water) and most likely thyroid hormones and the natriuretic peptides (Hirano, 1992). Although less studied in the eel than in other teleosts, cortisol-binding sites, partially characterised as glucocorticoid receptors, have been described in the gills and intestine of the eel.

1.8.8 Chromaffin tissue

The homologue of the mammalian adrenal medulla, the chromaffin cells, are present in and around the walls of the posterior cardinal veins of the eel and secrete dopamine, norepinephrine and epinephrine. It seems that several cell types are responsible for each of these secretions and in addition other neuropeptides including vasoactive intestinal polypeptide, neuropeptide Y and peptide YY as well as serotonin are also present (Hathaway and Epple, 1990; Reid *et al.*, 1995). For example, epinephrine and norepinephrine are secreted by innervated chromaffin cells while dopamine is released from

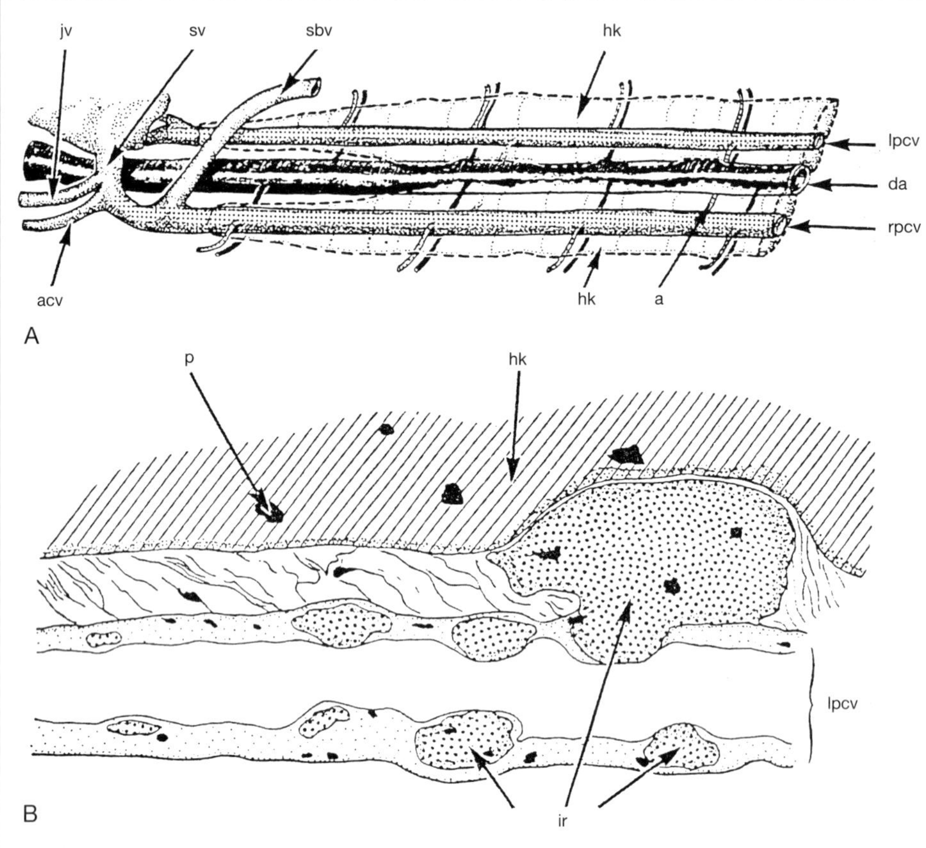

Fig. 1.33 The sites and arrangement of the adrenocortical homologue (interrenal tissue) of the European eel (from Chester-Jones *et al.*, 1964)

A the general arrangement of the blood vessels in relation to the head kidney and heart, viewed from the right side. The adrenocortical cells within and around the walls of the two posterior cardinal veins and to an extent in the anterior cardinal veins.

B diagramatically in longitudinal section, the islets of adrenocortical cells in relation to the head kidney and cardinal vein.

acv	anterior cardinal vein	jv	jugular vein
a	intercardinal anastomosis	lpcv	left posterior cardinal vein
da	dorsal aorta	p	pigment cells
dc	ductus Cuvieri	rpcv	right posterior cardinal vein
hk	head kidney	sbv	swim bladder vein
ir	adrenocortical (interrenal) islets	sv	sinus venosus

components of the vascular wall (Al Kharrat *et al.*, 1997). Basal blood levels of norepinephrine and epinephrine are closely correlated but have no apparent association with dopamine. However, epinephrine does promote the release of both norepinephrine and dopamine (Epple and Nibbio, 1985) and there is also evidence for cholinergic control (Reid and Perry, 1995; Al Kharrat *et al.*, 1997). Although the chromaffin tissue resides predominately in the cardinal veins, other sources may contribute to plasma catecholamines (Hathaway and Epple, 1989). The relationships between adrenomedullary function and other systemic neuroendocrine systems are not known for the eel, but associations with the corpuscles of Stannius are absent while possible interactions with insulin (*via* glycemic state) and interrenal function await study.

1.8.9 The gonads

The basic anatomy of the eel gonads is described elsewhere in this volume (Section 1.7.2). The principal gonadal steroids of *Anguilla* sp., include androstenedione, testosterone, 11-ketotestosterone, 17-hydroxyprogesterone, oestradiol-17β and 17,20β-dihydroxy-4-pregnen-3-one. These hormones fluctuate prior to sexual maturation and in naturally migrating (non-captive) eels there are clear associations with gonadal growth and gametogenesis (Dufour, 1994). Lokman and Young (1998) and Lokman *et al.* (1998) give clear analyses of the likely and actual interactions. The regulation of gonadal steroid production and gametogenesis come under the influence of the gonadotropins GtH-I and GtH-II. The process of vitellogenesis, a central part of the reproductive activity is induced by ovarian oestrogen, in particular oestradiol-17β and interactions with pituitary hormones especially prolactin and growth hormone have been suggested (Burzawa-Gerard and Dumas-Vidal, 1991; Kwon and Mugiya, 1994).

1.8.10 The gastro-entero-pancreatic system

The gut is a repository of many peptide hormones, often co-existing with amines. The gut hormones are produced in the so-called clear cells (enterochromaffin, agyrophil or argentaffin cells), which are stained by certain salts of silver. These cells are scattered throughout the length of the gut. This dispersed, or diffuse, system has been termed the gastro-entero-pancreatic endocrine system (Bloom and Polak, 1981).

The eel pancreas is in many respects unlike the typical aggregated teleost type, and indeed resembles that of the typical more dispersed tetrapod vertebrate arrangement (Fig. 1.34, Epple and Brinn, 1986); the endocrine islets of Langerhans are embedded in the exocrine component of the pancreas (Fig. 1.19). The functions of this whole system include the control of the passage of ingested materials, their proper treatment with regard to digestion, absorption and assimilation at appropriate sites and rates and the eventual elimination of faecal matter. The major hormones described in the eel include insulin, glucagon, pancreatic polypeptides and somatostatin, some of the sequences of which are now known. In the case of insulin, the sequences of Japanese and European eel insulins differ (Conlon *et al.*, 1991a). The distributions of immunoreactive insulin, somatostatin, glucagon, met-enkephalin and serotonin have been described in leptocephalus, glass eel and adult eels (l'Hermite *et al.*, 1985). Insulin, somatostatin and glucagon were present in the pancreatic islets while serotonin was present throughout the gut. Somatostatin cells were predominantly in the stomach and less in the intestine. Metenkephalin cells were mostly present in the pyloric caecum. Glucagon cells were also seen in the intestine and insulin was notably absent from the gut. The pancreatic islets of the leptocephalus gave strong somatostatin reactions, a weak reaction to glucagon and there were no immunoreactions at all to antisera against insulin, metenkephalin and serotonin. Their

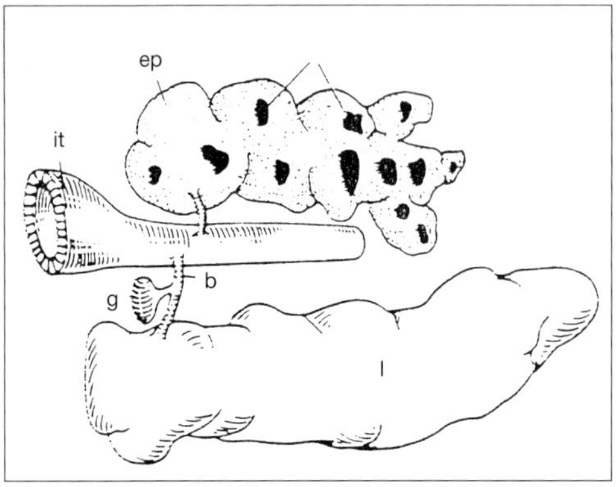

b bile duct
ep exocrine pancreas
g gall bladder
i islets of Langerhans
it intestinal tract
l liver

Fig. 1.34 The tetrapod type of pancreas found in the eel, *Anguilla*: there are scattered endocrine islets within the exocrine pancreas (from Epple and Brinn, 1986)

foregut contained small amounts of metenkephalin and glucagon; serotonin, somatostatin and insulin were absent. The distributions of the endocrine cells were similar in adult and glass eels. The functions of the GEP system have not been widely explored in the eel although somatostatin has been shown to act directly on enterocytes to stimulate water and salt absorption, or perhaps more exactly to antagonise inhibitory actions of serotonin and acetylcholine (Uesaka *et al*., 1994). The systemic/metabolic actions of insulin and glucagon are not dissimilar to those seen in other vertebrates (Inui and Ishioka, 1983a, b; Lewis and Epple, 1984). The anterior, middle and posterior regions of the intestine of the eel appear to utilise locally synthesised atrial and ventricular natriuretic peptides to regulate water and solute transport under seawater conditions (Loretz *et al*., 1997).

1.8.11 The heart

The heart of all vertebrates is now recognised as a major source of cardiac natriuretic peptides. The eel heart (see Section 1.6.1) contains the typical granular atriocytes in both the atrium and ventricle (Broadhead *et al*., 1992). The sequences of the peptides have been determined. In the eel there are at present three members of the natriuretic family of peptides: atrial natriuretic peptide, ventricular natriuretic peptide and C-type natriuretic peptide (Takei *et al*., 1994). They function to regulate branchial and gastro-intestinal solute transport and possibly water permeability as well as renal function. Specific receptors for several of the peptides have been observed in the gills, gastro-intestinal system, brain, kidney, head kidney (interrenal gland?), posterior kidney and urinary bladder and in the red body of the swimbladder; adaptation of eels to sea water decreased receptor densities in most tissues (Mishina and Takei, 1997). ANP and VNP appear to be differentially released when eels move between environments hyper- to hypo-osmotic to their extracellular fluid. Plasma and extracellular fluid volume, although influential, are not primary stimuli to the release of the peptides; rather it is the osmotic stimulus of entry into sea water that promotes their differential secretion (Kaiya and Takei, 1996, 1997). The eel intestine may produce atrial natriuretic peptide to have local, paracrine actions, on mucosal epithelial transport.

1.8.12 Kidney

The eel kidney (Section 1.7.1) contains renin and the component parts of the renin–angiotensin system have been elucidated, with eel angiotensin I having been sequenced. The functions of angiotensin include regulation of renal function, stimulation of the interrenal production of cortisol and the control of drinking behaviour and possibly blood pressure and flow to particular organs or specific vascular beds (Brown *et al.*, 1980; Henderson and Deacon, 1993; Butler and Oudit, 1995; Tierney *et al.*, 1995).

A renal enzyme responsible for the production of 1,25-dihydroxycholecalciferol (calcitriol) has not been detected in the eel. This sterol together with other metabolites (25-hydroxy, and 24,25-hydroxycholecalciferol) are present in eel plasma and in one study calcitriol was shown to increase from near zero values in freshwater-adapted specimens to over 10 pg/ml on adaptation to sea water (Henderson, 1997). Effects of calcitriol on gut calcium absorption and renal phosphate handling and bone metabolism have been described in eels (Lopez *et al.*, 1977; Fenwick *et al.*, 1984; Fenwick and Vermette, 1989).

A plasma kallikrein system has been described and a bradykinin-like peptide sequenced. Although likely, whether there is a renal tissue kallikrein–kin system in the eel awaits definitive study.

1.8.13 The corpuscles of Stannius

The corpuscles of Stannius are paired (usually) lying on and just below the surface of the posterior kidney of the eel and embryologically are derived from the pronephros. They are well innervated and have a rich vascular supply from the dorsal aorta and venous drainage into the posterior cardinal vein *via* parts of the renal circulation (Fig. 1.35).

At least three cell types are present. The early controversies about whether they were homologous with the adrenal cortex, whether they produced steroid hormones, and whether they produced renin or a substance akin to parathyroid hormone have now to an extent been resolved at least with respect to the predominant secretion. Much remains to be understood about the function of this product and the nature of their secretory products. Most interest has focused on what is a new hormone, perhaps unique to teleosts, stanniocalcin (also termed teleocalcin and hypocalcin) which has been sequenced. It is of interest that stanniocalcin and bovine parathyroid hormone have similar hypocalcaemic activities when administered to eels rendered hypercalcaemic by removal of the corpuscles of Stannius (Lafeber *et al.*, 1988a, b; Hanssen *et al.*, 1989). Arguments have been presented that, at least functionally, the corpuscles of Stannius represent the elusive parathyroid glands of teleosts (Tisserand-Jochem *et al.*, 1987). Relevant to this argument are the cardiovascular changes that follow removal of the corpuscles of Stannius, with decreased dorsal aortic blood flow and pressure, reductions in cardiac output and stroke volume, increased branchial shunting of blood; the net effect is to disturb organ perfusion especially to osmoregulatory organs including the kidney and gills (Butler and Oudit, 1994, 1995). The mammalian parathyroid gland contains alongside its calcium-regulating hormone a factor that has vascular actions (Pang *et al.*, 1980). Evidence for the presence of a parathyroid hormone-like substance (Lopez *et al.*, 1984), which seems to act upon calcium homeostasis (Milet *et al.*, 1989) remains to be fully explored. Stanniocalcin secretion

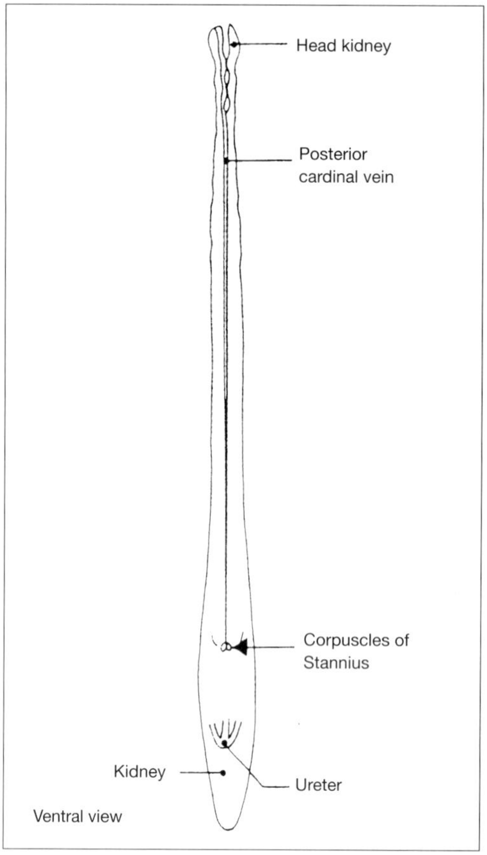

Fig. 1.35 Position of the corpuscles of Stannius in relation to the kidney from Ogawa,1992)

seems to be influenced directly by local calcium concentrations as well as by cholinergic innervation (Fenwick and Brasseur 1991; Mayer-Gostan *et al.*, 1992; Cano *et al.*, 1994; Fenwick *et al.*, 1995). However, the most telling experimental evidence points to the conclusions that teleocalcin is the major hormone produced by the corpuscles of Stannius and its primary functions are associated with calcium homeostasis.

1.8.14 The urophysis

The urophysis produces at least two peptides, the urotensins which belong to the CRH family of peptides. Their functions in the eel have been connected *inter alia* with ACTH release, the regulation of renal function, and caudal lymph heart activity.

1.8.15 Closing remarks to endocrine system

This brief outline of the eel endocrine system has concentrated on the more recent advances in knowledge of this area; for the earlier literature the reader is referred to the earlier edition of this volume (Tesch, 1977, 1983). The most significant contributions in recent years have been the identification and chemical characterisation of a number of eel hormones and the discovery of several new ones including somatolactin, hypocalcin (teleocalcin) and the cardiac natriuretic family of peptides. An area for future study will be the vascular endothelium of the eel in an endocrine context. In addition, the roles of cytokines and growth factors are at present somewhat neglected realms, but it is certain that they will provide experimental subjects for many biological phenomena from pubertal control to body fluid homeostasis and calcium metabolism.

1.9 Nervous system and sense organs (updated by S. Appelbaum)

1.9.1 The brain

Information on the brain of eels is limited and is available only for particular parts. Lissner (1923), in a general morphological description (Fig. 1.28) concisely outlines the structures as follows.

'The brain of the eel is as remarkable as its life. The very thick olfactory nerves gradually swell to terminate in two extremely large, egg-shaped olfactory bulbs situated on ridged corpora striata, which, particularly in the ventral caudal region, are enormous. The epiphysis is a short, cone-shaped structure wider distally. Caudal to this are flat roundish, optic lobes, which are smaller than the corpora striata. The optic lobes have a ridge that runs laterally, from the upper region behind, diagonally to the lower regions in front. The part of the optic lobe that lies caudally under this ridge is slightly flatter than the bulk of the lobe. This and the origin of the ridge can be explained by the fact that the large nerve bundles of the trigemino facialis group (V and VII), which run diagonally forwards, are pressed against the optic tectum. The cerebellum has a curious shape. It is a smooth rectangular structure with rounded edges that widen posteriorly, and lies flat on the medulla oblongata, revealing a deep, longitudinal furrow on the side thus creating the impression that the cerebellum has been folded on itself as a result of insufficient space, and that areas of contact have merged with one another. According to Victor Franz, the cerebellum of all teleosts has a ridge on the side which indicates the line of embryological dorsoventral fusion between the granular layer and the ventral part. This fusion does not occur in *Anguilla*, and the section of the corpus cerebellum under the ridge is the border of a granular layer, which stands out quite clearly. Also there are no eminentiae granulates in this region because their homologue is to be found in the 'granular border'. Franz describes the actual details very clearly. However, the brain of *Anguilla* is so unusual in its entirety that one cannot resist thinking that extensive secondary changes have taken place. The valvula is rather small. In the region of the medulla oblongata, under the cerebellum, there is a short fissure the rhombencephalon, which is surrounded by the two ledge-like protrusions that rapidly become narrower towards the rear. Posteriorly the medulla increases slightly in size, and then continues into the spina dorsalis. Ventrally, long, thin optic nerves, round in cross-section, and sturdy discrete ventral inferior lobes occur on either side of the midline. The lateral lobes are separated by a shallow oculomotor lobe groove. The hypophysis is large and flat with its front surface forming an obtuse angle while its caudal surface is rounded.'

The brain is very similar to that of the moray eels (Muraenidae) also described by Lissner (1923). Ito (1974) has compared the fine structure of the torus semicircularis in eels with the same organ in carps (*Cyprinus carpio*), loach (Cobitidae) and rainbow trout (*Oncorhynchus mykiss*). Pravdina and Chebotareva (1973) compared phospholipids in the brain of eels and rays (*Dasiatis pastinaca*) and found differences, which could be explained on the basis of different ecological and phylogenetic relationships.

The neuroanatomy and the relationships of the medulla oblongata have been described by Ishida (1958). The parasympathetic nervous system affects osmoregulation, although in the eel, in contrast to other freshwater fishes, hormonal control dominates neural mechanisms (Pequignot *et al.*, 1968; Section 1.8.3). Cutting through the IXth and Xth nerve of the brain was of importance for sodium balance (Mayer-Gostan and Hirano, 1976). Richard (1955, 1956) studied neural influences on the different locomotory reactions of the eel (Bickel, 1897; Gray, 1936; Blancheteau, 1972, 1973). Eels continue to show movements despite such manipulations as skin removal, brain destruction or indeed sectioning of the posterior nervous system. The latter contributes to neuromuscular activities. If the spinal cord is severed 1 cm behind the pectoral fins, electrical stimulation produces rhythmical, snake-like movements. It appears that the brain has an inhibitory influence on movement in intact eels. In fact, excitation appears to be controlled by the tail section. In any event, eels may be immobilised by wrapping them in a towel, particularly around the tail region. In addition, optic lobe activity inhibits mobility, as illustrated by the fact that feeding and migration take place mostly at night. This is confirmed by activity studies of Edel (1975b).

The eyes are not the only regions of the eel sensitive to light. Blinded animals will show flight reflexes when the head or tail is exposed to light, and this reaction still occurs when afferent nerves to the skin have been severed (Motte, 1963, 1964). Van Veen *et al.* (1976) confirmed that blinding as well as elimination of the pineal organ could not cut out light sensitivity of the head. From the citation of the authors (Oksche and Hartwig, 1975 in van Veen *et al.*, 1976) it is likely that cell types in the area of the third ventricle could be responsible for light sensitivity.

The commissural organ and Reissner fibre complex in the eel are thought to transmit optic signals to the motor nucleus of the spinal cord and have been described by Leatherland and Dodd (1968) who showed that changes in the brightness of the background stimulated the complex, but that other environmental factors had no such effect.

1.9.2 The eye

The eye of the eel shows poor development at hatching which is known also from other marine fishes; full visual capabilities are obtained during the following larval stages (Prokhorchik *et al.*, 1987).

Literature on the most important part of the eel's eye, the retina, is available from as far back as 1882 (Denissenko, 1882; Virchov, 1882). Information from more recent research on the retina is considered below.

A peculiar feature of the eel's eye, in comparison with that of other fish species, is its iris. In contrast to other species, the iris is relatively well developed and, furthermore, is able to regulate the amount of light entering the eye by appropriate degree of expansion and contraction (von Studnitz, 1933). In bright light the pupil contracts, in dim it expands; in other words it simply reacts according to the amount of illumination present – in contrast to humans where the diameter of the pupil has a much more complicated form of adaptation. As also has been shown in the isolated iris of the eel, the adjustment of the iris muscle occurs as the result of the direct influence of light on the iris. The greatest reaction is seen by using light that corresponds to the absorption maximum of rhodopsin (Seliger, 1963). As will be described below, rhodopsin makes up most of the pigment in the rods of the eel's retina. It is, therefore, assumed that, in order to activate the iris muscle, the same pigment composition is present in that muscle.

The ventral attachment of the lens is primitive in the eel as compared with that in other teleost fishes (Stramke, 1972); the lens is not attached by means of a musculus retractor lentis based on the processus falciformis, but by a broad ligament. Due to the absence of lens muscles, accommodation is not possible. Furthermore, the eel does not have a corpus chrioidae nor a musculus ciliaris. The blood supply to the vitreous humour and the retina also differs from those in other teleosts.

Biometric studies on the nucleus of the oculomotor nerve have shown that the yellow eel probably makes relatively little use of its eyes (Kirsche, 1966). All other species of fish studied, including the burbot (*Lota lota*) – which is well known as a 'non-visual' animal – have a larger nucleus than the eel. Just before its spawning migration, in comparison with other fish, the eel has relatively small eyes (Wunder, 1936). This suggests a similarily reduced capacity and restricted use just before the eel's spawning migration. With metamorphosis into a silver eel the diameter of the eye increases in size (Figs 1.36 and 1.37: Matchenis, 1965); this growth should be accompanied by an increase in the efficiency of the eye. The size of the eye increases by a factor of 1.2–2 (Todd, 1981a, 1982).

Simultaneously with this increase in eye size, the degree of maturity of the gonads also increases (Fig. 1.37; Olivereau and Olivereau, 1985). Thus, it appears that the degree of maturity of migrating eels caught in the sea can be determined from the size of their eyes. A kind of sex dimorphism also seems to be involved in the shape of the eyes: 98% of the males and only 23% of the females had 'protruding eyes' (Holmgren and Wickström, 1993). Knight (1982) showed that the size of the eye of larger eels was positively correlated with their body weight. However, one also finds enlarged eyes in eels, which have been prevented from migrating seawards (Wundsch, 1953). In this case it is highly questionable whether an increased degree of sexual maturity is also attained (Schnakenbeck, 1953b; Section 1.7.2). In addition, abiotic factors can have influence on the size of the eyes: in a water which was poorly penetrated by light, Caruso *et al.* (1988) found juvenile *A. rostrata* with enlarged eyes. Williamson and Castle (1975) reported a silver eel (*A. bicolor*) with large eyes, captured in a well on the Isle of Celebes, showing only poorly developed gonads. A similar finding, in Italy, of a yellow eel (*A. anguilla*) with large eyes gave rise to the suggestion that this development could have taken place as an aid to better orientation in the darkness of the well (Grassi and Calandruccio, 1897b). Also, in fish culture, eels, untreated with hormones, have been found with enlarged eyes (Beullens *et al.*, 1997). So enlargement of the eyes is not only the consequence of development to a 'deep sea eel'. It can also occur earlier, under the influence of appropriate environmental factors.

Fig. 1.36 Head of the eel (photo: Marschall)
Above: Female eel with enlarged eyes, caught in Mediterranean coastal waters near Martigue/Marseille in February, 1969
Below: normal silver eel

At the same time as the external morphological changes take place, morphological and physiologcal changes are occurring within the eye. Despite the increased size of the eye in the silver eel, the retina still retains its disproportionately large size – i.e. it does not become smaller in relation to the surface area of the body as is the case in all other species of fish that have been studied (Teichmann, 1954). At advanced stages of maturity and with large increases in eye diameter, the retina shows a 4-fold increase in its surface area. However, the total number of rods does not remain constant as was suggested by D'Ancona (1927, 1929), but increases considerably, that is, the original rod density is actually retained (Stramke, 1972). Stramke concluded that there was an increase in photosensitivity in the silver eel eye at the expense of resolution. Pankhurst's (1982) studies seem to confirm this. He observed a striking increase of the total number of rods in the retina of maturing eels (*A. anguilla*), although the density of light receptors remained constant. In contrast, the number of cones decreased. The changes in the eye of the eel seem to be accomplished relatively early during the maturing process.

Wunder (1936) noted that, in comparison with other fishes, which usually live in well-lit areas, the yellow eel stays in dimly lit or dark places and possesses a retina which is very sensitive to half-light. On an 80 μm strip of yellow eel retina there were 143 rods. Fish which are active in daylight – e.g. carp, pike, perch and many others – have only between 10 and 50 rods on a similar sized strip; only nocturnal fishes and fishes that live on the bot-

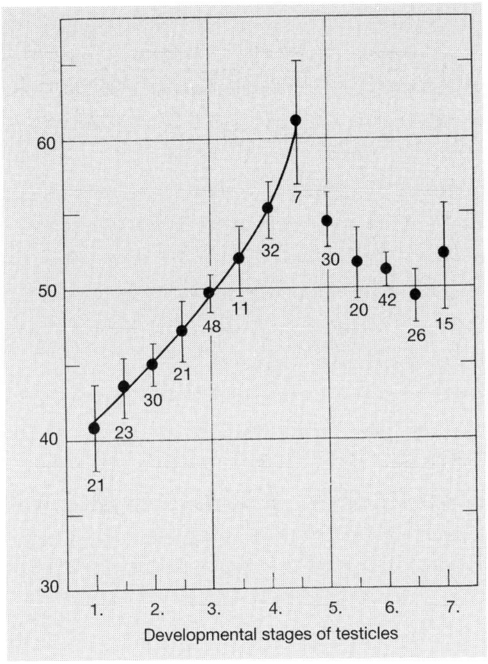

Fig. 1.37 Eye size in male silver eels, which were treated with hormones (see Section 1.7.2), in relation to the develomental stages of the testes. Vertical lines: standard deviation; figures: number of specimens (after Boëtius and Boëtius, 1967a)

tom, such as the burbot (*Lota lota*) and the barbel (*Barbus barbus*) have more rods than the eel. The cones in the eels retina are extremely small, and their ratio to rods is very small at 1 : 150 (Gordon *et al*., 1978). This indicates reduced capabilities where orientation depends on the visual system in daylight. Negative phototaxis increases with development from the yellow eel to the silver eel stage, as seen with migrating eels at light barriers (Section 5.9). An experiment to test the eel's colour vision produced negative results – in contrast to similar experiments with other species of fish (Betge *et al*., 1965).

The ability to see well in crepuscular conditions is not equally developed at all ontogenetic stages. For example, eels living in caves develop proportionately more rods in the lower side of the retina than on the upper side (Vilter, 1951). This is completely reversed in eels living in surface waters. Also, the number of rods in cave-dwelling eels is about seven times that of surface-dwellers. However, the rods in the eye of a moderately pigmented glass eel are distributed as in cave-dwelling adults. Vilter (1951) believed that, in cave-dwelling individuals, uninfluenced by light, the number of rods would remain constant during development into the yellow eel stage. But in surface-dwelling individuals, the number of rods on the under side of the retina must have decreased because this part of the retina lies in the pathway of light coming from above, and light inhibits the development of rods. On the upper side of the retina the rods develop in relatively large numbers because this region is less well illuminated.

Similar correlations with light can be observed between the centrifugal and centripetal migration of melanin in the retina, and the pigmentation stages in the young eel. Pigmentation and sensitivity of the retina increase in parallel. At the same time the migrational activity of the young eel decreases (Vilter, 1943, 1946c; Cahour and Chèze, 1973). Vilter (1942, 1946c) showed that the hypophysis influenced the distribution of melanin in the retina. In glass eels centrifugal migration of the pigment took place under the influence of the pars intermedia (Fig. 1.31) of the hypophysis. This change can be revoked by adaptation – i.e. through the sympathetic nervous system. According to Thornton and Howe (1974), visual factors appear to influence the hypophysis. These authors found structural changes in the secretory cells of the pars intermedia in the hypophysis when the background colour changed between black and white.

In addition to these morphological changes, during the second metamorphosis of the eel biochemical–physical peculiarities and transformations have been observed in the eye. Animals develop retinal pigments that are best suited for the absorption of the predominant wavelength in their particular environments. Fishes that live on the sea bed and in deep water are mainly provided with rhodopsin, which absorbs principally the kind of light that penetrates to greater depths; the wave length of this light is in the range 480–400 nm. In contrast, in freshwater fishes, porphyropsin predominates; this pigment absorbs light of >500 nm (Wald, 1958). The presence of vitamin A_1 in deep-sea fishes and of vitamin A_2 in freshwater fishes corresponds to the different light absorbing properties of the retinas in these two types of fishes. Both vitamins are found in anadromous and catadromous fishes – e.g. trout, charr, salmon, and of course the eel. However, in the eel, in contrast to the salmonids, vitamin A_1 and rhodopsin predominate, which can be attributed to the eel's marine origin.

Research on the wavelengths of rhodopsin in the European eel during different stages of development has shown that the colour of pigment in the retina changes during development from the yellow to the silver eel stage to still shorter wave lengths (Carlisle and Denton, 1959). Yellow eels have a purplish coloured retina, which is not unlike that seen in other freshwater fishes. Silver eels have a golden shimmer to the retina, which is similar to that of deep-sea fishes. These authors call this pigment chrysopsin. Comparison with the retina of the conger eel (*Conger conger*) reveals a great similarity. The results of Carlisle and Denton (1959) also apply to the American eel (Beatty, 1975). *A. rostrata* yellow eels have a mixture of rhodopsin P501 and porphyropsin $P523_z$, silver eels at an early development stage have a mixture of the rhodopsins $P482_1$ and P501 as well as a small quantity of porphyropsin $P523_2$. More advanced silver eels appeared to have no porphyropsin, but showed a predominance of rhodopsin $P482_1$ (Gordon *et al.*, 1978). *A. japonica* originating from fresh water showed no change of spectral sensitivity with increasing salinity (Niwa, 1979). Another fish species migrating to the sea (*Plecoglossus altivelis*) was also not affected by increasing salt content.

In conclusion, the enlarging eye and the correspondingly enlarging pupil make the eye more sensitive to light. The influence of light is rendered more effective because the photosensitive pigment of the retina becomes better suited to the deep sea conditions.

Pathological changes and anomalies in the eel's eye have been described. Macrophthalmus, for example, is known in the eel (Mercier and Poisson, 1927). In another case, the left eye had developed in the lower jaw (Drooglever, 1917; Vladykov, 1973).

1.9.3 Olfaction and taste

In contrast to its visual capacities, the eel has an extremely well-developed sense of smell. While in most species of fish the two nostrils of each nasal cavity lie close to one another, in the eel they are unusually far apart (Fig. 1.38). This is due to the very elongate shape of the olfactory rosette which, in other species, is virtually round or forms only a slightly elongated oval. The rosette stretches almost from the anterior margin of the upper jaw to the anterior border of the eye, where the posterior nostril is situated. The anterior, inflow, nostril lies at the anterior margin of the nasal cavity; it is tubular in shape and joins downwards.

The olfactory rosette is formed from a great number of olfactory folds. The burbot (*Lota lota*), which also has a well-developed sense of smell, has about 30 folds, and the rainbow trout (*Oncorhynchus mykiss*), which has a mediocre sense of smell, has about 15 olfactory folds, but the yellow eel possesses between 50 and 70 such folds, and, in some cases may have as many as 100 (Laibach, 1937; Teichmann, 1954; Kaluza, 1959; Gorovaya, 1973). Also, other anguilliform fish have an obviously similar olfactory capability; their olfactory rosette is similar to that of the eel (Yamamoto and Ueda, 1978). *Muraena undulata* even possesses 130–160 olfactory folds. In the glass eel, and especially in the leptocephalus, these laminate structures are considerably fewer (Fig. 1.39) and are of a different shape. In other words, the olfactory rosette and the nasal cavity in the early stages of the eel's development are similar to those of the adults in other species of fish, but, as early as in the eel larva it shows a rosette, resembling a feather-like leaf.

A further noticeable characteristic of the nasal cavity of the eel is that it and particularly the posterior part of the rosette, are well provided with cilia; these enable a strong flow of water to pass through the cavity, and, in conjunction with the special shape of the latter, also to increase contact between the olfactory epithelium and the water (Teichmann, 1959). The histology of the epithelium at various stages of development has been described by Laibach (1937) and Schultze (1972), who found increases of the various cell

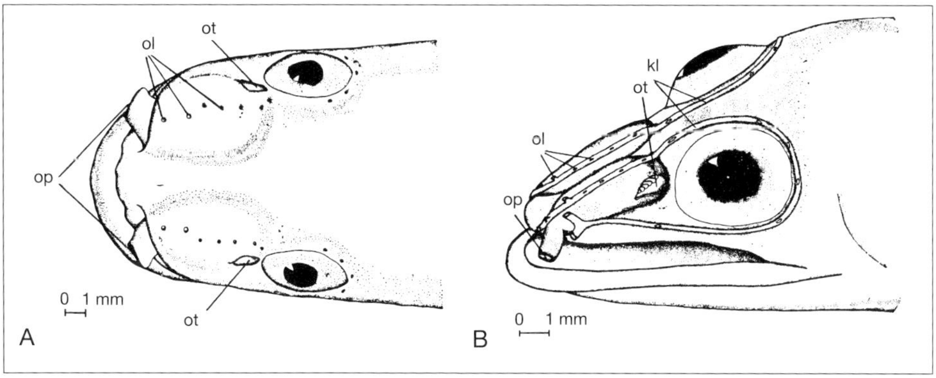

Fig. 1.38 Snout of the eel (after Kaluza, 1959)

A Head from above
B Head from the side; the nasal cavity exposed to show to olfactory rosette

kl Lateral line canal
ol Lateral line canal opening
op Anterior nostril
ot Posterior nostril

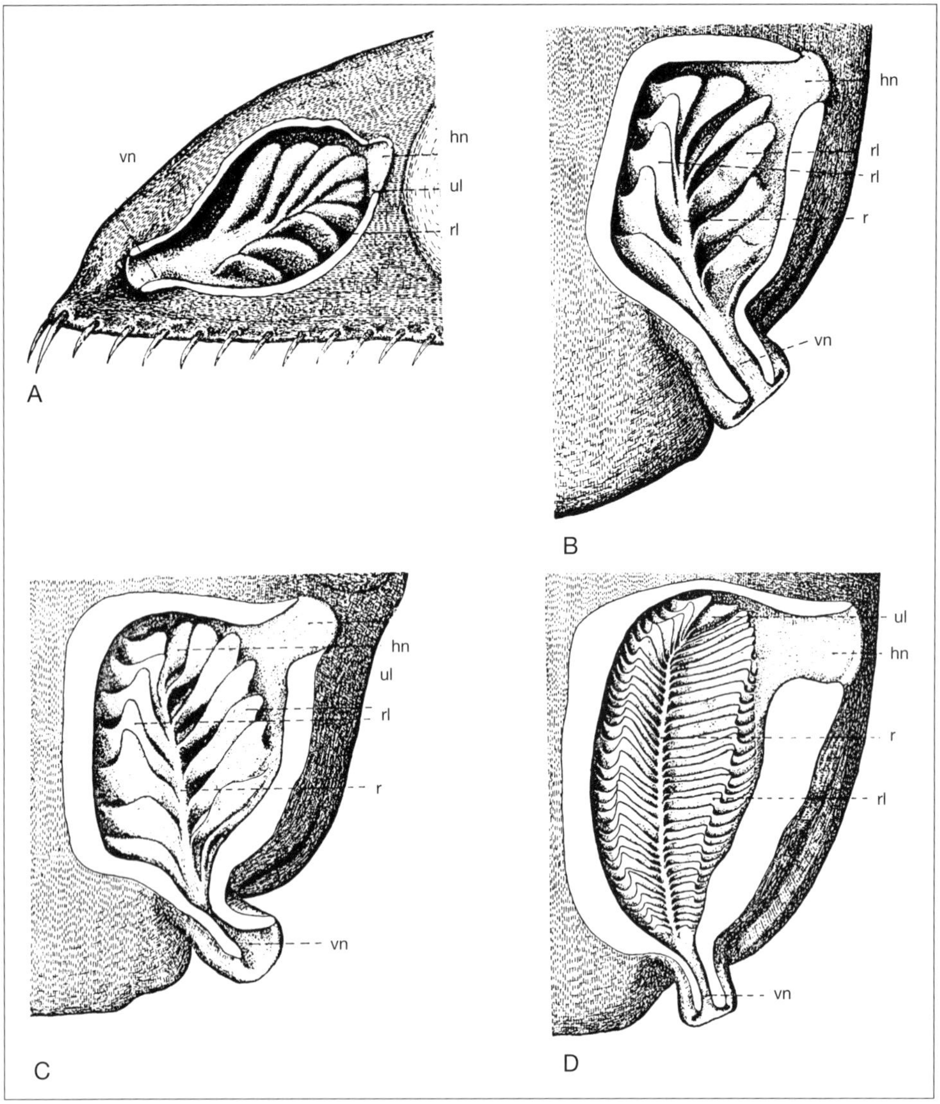

Fig. 1.39 Olfactory cavity at different developmental stages of the eel (after Laibach, 1937)

A Leptocephalus
B Glass eel
C Pigmented eel 8–12 cm long
D Yellow eel 45 cm long

hn posterior nasal orifice
r raphe
rl olfactory lamella
ul unpaired lamella
vn anterior nasal orifice

elements during ontogeny. Kaluza (1959) investigated the blood-vascular system of the nasal region and of the olfactory folds (Gorovaya, 1973). Studies of the mucous cells in olfactory lamellae of immature and artifically matured *A. anguilla* showed atrophy of these cells in specimens of advanced development: they decreased from a maximum of 443 /mm^2

in sexually immature eels to as low as 19/mm^2 in sexually maturing eels (Pankhurst and Lythgoe, 1983).

Comparison with other species of fish shows just how much the immature eel must rely on olfaction for sensory perception. In most species of freshwater fish studied, the olfactory surface area, measured as a percentage of the retinal area in the eye, is between 14% and 140%; in the eel, however, this proportion reaches 623% (Teichmann, 1954). The eel, during its continental life, is the most extreme example of an 'olfactory animal' yet found among fishes. A further indication of the eel's well-developed sense of smell lies in its relatively large olfactory bulb, which, because of its position, belongs to the 'salmonid' type; the eel also has large corpora striata, which play a part in the processing of olfactory stimuli (Fig. 1.28; Harder, 1964; Gorovaya, 1969, 1973).

This dependence on a sense of smell accords well with the eel's almost unbelievably refined ability to differentiate between various scents. In painstaking conditioning experiments it was shown that the eel can perceive the scent of roses (β-phenylethyl alcohol) even when the latter is diluted by $1:2.857 \times 10^{18}$ parts (Teichmann, 1957). Such a degree of dilution corresponds to a solution of one ml of scent in a volume of water 58 times that of Lake Constance (Bodensee). In addition, there are only about 1800 scent molecules in 1 ml of water. The minnow (*Phoxinus phoxinus L.*) is also very sensitive to smell; it needed a concentration of $0.75–2.2 \times 10^{14}$ molecules/ml of rose scent and, therefore, much more than the eel (Teichmann, 1957; Neuhaus and Appelbaum, 1980). Teichmann (1959) has calculated 'that at the threshold concentrations indicated there is only a single scent molecule in one of the two olfactory organs at any one time. The specialised structure of the olfactory organs appears suited to act as a quantitative filter for the scent molecules in the water that flows through the nasal cavity. At threshold levels of concentration, central excitation of the olfactory area is only obtained when the arrival of two or more stimuli in the nose are registered. The temporal interval between individual stimuli can probably be as much as three seconds or even slightly longer. However, if this interval is exceeded, then central excitation does not occur. The stimulus is subliminal.' The eel is thus almost as sensitive to smell as the dog, which is not surpassed by any other animal.

Other scents are just as effective if they are detected in slightly higher concentrations. Among these is the scent of prey animals, which the eel locates by smell rather than by sight. One of the eel's preferred prey is the tubicolous blood worm (*Tubifex* spp.). Its smell was still perceived by the eel even when five *Tubifex* with a total weight of 25 mg were finely ground and diluted in 6.87×10^{12} ml of water. In view of the eel's olfactory capabilities it is not surprising that hungry eels would, after a short time swim up to and bite at a hand that had previously been in contact with meat (Schiemenz in Wunder, 1936). If the other hand, which had not come in contact with meat, was held in the water, it excited no interest.

The eel's ability to detect such low concentrations of scent is most marked in late winter and high summer. In late summer and in early winter the concentration of scent must be raised by 10^6 before the eel can perceive it (Teichmann, 1959).

The eel's sense of smell enables the fish to follow a scent trail. 'Orientation is achieved by trial and error (through the successive perception of differences) and not by a topotactic adjustment (the simultaneous perception of differences). In addition to orientation in the horizontal plane, orientation in the vertical plane plays a very significant role in the life of the eel' (Teichmann, 1959).

The eel's use of smell in orientation has frequently been exploited by baiting fishing gear. The eel may perceive the bait on an angler's hook, on a long line (eel line), or at naring (Section 5.3), by its sense of smell. The effectiveness of wicker eel baskets is increased several fold when bait is placed inside them. In this type of trap, the eel is not led in by the channelling effect of a barrier nor as a result of protection from light, nor by the need for contact (i.e. negative phototaxis and thigmotaxis), but as a result of osmotaxis (attraction by smell) (Tesch, 1970; Section 5.4).

Attempts have been made to develop artificial baits, which could appear attractive to the eel without its being specifically conditioned to them (Schoeniger, 1951). However, none of the many substances tried, most of which were similar to camphor and aniseed in smell had any decisive effect. Only Cibeton can be considered as being at all attractive, but it is used only occasionally because of its high price. There are also substances that serve to drive the eel away – such substances are of practical importance, for example, in protecting nets (Mohr, 1966, 1969). Carbolineum has a repellent effect as long as it is not leached out of the netting; diesel oil used as an emulsion, tobacco smoke, ammonia (>0.0002%), copper acetate (>0.001%) and phenacyl chloride (>0.0002%) also serve to repel eels. Net preservatives such as Racorit, Jolasteen and black varnish have no effect once they have hardened.

Since many of the eel's prey animals, for example, the smelt (*Osmerus* sp.) and various crustacea, frequently act as a strong lure not only to yellow eels but also to elvers, it is obvious that there are certain scents, to which the eel responds, not as a result of experience or conditioning, but through innate reactions. Proof of this is seen in the fact that even glass eels show reactions that suggest a specific olfactory sensitivity (Creutzberg, 1961; Miles, 1968b). Glass eels swimming around in sea water show increased activity when fresh water is added to their container. But if one adds tap water, there is no such response. On the other hand, natural surface water from a lake proved to have great effect. If one passed the water through a carbon filter, it lost its 'attractiveness'. Glass eels, arriving straight from the sea, cannot yet have had any experience of the various scents contained in fresh water. The capacity to detect these scents, despite their marked dilution by sea water, must be innate.

In a study of four rivers, each had water that was differentially attractive to glass eels. One river in particular had an extremely good luring effect. Oddly enough the glass eels from this river were particularly easy to activate with water from other rivers. The strength of attraction increased with the pH of the water. However, merely altering the pH alone had no effect; but a higher pH value coupled with a higher degree of fertility in the area from whence the water came, resulted in increased 'attractiveness'. River water from an area consisting of predominantly granitic rocks was the least attractive. Water in which glass eels have been kept appeared to be less attractive than water which had not been in contact with glass eels; on the other hand, water in which older eels lived, increased its attraction for glass eels (Miles, 1968b). Nordeng (1971) has suggested that these results indicate that glass eels may even be attracted by substances (pheromones) given off by older members of their species.

Substances attractive to eels contain dissolved and undissolved (particulate) organic material, can be destroyed by bacteria and thereby made ineffective, may be unaffected by heat, and finally, are not volatile (Miles, 1968b). Scents attractive to eels are in no way similar to those attractive to salmon. Homing salmon orient by means of substances which are mostly volatile and which are destroyed by heat (Hasler and Wisby, 1951; Fagerlund *et al.*, 1963).

Japanese studies on the nature of these attractants have produced more conclusive results (Hashimoto *et al.*, 1968). Among amino acids, quaternary ammonium bases and nucleotides of a mussel, only amino acids proved to be attractive. However, they were only attractive when used in combination with one another. None of the 18 amino acids could act as a lure by itself. Among the effective amino acids were taurine, aspargine acid, threonine, serine, glutamic acid, glycine and alanine.

The eel's extremely refined sense of smell has naturally raised widespread interest and has led to the assumption that the eel uses its olfactory organs not only to locate food but also to locate different places. In land animals this is a well-known phenomenon – for example, in the dog, which is considered equal to the eel in its olfactory capabilities (Teichmann, 1959). Other species of fish, such as the salmon, provide good evidence that aquatic animals, too, can follow scents. Indeed, the orientation of glass eels towards fresh water is thought to be a form of orientation by smell. However, this is only a general orientation to scents from any natural source of inland water flowing into the sea. It is highly unlikely that individual inland waters have any specific effect (Miles, 1968b); in other words, glass eels are attracted by the smell of any river. But it could still be argued that eels migrating to their Atlantic spawning grounds use olfactory cues for orientation, and that yellow eels transported some distance away find the way back to their old home by using the sense of smell (Mann, 1965; Tesch, 1967a; Deelder and Tesch, 1970; Tesch, 1970). When discussing the use of smell for orientation by salmon in rivers and also by the eel in running water – if, that is, the processes involved are the same for the two species – it should be noted that a 'scent trail' is followed and that the possible direction of this trail is narrowed down by the sides of the river, thus making it easier to follow. A difference in scent only exists for a very short while at the junction between the two rivers where also the scent of the confluent rivers is sharply contrasted. Such clearly contrasted trails are not available in the sea. A displaced eel that is homing many kilometres to some point in Heligoland, for example (Tesch, 1967a, 1968b), would be confused if it came across widespread 'scent waves' from Heligoland, presuming such scent waves were present. Keeping this image in mind, and also if one remembers the difficulty eels have in following a scent trail a few centimetres long in an aquarium, then the idea of their differentiating between huge, diffuse waves of scent becomes improbable. Besides, hydrographic factors make it unlikely that scent from Heligoland would reach Cuxhaven, for example, which is 65 km away. Locating the Gulf Stream would be even more difficult using a trial and error approach.

Experiments in which eels with blocked nasal cavities were transported from one North Sea coastal area to another, showed that such eels display the same homing trends as control eels (Tesch, 1970). Similar experiments, conducted in the Bay of Gdansk exhibited no evidence for an olfactory orientation (Karlsson *et al.*, 1988). Thus, location by olfaction is unlikely, at least as far as the detection of distant goals in the sea is concerned. Evidence that olfaction may play a role in orientation in the sea and in large inland lakes is, as yet, not available for other species of fish either (McCleave, 1967; Royce *et al.*, 1968; Jahn, 1969).

Taste is considered to be a competing sense in fishes. Generally, it is not as sensitive as olfaction, but documentation is not available. Taste buds of the eel's oral cavity, lips and the epithelium are known from electron-microscopical investigations (Pevzner, 1978). They exist in different numbers and sizes. *A. anguilla* probably dislikes bitter tastes (chinin-hydrochloride), but, accepts salty ones. Sweet tastes seem to produce no negative reactions; sour ones are accepted or rejected depending on concentration (Neuhaus and Appelbaum, 1980).

1.9.4 Hearing and the lateral line sense organs

The Anguillidae do not belong to the Ostariophysae – i.e. eels do not have a Weberian apparatus connecting the swimbladder with the inner ear; therefore, direct transmission of vibrations from the swimbladder to the auditory organs probably cannot occur. However, an acoustic function of the swimbladder cannot be excluded completely (Jerkoe *et al.*, 1989).

Mathiesen (1984) described the structure of the inner ear and the site and innervation of hair cells. As generally agreed, non-ostariophysian fishes have less acute auditory perception. This is certainly the case in the eel (see below). Attempts have been made to determine the centres of sound perception in *A. anguilla* (Diesselhorst, 1938). Destruction of the two auditory labyrinths resulted in a reduction of acuity from about 600–400 Hz. Above 600 Hz, according to these studies, the eel's perceptual capacity shows a sharp drop anyway. The same apparent loss of sensitivity was produced not only by the removal of the pars inferior of the inner ear but also in contrast to other species, by removal of the pars superior. Thus in the eel, the site of auditory perception in the frequency range already mentioned cannot be located more precisely and it appears that the eel is able to detect sounds below 400 Hz also through some other sense organ.

Diesselhorst (1938) suggests the tactile sense of the skin is the receptor mechanism for sounds below 400 Hz. The lateral sense organs of the eel, which are also able to detect vibrations, are not activated by tones above 150 Hz (Schrieven, 1935).

More refined methods (Stepanek, 1968) have shown that the eel's capacity for detecting higher sound frequencies is not quite as limited as was suggested earlier. If one makes a distinction between a weak response and an abnormal reaction, then the 600 Hz frequency mentioned above as the upper limit of hearing should be taken to be an abnormal reaction for young eels 8 cm in length. At 800 Hz weaker responses can still be seen in such animals. Older eels (30 cm in length) show a strong reaction at 2500 Hz and will even show a weak response to 9000 Hz. In addition, eels display an escape reaction to sounds of only 50 Hz, while at sounds of 400 Hz they merely become restive. Whether this is evidence for the existence of different receptor organs for different frequency ranges is a hypothesis that needs closer investigation.

A comparison with other species of fish shows that as far as auditory acuity is concerned, the eel is relatively insensitive. Out of a total of 12 species studied, four – the trout, the minnow, the grayling and the pike – show much greater sensitivity, particularly in the higher frequency ranges (Stepanek, 1968). In the lower ranges, too, the eel is less sensitive; its lower limit of perception is between 50 and 70 phon, while the minnow (*Phoxinus phoxinus*) for example, and the mormyrids can perceive sounds as low as 20 to 30 phon. Similar findings are recorded for interval discrimination; this stands at an octave in the eel and is, therefore, rather crude (Diesselhorst, 1938).

Electron-microscopical–histological studies on the lateral line organ are available from *A. japonica* (Yamada and Hama, 1972). Its structure corresponds to those organs which are known to receive mechanical stimuli. However, in the sense cells of the eel, crystalline particles have been found which resemble those of the electro-sensitive lateral line organ of the catfish (*Ictalurus nebulosus*). In the deep-sea eel *Cyema atrum*, lateral line organs could be identified showing still more the character of electro-receptors (Meyer-Rochow, 1978). That the eel can sense extremely weak electrical currents, which resemble those of electric fishes, is documented by McCleave *et al.* (1971) and Enger *et al.* (1976). Sensitivity to electricity and magnetism is probably important for many deep sea fishes and the eel, considering its oceanic migration and orientation to its spawning grounds, should belong to these (Tesch, 1978a, b, 1989, 1995; Tesch *et al.*, 1979).

During transition from the yellow to the silver eel, the lateral line organ develops further histologically (Zacchei and Tavolaro, 1988). This includes the extension of the head channels caudally and should favour the increased motile activity of the silver eel. Alnaes (1973a–c) attempted to identify the complicated system of sensory cells and their afferent and efferent nerves by using an electrophysiological and histological approach.

2 Developmental stages and distribution of the eel species

2.1 Introduction

The eel species are sometimes hard to distinguish from one another. This is especially so for larvae. They continually change body shape and pigmentation, so that the characteristics of various ontogenic stages are not comparable. Very few characters persist through all stages before and after metamorphsis. Even such a fundamentally important character as colouration or spotting pattern of the skin is essentially valid only after transformation from glass eel to yellow eel and until transformation to silver eel. For example, the so-called marbling first becomes clearly evident in animals >20 cm long. At the silver-eel stage it diminishes (Ege, 1939; Jubb, 1961). Characters that typify all stages from larva to silver eel are the vertebral and myomere counts (the latter being one unit higher than the former), as well as the relative lengths of the dorsal and anal fins (Figs 2.1 and 2.2). All other characters can be compared only within developmental stages of about the same age. Besides the characters represented in Tables 2.1 and 2.3 and in Figs 2.1 and 2.11 mainly for older stages, the following are suitable for larvae:

- position of the melanophores;
- position of the vertical intestinal blood vessels (Fig. 2.2);
- dentition (Fig. 2.11);
- anal fin shape.

However, these criteria are not applicable unless complete knowledge of metamorphosis in the individual species is available. In this regard, deficiencies exist especially for the numerous and often sympatric Indo-Pacific species (Castle, 1963). Myomere counts, fin lengths, and geographic distribution may leave the choice between two or three candidate species undecided.

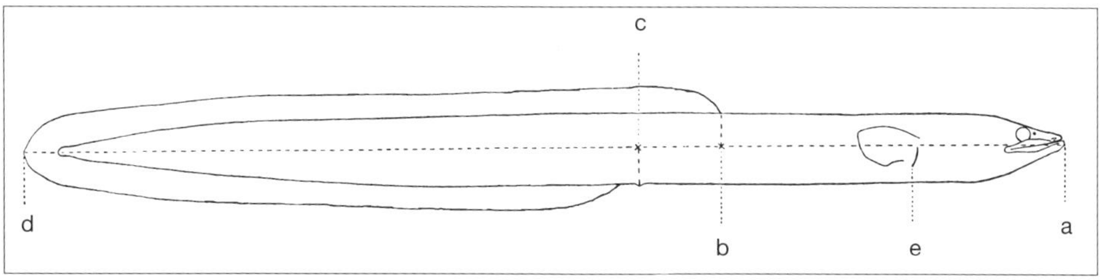

Fig. 2.1 Body measurements of the eel (after Ege, 1939, reproduced from Dana Report **16**)

a–d Total length   a–b Predorsal length   a–c Preanal length   a–e Head length

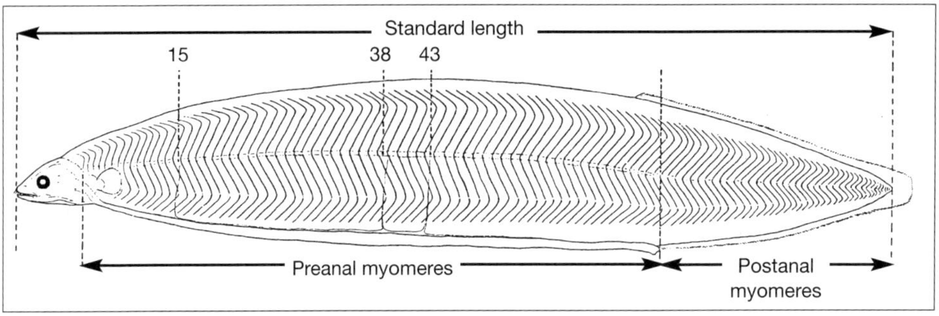

Fig. 2.2 Body measurements of the eel larva (Leptocephalus), showing the standard length, positions of the preanal and postanal myomeres, and three vertical intestinal blood vessels (15, 38 and 43). In this case, the subject is a short-finned eel that shows only a few preanal myomeres (3) from the beginning of the dosal fin fold to the most prominent point of the anus (after Jespersen, 1942, reproduced from Dana Report **22**).

The genus *Anguilla* Shaw (*Gen. Zool.* 4, 15, 1803) is systematically classified within the series Pisces as follows:

- class Teleostomi;
- subclass Actinopterygii;
- order Anguilliformes;
- suborder Anguilloidei;
- family Anguillidae.

The family Anguillidae has the following characters in common with the rest of the Anguilliformes, insofar as the organs in question are present or developed:

- swimbladder connected to the gut;
- no spiny fins;
- cycloid scales;
- no mesocoracoid and no post-temporals;
- premaxillae and mesethmoid fused;
- paired orbitosphenoid;
- teeth present on the maxillae;
- no basisphenoid and sympleticum;
- parapophyses and sometimes also the arches fused with the vertebrae;
- no myodome;
- gill slits narrow;
- dorsal and anal fins very long and fused with each other caudally;
- bones contain bone cells.

For further information about the skeleton, see Sections 1.1 and 1.2 and Table 2.1. A total of 28 recent and fossil families are included in the Anguilliformes (Berg, 1958). Distinguishing characters of the family Anguillidae and thereby also of the genus *Anguilla* are:

- special character: pectoral fin supported by 7–9 radials, 11 in juveniles;
- mouth terminal;
- lower jaw somewhat longer than upper jaw;

Table 2.1 Characteristics of different species and sub-species of the genus *Anguilla* Shaw (after Ege, 1939). Principal diagnostic features are italicised.

Species	No. of vertebrae: Total	No. of vertebrae: Pre-haemal	Distance between the verticals from beginning of dorsal fin to anus (% of total length)*					Length of vomerine tooth band as a % of the length of the maxillary tooth band: ♂♂	♀♀	♂♂+♀♀
A. celebesensis	103.4	39.6	9.0							86–67
A. megastoma	*112.3*	41.7	11.1			Vomerine tooth band than the narrower band of teeth on the maxilla				80–81
A. interioris	105.4	40.7	13.0							
A. ancestralis	*103.3*	38.8	10–11							
A. nebulosa nebulosa	109.1	40.7	11.7		Skin marbled		Tooth rows on the maxilla even			70
A. nebulosa labiata	111.3	40.9	11.9							73–74
A. marmorata	105.6	41.1	*16.3*							70
A. reinhardti	107.8	42.5	10.8	Dorsal fin long			Uneven			75
A. borneensis	105.5	40.6	11.5							
A. japonica	*115.8*	43.6	9.2					84	82	
A. dieffenbachii	112.7	44.3	11.1			Vomerine tooth band wider than the band of teeth on the maxilla				81
A. mossambica	*102.9*	40.5	*14.6*							73–74
A. rostrata	*107.2*	42.8	9.1					*69–70*		
A. anguilla	*114.6*	45.2	10.2		Skin uniformly coloured		Vomerine tooth band narrows before the middle of the plate	*80*	*79*	82–83
A. bicolor pacifica	107.1	43.1	0.2							85–86
A. bicolor bicolor	109.5	43.3	0.8						71	
A. obscura	*104.0*	41.5	3.6	Dorsal fin short			Vomerine tooth band narrows after the middle of the tooth plate			70
A. australis	*112.2*†	*46.2*	1.9†							70

*See Fig. 2.1b–c, †mean between the earlier two suggested subspecies.

- teeth small and in several rows on jaws and gums;
- lateral line well developed;
- gill openings clearly separated from one another;
- inner gill openings wide;
- tongue present;
- lips thick;
- frontal bones paired;
- palatopterygoids well developed (Berg, 1965).

The characters of the individual species are shown in Tables 2.1 and 2.3, as well as in Fig. 2.11. Tabeta *et al.* (1976) compared nine of the species described in Table 2.1 and, to the extent that a sufficient number of animals were available for examination, arrived at similar results. Van Utrecht and Holleboom (1985) found the same for *A. obscura*, and they also presented many biometric data. A remarkable deviation with respect to the vertebral count was evident for *A. japonica*. Certain minor differences were also found for *A. anguilla*, when glass eels of different developmental stages or from collection points at various northern or southern distances were compared (Boëtius, 1980). Ege (1939) compiled information about the synonyms. In Tesch (1991), 33 synonyms are given for *A. anguilla* alone.

2.2 European eel (*A. anguilla*) and American eel (*A. rostrata*)

2.2.1 Spawning areas and larvae

In 1856 (Kaup), a fish from the Strait of Messina was discovered, which, because of its small and slender head, was named *Leptocephalus brevirostris* (Fig. 2.3). The body had a thin, high-backed form resembling a willow leaf. Forty years later the Italians, Grassi and Calandruccio (1897; Grassi, 1896), found that the fish called *Leptocephalus* was a larval stage of the river eel. Further investigations led to speculation that the spawning sites were in the depths of the Mediterranean Sea. There, *Leptocephalus* was supposed to live on the sea bed and be drawn to the surface by the Strait of Messina's turbulent currents. But they could not prove this hypothesis, primarily because they did not succeed in catching early-stage larvae, except those >6–8 cm. Likewise, they could not collect any eels that were ready to spawn, an effort that remains unsuccessful to the present elsewhere, as well. Today the Azores are still the closest area to the spawning grounds where spawning-migrant eels are caught (Bast and Klinkhardt, 1988).

Knowledge expanded suddenly when the Danish research ship 'Thor' caught a *Leptocephalus* specimen in the Atlantic Ocean west of the Faroe Islands, the first outside the Mediterranean Sea (Schmidt, 1912). A further catch the same year by an Irish ship stimulated the Danes to start systematic investigations into the matter of larval distribution in 1905. In June 1905, Schmidt (1906) was able to collect hundreds of *Leptocephalus* larvae off the 1000-m depth line with the Petersen young fish trawl, a relatively new technique in that day. But he had no success in catching younger animals the next year. On the contrary, at a later season in August–September and at a lesser distance from shore, even older lar-

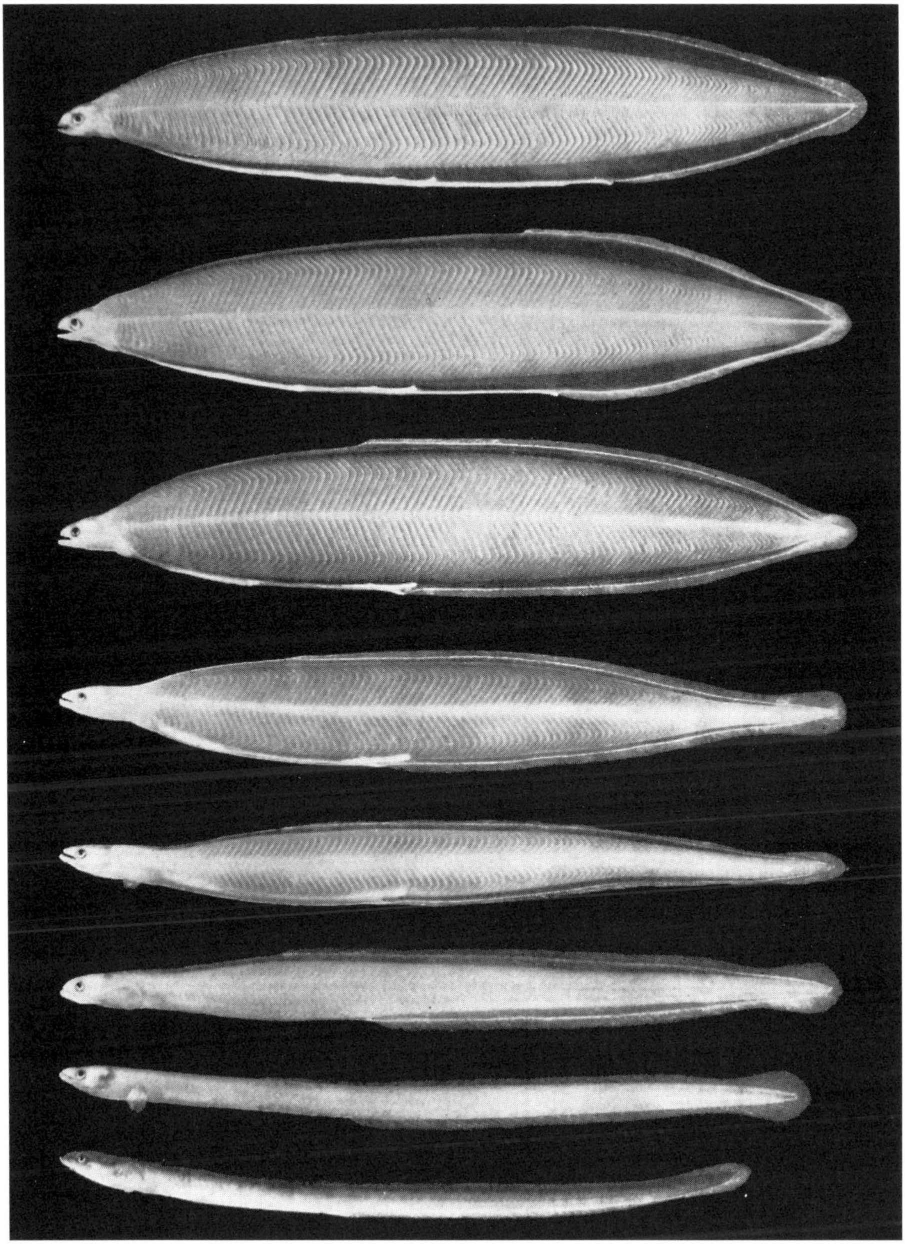

Fig. 2.3 Development of the eel (*A. anguilla*) from the larva (Leptocephalus) to the glass eel (after Schmidt, 1909a)

1 Fully grown larva 75 mm long (Stage I)
2–6 Metamorphosis of the larva into Stage IV, shortly before reaching the typical eel shape
7 Glass eel at its earliest developmental stage (Stage V)
8 Eel at the beginning of pigmentation (Stage VI); on pigmentation during larval development, see Table 1.2

vae were caught, which approached the glass eel stage. The youngest stages came from farthest out – over a water depth of 4000 m. These and later investigations (Schmidt, 1909a) indicated that the spawning sites had to lie far out in the Atlantic Ocean, north of the equator, and that the larvae were not to be found on the sea bed but near the ocean surface.

To support the idea that the spawning grounds of the European eel (*A. anguilla*) could not exist in the Mediterranean Sea, Schmidt (1912) put forth convincing results:

1. Vertebral counts of 1700 Atlantic eels (Iceland and Madeira) averaged 114.736, and 1000 Mediterranean eels averaged 114.731 vertebrae. Thus, the two groups did not differ significantly in this classic taxonomic character.
2. In the Mediterranean Sea in 1908–10, specimens <70 mm, thus young ones, made up only 5% of the larval eel catch in the area east of 3° W, but west of this longitude, this size group composed 60%. Thus, the farther east of Gibraltar, the older the larvae. Moreover, the number of larvae near Gibraltar was significantly larger than further eastward. In the eastern Mediterranean basin, that is, east of Italy, no larvae were encountered at all.
3. The larvae were present at Gibraltar mainly in winter, but were in the mid-Mediterranean Sea near Messina, in spring and summer. This time difference corresponded to the period the larvae could have needed to drift from Gibraltar to Messina. The drift resulted from the surface current that flowed from Gibraltar eastward, despite the fact that an active migration also could have gone on.
4. Just as in western Europe, one found Italian glass eel fisheries only on the west coast. In contrast, there was none on coasts further to the east (lack of fisheries for them does not rule out glass eel occurrence, such as on the Adriatic coasts – see Heldt and Heldt, 1929b).
5. In the Strait of Messina a lantern fishery takes place, in which migrating eels were caught in hand-operated nets. These eels, their swimming direction easily seen, were migrating northward, that is, toward the Strait of Gibraltar. At Messina silver eels were also caught later in winter or spring than they were in the eastern Mediterranean Sea, where they would have to start migrating in order to pass Messina by the turn of the year.
6. Eels seldom occur in the Black Sea. They would probably be more numerous there if spawning sites were available for them in the eastern Mediterranean Sea.

Arguments for presence of spawning grounds within the Mediterranean Sea are attributable mainly to Grassi and Calandruccio (e.g. 1897) (Schmidt, 1912). They established that a great many eels having enlarged eyes occur in the Strait of Messina. In more northerly Europe such specimens, which lead one to suspect an advanced maturity, are encountered only rarely (Section 1.9.2). Schmidt (1912) found the frequent observation of large-eyed eels at Messina to be borne out. The present author's own unpublished observations of eels supplied commercially from the Mediterranean Sea to Hamburg even showed that the frequent occurrence of large-eyed eels is not restricted to the Strait of Messina.

In a shipment of eels from the Rhône delta near the Mediterranean coast in February–March 1968, most of these fish had more or less enlarged eyes (Fig. 1.36), but the state of ovarian development hardly differed from that of silver eels from northwestern Europe;

the same seems true for eels from Messina (Schmidt, 1912). Undoubtedly, enlarged eyes signify rather advanced maturity, as experimental observations have shown (Section 1.2.9), but they are only a secondary character that presumably can develop faster at higher temperatures and correspondingly increased thyrotropic hormone production (Fontaine, 1961) than can the gonads, which depend on gonadotropic and oestrogenic hormones.

A further argument for an isolated spawning area in the Mediterranean Sea could be the difference in protein composition between Mediterranean and Atlantic eels (Drilhon *et al.*, 1967; Drilhon and Fine, 1968, 1971). The authors found differences in the composition of serum transferrins from the Atlantic coast and Mediterranean Sea. Also, differences were found between areas within the Mediterranean Sea, for example, between eels from the French, Greek and Turkish coasts. This could be taken to mean that various spawning aggregations are possible within the Mediterranean Sea, but existence of genetically different eel groups has not yet been shown (Pantelouris *et al.*, 1970, 1971; Section 2.5).

Furthermore, the Mediterranean Sea catch of 30-mm-long leptocephali should bolster the hypothesis that one or more spawning areas exist there. Beyond that, it should be mentioned that in view of relatively large catches of eels within the Mediterranean Sea at its outlet in the Strait of Gibraltar, hardly any silver eels are apparently included (Tucker, 1959). This is also not surprising, as only a very small percentage of the silver eels migrating on the German Baltic Sea coast are caught because the fishing gear reaches a depth of little more than 10 m (Martinköwitz, 1961). Most of the Strait of Gibraltar, however, is deeper than 200 m. Moreover, few eels can be caught at Gibraltar because no suitable fishing gear can be used (Jones, 1959).

Ekman (1932) raised the possibility that if no outward migration takes place at Gibraltar, then the Mediterranean Sea could be considered as a sort of gigantic trap.

The comprehensive studies by Schmidt (1923, 1925a) eventually led to the area in which larvae as small as 5 mm (i.e. not much larger than the 3–4-mm hatching size) occurred. Thus, the spawning ground had to lie very close by. The centre of the area in which larvae of 10 mm and less were found is at 26° N and 60° W (Fig. 2.4).

Fricke and Tsukamoto (1998) doubt that eels could assemble in this 5000–6000 m deep area without a solid reference point for orientation. They suggest that an underwater mount, the so-called Echo Bank, which lies far more than 100 km southeast of the main concentration of larvae, could be the spawning ground. The spawning area of Japanese eels is reported to have analogous conditions (Tsukamoto, 1992). For the Echo Bank there would also be clues for orientation of European eel spawners toward it (Fricke and Käse, 1995). However, there are significant arguments against these assumptions. In original studies by Tesch and Wegner (1990), and also by American investigators (Castonguay and McCleave, 1987) in catches of larvae at about 26° N, freshly hatched yolk fry were caught that could have drifted a few kilometres at most. Moreover, the spawning area extended more than 2000 km from east to west (Schoth and Tesch, 1982), which would require drift paths to be much wider and directionally much more diverse.

All ships participating in this sampling of the western Atlantic Ocean usually caught American eel (*A. rostrata*) larvae along with the European eel larvae. These could be distinguished easily from *A. anguilla*. *A. rostrata* averages 107.2 vertebrae (Ege, 1939), 7.5 less

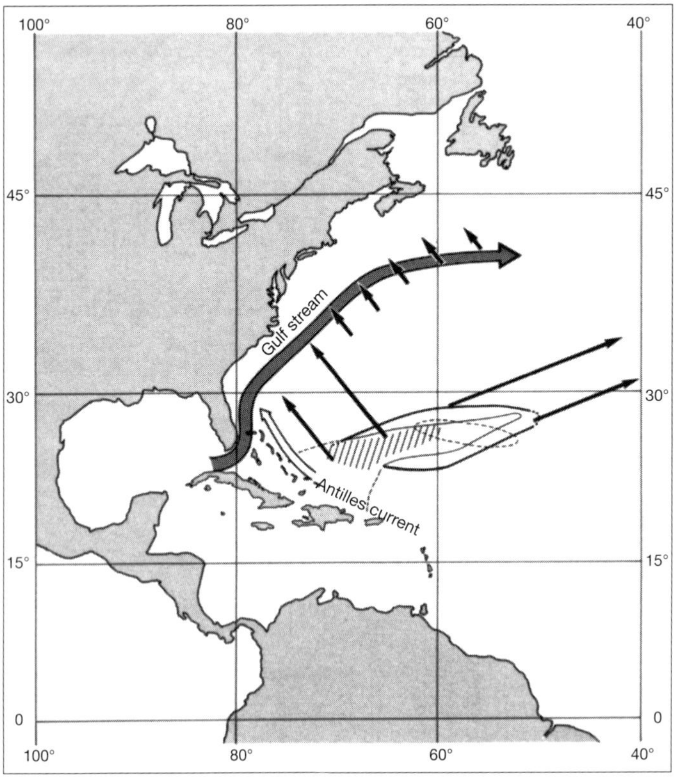

- Dashed-line ellipse: the distribution of European eel larvae up to 7mm long as found by J. Schmidt (after Tåning, 1938)
- Thick boundary line: *A. anguilla* <7 mm
- Thin boundary line: *A. rostrata* <10 mm
- Hatching: *A. rostrata* <7 mm
- Arrows: Direction of presumed active larval migration (Section 3.1).

Fig. 2.4 Limits of occurrence until the larvae reach sizes of 7 and 10 mm for both Atlantic eel species, according to results of the 1979 German Sargasso Sea expedition (Schoth and Tesch, 1982)

than the mean of 114.6 for *A. anguilla* (Table 2.1). Overlap occurs in only 1.5% of individuals at issue (Fig. 2.5). But American eel larvae, termed *Leptocephalus grassi* by Eigenmann and Kennedy (1902), were represented by only about 2300 specimens. In contrast, 12,000 *Leptocephalus brevirostris*, European eel, were caught. This could have been partially or entirely because Schmidt's (1923, 1925a) investigations concentrated on the spawning area for *A. anguilla*, and that the season of sampling was unfavourable for *A. rostrata* (Vladykov, 1964).

Schmidt (1923) assumed on the basis of his sampling results that the American eel's spawning areas are to be expected farther west for the most part – about 15°, according to his graphical representation – and somewhat farther south. In fact, Schmidt's catch ratio of almost 1 : 6 resulted mainly because too little sampling was done in the most westerly area.

Kleckner and McCleave (1980) undertook detailed analysis and summary of the entire larval collection by North American ships not included in Schmidt's (1923) collections. They recorded 2009 *A. rostrata* and 899 *A. anguilla*, thus, a ratio definitely skewed toward American eels. Schmidt bolstered his observations with the fact that the 2000 t/a commercial catch of *A. rostrata* on the continent amounted to only a fifth of the corresponding 10,000 t/a for *A. anguilla* (Schmidt, 1923; Vladykov, 1966).

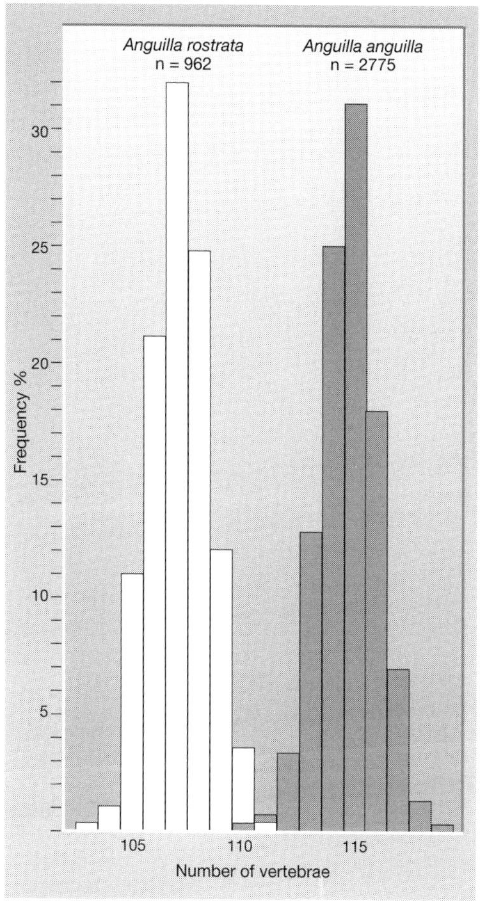

Fig. 2.5 Frequency distributions of vertebrae for *A. anguilla* and *A. rostrata* (data from Schmidt, 1913 and Ege, 1939)

Later results did not change the basic conclusions from Schmidt's studies and interpretations concerning *A. rostrata*. July 1966 sampling in the southern Caribbean Sea between Panama and Trinidad found no *A. anguilla* among many other anguilliform larvae, even though *A. anguilla* larvae abounded at that time in their known range to the north (Smith, 1968). Also further west in the Gulf of Mexico, there is no sign of a quantitatively strong occurrence of *Anguilla* species (Eldred, 1971). According to Kleckner and McCleave (1980), the southern-most catches of small larvae stem from the transitional area between the Gulf of Mexico and the Caribbean Sea. On the basis of Schmidt's (1925b) reports, only 2% of the American commercial catch of eels came from the coasts or the catchment basin of the Gulf of Mexico, and more recently it has been less (Vladykov, 1966). All other eels were caught on the American east coast. The larvae are apparently carried there by the strong, dominant current that has its source in the north equatorial current and extends through the Strait of Florida into the Gulf Stream (Fig. 2.4; Kleckner and McCleave 1980). Thus, the spawning area of the American eel extends somewhat further than Schmidt (1923, 1925a) suggested. But significant parts of it overlap with that of the European eel (Fig. 2.4) (Tesch *et al.*, 1979; Schoth and Tesch, 1982). However, the question of

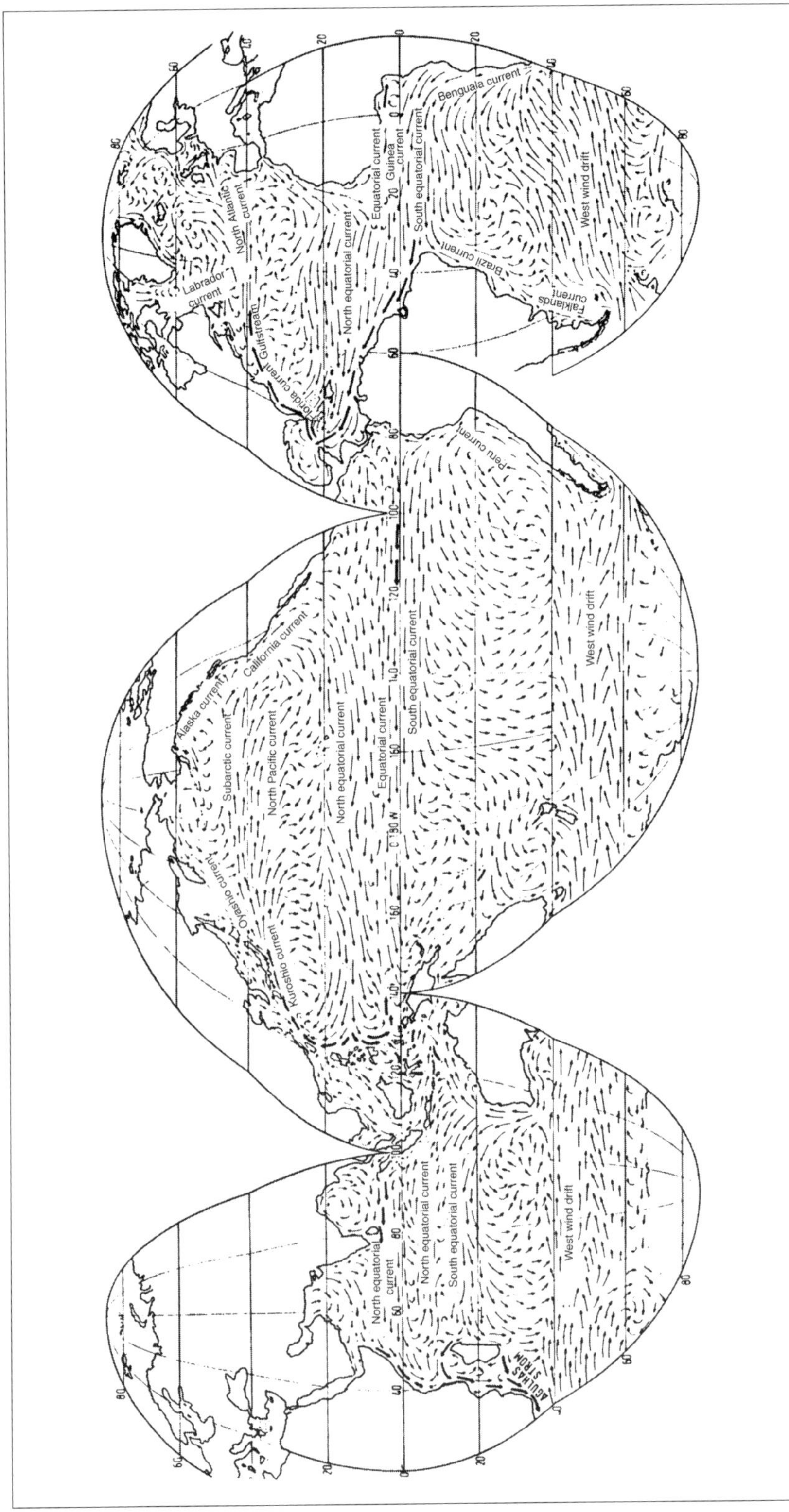

Fig. 2.6 Surface currents of the world's oceans in February/March (after Sverdrup *et al.*, 1942)

how *A. rostrata* can reach the southern-most points of its distribution near Trinidad and Guyana, despite the northward current, remains unexplained (Vladykov 1964).

The Florida current and Gulf Stream transport the American eel larvae northward. Whereas abundances of small leptocephali (stage 0) are found in the assumed spawning area in February-April, there are but few larger leptocephali (stage I) in the entire North Atlantic Ocean at that time (Kleckner and McCleave, 1980). In August, on the other hand, large larvae (40–67 mm) occupy the Gulf Stream's whole area up to the Gulf of Maine. Then, from October to December, only a few large *A. rostrata* larvae (Kleckner and McCleave, 1980) or even glass eels (Kracht and Tesch, 1981) remain far out in the West Atlantic Ocean. At this time onward they approach the American continent and Greenland as glass eels. In contrast, the European eel larvae continued to be caught in autumn (Schmidt, 1923, 1925a). The following spring, having reached 45 mm, they still showed up in large numbers in the northern half of their spawning area and were to be found in large or small amounts as far away as the Azores (Tesch *et al.*, 1979; Kracht and Tesch, 1981), while the newly hatched year class was already developing in the Sargasso Sea. Figure 2.8 shows the size difference between larvae of the two-year classes. Larvae of 60–80 mm (Tesch, 1980, 1998), the so-called stage II, appeared in all catches east of the Azores, right up to the European continental shelf especially in autumn but also in smaller quantities at other seasons (Schmidt, 1909a; Tesch, 1980, 1998; Tesch *et al.*, 1986; Tesch and Niermann, 1992; Kracht, 1982; Kracht and Tesch, 1981; Antunes and Tesch, 1997a; Kleckner and McCleave, 1980). Only occasionally did stage II eels also occur in catches west of the Azores, particularly west of 35° W.

Schmidt (1923) constructed growth curves based on the eastward change in size of *A. anguilla* and on the seasonally increasing size of *A. rostrata*, which the representation in Fig. 2.7 more or less follows. Boëtius and Harding have not yet confirmed this seasonally synchronised size progression from west to east on the basis either of the material men-

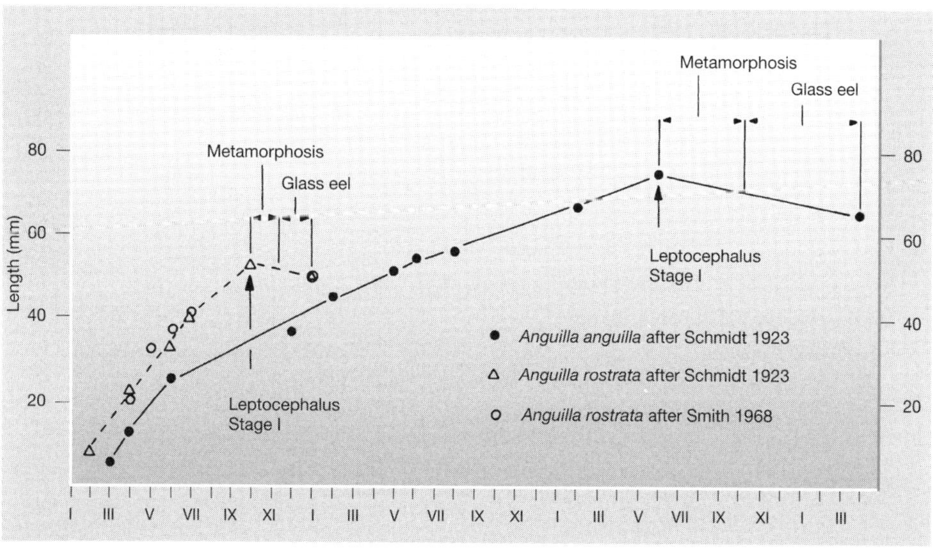

Fig. 2.7 Growth in body length for *A. anguilla* and *A. rostrata* according to data from Schmidt (1923) and Smith (1968), respectively

tioned or, in particular, of their revision of Schmidt's material (Boëtius and Harding, 1985). In this connection, perhaps a role was played by the studies of daily ring counts (Lecomte-Finiger, 1994), according to which the Atlantic crossing should already be finished after a year. These counts on glass eel otoliths have proved to be unusable, at least for the European eel (Antunes and Tesch, 1997b). Subsequent large collections of eel larvae (Fig. 2.8) have made Schmidt's interpretation more convincing (Tesch, 1998). According to the somewhat problematic annuli counts of van Utrecht and Holleboom (1985), which more closely approach our ideas about growth (Antunes and Tesch, 1997b), European glass eels are 2–6 years old and average 3–4 years old. Temporal progress of American eel growth can be confirmed by the more recent studies. The larval development period is estimated at 8–9 months, even if individual larvae stay in the marine environment longer than that.

How the American eel larvae reach their respective continents may be explained to a certain degree by Gulf Stream drift. For the European eel this hypothesis can no longer be accepted for the following reasons:

1. The Gulf Stream was detectable only to a point not far beyond the edge of the North American continental shelf, that is, at about 40° W. Beyond that, significant parts of its water mass deviate to the right and flow back southwestward (Fig. 2.10). Even before that, eddies break off to the sides and do not continue northeastward.
2. Experiments with drifting cards, the results of which were used to demonstrate the duration of leptocephalus drift, account for surface currents exclusively (Harden Jones, 1968). These are generated by the prevailing westerly winds and are independent of the Gulf Stream and North Atlantic Current, which reach depths of several hundred metres. The fully-grown larvae of the European eel in the North Atlantic Ocean prefer a depth of 350–550 m in daytime and 30–120 m at night (Tesch, 1980; Tesch *et al.*, 1986). The author's sampling in the Sargasso Sea revealed a preferred depth of 160 m during day and of 60 m at night (Tesch *et al.*, 1979; Tesch, 1979b; Schoth and Tesch, 1982, 1984). No eel larvae were caught with surface (neuston) nets in the Sargasso Sea in 1979 (H.-C. John, personal communication). Thus, it is hard to imagine how the larvae could reach Europe *via* a surface current that is only about 20 m deep.
3. Netting by the 1979 German Sargasso Sea expedition showed relatively rich larval occurrence southwest of the Azores, that is, in an area that the Gulf Stream and Atlantic Current do not touch. North American sampling also recorded abundant eel larvae from this area to Madeira (Kleckner and McCleave, 1980). The same went for Danish catches (Boëtius and Harding, 1985).
4. So-called stage II European eel larvae in 1979 were significantly smaller off Gibraltar than between the Azores and the Bay of Biscay (Kracht, 1982; Kracht and Tesch, 1981; Tesch at al., 1986; Tesch and Niermann, 1992). According to the investigations with surface drift cards (Harden Jones, 1968), the larvae off Gibraltar would have to have been older than between the Azores and the Bay of Biscay. Thus, it is unlikely that the larvae at Gibraltar arrived *via* a detour through the Biscay area. Surely, they had followed a more direct route from the Sargasso Sea to Gibraltar.
5. It is known that fish larvae drift with the current, and that locational changes happen in that way. Leptocephali also drift, and even the adults can undergo movement by drift. Moreover, the leptocephalus is regarded as 'a current-wafted willow leaf,' even

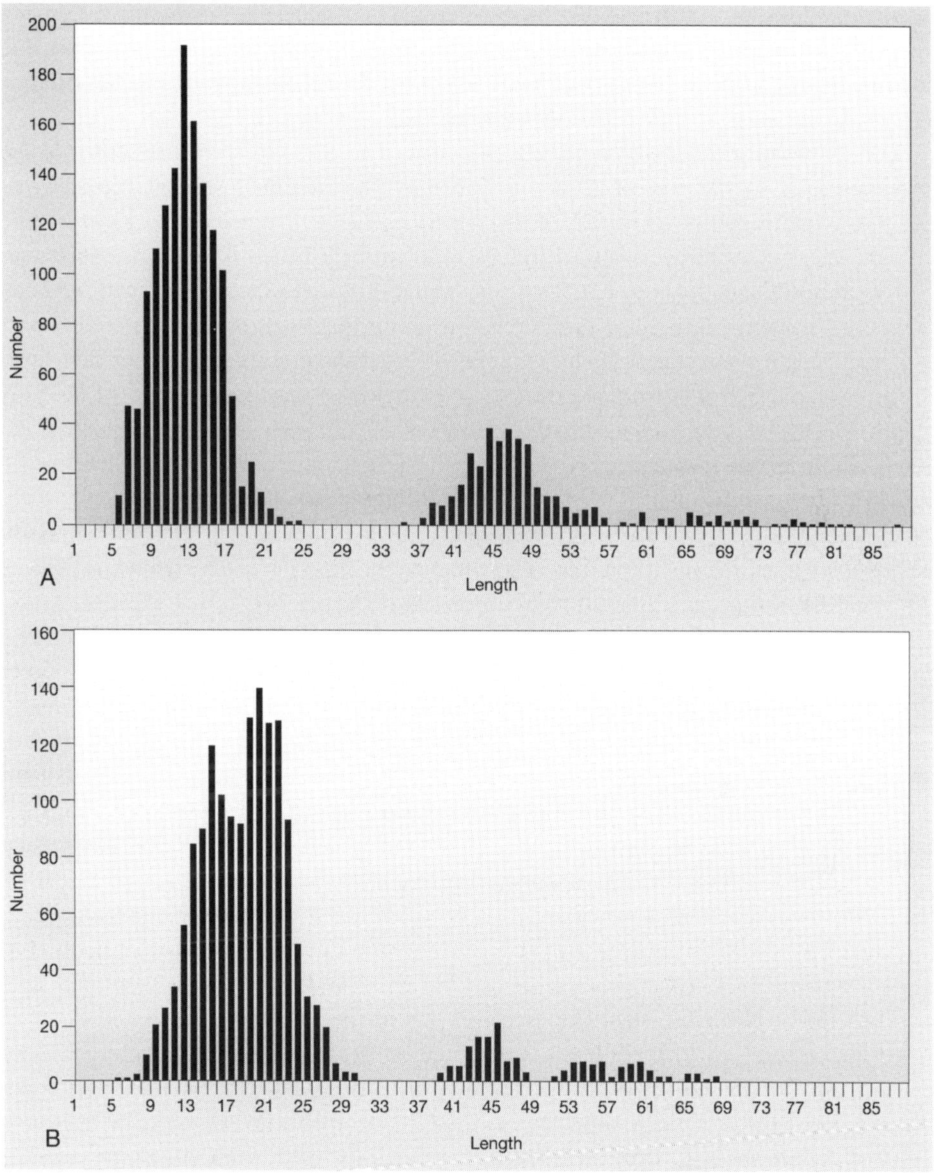

Fig. 2.8 Length frequency distributions of larval eels (*A. anguilla*) in the North Atlantic Ocean (Tesch, 1998)

A Composite from data of Schoth and Tesch (1982) and of Kracht (1982) from catches between 22 March and 6 May 1979 in the area between the central spawning grounds and 35–48° N, 7° W (Tesch, 1982)

B According to catches by Schmidt on 'Dana II' from 22 April to 14 June 1922 between the central spawning grounds and 37°40′ N, 26°00 W (based on data revised by Boëtius and Harding, 1985)

less able to proceed under its own power than other fish larvae are. Spärck's (1930) studies of the eel larva's basal metabolism contributed in no small measure to this view (Section 1.4). Unfortunately, they were conducted on only a very small sample of animals. Freshly caught, uninjured leptocephali are not at all so inactive and slow as previously assumed.

Observations by Kracht and Tesch (unpublished) revealed that furthermore they are very reactive to environmental stimuli. Because they are larger than other fish larvae, and therefore could swim farther during the same period, they probably can cross the Atlantic Ocean at least partially under their own power.

The following calculation makes this plausible. The distance from the centre of the spawning area to the Bay of Biscay is 6000 km. The migration period according to Schmidt's interpretation (Fig. 2.7; Tesch, 1998) could be 2 years and 8 months. The straight-line course of travel would then have to be swum at 7 cm/s. At a mean larval length of 5 cm, this would be 1.4 body lengths/s. The migration speed of fishes is around 2–3 body lengths/s (Blaxter, 1969). Silver eels fitted with ultra-sound transmitters covered almost one body length/s. Juvenile fish can swim faster in terms of body lengths per unit time. Volga river roach (*Rutilus rutilus*) 24 mm long attained a maximum speed of 30 body lengths/s, whereas fully grown salmonids reached only 10 body lengths/s (Bukhanevich, 1969). The calculated over-all speed of 1.4 body lengths/s for eel larvae would therefore lie within the realm of possibility, even though the increased distance imposed by daily vertical migration would mean that they would have to swim a bit faster yet. Orientation would be possible on the basis of a compass course that depends on the earth's magnetic field, as discussed for the oceanic migration of silver eels to the Sargasso Sea (Section 3.4.6). Salmon crossing the Pacific Ocean presumably also orient to the earth's magnetic field. Juvenile sockeye salmon (*Oncorhynchus nerka*) choose a geomagnetically oriented direction corresponding to that in open, natural water bodies (Quinn *et al.*, 1981).

The distribution and geographical separation of the two Atlantic species was previously based on the differential growth and age of the larvae. The American eel larvae would not be able drift as far across the Atlantic Ocean as European eel larvae do because they are thought to finish adapting for continental life by the age of 8–9 months, which would be too early. The European eel is not supposed to be able to become 'continental' in America or Bermuda because it would not yet have developed enough for metamorphosis. Greenland would not be touched by the North Atlantic Current, thus would not undergo any colonisation by European eels. On the other hand, Iceland would, because that current brushes it. However, how American eels get across the cold Labrador Current to Greenland is hard to explain in terms of drift and larval age.

Active migration and a genetically fixed compass course make the interpretation seem simpler. Even for American eels, an active migration with a westward- to northward-oriented compass course is not excluded, even though surely the Gulf Stream plays a crucial role in distributing them. This would compensate for the possible eastward drifting if eddies split off of the right side of the Gulf Stream, and a better supply of larvae to southeastern North America would be guaranteed. A northeastward compass course seems reasonable for European eel larvae if self-propulsion would play the main role; it would be the reverse of the course that silver eels take in spawning migration (see below). However, apparently drift with the current and self-propulsion both play roles (Käse and Tesch, unpublished). Simulated drift of eel larvae shows that they would not get out of the spawning area without supplementary directional swimming.

The breadth of the east–west-oriented spawning area is at least 2200-km – possibly 2500-km (Tesch *et al.*, 1979; Schoth and Tesch, 1982), ensuring a broad distribution along the European and African coasts, even if all larvae were to take the same compass course.

Presumably, the Gulf Stream also has a distributional function, by which it transports larvae northward and facilitates a much more northerly start for those that happen to get caught in it; thereby, Europe's northernmost parts are supplied with eel larvae. The recently studied Azores Current (Käse and Zenk, 1996) seems more important though, as it could transport the larvae to Gibraltar and North Africa. Therefore, a role of active, oriented swimming in the migration is very probable, which is also confirmed by a more precise geographical analysis of the larval samples (Tesch, 1998).

In this connection, it is of interest that significant numbers of American glass eels arrive in North Europe. Boëtius (1976) found that at two freshwater entry locations in Denmark, 3% of the eels were identifiable as *A. rostrata*. Of all the 15,000 eel specimens examined in Europe to date, 0.16% of the glass, yellow, or silver eels had 109 or fewer vertebrae (Boëtius 1980) – indicating that they were *A. rostrata*. In North Europe, the portion amounted to 0.25%, in Central Europe up to 0.06%, and in South Europe zero. Thus, fewer 'strays' occurred in the south than in the north, where the two species overlap the most. There, genetic studies have shown that 2–4% of the gene pool is attributable to *A. rostrata* (Avise *et al*., 1990; Section 2.5).

A possible explanation, among others, would be hybridisation where spawning sites coincide. Comparini (personal communication) ruled that out, based on his electrophoretic studies of larvae in the Sargasso Sea (Comparini and Rodino, 1980). It is additionally noted here that glass eels entering fresh water of a North European locality early in the season averaged fewer vertebrae (about 0.5) than those entering later; also, of eels entering fresh water at any one time, the larger ones had more vertebrae (Boëtius, 1976). The extent to which developmental processes during or after the larval period play a role is unclear; a slight increase in vertebral count continues even into the first or second freshwater year.

The preferred depth of eels that are readying to spawn and the spawning depth itself remain in question because no spawners or freshly deposited eggs have yet been found. No spawners at all have been caught yet in the Sargasso Sea despite intensive efforts (Post and Tesch, 1982). Fish (1927) claimed to have collected eel eggs at 700-m depth near the Bermuda Islands, but later this was severely questioned after comparing the exact descriptions (Schmidt, 1929; Tåning, 1938; Bigelow and Schroeder, 1953; Fontaine *et al*., 1964), as the compact oil globule was absent, which is still present in the youngest larval stages (Fig. 2.9). The discoveries of adult eels at sea have come mainly from predator stomachs so far, and involve eels that apparently had just begun migration to the Sargasso Sea. Such was the report of Vaillant (1896) about an eel in the stomach of a sperm whale (*Physeter catodon*) near the Azores. As a whale's diving depth is limited to about 700 m, this leads us to assume that migration of spawner eels takes place at moderate depths. The trawl catch of a female eel likewise east of the Azores also confirms this (Bast and Klinkhardt, 1988); her oocytes were 0.25 mm in diameter. Thus, she was far from being mature. Further remains of migrating eels came from swordfish stomachs in the Strait of Messina, that is, still very far from the Sargasso Sea (Grassi and Calandruccio, 1897a, b). This also indicates pelagic migration of eels. Two further specimens came from predatory fish caught in trawls at 730 m (Reinsch, 1968). The two predators, *Mora mora* and *Aphanopus carbus*, which each had eaten a 44-cm female eel, inhabit the deep-sea benthos, so the eels probably were close to the sea bed, too. The catch location lay about 100 sea miles northwest of the

Hebrides, also quite far from the Sargasso Sea. The eels had oocytes of 0.15 and 0.12 mm diameter, respectively, therefore, were no further developed than than silver eels from the continental region. Much the same could be concluded from the eye diameters. There is further direct evidence from pelagic trawl catches at 300–400 m in the Faroe Islands vicinity (Ernst, 1975; Boëtius, personal communication). They confirm an eel migration in open water and at moderate depth.

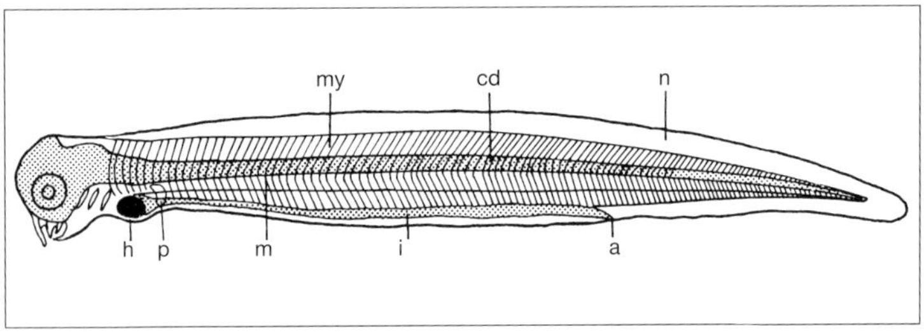

a Anus
cd Neural tube
h Oil globules in the yolk
i Gut
m Notochord
my Muscle segments
n Embryonic fin fold
p Pectoral fin

Fig. 2.9 Pre-leptocephalus larva of *A. anguilla*, 6 mm long (after Schmidt, 1923, completed by Bertin, 1956)

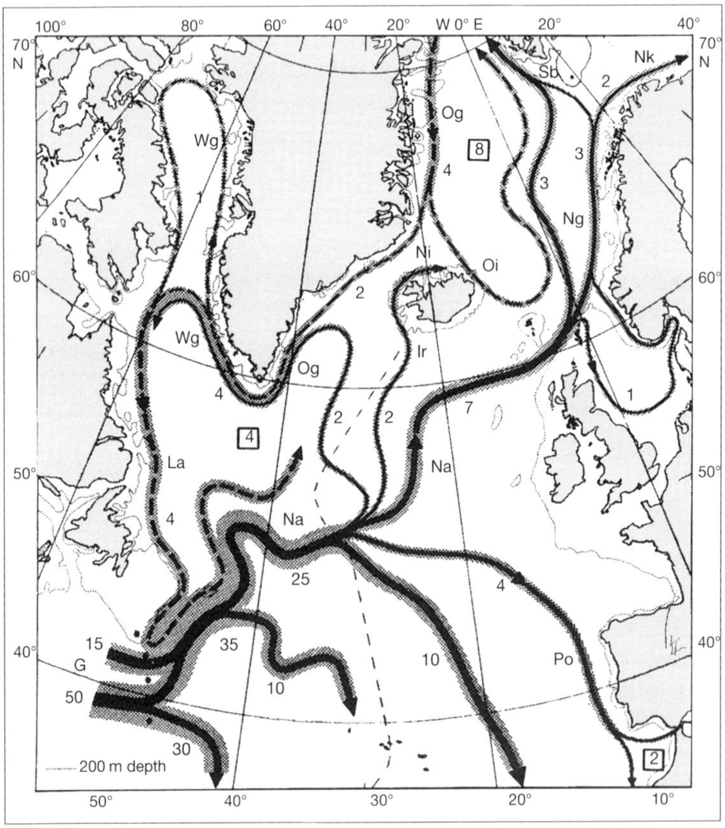

G Gulf Stream
Ir Irminger Current
La Labrador Current
Na North Atlantic Current
Ng Norwegian Current
Nk North Cape Current
Ni North Iceland Current
Og East Greenland Current
Oi East Iceland Current
Po Portugal Current
Sb Spitzbergen Current
Wg West Greenland Current
---- Mid-Atlantic Ridge

Fig. 2.10 Movements of water masses in the North Atlantic Ocean above 2000 m depth (in 10 million m^3/s) from the point where the Gulf Stream splits into several northern and southern components (after Dietrich *et al.*, 1975)

Marking experiments established the migrational depth in the Bay of Biscay on the continental slope of the Spanish–Portuguese west coast and in the Mediterranean Sea. Those studies revealed maximum depths of migrating eels to be 600 m (Tesch, 1979a, b, 1989, 1995). Little was learned from the first tracking of hormone-treated but not completely spawning-ripe eels in the Sargasso Sea. Depths of 700 m were occasionally recorded; but between 100 and 600 m no depth preferences were evident (Tesch, 1989). Fricke and Käse (1994) were able to track two mature eels in the Sargasso Sea and determined that the animals already went to depths of 200–300 m shortly after release. The observations from a bathysphere by Robins *et al.* (1979) on presence of an 'eel' at 2000 m are still to be critically evaluated.

Therefore, conclusions about the depth of spawning sites still can be made only on the basis of data about where the youngest larvae occur. The yolk fry came from 200 to 300 m (Schoth and Tesch, 1984; Kleckner and McCleave, 1988). As mentioned above, the somewhat older stage 0 larvae were found at depths of 60–160 m, depending on the time of day. But the eggs must not fall below a certain depth, to stay buoyant. The same is to be expected of larvae that have large yolk sacks. Therefore, the occurrence of the eggs should not deviate essentially from the depths of 200–300 m where yolk sac larvae exist. Boëtius and Boëtius (1967a) assume, on the basis of their experiments in obtaining spawning readiness in male eels, that the spawning act had to take place about 150 m deep. It is very unlikely that the eels spawn near the sea bed. Water depth is about 5000 m at this place. Here also, the water temperatures of <5°C are too low for eel spawning. Fairly high temperatures were required in order to prolong experimentally the eels' increased maturity or even readiness to spawn – initially 24–25°C in trials with female eels, and 20°C during spawning (Fontaine, 1961; Fontaine *et al.*, 1964; Boëtius and Boëtius, 1967b).

A trough having relatively warm water south of Bermuda characterises the spawning area in the Sargasso Sea (Ekman, 1932). This area is bounded on the north by a thermal front, where, under favourable conditions, the surface water rises 2°C within a few kilometres from north to south (Katz, 1969). The authors also found this front in 1979 (Wegner, 1982) and in 1981 (Tesch and Wegner, 1990), but in 1979 there were only a few places where it was as strongly pronounced as in the earlier studies. Although north of this thermal front the water temperature decreased only from 19.5°C at the surface to 18.5°C at 200 m, south of the front it fell from at least 22.5°C at the surface to 18.5°C at 200–300 m. At the same time, the authors found an elevated salt concentration reaching to a depth of about 100–150 m in the south of the front (Wegner, 1982; Tesch and Wegner, 1990). This area in which small larvae exist is characterised by elevated temperature and elevated salinity (Kleckner *et al.*, 1983; Kleckner and McCleave, 1988). South of the Atlantic spawning area, water temperature rises further, and at the same time the salinity decreases, a situation unfavourable for buoyancy of the larvae.

At the same time the temperature increase in the direction of the Sargasso Sea could aid orientation of the migrating eels (Ekman, 1932), but surely it is only one of the cues, if one at all. Orientation toward higher water temperature would hardly be possible for the Mediterranean eels because temperatures there already resemble those of the Sargasso Sea.

Salinity, which increases in passing southward across the thermal front, could also guide the eels in their migration. Similarly, the spawning area of Japanese eels borders on a salinity front (Tsukamoto, 1992). Laboratory studies showed that the eel can smell the salinity (Hain, 1975).

Further factors have been studied with regard to their influence on the orientation of eels in their migration toward the spawning area. For example, the Gulf Stream could spread odour particles into northern regions. Then, on the edges of the Gulf Stream perception of the difference between the Gulf Stream odour and that of the adjacent water mass would give the eel the cue for swimming in the direction of the Sargasso Sea (Hasler, 1956). Presumably odour is not the decisive cue in long-distance orientation (Deelder and Tesch, 1970; Tesch, 1970). Obviously, olfaction played no role for yellow eels in finding the way back to their home area. Successive differential perception (Teichmann, 1959) would take up too much time for the enormous migration route to the spawning area. Moreover, the boundaries between the Gulf Stream and adjacent water masses are hardly definite, so a constant direction could not be maintained, especially if one considers the eddies that are breaking away. Also, the Gulf Stream presumably reaches the European continent in a very weakened form (p. 84). Even in an aquarium, finding the source of an odour *via* orientation to higher concentrations of it is out of the question (Teichmann, 1959; Hasler, 1966).

Finally, olfactory orientation in combination with rheotaxis, that is, a matter of odour and simultaneously occurring current, is improbable because the eels certainly cannot detect current differentials within the huge water masses. For eels that have reached the Sargasso Sea, it is to be considered whether pheromone perception aids them in finding each other. There are indications that the progress of glass eels entering fresh water is influenced by odours of individuals of the same species (Miles, 1968b).

In addition, an optical orientation by detection of stars or the sun must be disputed. The improved eyesight of migratory eels (Section 1.9.2) is surely to be understood more in terms of a change to cope with diminished light in deeper water. Furthermore, how would a star pattern be recognisable by fish that are migrating at a depth of, for example, 600 m?

The eel's proven directional sense remains as the most probable means of orientation. American silver eels swim southward (Miles, 1968a; Souza *et al.*, 1988), European silver eels westward to northward (Määr, 1947; Tesch and Lelek, 1973a, b; Edelstam, 1965).

For this, the operative environmental factor and the fish's means of reception are still unknown. It is known from tracking of eels fitted with ultrasonic transmitters that migrating eels can maintain a compass course (Tesch, 1974a, 1975b, 1979a, 1992). For *A. anguilla*, presumably this course is west- to southwestward across the deep-sea region (Tesch, 1978a, 1989). In addition, European and American eels tested in the laboratory proved to be sensitive to the earth's magnetic field (Branover *et al.*, 1971; Gleiser and Khodorkovskii, 1971; Tesch, 1974b, 1975b; Souza *et al.*, 1988; Tesch *et al.*, 1992). Influence of the earth's magnetic field on compass direction was also shown in Pacific salmon juveniles during lake migration (Quinn, 1980). Therefore, there is at present no explanation other than the earth's magnetism for the directional information of eels in crossing the Atlantic Ocean and other marine areas, aside from the fact that, *via* drift, water current plays a role in the migrational direction of the headway over the ground (Fricke and Käse, 1994). On reaching the spawning area, other environmental stimuli, for example, pheromones, can ensue for the congregation of spawning populations of eels.

Certainly, the spawning area of American eels extends farther southwestward than previously assumed (Kleckner and McCleave, 1980). Contrary to the previous ideas, major parts of it overlap with the spawning area of *A. anguilla* (Fig. 2.4; Tesch *et al.*, 1979; Schoth and Tesch, 1982). The ideas of Tucker (1959) that the European eel cannot negotiate the distance to the Sargasso Sea, and that the eels occurring in Europe are environmentally modified offspring of American eels is refuted on the basis of evidence (D'Ancona, 1959a; Jones, 1959; Deelder, 1960a; Fontaine, 1961; Bruun, 1963; Sinha and Jones, 1967a). Besides taxonomic differences, such as vertebral count (Schoth, 1982; Strehlow, 1996). and arrangement of the anterior teeth (Fig. 2.11), there are differences of a biochemical nature (Fine *et al.*, 1967; Sick *et al.*, 1967; Ligny and Pantelouris, 1973; Williams *et al.*, 1973; Rodino and Comparini, 1978a, b; Jamieson and Turner, 1980; Comparini and Schoth, 1982; Avise *et al.*, 1990; Tagliavini *et al.*, 1996a; Section 2.5) and with respect to chromosomal morphology (Ohno *et al.*, 1973; Passakas, 1981). This will be discussed in Sections 2.4 and 2.5.

2.2.2 Continental occurrence

Rather definite ideas about the European eel's geographic distribution exist today (Table 2.2; Fig. 2.12). The northern and southern limits are of particular significance because, nearer the centre, little doubt can exist about the eel's occurrence. As eels were

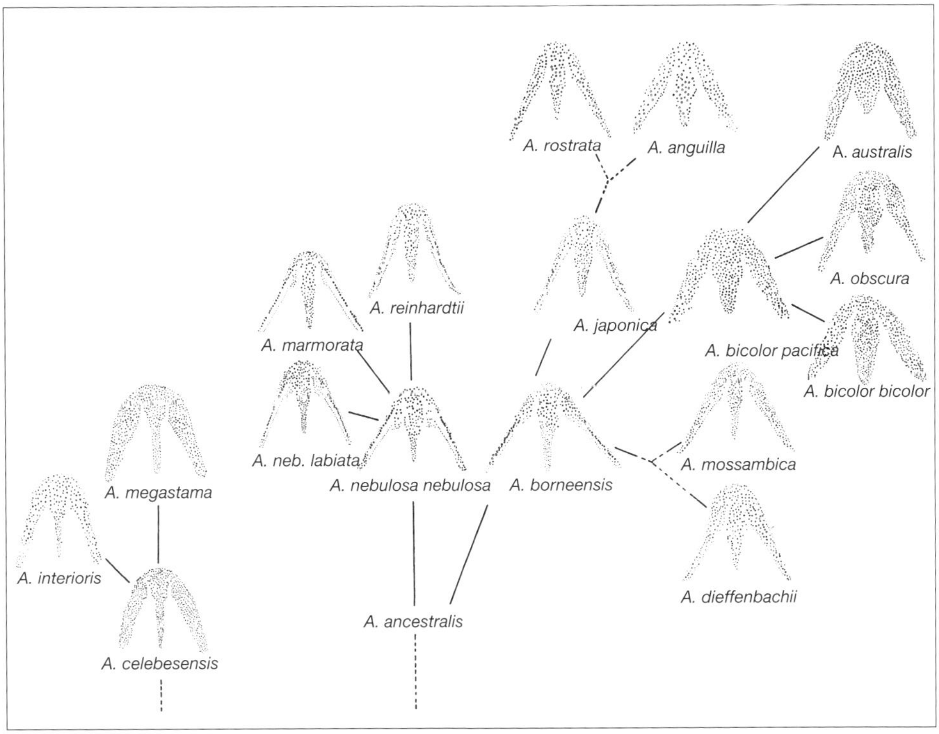

Fig. 2.11 Tooth patterns from the upper jaws of various *Anguilla* species, and the phylogenetic relationships (after Ege, 1939, reproduced from Dana Report No. 16)

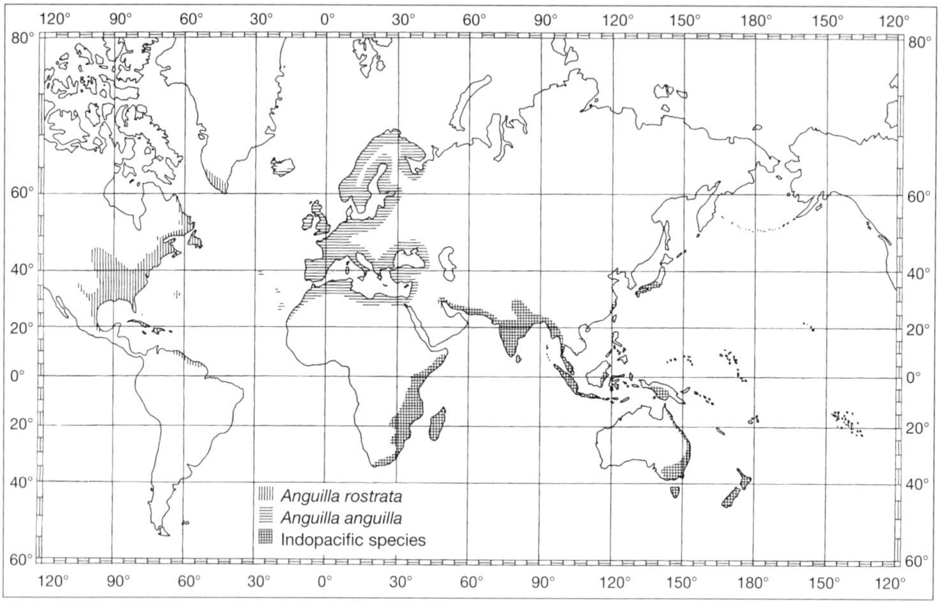

Fig. 2.12 Distribution of *Anguilla rostrata*, *Anguilla anguilla*, and the *Indopacific Anguilla* species

Table 2.2 Postmetamorphic distribution of the various species of the genus *Anguilla*

Geographic region	Species	Geographic distribution	Author
North Atlantic	*A. anguilla* Linnaeus	Europe and North Africa. Northwards: as far as North Cape and to the east of it as far as the Murmansk coast, in Kola Bay and in Northern Dvina. Southwards: as far as coast of Morocco and Canary Islands. Eastwards: the whole of the Mediterranean region, then Black Sea. Westwards: Iceland. Madeira, Azores	Schmidt, 1909b Sorokin and Konstantinov, 1960
	A. rostrata (LeSueur)	North and South America. Northwards: southern Greenland, Newfoundland, Labrador. Southwards: Gulf coast of Mexico as far as Tampico, and in Panama, Greater and Lesser Antilles, e.g. Cuba, Jamaica, Puerto Rico, St. Croix, Grenada, Dominica, St. Vincent and on the South American mainland as far as Guyana. Eastwards: Bermuda Islands	Schmidt, 1909b; Jensen, 1937
North East Pacific	*A. japonica* Temminck and Schlegel	Japan and China. Northwards: Hokkaido, coast of Manchuria, Liao-ho river. Southwards: Hainan, Gulf of Tonkin. Westwards: Bonin Islands	Ege, 1939; Matsui, 1952
Northern Indian Ocean	*A. nebulosa nebulosa* McClelland	Ceylon, Indian hinterland, Myanmar (Burma), Andaman Islands, upper drainage region of the rivers	Ege, 1939

Table 2.2 *(Cont'd)*

Geographic region	Species	Geographic distribution	Author
Northern to Equatorial Indian Ocean	*A. bicolor bicolor* McClelland	Northwards: Indian hinterland, Myanmar (Burma). Eastwards: Sumatra, Java, North-west Australia. Westwards: Madagascar and the smaller islands of the Seychelles, Reunion, coast of East Africa, i.e. Kenya, Dar es Salaam, and isolated specimens further south as far as the Cape coast mainly in coastal regions	Ege, 1939; Jubb, 1961; Kiener, 1965
	A. nebulosa labiata Peters	East African coast of Kenya, where it is the most common eel species, as far as southern Mozambique, in Lake Malawi, isolated specimens as far as the Cape coast, rare on Madagascar, mostly on the upper stretches of the rivers, near Nairobi at a height of 2000 m	Ege, 1939; Frost, 1954; Someren and Whitehead, 1959; Jubb, 1961; Kiener, 1965
	A. mossambica Peters	Coast of East Africa, most southern species, from Zanzibar to the Cape of Good Hope, slightly less frequent in Kenya, the most common species on Madagascar	Ege, 1939; Frost, 1954; Kiener, 1965
Indian and Pacific Oceans; equatorial regions	*A. marmorata* Quoi and Gaimard	Most widely distributed. Westwards: Natal, Madagascar (not very common) and Islands near Madagascar, Mauritius and Johanna. In South Africa from Port Elizabeth and beyond. Northwards: Sumatra, Hong Kong, Taiwan, Marianas Islands, the southern Japanese islands of Yaku, Philippines. Eastwards: Islands in the Western Pacific, e.g. from Samoa to the Marquesa Islands. Southwards: New Caledonia	Ege, 1939; Jubb, 1961, 1964; Kiener, 1965; Matsui, 1952; Nishi and Imai, 1969
Pacific equatorial regions	*A. bicolor pacifica* Schmidt	Insular, eastern part, particularly Celebes and New Guinea.	Ege, 1939
	A. celebesensis Kaup	Northwards: Philippine Island of Luzon. Southwards: Sunda Islands Roti and Timor. Westwards: Nias Islands west of Sumatra. Eastwards; Molluccan Islands of Halmaheira and Geelrink Bay in the eastern region of New Guinea	Ege, 1939
	A. ancestralis Ege	Menado on the eastern part of Celebes	Ege, 1939
	A. interioris Popta	River Bo, tributary to the river Mahakam on Borneo	Ege, 1939
	A. obscura Günther	North-eastwards: Molluccan Islands Halmaheira. Central region: islands north-east of Australia, e.g. New Caledonia, Fiji Islands and Samoa. South-eastwards: Tahiti, Cook Islands. South-westwards: north-east Australian coast 11° to 21° N	Ege, 1939 Beumer *et al.*, 1981
South-eastern to southern Pacific	*A. megastoma* Kaup	North-eastwards: Solomon Islands, from there in a south-easterly direction to the New Hebrides, New Caledonia, Fiji Islands, Society Islands and Cook Islands as far as Pitcairns in the South Pacific	Ege, 1939

Table 2.2 *(Cont'd)*

Geographic region	Species	Geographic distribution	Author
	A. reinhardtii Steindachner	East coast of Australia from Cape York in the north to Port Phillip in the south, northern New Zealand.	Ege, 1939; Jellyman *et al.*, 1996; McDowall *et al.*, 1998
	A. australis (authors of the previously suggested sub-species: Richardson and Phillips)	South-eastern Australia from Cape Byron on the east coast to Warrnambool on the south coast, and Tasmania, Lord Howe Islands, New Zealand, north of Norfolk Island, New Caledonia: Eastwards: Chatham Islands. Southwards: Auckland Islands	Ege, 1939; Dijkstra and Jellyman, 1999
	A. dieffenbachii Gray	New Zealand, Chatham and Auckland Islands.	Ege, 1939

caught on the Murmansk Coast and the North Cape, they are certain to be found everywhere to the west and south of that. Likewise, the southern limit is known from Schmidt's (1909b) investigations. The Canary Islands represent the most southerly collection sites to date. The present author found recent confirmation of this in reports of fishers from Gran Canaria and La Palma. In addition, reports of *Anguilla* catches often appear in angling periodicals there. Therefore, the question arises about whether eels occur further south on the West African coast.

The western occurrence of *A. anguilla* at some Atlantic islands like Canaries, Azores and Iceland appears to represent the boundaries of the distribution of the European eel and the eastern border for the American eel. The eastern distribution is more problematic. Although little doubt exists about the Baltic Sea and its catchment basin, statements about the Mediterranean Sea's eastern inlets must be considered very critically. This applies especially to the seas lying southeast of the Suez Canal. Even though it is possible for an immigration to get there, the occurrences reported there are results of installing the Suez Canal. Similarly, eels were variously reported in the Volga river, which may have immigrated *via* canal connections with the Baltic Sea (Schmidt, 1909b). Hardly any doubt underlies the occurrence reported by Schmidt (1925b) at Massaua on the west bank of the Red Sea. Less certain are the discoveries of European eels that Ege (1939) reported from East Africa, of which three specimens are supposed to have come from Nairobi; confusions with *A. mossambica* (Jubb, 1964) are possible. Studies by Frost (1957b) in the same area could not confirm the occurrence of *A. anguilla* in Kenya.

There can be little doubt that *A. anguilla* enters the Black Sea naturally. Schmidt (1909b) reported numerous eel catches from that catchment basin in the 19th and even in the 17th and 18th centuries – an era when it was not yet possible to transport eels over such great distances for stocking. For example, he cited statements by Massili, according to which eels were caught at Linz, Krems, and Vienna. Massili's quotations about catches in the upper and lower Dnepr river were numerous. The quotations came for the most part from the 19th century and, in part, told of consistent annual catches.

Catches were also reported from three different places in the Sea of Azov in the 19th century. After World War II eels were found in the Black Sea catchment rarely, as documented in the publications of Drapkin (1964), who reported on eleven eel catches in the northeastern Black Sea, as well as in Telegut (1955) and Gyurko (1961) from the lower Danube catchment. Occasionally, amounts of eels were found, for example, in the Theiss and Danube rivers in 1960 (Sterbetz, 1960), but it is not clear whether they originated from artificial stocking or from natural development of a particularly strong year class. Individual eels can be caught in the Bulgarian rivers that flow into the Mediterranean Sea, for example, the Struma, Mesta and Maritza (personal communication, Michailowa, Bulgarian Academy of Sciences, Sofia). However, they are caught more rarely in the Black Sea tributaries. Likewise, eels sometimes occur in the lakes on the Black Sea coast; these probably had intruded from the Black Sea. However, the eel does not occur in Caspers' (1951) faunal lists for the Bulgarian coast, leading one to suppose that these eels can represent only isolated cases.

Thus, in the Black Sea the eel is at the edge of its range. That eels are relatively abundant today in European–Russian water bodies is entirely attributable to stocking (Orlov, 1966; Kokhnenko, 1958, 1967). For incomprehensible reasons, even *A. rostrata* was stocked in central and south Russian waters (Kokhnenko, 1975). Stocking played a greater rôle still in the Danube catchment's middle and upper areas (Volf and Smisek, 1955; Meyer-Waarden, 1964; Schmid, 1962, 1964, 1966; Lassleben, 1966; Wiesner, 1966).

Eels have been stocked in southern Germany since 1881 (Wiesner, 1966; also Hofer, 1897; Haack, 1879, 1881). These statements reveal that already at the end of the 19th century all occurrences of eels in the Danube basin could have been due to stocking. After World War II, eel stocking increased substantially in southern Germany. In the Danube river near Dillingen, there were eel harvests of 3.5 kg/ha in each of the years 1959 and 1960, 5.5 kg/ha in 1961, and 8.0 kg/ha in 1962. This exceeded the yield of many north German water bodies. There are also considerable eel populations in the smaller lakes of Upper Bavaria. Well over 3 kg/ha were caught annually from the Hartmannsberg lakes of the Rosenheim district (Schmid, 1966). In the Upper Palatinate, most of which lies in the Danube catchment, the annual eel yield was around 3000–5000 kg (Dorfner, 1966). In Lake Balaton since 1961, 60,000–400,000 glass eels have been stocked annually, and consequently considerable amounts of eels for human consumption have been caught there (Koops, 1967a; Biró and László, 1970).

It is all the more astounding that so few eels occur in the lower Danube area. One would expect to find at least the downstream migrating silver eels there. There are indications from the upper Danube that the eels indeed migrate downstream in considerable numbers and do not choose the western direction as is said in popular publications (e.g. *Fisch und Fang* 7/1966, p. 196; *Elbe-Jetzel Zeitung*, 8 September 1963, p. 3). Likewise, it was said in a report of the Danube Fishery Association that turbine-caused injuries (Section 7.4.4) are discovered on about 10 eels per year, and that many times that number of eels perish without being seen. Without exception, large specimens are involved in these cases, so it can be assumed that these are emigrating eels (Wiesner, 1966). 'As experience shows, the silver eel emigrates from the individual Upper Bavarian lakes at quite different seasons' (Schmid, 1966). These statements create the impression of a quite normal silver eel migration in the Danube river area. Therefore, the reason eels are so rarely caught in the lower Danube area presumably lies more in deficiency of the fishing operation there than in scarcity of eels that pass through the area.

The Bermuda Islands are also considered as a boundary between the ranges of both Atlantic eel species. Although larvae of the European eel are more abundant than those of *A. rostrata* in the vicinity of these West Atlantic islands (Schmidt, 1925a; Tåning, 1938), only the American eel occurs on the islands themselves (Boëtius and Boëtius, 1967b). Greenland in addition lies at the border between the two ranges. The few samples known thus far (Schmidt, 1909b; Jensen, 1937) are based on seven specimens. Further sampling of this would be important. 'Strays' of the other species are very possible, as Bruun (1937) determined in counting the vertebrae of several hundred glass eels from the Spanish north coast. In this sample, *A. rostrata* were identified clearly: from later investigations on 6460 Danish juvenile eels, 0.3% were probably classifiable as American eels (Boetius, 1976, 1980). The probability of finding *A. rostrata* would be greater on the Azores, far to the west. However, Schmidt (1909b) identified 34 eels from these islands as *A. anguilla*. The present author's investigations on a sample caught by electrofishing likewise did not identify any eel for sure as *A. rostrata*; 30 counts on X-rayed spinal columns showed 114.0 (111–118) vertebrae – therefore, perhaps at best rather insignificant amounts of American eels (Tesch, unpublished). The sample's only questionable eel would be the one having the very rare vertebral count of 111 (Fig. 2.5).

The American eel's northern continental range is seen in Fig. 2.12 and Table 2.2. In Canada, Quebec, with the St Lawrence current, shows the greatest abundance of eels (Fig. 2.13). Presumably, this is attributable to favourable conditions for operating weirs and draw-nets, as well as to the largest freshwater drainage basin on North America's east coast. Toward the north, for example, in Newfoundland, the catches decline sharply (Eales, 1968). However, distributional studies showed that *A. rostrata* occurs in almost every river system of Newfoundland's coast (Flechtner and Anderson, 1972). On the southwest coast, eels are so abundant that a commercial fishery developed. It is concentrated on St George's Bay and harvested more than 36 t in 1972. In Labrador at that time

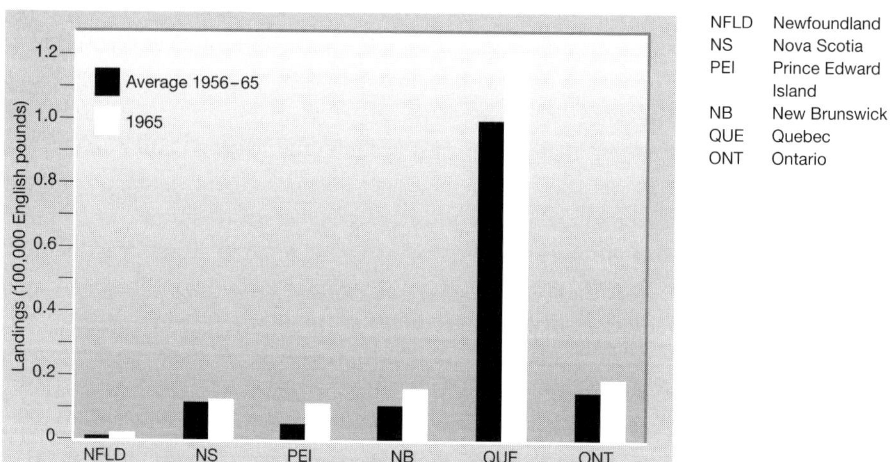

Fig. 2.13 Mean annual eel landings (weight in pounds) over the 10-year period, 1956–65, compared with those for 1965 in Canadian provinces (after Eales, 1968)

the eel population could hardly support commercial fishing. As mentioned, the reason for this could be that the cold, southward flowing Labrador Current (Fig. 2.6) may hinder the spread of larvae or glass eels.

Natural immigration of eels into the Great Lakes system is possible only as far as Lake Ontario, where a considerable commerical fishery used to take place, with annual harvest reaching 136.2 t in 1959 (*Canad. Fisherman* **51**, No. 3, 1964). In Lake Erie and the Great Lakes lying above it, eel immigration is blocked by Niagara Falls. But apparently, a few eels manage to get past them (Schmidt, 1909b). They are assumed to have entered *via* the Welland Canal. An eel ladder in the Moses Saunders Dam also promoted their passage into the Great Lakes (Whitefield and Kolenosky, 1978).

Finally, eels immigrate into the Mississippi river system, but not in very great numbers anymore. They are better known from the Chicago area of Illinois, for instance, than from adjacent Lake Michigan. In a lake on the Illinois river, a Mississippi tributary, anglers caught 21 eels in 5 years, and commercial fishing took 16 more (Starrett and Fritz, 1965). Eels are scarcer yet in the more northerly tributaries, for example, in the Wisconsin river system. Compared with Schmidt's (1909b) statements, a significant later decline in abundance was seen. This was caused principally by dams. Around the turn of the century eels were well distributed in the upper Mississippi and Missouri rivers. In the state of Illinois, for example, 20,000 pounds of eels were still being landed. In 1902, commercial fishers on the Mississippi river in Wisconsin harvested 1745 pounds, and that river yielded 900 pounds in Minnesota even further to the north.

Today, eels are still relatively abundant in the lower Mississippi river and other Gulf of Mexico tributaries. However, already at the turn of the century harvest was less than on the continent's east coast. As previously described, this disparity was presumably attributable to the ocean currents, which carried few eels into the Gulf of Mexico. Eels are, therefore, scarce on more southerly coasts (Section 2.2.1). On the other hand, the Suwanee river, which enters the Gulf of Mexico in Florida, has a tributary in which considerable numbers of eels are caught year-round (Mellier, 1967). Of course, the lower Mississippi area's streams are clearly recognised as populated with eels (Gunning and Shoop, 1962).

Stocking introduced *A. rostrata* not only into remote areas of North America, for example, Alberta's Saskatchewan river (Radford, 1972), but also into natural waters of other parts of the earth. For example, Kokhnenko (1975) reported that *A. rostrata* were stocked in waters of central and southern Russia, and that specimens were recovered which had grown well. Even in East Asia, at present *A. rostrata* may be common in natural waters, from aquacultural stocking.

2.3 Indo-Pacific Eel species

Geographically distant from the two Atlantic forms are the Indo-Pacific eel species. The genetic autonomy of the Indo-Pacific eel species is undoubted, even though they resemble closely the Atlantic species in various respects (Sections 2.4 and 2.5). However, the systematic separation of the Indo-Pacific species from each other is by no means as definite as in that between *A. anguilla* and *A. rostrata*. Schmidt's papers and their completion by Ege (1939) served to reduce about 50 assumed species to 16. Nevertheless, an open field remained for investigations, although for the main species genetic knowledge has improved (Section 2.5).

In considering the world distribution of eels (Fig. 2.12), one tends to infer, besides the aforementioned split, a certain segregation between southeast African and Indo-Pacific taxa. For reasons of overview, this grouping of the forms will be made in what follows. But according to present opinions, this classification includes several associations in which some species occur simultaneously in areas located very distant from each other. Most remarkable is the example of *A. marmorata*. This species occurs throughout almost the whole Indo-Pacific area that eels inhabit (Figs 2.14 and 2.20). *A. nebulosa* and *A. bicolor* also occur in both the Indo-Pacific and southeast African regions (Figs 2.14, 2.20 and 2.21). The division of *A. nebulosa* into one subspecies designated for each of these major regions seems invalid. The extent to which acceptance of species, subspecies, or even local forms is justified is among the questions that are easiest to resolve by molecular-genetic methods (Section 2.5).

*2.3.1 Southeast African Ranges (*A. marmorata, A. nebulosa labiata, A. mossambica *and* A. bicolor*)*

Fig. 2.14 shows the larval and adult distributions of the four eel species that occur on the southeast coast of Africa: *A. marmorata*, *A. nebulosa labiata*, *A. mossambica* and *A. bicolor*. In addition, the presumed spawning area is shown in the South Equatorial Current northeast of Madagascar. In this area there is also a zone of warmer water (Jespersen, 1942), similar to the situation in the North Atlantic spawning area. Water of 15°C extends about 400 m deep in an area lying near 60° E, between 10° and 20° S. The spawning area is inferred from that fact, from the predominantly westward oriented current, and from the presence of relatively small, 37-mm-long larvae north of Madagascar. Many fewer clues exist here than in the Atlantic Ocean, near the Mariana Islands in the Pacific Ocean, or southwest of Sumatra in the Indian Ocean (Fig. 2.18); therefore, the speculations involve major uncertainty. Judging by the maps of currents and larval distribution (Figs 2.6 and 2.14), the larvae are swept toward Madagascar and the African east coast, insofar as no active migration occurs (Section 2.2.1). Some of them are carried around the island's north end, others around the south end. Due to its exposed position, Madagascar is inhabited by all four species of eels. The northward drifting larvae populate the African coasts mainly. Some of them pass between Africa and Madagascar, continuing southward in the warm Mozambique Current, and some of these, in turn, reach the coasts of Mozambique and Natal. Likewise, the leptocephali that are carried south around Madagascar enter the Mozambique Current and the Agulhas Current, which continues further southward. From Port St Johns onward, cold counter-currents on the coast and the Atlantic west wind drift block the drift from continuing to the west, so westward distribution of the various species is hardly possible much further than Cape Town. It is assumed that the eels that exist further south than 31° S, and that, for example, reach as far as Cape Town, are definitely not glass eels but rather had already reached the coast a year earlier and had proceeded further by active migration.

Like the Atlantic species, the South African species have a definite time for transition from the coastal waters into the fresh water of the rivers. For the two most abundant

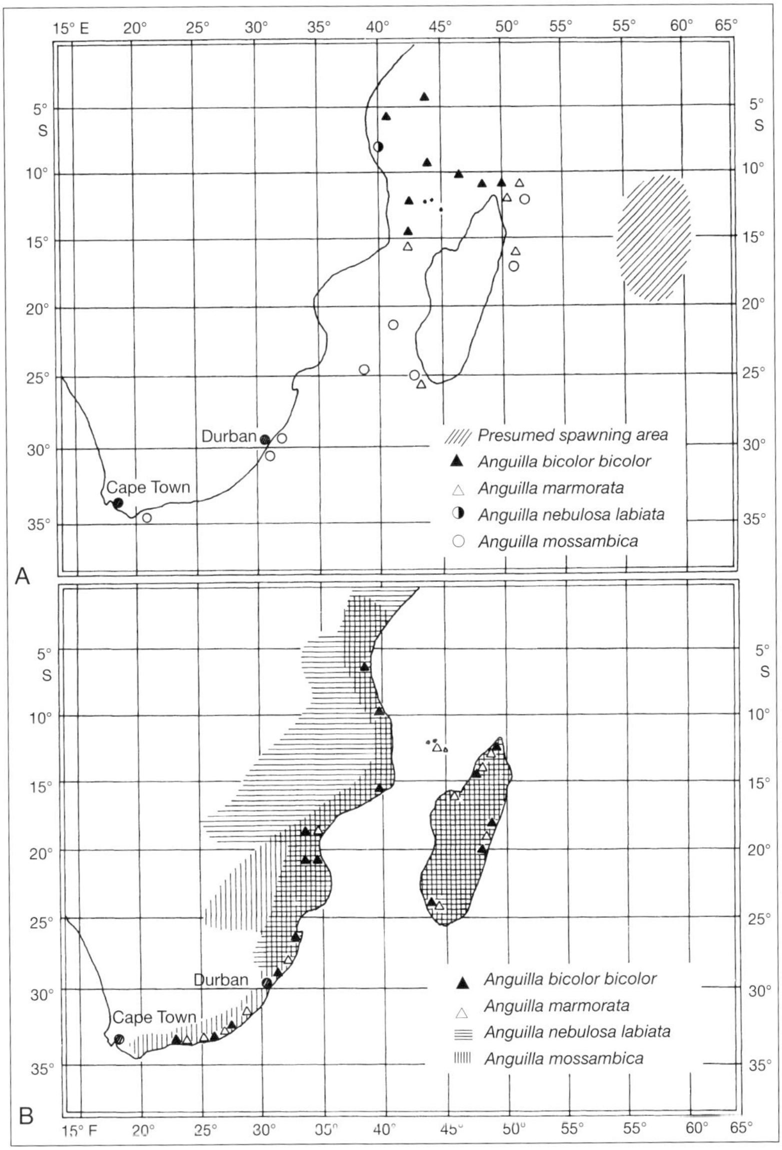

Fig. 2.14 Distribution of southeast African eel species (modified after Jubb, 1961)

A Larval occurrence and presumed spawning area
B Coastal and inland distribution

species, *A. mossambica* and *A. nebulosa labiata*, this is in the summer months of January to March (Frost, 1957a; Jubb, 1961). Therefore, they probably have a seasonally determined spawning period, in contrast to the tropical species of the Indo-Pacific region. This hypothesis is bolstered by the fact that the silver eels of all four species emigrate during a distinct season, namely November until March. If the eels spawn in the area lying just off Madagascar to the northeast, then it is also to be assumed that larval development lasts less than a year. This assumption is supported by the fact that the glass eels of *A. mossambica* and *A. nebulosa* are only a little longer than 50 mm, thus contrast, for example, with the 70-mm *A. anguilla* 'short-migrators.' Jubb (1961) assumed that the three species, *A. marmorata*, *A. nebulosa labiata*, and *A. bicolor bicolor* spawn in the northern part of the supposed area. This suggests that the observed distribution, especially of the glass eels and larvae (Fig. 2.14) in contrast to *A. mossambica*, covers primarily the northern part of the range, namely East Africa.

With its four eel species, the Isle of Madagascar presents the opportunity to ascertain various requirements that these species have for their inland migration and range-expansion. *A. bicolor bicolor* penetrates the least distance inland; this species is found in fresh water but does not get beyond the coastal waters. An exception is known from Zimbabwe (Frost, 1957a). *A. marmorata* immigrates a bit further inland – to about 500 m elevation. *A. mossambica* and the type form of *A. nebulosa labiata* found on Madagascar reach 1000 m and higher (Kiener, 1965). Similar distributional differences, which emphasise vertebral count and fin length (Table 2.1), were also found for the two New Zealand species (Section 2.3.3). They might be attributable to differing environmental requirements. But the inland extent of the range, or even the vertical distribution, cannot always be determined simply on the basis of individual records of occurrence. As Jubb (1961) wrote, *A. nebulosa labiata*, for example, is found at lesser distance inland as one goes farther from the main area of its distribution, until finally it is only on the coast. Normally this species occurs far inland. In Kenya, *A. nebulosa labiata* is the only species still found at elevations over 1000 m (Frost, 1954). This species inhabits the Zambesi river system for 1000 km inland, in the process surmounting a series of waterfalls, having a total height of 60 m (Frost, 1957b). Jubb (1964) even reported a single specimen that was caught above Victoria Falls. Even the 128-m-high dam erected in 1958, creating the 532,400-ha Kariba Lake, seems not to have blocked *A. nebulosa labiata* from populating the area above it in considerable numbers (Balon, 1971, 1975). *A. mossambica* might occur there, as well. *A. nebulosa labiata* overcomes similar obstacles in ascending into Lake Nyasa. The outflow of this lake, with many waterfalls and rapids, presents an elevational change of 300 m in a 50-km reach. The lake's endemic fish species number 185, which this isolation protects from immigration of further species of the Zambesi river system. *A. nebulosa labiata* gets over this barrier, except when drought dries up the outlet (Schmidt, 1925b; Jackson, 1959). Further to the south at the edge of *A. nebulosa labiata*'s main range, *A. mossambica* is the predominant species inland. In Southern Rhodesia, for example, they were caught 750 km from the sea at about 1400 m elevation; *A. nebulosa labiata* was not present until 150 m lower, and then only as a few specimens (Jubb, 1961). The species *A. marmorata* is also at the edge of its range along the whole southeast African coast, so the inland distribution of this species is minor there. There are exceptions though. Thus, *A. marmorata* was a very adaptable species in Zimbabwe, 1000 km inland (Frost, 1957b), and at another place 1530 m high (Here, after Ege, 1939).

*2.3.2 North Pacific temperate zone (***A. japonica** *and* **A. marmorata***)*

The Pacific eel species that corresponds to that of the northwest Atlantic Ocean is the Japanese eel (*A. japonica*). The individuals of this species, presumably like those of the American eel, come up out of the North Equatorial Current into the Kuroshio Current, a strong current that the coast directs northward (Fig. 2.6), which corresponds to the Atlantic ocean's Florida Current and Gulf Stream. By this means, the Japanese eel's range extends from 18° N in the south to 42° N in the north (Fig. 2.15). As in the case of the Atlantic eel species, active swimming by the larvae to get out of the currents cannot be ruled out.

Japanese scientists tried to discover the spawning area of *A. japonica* after they had caught only a few larvae (Matsui *et al.*, 1968; Matsui, 1971; Tabeta and Takai, 1973, 1975; Nishiwaki, 1974; Tanaka, 1975; Nakao, 1976; also Matsui, 1957). In 1991, the 'Hakuho Maru' succeeded in reaching the spawning area (Tsukamoto, 1992). A total of 911 *A. japonica* larvae were caught. Among these, at 15° N, 137° E, were 239 leptocephali, 7.9–24.5 mm in length and estimated to range in age from 10 to 40 days. The sampling sites were on the north edge of the North Equatorial Current (flowing 20 cm/s westward in that area) and, like the Atlantic eels, mainly 75 m deep. Accordingly, the main spawning area must lie still further to the east, nearer the Marianas Islands. South of the primary sampling area was a front that had higher salinity, which corresponded to the spawning area of Atlantic eels. A repeat of the sampling in 1994 confirmed this result (Tsukamoto, unpublished). A consolidation of the result is being made. Based on the data, it is presumed that, as hypothesised for the Atlantic eels, the spawning could be taking place on ocean mounts that rise from the ocean depths east of where the larvae were discovered (Fricke and Tsukamoto, 1998), even though no small larvae were found right at those places.

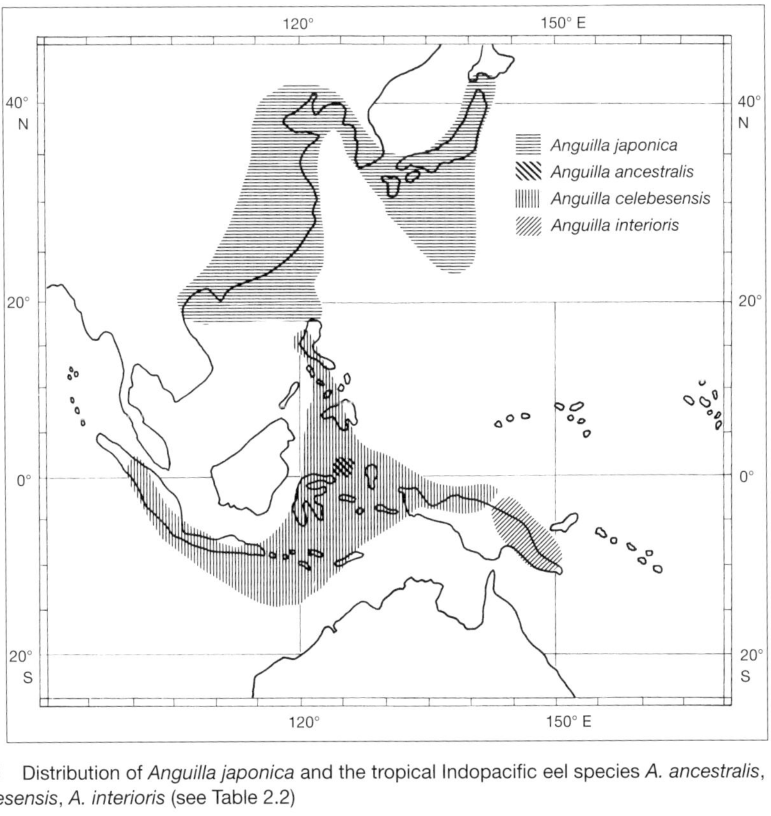

Fig. 2.15 Distribution of *Anguilla japonica* and the tropical Indopacific eel species *A. ancestralis*, *A. celebesensis*, *A. interioris* (see Table 2.2)

Emigrating Japanese eels were caught with various gear over ocean depths of 70–300 m in the Yellow Sea and Chinese Sea, in the Strait of Korea, and in the area surrounding Japan (Matsui, 1957). Obviously, they were swimming primarily at rather shallow depths, and several specimens could even be caught with hand-nets. Their oocyte diameter range, 0.17–0.27 mm, showed well progressed maturity, which indicated that the spawning sites were closer to the continent than those of the European eel.

In the opinion of Matsui (1952, 1957), *A. japonica*'s glass eel migration in the lower reaches of rivers starts 2 months earlier than that of *A. rostrata*. It begins in October and ends in May (Tzeng, 1985), and therefore, hardly differs from that of *A. anguilla* (Gandolfi-Hornyold, 1934; Heldt and Heldt, 1929a, b; Schmeidler, 1963; Tesch, 1965; Elie, 1979; Fig. 3.4). But the seasons of entry into fresh water are hard to compare for various eel species because large local differences occur within species. Speculations about the spawning season based on immigration times (Matsui, 1957), or about the spawning area distance, are therefore illusory; the same applies to comparison with the tropical eel species, which presumably spawn year-round or for a major part of the year (Jespersen, 1942). However, *A. marmorata*, which also occurs within *A. japonica*'s range (see below), enters fresh water only in the winter half of the year (Nishi and Imai, 1969).

In the Japanese Islands, most eels are caught on the southeast coast, that is, the oceanward side, so the main immigration of eels is to be expected on that side and *via* the Kuroshio Current (Fig. 2.16). Catches are considerably less on the continental side. Eels are almost absent on the northern island of Hokkaido, where they reach the northern limit of their range (Fig. 2.15). The reason for the range's relatively weak northward extension is perhaps the cold water of the southward flowing Oyashio Current. *A. japonica* is sup-

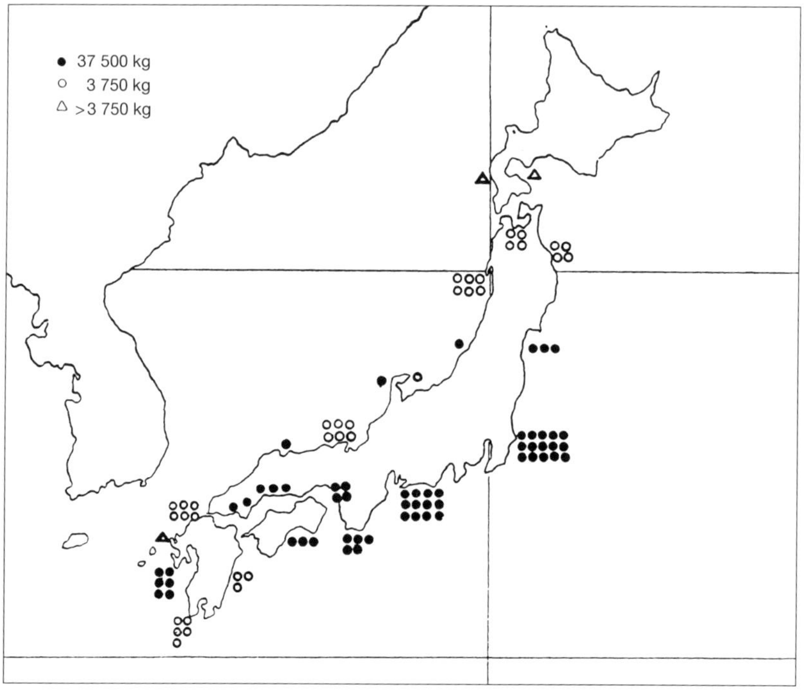

Fig. 2.16 Annual catches by the Japanese fishery in various districts (modified after Matsui, 1952)

posedly quite numerous in Korea, and likewise in China it is still caught in marketable amounts. But, on the Asian mainland it is much less frequent than on the Japanese islands. It reaches the southern limit of its continental range in the Canton river. In the Yangtse river it is found for about 600 km upstream (Schmidt, 1925b).

In contrast to the Atlantic species, which are allopatric, *A. japonica* overlaps, at least in the southern half of its range, with *A. marmorata* (Section 2.3.4; Fig. 2.20). The Isle of Quelpart to the south of Korea, as well as the southern coast of the Japanese islands show such occurrences (Nishi and Imai, 1969). No difficulties of identification are involved. As opposed to *A. japonica, A. marmorata* has marbled colouration and has 10 less vertebrae. Also, the distance between the beginning of the dorsal fin and the anus is significantly shorter in *A. japonica* than in *A. marmorata*. Although a mean vertebral count of 105.6 has been established for *A. marmorata* (Ege, 1939; Table 2.2), specimens of that species on a south Japanese island averaged 104.7, with a range of 101–107 (Nishi and Imai, 1969).

Certainly, the differentiation between *A. japonica* and *A. anguilla* is meaningless from a strictly taxonomic standpoint, but it becomes crucial in terms of human activities conducted for economic purposes, for example, when European glass eels are imported into Japan for stocking, and then a few of them escape into natural water bodies (Tabeta *et al.*, 1979). The two species are very similar with respect to meristic characters, except for the ano-dorsal vertebral count (Table 2.1). There is, however, great difference between them in terms of electrophoretically determined genetic characteristics (Sick *et al.*, 1962; Taniguchi *et al.*, 1972; Taniguchi and Morita, 1979; Tagliavini *et al.*, 1996b). These show that species having almost equal vertebral counts, such as *A. anguilla* and *A. japonica*, can be more distant genetically than some that have very different numbers of vertebrae, such as *A. anguilla* and *A. rostrata* (Section 2.5). Likewise, other species from the Pacific region, for example, *A. australis*, were imported and later captured in natural waters. But the identification of the short-finned eel in question was easy (Tabeta *et al.*, 1977); of course, a reproductive life stage is impossible for these animals.

*2.3.3 Southwest Pacific eel species of the temperate zone (***A. australis** *and* **A. dieffenbachii***)*

The uncertainty about the location of the southwest Pacific eels' spawning area resembles the situation for the southeast African species. Only a few leptocephali were found in this area, in which, besides the temperate zone species, four others from the tropical Pacific have shares (Castle, 1963). The smallest larvae are about 25 mm long. Most of the larvae originate from the area northwest of the Fiji Islands. South of the equator, a constellation of currents exists that mirrors the Sargasso Sea; the prevailing South Equatorial Current there is constantly oriented to the west (Fig. 2.6), so these larvae, too, must come from areas lying to the east. An elongated depression of warm water from the depths extends between 10° and 20° S and between 140° and 170° W (Jespersen,1942). According to Schmidt (1925b), this depression is somewhat further south, near about 20° S. All larvae except one were found west of this area. The great elongation of the water mass and of the larval distribution, as well as the species diversity, suggest that the spawning areas likewise line up, extending from west to east. Castle (1963) also assumes that the spawning area lies between the Fiji Islands and Tahiti. The core of this area would be 170° W by 18° S.

Jellyman (1987) discusses the literature data on oceanography, body size and time of freshwater entry for glass eels, otolith microstructure, and gonadal development stage for silver eels and arrives at a somewhat more northern spawning area (between 150° and 170° W by 5° and 15° S). He assigns *A. australis* to the area northeast of Samoa and *A. dieffenbachii* to the area east of Tonga.

If the freshwater ranges of *A. australis* and *A. dieffenbachii* are compared (Fig. 2.17; Table 2.2), then it is very probable that their spawning area extends very far to the east, perhaps the farthest of all. Based on this idea, it is likewise evident that *A. dieffenbachii* has the longest route to traverse, for the farther east the spawning area is located, the farther south the larvae can be drifted. This assumption is supported by the fact that the glass eels of *A. dieffenbachii* have the greatest mean length. At stage V A II (Fig. 1.2), they are 64–70 mm long. At the same stage, glass eels of *A. australis* are 49–62 mm (Ege, 1939).

The taxonomic differences (Table 2.1) of the subspecies *A. australis australis* and *A. australis schmidtii* are very small and, although it is true that they can be confirmed by means of Jellyman's (1987) vertebral count, this is not so with respect to electrophoretically determined genetic pool (Dijkstra and Jellyman, 1999).

The species *A. australis* and *A. dieffenbachii* are distributed in very restricted fashion on the mainland of Australia and the islands of New Zealand (Fig. 2.17; Table 2.2). *A. dieffenbachii* can even be designated as the 'New Zealand eel' because it occurs only there. In New Zealand it is called the 'long-finned eel' in contrast to *A. australis*, which is known as the 'short-finned eel' and is, likewise, if one uses the obsolete subspecific designation, a 'New Zealand eel' (for distinguishing characters, see Table 2.3).

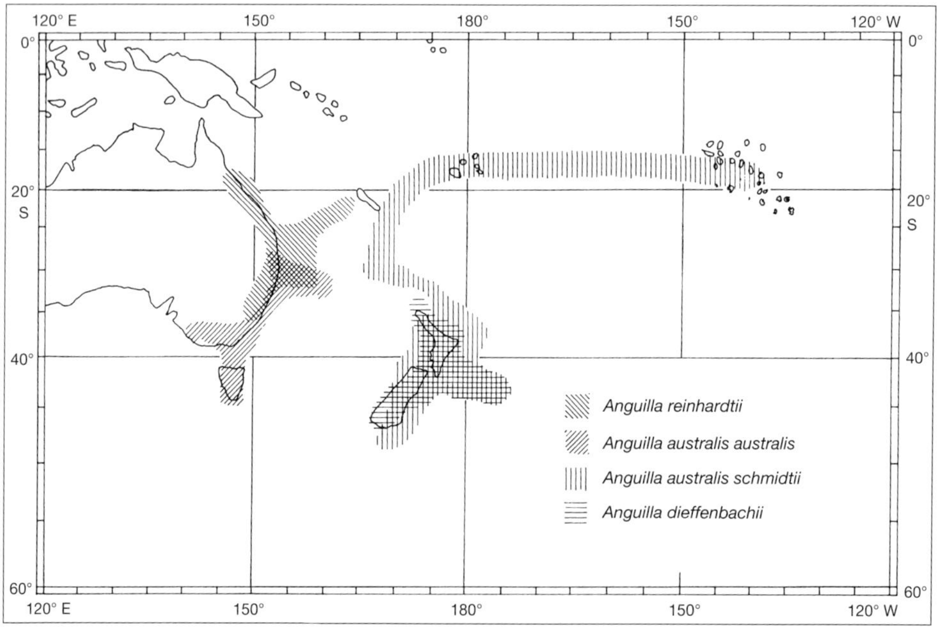

Fig. 2.17 Distribution of *A. reinhardtii*, *A. australis australis*, *A. australis schmidti* and *A. dieffenbachii* (after Ege, 1939). The differentiation between the two subspecies of *A. australis* is no longer reasonable (Dijkstra and Jellyman, 1999)

Table 2.3 Differences between the two eel species of New Zealand (after Cairns, 1942a) (see also Table 2.1)

A. dieffenbachii: Long-finned eel	*A. australis*: Short-finned eel
1. Dorsal essentially longer than anal fin	1. Dorsal only slightly longer than anal fin
2. Vomer teeth form a narrow band	2. Vomer teeth form a compact club-shaped band
3. Eye in front and above the corner of the mouth	3. Eye immediately above the corner of the mouth
4. Lips thick	4. Lips thin
5. Head broad	5. Head narrow
6. Nostril organ large	6. Nostril organ small
7. Mouth opening large, jaw bones strong	7. Mouth opening small, jaw bones small
8. Tail broad, caudal fin well developed	8. Tail narrow, caudal fin weakly developed
9. Pectoral fin large	9. Pectoral fin small
10. Maximum length >180 cm, weight >18 kg	10. Length seldom >90 cm, weight seldom >1.8 kg
11. Male silver eels length 55–65 cm, weight 0.9–1.1 kg	11. Male silver eels length 35–45 cm, mean weight 0.25 kg
12. Scales not very distinct	12. Scales distinct

Three further Indo-Pacific eel forms deserve designation as 'short-finned' (Table 2.1). All other eel species are long-finned. In contrast to the short-finned species, their dorsal fins begin relatively far forward, so that the distance from the start of the dorsal fin to the anus appears very large relative to body length. Extremely long fins characterise the conger eel (*Conger conger*), which does not belong to the Anguillidae; it is distinguished from the Anguillidae primarily by this feature.

Besides New Zealand, the Auckland and Chatham Islands are within the ranges of *A. australis* and *A. dieffenbachii*. Moreover, the short-finned New Zealand eel (*A. australis*) is found in abundance north of New Zealand on the Norfolk Islands and New Caledonia. It also occurs on the Fiji Islands, where there are three further eel species: *A. obscura*, *A. marmorata*, and *A. megastoma* (Ege, 1939; Beumer, 1985). *A. australis* is not confirmed for Tahiti (Marquet, personal communication).

In New Zealand, *A. dieffenbachii* and *A. australis* were caught in nearly equal amounts: 547 and 499 t, respectively, in 1990/1 (Jellyman, 1993). Their abundances differ among various water bodies. Judging by catch statistics of Todd (personal communication), *A. australis* dominates; it accounted for 64% of the total catch in 1979. Contributing greatly to this was coastal Lake Ellesmere on the east side of the South Island; in 1976, 647 t, mainly *A. australis*, about a third of the total New Zealand eel catch, were landed from this lake (Todd, 1977). *A. australis* also seems to dominate in a lake on the south coast of the North Island (Todd, 1980). In contrast, *A. dieffenbachii* alone is caught in some deep, relatively large mountain lakes on the South Island (Jellyman and Todd, 1980). The long-finned eel (*A. australis*) also dominates in the upper reaches of two small rivers near Wellington; downstream from that, however, both species have good living conditions (Todd, 1980). Both species were also found in very similar abundance (Burnet, 1959, 1969a). The proportions of *A. australis*, in terms of numbers of individuals caught in three streams in the south of the North Island, were 11–88%, 1–21%, and 32–37% among different sections of each stream. In terms of biomass, the percentages of *A. australis* were in most cases less because this species is small-bodied. Therefore, corresponding to the larger proportion of short-finned eel in the national harvest data, the numerical proportion of this species is substantially larger yet.

According to present ideas, the long-finned eel migrates farther inland, whereas the short-finned eel stays mainly in the coastal region. This is true in general, but this differential distribution is presumably not so much a result of the migrational characteristics of the two species as it is a consequence of different environmental requirements. Concerning this, Burnet (1968) writes: 'The short-finned eel is selective in its search for habitat, and is the exclusive inhabitant of some coastal waters. It generally prefers lakes and spring streams. The long-finned eel is widely distributed, and it is found that almost every water has either one or both species'. Thus, the short-finned eel absolutely does also inhabit the upper courses of rivers, but these are for the most part very small individuals; only when electrofishing was developed did investigators discover the occasionally major eel abundance there. Burnet (1952b) assumes that with increasing body size, the short-finned eels occupy slower-flowing rivers, lakes, and near-shore waters.

In eastern Australia, besides *A. australis*, the more tropical long-finned and marbled species, *A. reinhardtii*, is present. This species was also discovered in the north of New Zealand (Table 2.2). In the southeast of Australia, *A. australis* seems to predominate. According to a newspaper item, 4.5 t of this species was harvested in the western part of the State of Victoria and exported to Europe (*Trade News*, Ottawa 18, 10–11). Beumer (1979) reported that in Victoria from 1975 to 1977, 200 t of eels were harvested annually. In a lake in the southern part of the country, near the most southerly point on Australia's coast, *A. australis*' proportion amounts to 95%. The diets of *A. australis* and *A. reinhardtii* in this lake differ little.

2.3.4 Tropical eel species (**A. celebesensis, A. megastoma, A. interioris, A. ancestralis, A. nebulosa, A. marmorata, A. reinhardtii, A. borneensis, A. bicolor** *and* **A. obscura**)

Schmidt and his fellow workers were more successful in locating the spawning areas of Indo-Pacific tropical eels than those of temperate zone species. About 1300 larvae were caught in a relatively limited area immediately southwest of Sumatra; another 235 northwest of New Guinea. These two areas, as well as that described for the southwest Pacific species (Section 2.3.3) presumably provide the major spawning areas of the tropical eels.

The distributional pattern of eels in south-east Asian islands (Fig. 2.18) already gave a valuable clue toward locating the spawning area. Although the deep-sea-facing south-west coasts of Sumatra and Java and the rivers that drain them are inhabited by eels, the north-east coasts that front on the shallow shelf area harbour no eels. Eels do not inhabit Borneo's southwest side, which is oriented toward this shelf area, but ample eels are found on its east side; the same applies to the whole Celebes Island that lies before it. On this side of Borneo, the ocean is over 1000 m deep in spots, a condition essential for the presence of eel larvae. The East Indian Archipelago exhibits a situation similar to that of the Japanese islands, on which most of the eels are likewise found on the side toward the continental slope (Fig. 2.16). Thus, the East Indian Archipelago's spawning areas are presumed to be on the southwest and east sides.

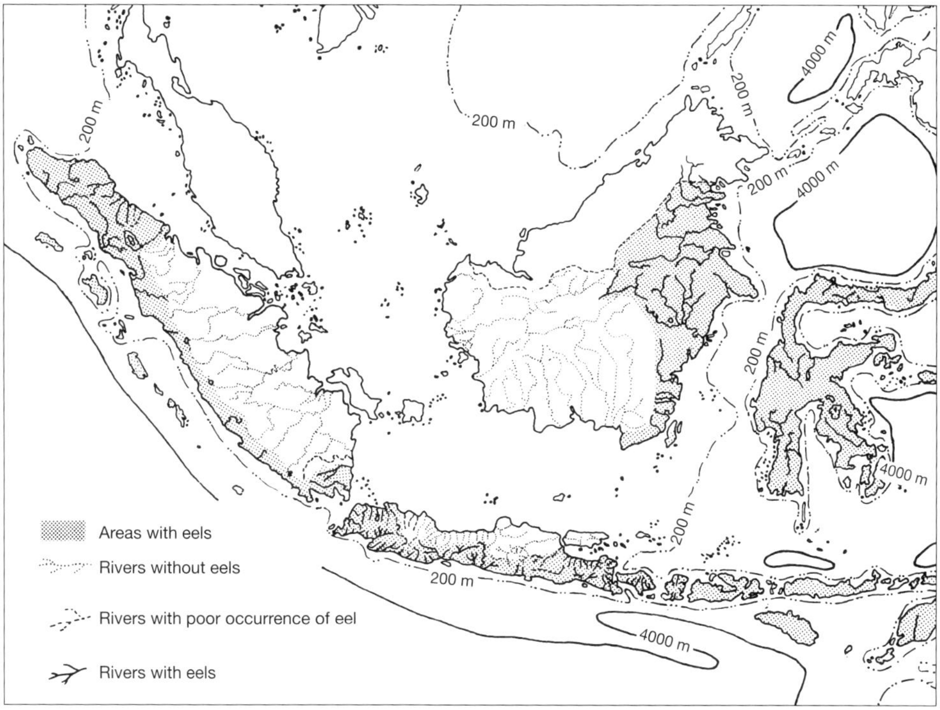

Fig. 2.18 Distribution of *Anguilla* species on Sumatra, Java, Borneo, Celebes, and neighbouring islands (modified after Delsman, 1929)

Figure 2.19 shows the number of eel larvae at each of Jespersen's (1942) sampling stations off Sumatra. According to this, a maximal concentration exists directly on the continental slope between 2000 and 5000 m deep in front of the central part of Sumatra's southwest coast. Here too, as in the Sargasso Sea, high concentrations of larvae are found – particularly leptocephali <20 mm long – in conjunction with high temperatures in deep water. Here, the water was 12°C at the 300-m depth, and it was only 11°C at that depth in the surrounding area. Of this spawning area's 1300 larvae, at least 950 were recognised as *A. bicolor bicolor*. This species is also short-finned as larvae, thus, easy to distinguish from other larvae. The rest of the larvae, that is, long-finned eels, were divided among *A. marmorata*, *A. nebulosa nebulosa*, and *A. celebesensis*. They usually could not be distinguished from each other. The species most possible to identify was *A. nebulosa nebulosa*, with its relatively high numbers of vertebrae and myomeres. The identification, therefore, depended substantially on knowledge about the distribution of the involved eel species on the adjacent land areas (Figs 2.15, 2.17, 2.20 and 2.21). Larvae of *A. nebulosa nebulosa* and *A. bicolor bicolor*, caught by the Japanese, confirmed the range indicated for them in the figures drawn from earlier studies (Table 2.2).

East of the Indo-Malayan island archipelago, the situation with regard to delimiting the spawning area is not so favourable. Only 235 larvae were caught. The distributional image shows a maximal concentration of larvae east of the Moluccas and directly on the north coast of western New Guinea (Jespersen, 1942). In this area, the distribution of eel larvae comes close to the shelf area, and between the places where larvae occur lie islands and

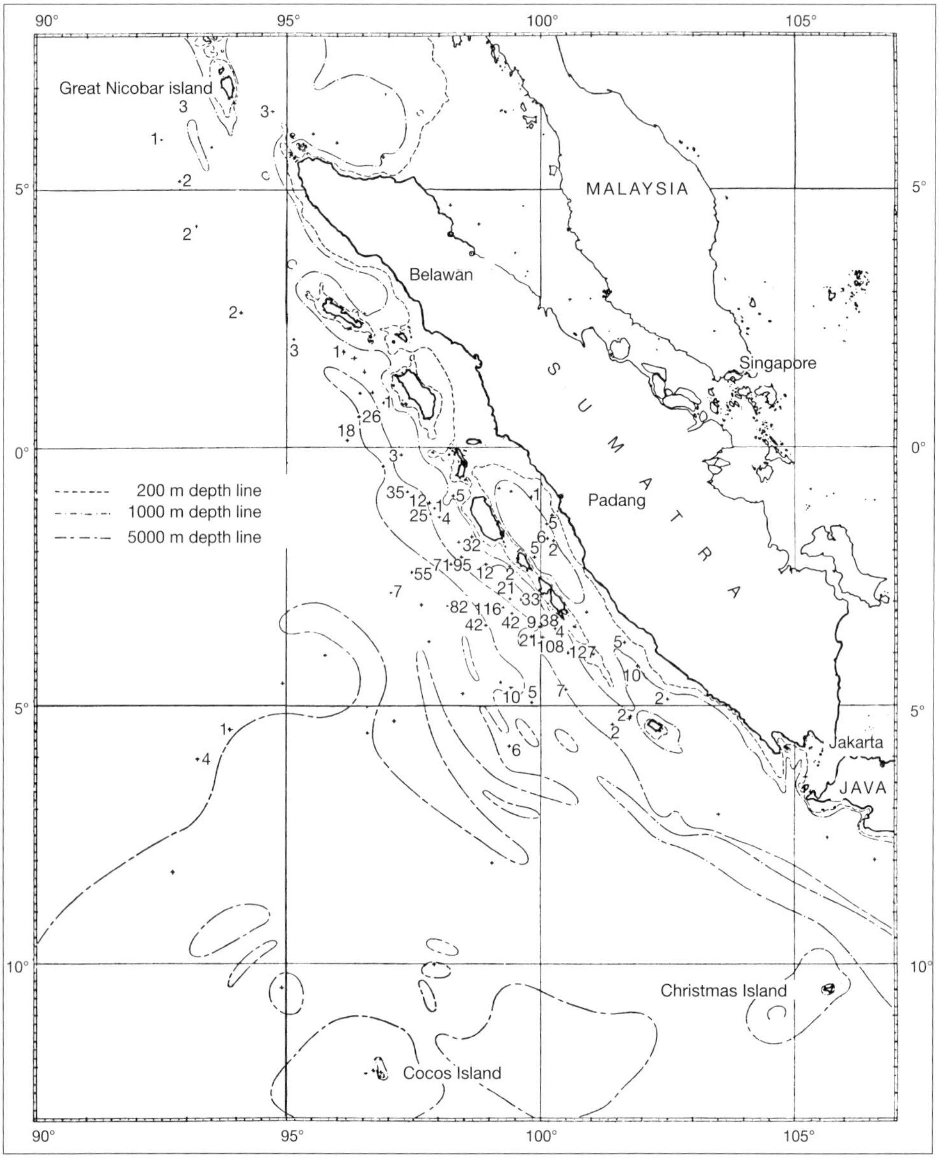

Fig. 2.19 Numbers of eel larvae caught at various stations near Sumatra, calculated per 10 trawl hours (after Jespersen, 1942, reproduced from Dana Report **22**)

shallower parts of the ocean, so existence of several interdependent spawning sites can be presumed. Here, systematic sorting of the sampled animals into long- and short-finned specimens was possible. From that it seems the long-finned species occur closer to the coast than do the short-finned species. Furthermore, the northern part of the short-finned species' range extended into the area south of Taiwan. The northernmost station where long-finned species were found was in the area south of the Philippine Islands.

In the case just described, *A. bicolor biclor* is presumably the species most frequently found; most of the short-finned larvae probably belonged to this species, which amounted

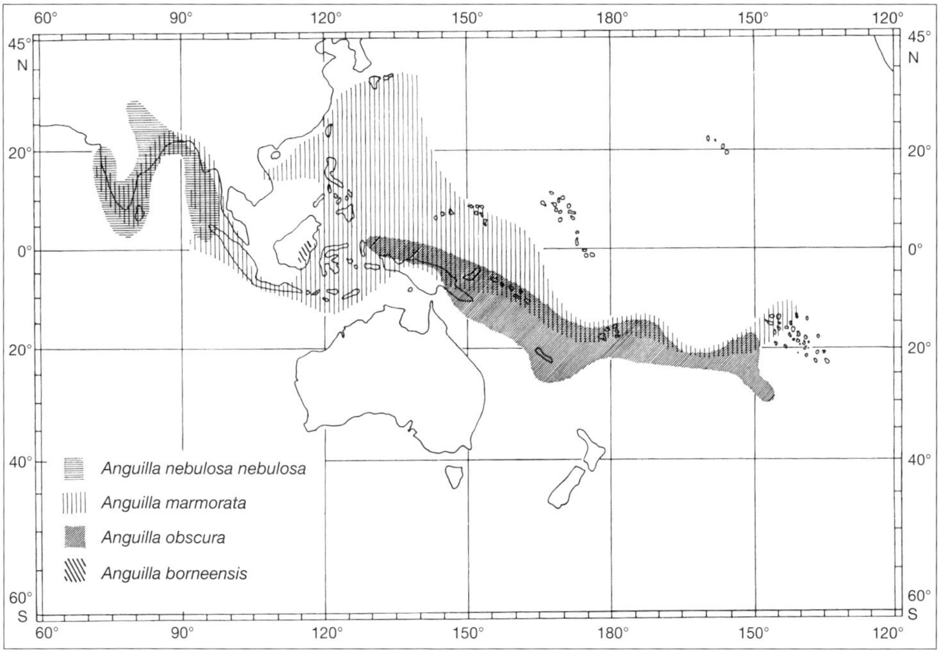

Fig. 2.20 Distribution of *A. nebulosa nebulosa*, *A. marmorata*, *A. obscura* and *A. borneensis* (see Table 2.1)

to 129 of the collected larvae. *A. obscura*, the second short-finned species represented in the area, has only a minor proportion, as taxonomic studies of Jespersen (1942) make obvious. Supporting this is also the minor inland distribution of the species in the western part of its range, which is limited to a few occurrences in the northern part of New Guinea (Fig. 2.20). Some *A. obscura* larvae in the stomachs of tuna (*Katsuwonus pelamis*) confirm their presence in this area (Matsui *et al.*, 1970).

Among long-finned species, according to the inland distribution of these eels, are *A. marmorata*, *A. borneensis*, and *A. interioris*. Taken together, they compose the largest proportion of all the long-finned eel larvae collected. They were distinguishable from the rest of the larvae by their vertebral count (Table 2.1). The remaining larvae must have included *A. celebsensis* and *A. ancestralis*. As Jespersen (1942) writes, *A. marmorata* is the one found most abundantly in the area, and therefore, is the one predominantly represented among the long-finned larvae. However, because *A. celebesensis* is also distributed throughout almost the whole inland larval area, this species must be represented to some extent in the larval collections, as also seems evident from the morphological characters. In the entire island area, the remaining species have only a tightly restricted distribution (Fig. 2.15), so, presumably, they are likewise hardly represented in the larval samples (Fig. 2.21).

Favourable hydrographic conditions, such as are apparently required for eel spawning, were also found in the area of elevated larval concentration northwest of New Guinea (Jespersen, 1942). The temperature at the 300-m depth was 15°C, and the salinity was higher than in the surrounding water masses.

In both assumed spawning areas, for example, near Sumatra and near New Guinea, the currents are not so clearly arranged as in the presumed spawning sites of the eels that live

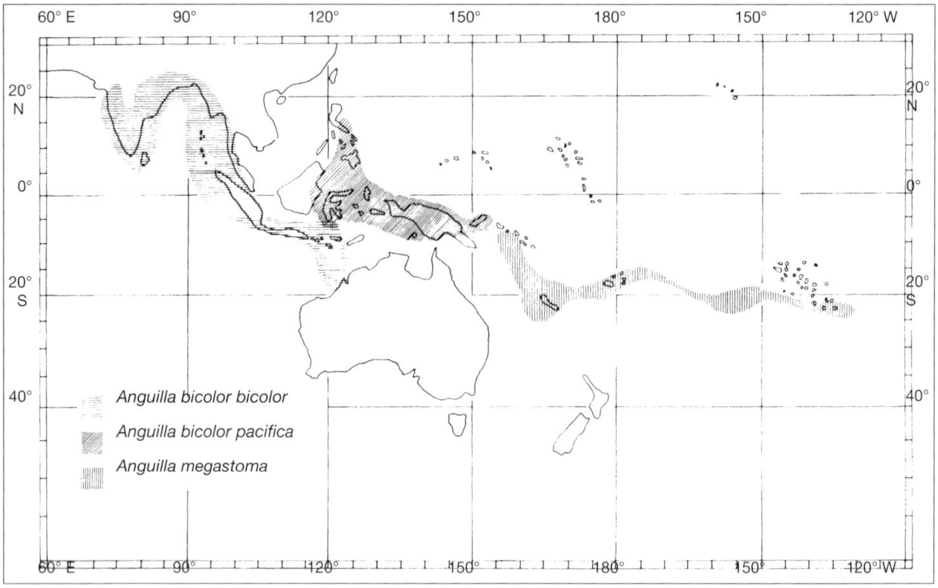

Fig. 2.21 Distribution of *A. bicolor bicolor*, *A. bicolor pacifica* and *A. megastoma* (modified after Ege, 1939)

in temperate climates (Fig. 2.6). Perhaps in the area of highest larval concentration near New Guinea, an eddy exists in summer, which is small but quite strong (Morskoi Atlas). It seems as if, especially near Sumatra, the larvae can be drifted in several directions. This is especially important for such widely distributed species as *A. marmorata* and *A. bicolor. A. marmorata* is found, for example, far west, as well as, far east of the Sumatran spawning area. However, according to present knowledge, it has not yet been clarified whether still further spawning areas exist northwest of New Guinea for these and other species. For *A. marmorata*, the occurrence of suggested larvae of this species indicate a further spawning site in the above-described area of southwest Pacific eels (Castle, 1963). Presumably, the spawning areas described in this section are out of the question for the somewhat more southern species, *A. megastoma* and *A. reinhardtii*. Surely, they reproduce in the spawning area of the South Equatorial Current (Section 2.3.3), as discoveries of larvae from that area indicate (Castle, 1963).

The water depths of the spawning sites can be ascertained from the vertical distribution of the larvae. According to depth distribution studies northwest of New Guinea (Fig. 2.22), most short-finned larvae occur at 100–200 m, and long-finned larvae at about 200 m, provided that the depth calculations, based on cable length of the deployed gear, are valid. For the Sumatran area, a maximal larval concentration was evident at 120–300 m depth. This applied to the older larvae, as well as the younger ones (<20 mm long). There, the densest concentration of larvae >20 mm long was presumably at a depth of 200–250 m. It is, therefore, to be assumed that the spawning sites are to be estimated at similar depths (Section 2.2.1) But the depth preferences that are quoted should be regarded with caution, because it is known that the preferred depths differ between day and night, as Jespersen (1942) himself determined, and as the studies of Atlantic eels have shown (Schoth

- Ordinate: number of larvae per 10 hours of trawling
- Abcissa: Length of the trawl tow line paid out; fishing depth equals about half that distance

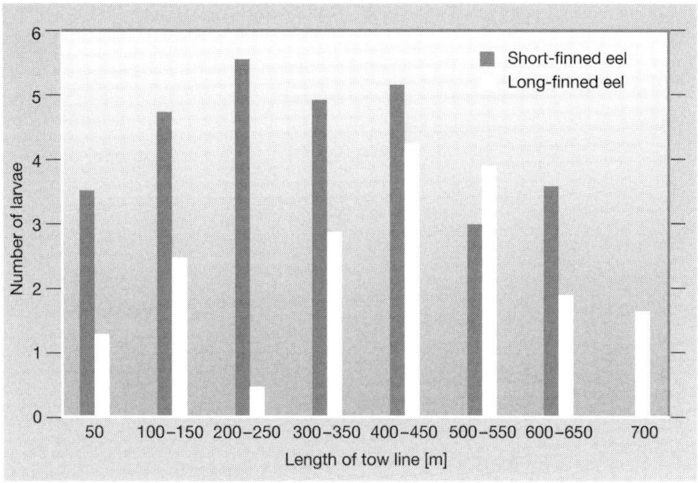

Fig. 2.22 Depth distribution of larvae of short- and long-finned eels northwest of New Guinea (after Jespersen, 1942, reproduced from Dana Report **22**)

and Tesch, 1984). The night-time depth of larvae in the Sargasso Sea, as well as in the area where Japanese eel larvae occur is around 75 m (Section 2.2.1).

Little can be said about the season of spawning, based on the Danish investigations. The collection of glass eels indicates that the entry into freshwater, and accordingly also the spawning period does not tend to be limited to any one time of the year (Ege, 1939; Jespersen, 1942). During the 2-month larval sampling in the Sumatra area, three size groups of *A. bicolor bicolor* were found. In contrast, Indian studies (Pantulu, 1956) on *A. nebulosa nebulosa*, which presumably also spawns near Sumatra, found that time of freshwater entry is seasonally very restricted. There, as in Europe and North America, freshwater entry of glass eels was limited to the winter half of the year. As temperature barriers are hardly to be expected while the larvae are drifting toward the continent, the observed annual rhythm can hardly materialise. However, a correlation with this region's dry and rainy seasons is possible. The streamflow discharge regime of the rivers determines the emigration of silver eels to the sea, such that the high-water rhythm can cause a rhythm in the spawning times. But in Calcutta, for instance, there is a summer period and a winter period, by which, actually, a semi-annual rhythm would have to develop. Therefore, an influence on the freshwater entry of glass eels, which would likewise be conceivable, seems improbable. A dependence of the glass eel migration, as well as that of silver eels and accordingly the seasonal spawning time also seems questionable, in so far as the progression of weather is not the same north and south of India, so, presumably, no consistent rhythm of spawning and migration can develop. Much more probable, therefore, is an influence that is likewise caused by the trade winds but that is much more regular: the different winter and summer currents of the northern Indian Ocean. They flow westward in winter, that is, from Sumatra toward the Indian coasts (Fig. 2.6). But in summer they go in the opposite direction. Then, neither the leptocephali nor the glass eels would succeed in overcoming the at least 2000 km distance from Sumatra to India, which in places flows

Table 2.4 Mean length of different tropical species of glass eels of the developmental stage VI A (Table 1.1) of different catch seasons and locations, after Ege (1939), compared with data presented by Tabeta *et al.* (1976) on the same and different species.

Species	Total length (mm)	
	after Ege	after Tabeta *et al.*
A. celebesensis	48	48–57
A. ancestralis	49–54	
A. nebulosa nebulosa	49–54	
A. marmorata	49–50	45–53
A. reinhardti	50–58	
A. bicolor bicolor	52–55	52–58
A. obscura	49	
A. anguilla		61–77
A. rostrata		57–71
A. japonica		54–89
A. australis		56–65

1 nm/h (Morskoi Atlas). Thus, detailed studies are needed to clarify the question of tropical eel spawning seasons.

At the glass-eel stage, tropical species reach lengths of about 50 mm (Table 2.4). Thus, they appear, judging by the two Atlantic species, to have a relatively short developmental period (Fig. 2.7) and, therefore, a short distance of migration. If one analyses the growth values reported by Jespersen (1942) for *A. bicolor* larvae, then a development time of 3 months from egg to larval stage I (Table 1.2) seems possible. For example, this species needed only a month and a half to increase its mean length from 39 to 50 mm. Development to the glass-eel stage had to be completed in 4 months at the latest. In consideration of the sometimes very close proximity of the spawning sites to the coast, these calculations of time spans were probably too long rather than too short.

2.4. Zoo-geographic relationships

The high correlation of homologous structures of the eel species on the one hand and their worldwide distribution on the other raise questions for research into their common ancestry. The problems are geologic, zoo-geographic, and genetic. It should be said at the outset that at present nothing final can be said about the place of their common origin. Therefore, the various questions are merely indicated here.

For instance, for many years there has been no doubt that the two Atlantic species are closely related. A hypothesis even arose that considered the European and American eels to be a single species (Section 2.2.1). Genetic studies have confirmed the close relationship of the two Atlantic species and the greater distance to the Japanese eel (Section 2.5). It was, therefore, obvious that the common origin of these two, surely very closely related species should be investigated first. How did such extraordinarily great spatial separation happen?

Von Ubisch (1924) and later authors (e.g. Ekman, 1932) drew a connection between the eel migration and Wegner's continental drift theory. According to this theory, the present American continents and Europe–Africa once formed a single continental mass. In the Cretaceous period, this large continental mass split apart. During the Tertiary period, an ever-wider trench developed, which finally formed the Atlantic ocean. The spawning grounds of that era's eel form had always been in the location of the present-day Sargasso Sea, in the Tertiary-period Atlantic, an ocean (or lake), which was then much smaller than the present Atlantic Ocean. While the spawning site and perhaps also its hydrographic conditions shifted westward, the feeding and growing areas, stayed – or drifted – to the east. To reach its spawning grounds, the European eel had to perform an ever-longer migration.

It is, therefore, probable that the two present Atlantic eel forms are developed from a single species because they originally lived together within a much smaller space. From genetic studies of DNA, we can infer that the divergence began about 2 million years ago, but they could not then have been called genuine species (Avise *et al.*, 1986). In addition, parasitologic findings indicate that the two Atlantic species are more closely related to each other than to the Indo-Pacific species (Sections 7.3.2 and 2.5).

Therefore, the question arises about the connection that must have existed with the Indo-Pacific eel species at some time or other. There is little probability that immigration took place across the area now occupied by Middle America, which did not always present a barrier between the Atlantic and Pacific Oceans. A connection of the Atlantic eels with the Indo-Pacific species is much more likely to have taken place toward the east. The present-day Mediterranean Sea represents the remainder of an earlier, larger ocean, the so-called Tethys, which lasted through almost the entire better-known part of geologic time, and which existed before America and Europe–Africa separated. That ancient ocean divided a northern and a southern continent and existed from the Mesozoic until the beginning of the Tertiary period. The Tethys connected present areas of the Atlantic Ocean with the Indo-Malaysian area. In the era of this Mesozoic ocean, the faunas in the Pacific and Atlantic Oceans had certain similarities, detectable today not only among the eels, but also other kinds of animals. The existence of eels is known from the sediments of the Tethys from the Cretaceous into the Miocene (Stinton, 1975). The eel population of that time was split up by geologic events, much as also happened later in the separation into east and west Atlantic eels.

The creation of the Indian Ocean resembled that of the Atlantic. India was originally associated with the southerly bordering parts of the ocean and Australia as its own plate that was drifting northward and had a southwestern connection with Africa (Bullard, 1969).In view of this, we can explain the divergence of Indo-Pacific and African species, which, from the taxonomic standpoint has, in some cases, not yet led to formation of their own 'species' to the extent represented by the difference between the European and Japanese eels (Section 2.5). For reasons already stated, one must be sceptical of the idea that the Indo-Malaysian area was the original eel-form's home range, from which all later species spread, thus also that the eventual Atlantic species had migrated eastward from the Pacific into the Atlantic Ocean across the then non-existent Middle American barrier (Ekman, 1932). This view was repeatedly defended on the basis of the Indo-Pacific's species richness. A species-rich area is often considered as the 'original home'. The aforementioned geologic events and palaeontological data suggest that the divergence into species took place only after the separation. As a result of unfavourable climatic conditions in the Atlantic Ocean, a species impoverishment took place, as was also the case for other marine animals.

Despite this, the hypothesis that eels originated in the Indo-Pacific cannot be completely dismissed. The following molecular-genetic consideration conveys further understanding about the ancestry of the eel species. Parasitologic aspects support these (Section 7.3.1).

2.5. Phylogenetic relationships and population genetics (revised by T. Wirth)

2.5.1 Molecular systematics of the genus Anguilla

The genus *Anguilla* is composed of 15 different species that are located mostly in tropical and subtropical regions. The recent development of molecular markers and the rise of molecular systematics as a new discipline made it possible to reassess the evolutionary relationships of the eels at the DNA level. This approach is a good alternative to previous phylogenies based on morphological characteristics (Ege, 1939) since gene trees are unlikely to be affected by analogies and since the use of a molecular clock gives the opportunity to calibrate and date the speciation events within this genus.

Recent contributions based on mitochondrial DNA (Aoyama and Tsukamoto, 1997) confirmed that ancestral eels originated in the western Pacific and that multiple radiation events had occurred in the Indo-Pacific region. Moreover, these authors claimed that eel speciation started at around 43.5 Mya. They also suggested that Atlantic eel speciation corresponds to a recent event which occurred 10.2 Mya ago and that their ancestors dispersed westward by transport of larvae through the northern edge of the Tethys Sea. Though these results were partially confirmed by another group (Bastrop *et al.*, 2000), the route of the Atlantic eels is still a matter of debate. Lin *et al.* (2001) studied mitochondrial cytochrome b and 12 S rRNA genes and ended up with a slightly different scenario for the evolutionary relationships of these catadromous species (Fig. 2.23) First, *Anguilla* radiated about 20 Mya ago. Second, the ancestors of the Atlantic eels probably trekked across the central American Isthmus to the Sargasso Sea. These conflicting conclusions are probably caused by differences in the number of taxa studied, the type of genetic markers and type of analyses that were made. The study of Lin *et al.* is the most complete, so their scenario is most likely. Moreover, all these contributions confirmed that morphological features are unstable, or that they appeared independently in different lineages and, therefore, that they perform poorly under an evolutionary perspective. Molecular phylogenies remain our best estimators of the true evolutionary history of *Anguilla*.

2.5.2 Inter-specific genetic differentiation and the Atlantic eel paradigm

Protein electrophoretic migrations boosted genetic studies on the evolutionary relationship of the Atlantic eels in the early 60's. The species status of *A. anguilla* and *A. rostrata* was always questioned and fish biologists clustered into two entities defending respectively

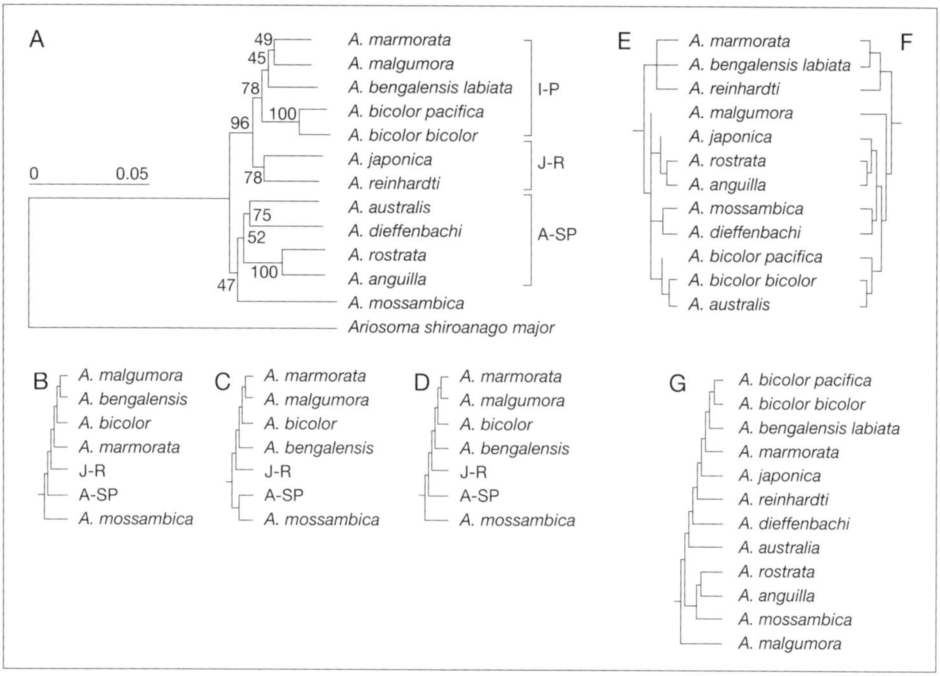

Fig. 2.23 The inferred topology based on the Tamura–Net distances and the neighbour-joining method. The numbers at the nodes are bootstrap values from 5000 replicates. The scale bar represents the branch length. The notations are I-P= Indopacific-, J-R= *A. japonica* and *A. reinhardti*, A-SP= Atlantic-South Pacific group (after Lin *et al*., 2001).

the single species model or the two species model. The former opinion was based on Tucker's view (Tucker, 1959) where he advocates that all European eels are the offspring of American parent stock and that they do not participate to former reproduction, and was supported by the works of Sick *et al.* (1962) on haemoglobin. However, most of the former allozyme works (Fine *et al.*, 1964; Drilhon *et al.*, 1966, 1967; Pantelouris and Payne, 1968) confirmed the two species status and succeeded in demonstrating that a genetic isolation does exist between the American and the European eel. Karyotype analysis (Passakas, 1981) showed that the two species had the same number of chromosomes ($2n = 38$) but differ with respect to the number of free chromosome arms (NF = 60 and 58 for *A. anguilla* and *A. rostrata*, respectively). The two species scenario was definitively settled by Avise *et al.* (1986) mitochondrial DNA studies. Samples of the European eel proved to be highly distinct from *A. rostrata* in mtDNA genotype. Aoyama *et al.* (1996) studied the mtDNA sequences of eight *Anguilla* species and found that *A. anguilla* and *A. rostrata* showed a lower level of divergence (0.024) than the others (0.051–0.085). Tagliavini *et al.* (1996a) also studied the mtDNA sequences of five *Anguilla* species and confirmed that the genetic distance between the Atlantic species was small.

Moreover, Avise *et al.* (1990) studied allozyme markers, mitochondrial DNA markers and vertebral numbers in Icelandic eels and concluded that the Icelandic population included, in low frequency, the products of hybridisation between American and European eels. Approximately 2–4% of the gene pool in the Iceland eel population was derived from American eel ancestry.

2.5.3 *Intra-specific genetic differentiation of* Anguilla *species*

Most of the works on intra-specific variations of *Anguilla* species are based on isozyme studies. Williams *et al.* (1973) found genetic differentiation of the American eels along the east coast of North America. Koehn and Williams (1978) further demonstrated that the

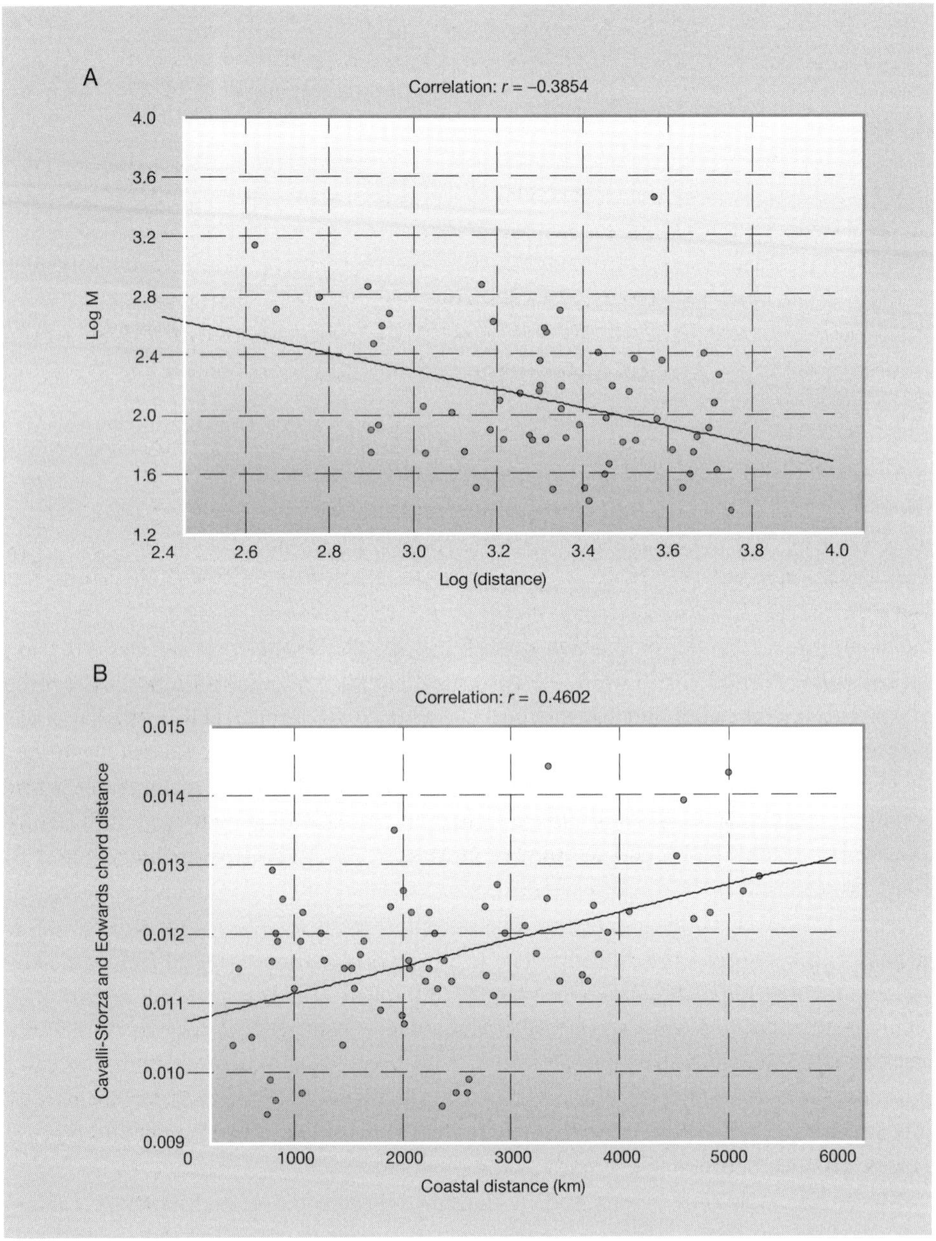

Fig. 2.24 Genetic distance (D_{CE}), based on seven microsatellite loci, *versus* coastal geographical distance for all possible pairwise combinations of 13 *A. anguilla* samples. Positive correlation (r) is significant, regression analysis: $y = 0.0107 + 4 \times 10^{-7}x$, $r^2 = 0.212$, $p = 0.0045$ (Mantel Test) (after Wirth and Bernatchez, 2001 and changed: Wirth, personal communication).

latitudinal cline in the two loci, PHI (=PGI) and SDH, were temporally stable. On the other hand, Comparini *et al.* (1977) found no significant differences in the PGI allele frequency between the Mediterranean and Atlantic samples of European eels. Rodino and Comparini (1978b) further confirmed that there was no genetic differentiation in the European eels. Taniguchi and Numachi (1978) studied two samples of the Japanese eel *A. japonica*, one from Fuchien, China and one from Kochi, Japan. They found no significant differences in the allele frequencies at each of the three loci: 6PGD, ICD and GOT.

Chan *et al.* (1997) studied the allele frequency of IDH (=ICD) and PGD in *A. japonica* and their results supported the existence of a latitudinal cline in allele frequencies along the east Asian coast for the Japanese eel.

The search for genetic differentiation strongly depends on the type of genetic markers that are used in a study. Enzymes that play a clear role in animal metabolism are putatively under selection pressure and one cannot exclude that some genes show a selective gradient, which is not the picture of population structure but the product of natural selection. Therefore, the jeopardised isozyme results should be interpreted cautiously.

Markers like mitochondrial DNA are more likely to be neutral, and both Sang *et al.* (1994) for the Japanese eel and Lintas *et al.* (1998) for the European eel failed to detect any geographic clusters in their samples.

Wirth and Bernatchez (2001) provide the latest stage on this topic. These authors used neutral and highly polymorphic genetic markers, which are used in forensics. Microsatellites show high mutation rates, and therefore, provide the opportunity to detect recent events and weak genetic signals. The authors analysed 13 samples of 50 individuals from the north Atlantic, the Baltic Sea and the Mediterranean Sea basins (Wirth and Bernatchez, 2001) and detected a global genetic differentiation. They also demonstrated that European eels exhibit isolation by distance (Fig. 2.24), which implies non-random mating and restricted gene flow among eels from different sampled locations. A tree based on genetic distances between the samples grouped the Mediterranean samples in a distinct clade, as well as the North Sea and Baltic Sea samples. Icelandic eels were intermediate between other *A. anguilla* samples and *A. rostrata*, which is congruent with the hybrid status of Icelandic eels. Based on these results, the hypothesis of panmixia in the European eel is refuted and consequently the reproductive biology of European eel must be reconsidered.

2.5.4 Perspectives

The development of population genetics algorithms and the use of genetic markers like microsatellites allow the elucidation of species demography (contraction, expansions and constant size). Effective population size could be estimated, changes could be dated, all kinds of information which are of first importance in the management of the declining eel stocks. The cohabitation of eels and molecular tools will certainly last for a while!

3
Post-larval ecology and behaviour

3.1 Glass eels during approach to the continental shelf and into fresh water

3.1.1 Migration in the ocean during completion of the larval stage and thereafter

'Glass eel' is defined here as all developmental stages from completion of leptocephalus metamorphosis until full pigmentation. This includes stages VA to VIB (Section 1.3.3). The continental slope represents a dividing line between the occurrence of larvae and transformed animals (Schmidt, 1909a; Antunes and Tesch, 1997b). At the start of the continental shelf, about the only stages encountered are those that have already taken on the external form of the adult eel, but their pigmentation still differs completely from that of adults. The continental slope represents a boundary only for the larvae, though. Stage-VA glass eels, that is, freshly transformed larvae are still found far from the shelf, for example near Bermuda, 1000 km off the North American coast. Antunes and Tesch (1997a) even found an American glass eel further north, 300 km east of the continental shelf at a position of >4450 m water depth. The depths that 'oceanic glass eels' prefer are obviously identical to those of the larvae (Tesch *et al*., 1986).

By which influences the larvae and early (oceanic) glass-eel stages are kept at the continental shelf including positions of 300–1000 m water depth, is not yet clarified.

Surely, it is hydrographic influences that hinder the eels that are not ready from prematurely invading areas that are still hostile for them. The locomotory mechanisms of larvae differ completely from those of glass eels (Dean, 1912). Whereas the larva glides along without significant expenditure of energy and without creating turbulence, the glass eel swims in a snake like manner with a more side-to-side motion. The larva's slim, willow-leaf-like shape should facilitate its planktonic drift, but the body shapes of the glass eel and fully developed eel show more adaptation for bottom dwelling (Wunder, 1963, 1964). The larva undertakes movements mainly to change location, whether this is in an innate compass direction or not. One body segment after the other glides along in a line in the water, like train carriages on a track. The head can be moved away from the line to find food. In contrast, goal-oriented forward motion dominates in the glass eel. With increasing frequency it swims actively against the current and from the coast into inland waters (Section 3.1.2).

Given the movement characteristic of larvae and early glass eels, for the time being it is still not understandable how they could resist, by their own activity, being drifted across the 1000-m depth contour into the shallower shelf area (see above). In this regard, Deelder (1970) took up the ideas of Westenburg (1952), who suspected that fish sense ocean depth on the basis of physical influences; they could, for instance, swim along the coast, constantly at a particular constant depth. They sense depth on the basis of reverberations from the ocean bed. The reverberations are brought about by waves on the sea surface, and their frequency, after reflection from the bed, increases to four times the water depth. The larvae could react to these frequencies, and then take certain action to avoid being drifted into shallower water. Observations of freshly caught leptocephali showed that they are not at all so motionless or passive (Deelder and Tesch, unpublished; Kracht, unpublished) as represented above. For example, according to shipboard aquarium observations, they were capable of quite rapid escape movements.

It is evident that glass eels already appear on the French (and Spanish) coasts in September (Fig. 3.1), whereas, they are first encountered at the outlet of the English Channel and in the German Bight in February. A longer time is necessary for reaching the eastern (continental) parts of the ocean than to cross the continental slope. Besides that, water temperature may also have an effect because the more north-easterly parts of the ocean are too cold to allow early entry by glass eels. Contrary to previous opinions (Van Heusden, 1943; Creutzberg, 1961), however, the invasion proceeds faster than it had been

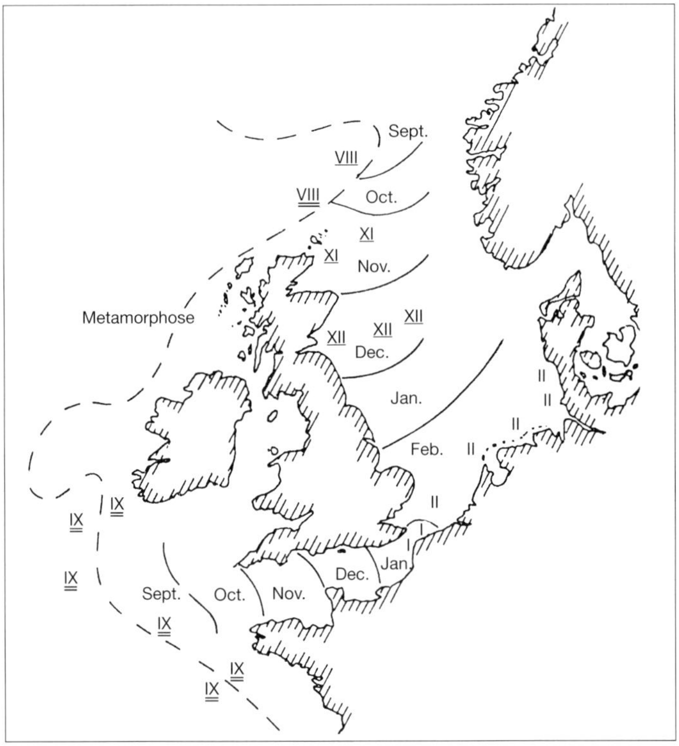

Fig. 3.1 Probable arrival times of glass eels on the coasts of northwestern Europe, according to various authors:

- Roman numerals with single underlining: months according to Bowman (1913)
- Roman numerals with double underlining: months according to Schmidt (1909a)
- Data for January and February (I and II): after Creutzberg (1961)
- In German coastal regions: after Tesch (1971)

assumed. First arrival in the German Bight is not in April or March, but in February (Tesch, 1971) and even, according to surface-net catches, in January (Nellen and Hempel, 1970). In the Skaggerak, glass eels were sighted at the surface around a submerged lamp or caught by net hauls at the surface at the beginning of April (Lindquist, 1972). One could also expect earlier arrival in the Baltic Sea, presumably already in February/March, as is reported for the river estuaries (Section 3.1.2). Arrival of glass eels in the open southern North Sea already in January/February is confirmed by the catches reported above for the North Sea off the estuaries of the rivers Ems and Elbe.

Glass eels appear on the Mediterranean coasts about as early as on the Atlantic coasts, and apparently even much earlier at Gibraltar than in the Bay of Biscay (Kracht and Tesch, 1981; Tesch *et al.*, 1979). Schmidt (1912) found fully developed and metamorphosing larvae in the central Mediterranean Sea in springtime, thus at about the same time as this would be expected west of the British Isles beyond the 1000-m contour. Thus, before metamorphosis is complete, these animals exist fairly close to the coasts, so that in the Mediterranean Sea's high temperatures, freshwater entry can already ensue in autumn (Fig. 3.2; Heldt and Heldt, 1929b). Also, in the southern part of occurrence of the Japanese eel, in Taiwan, from sampling in the estuaries it is evident that first entry takes place in autumn (Tzeng, 1985).

The transition from salt to fresh water is a life stage that creates physiological changes in juvenile eels that are just as fundamental as the effect of crossing the 1000-m depth

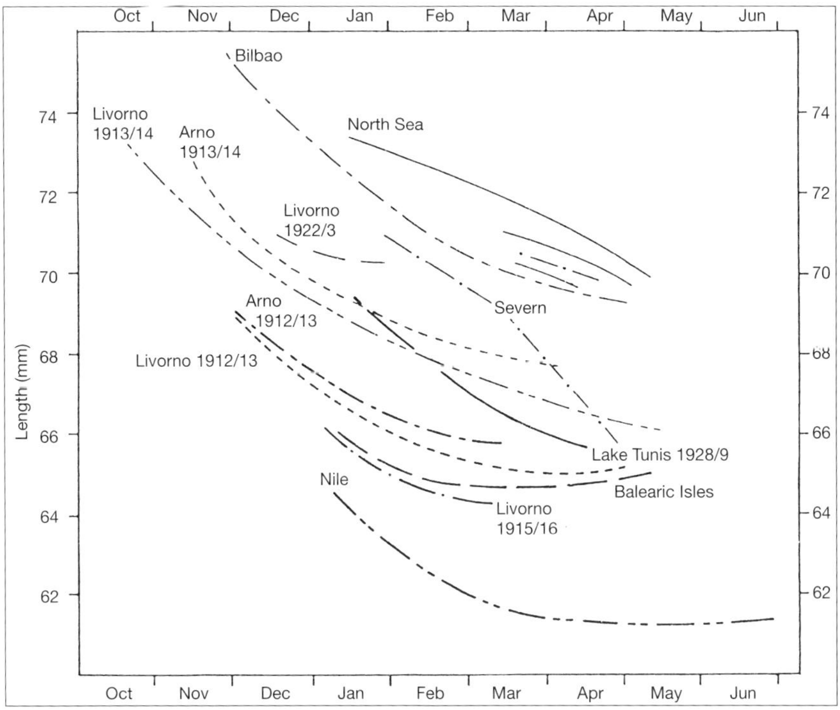

Fig. 3.2 Length of glass eels caught during various months in the Mediterranean region and western Europe (modified after Heldt and Heldt, 1929a)

contour is in transforming larvae and 'oceanic glass eels' into glass eels. Coincident with those events are basic behavioural changes and the onset of pigmentation (Section 1.3.3). In the interim between these two events, the glass eel's migratory behaviour is largely unexplained because until now sampling in the deep shelf area has been largely unsuccessful. Sampling was more successful on the shallower shelf, for example in the North Sea and in the English Channel. Based on the material in Fig. 3.1, Creutzberg (1961) inferred that glass eels in the North Sea are transported 7 km/day from north to south. As similarly strong north-to-east residual currents were observed, he assumed that the coastal migration of glass eels results mainly from passive transport. This is inconsistent with the glass eel's primarily active manner of propulsion. Deelder (1952, 1970) puts these data in doubt in that he assumes that the eels would have to migrate faster.

According to Schmidt (1909a), it is over the 1000-m-depth contour that metamorphosis is concluded and stage V reached. The glass eels arrive off the Holland–Belgium coast in January, and off the German coast in February (Tesch, 1971). Thus, to reach the Dutch coast, they traverse about 1000 km in about 4 months an apparent movement of >8 km/day. To the German Bight would be 1200 km within 5 months, therefore, about 8 km/day. This amounts to a somewhat faster migration than 7 km/day, but not very much more.

Nevertheless, there is a further argument for active glass-eel migration: how else should these developmental stages reach all the coasts of the Mediterranean Sea and the coasts of the Baltic Sea so early? Quite different currents prevail on the various sections of these coasts, and passive transport seems out of the question in many cases.

The remaining factor to consider is transport by tidal current, but this could only be extremely slow in the vicinity of the 1000-m-depth contour. However, near the coast where the currents are faster, this means of transport seems to have been proved for *A. anguilla* (Deelder, 1952, 1958, 1960b; Creutzberg, 1961) and *A. rostrata* (McCleave and Kleckner, 1982; Wippelhauser and McCleave, 1987). According to surface and deeper sampling in the tidal area, glass eels were to be found at all depths at night during flood tide, whereas, during ebb tide they existed only here and there in the pelagic layers (Fig. 3.3). In daytime plankton-net hauls, hardly any glass eels were caught during ebb tide, while substantial amounts were during flood tide, but only in the lower layers of water. These manifold observations by various parties lead to the conclusion that glass eels use landward tidal transport to reach inland waters. As they cannot be caught with plankton nets during ebb tide, we can infer that they stay on the sea bed then, where they are protected from excessively strong currents and, under certain circumstances, could make further headway towards the coasts despite the outrunning water by swimming against the current. Figure 3.3 moreover leads to the conclusion that glass eels are negatively phototactic and, therefore, stay in deeper, less well-lit water layers during daytime. From this behaviour, influenced by tidal current and daylight, we can infer a 14-day rhythm of activity, as is the case in the tidal area of rivers for all eel species that have been studied (see below). A 14-day rhythm was also evident in the daily rings in otoliths of type-A glass eels that were reared in aquaria after completing metamorphosis, a situation that rules out the influence of time-setting features in the immediate environment (Lee and Lee, 1989). But catches of *A. japonica* in the coastal area of Taiwan indicated that before entry into rivers eels in the ocean showed a 4-week rhythm, in that they were abundant in open water only at new moon (Tzeng, 1985). Thus, the greater brightness during full moon seems to drive glass eels into deeper areas of the ocean, whereby they escape predation. Kracht (1982) confirmed these behaviours for still younger developmental stages of *A. anguilla via* mid-Atlantic larval catches.

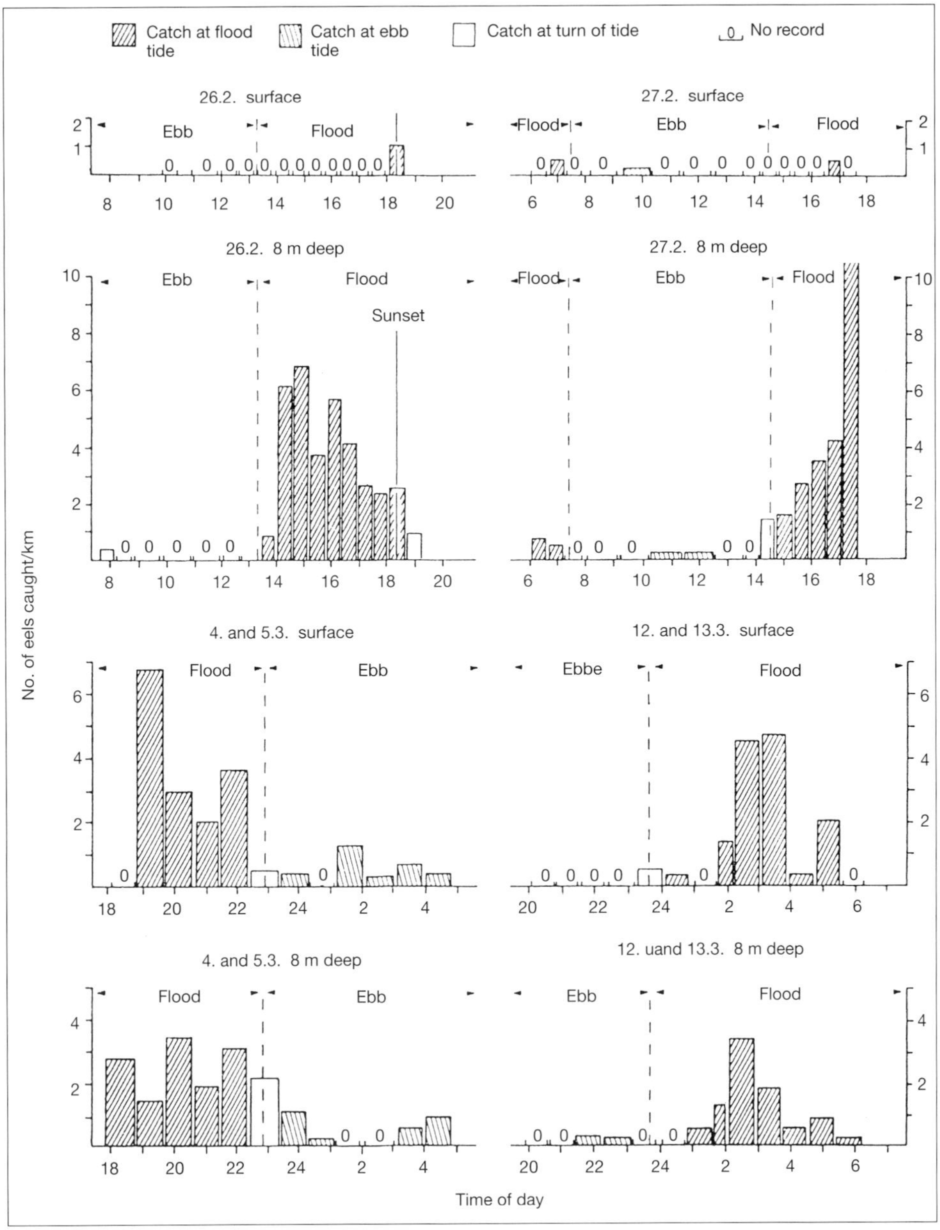

Fig. 3.3 Glass eel catches with plankton nets in the Texel Current, February/March 1957, indicating the various concentrations of these animals at the surface and in deeper water during high and low tide, and during day and night (modified after Creutzberg, 1961)

3.1.2 Entry into fresh water

3.1.2.1 Attractive substances in fresh water

As Creutzberg (1961) found experimentally, glass eels from the coastal area show a preference for the natural inland surface water, but not for charcoal-filtered fresh water. This treatment eliminates the typical odour substances from fresh water, such that it loses its attractiveness (Section 1.9.3). Thus, if, in nature, sea water mixed with natural fresh water from rivers and lakes flows towards the eels, they will react rheotactically. Creutzberg (1961) believed that this caused the eels to swim against the ebb tide in the direction of the inland waters, whereas during flood tide and its increasing seawater influence, they submit to the current and let themselves be drifted. Tesch (1965), on the basis of his investigations in the upper tidal area, questioned whether these changes in tidal rhythm behaviour could not also be brought about by an endogenous rhythm because all factors that could trigger a behavioural change, such as freshwater influence, altered turbulence and turbidity, very often do not coincide with flood or ebb tide. Maybe they could be supplementary zeitgebers and provide the necessary synchronisation. Calculations and laboratory experiments of Wippelhauser and McCleave (1987, 1988) on glass eels collected from North America's Penobscot river yielded confirmation of this.

In laboratory experiments, Tosi *et al.* (1990) established that the attraction to inland waters can be brought about in different strengths by three attributes of the water body: temperature, salinity and odour.

Distinction must be made between the effect of fresh water as a stimulatory medium on the one hand, and, on the other, its property as a medium in which eels reside (Schulz, 1975). Glass eels also prefer fresh water, when tap water is involved. They merely have to be given the opportunity to leave media of another salinity (18‰ and 36‰) again, that is, they must have time to change back and forth between the different media and to choose. The preference for fresh water is attributed to the low-salt concentration of the glass eel's interior substance; in contrast, young yellow eels have a higher concentration in their body fluid (Parry, 1966) and preferred brackish water of 18‰ (*source*: lower Elbe catchment basin; Schulz, 1975).

However, there is no irreversible preference of all the eel's post-larval development stages before spawning migration for fresh or brackish water. This can be shown by measuring the strontium : calcium content of the otoliths of eels from regions having various salinities (Tzeng *et al.*, 1997; Tsukamoto *et al.*, 1998). Commercial fisheries on eels in the North Sea have shown that eels can live in all salinities between freshwater and open sea conditions and can change between high and low salinity when challenged during a seasonal change of biotope (Peñaz and Tesch, 1970; Aker and Koops, 1976; Löwenberg, 1982).

3.1.2.2 Temperature barriers

The glass-eel invasion was observed at one place on the Dutch coast that had a rather more abrupt transition from sea or brackish water to fresh water than in lengthy estuaries, such as the Elbe (Tesch, 1971), the Loire (Elie, 1979) or the Sèvre Niortaise (Gascuel,

1986, 1987). In Holland, fresh water is conducted into the Waddensea through locks in the IJsselmeer's final dike, making a very sudden transition from the ocean into fresh water. The glass eels arrive there in temperatures that are very low, but generally at least 4.5°C (Deelder, 1952, 1970). During sudden temperature declines they are confronted with temperatures <4.5°C, if they have already penetrated into that area. In these very low temperatures they must undergo a certain physiological transformation before entry into fresh water. Therefore, they stop for some time in the sea water in front of the locks because the water temperature is probably too low for an immediate adaptation. Then they swim around individually and are not visible from the water's surface. Aquarium experiments indicated that these eels avoid fresh water. In a freshwater-fed aquarium, they did not swim into the freshwater flow, as did other eels that had been caught while migrating upstream in fresh water. It was established by colour marking of Japanese glass eels in a Taiwanese estuary that the immigration was stopped normally for 1 day, but in a few cases for as much as a week (Tzeng, 1984) even though the temperature difference certainly was not the main reason here for the delayed adaptation to fresh water.

Just below the IJsselmeer locks it was then observed that the glass eels gradually formed swarms. This was the beginning of readiness for freshwater entry. Various investigations have shown that this readiness starts at about 6–8°C (Deelder, 1952; Creutzburg, 1961; Tesch, 1971). In the fresh water of the river Ems they were expected for the first time at 9–11°C (Schmeidler, 1963). In western France's river Sèvre Niortaise glass eels actively enter fresh water when the water reaches 10–12°C (Gascuel, 1986). *A. japonica* was first caught with regularity during river entry when the water temperatures no longer fell below 8–10°C (Matsui, 1952). *A. australis* and *A. dieffenbachii* began to enter the estuary of a small New Zealand river in considerable numbers when the water temperature exceeded 9°C (Jellyman, 1977). These low temperatures are not now reached for the Japanese and European eels that enter fresh water further south in the northern hemisphere (see below). In Taiwan, it was even observed that the glass-eel catch reached its maximum at the lowest coastal water temperatures, for example 16.4°C on 2 January 1981 (Tzeng, 1985).

Freshwater entry is to be expected earlier in the lower latitudes than further north because temperatures are higher there:

- in estuarine areas of the Baltic Sea, eels were first caught at the beginning of April to the end of May (Herold, 1933; Herrmann, 1957);
- in the Ems from mid-April to the beginning of May (Schmeidler, 1963);
- in Italy's Tiber in February or January to March (Ciccotti *et al.*, 1995) or earlier (Fig. 3.2);
- in Portugal's Rio Minho, November to July (Weber, 1986);
- on the French Atlantic coast, November to April (Gascuel, 1986, 1987);
- in the Guadalquivir, that is, in south-westernmost Europe, they should, according to the present author's personal information, enter fresh water year-round, with a certain maximum in winter.

Figure 3.4 presents the seasonal distribution of glass-eel catches in Europe's major freshwater entry area, that of the Loire.

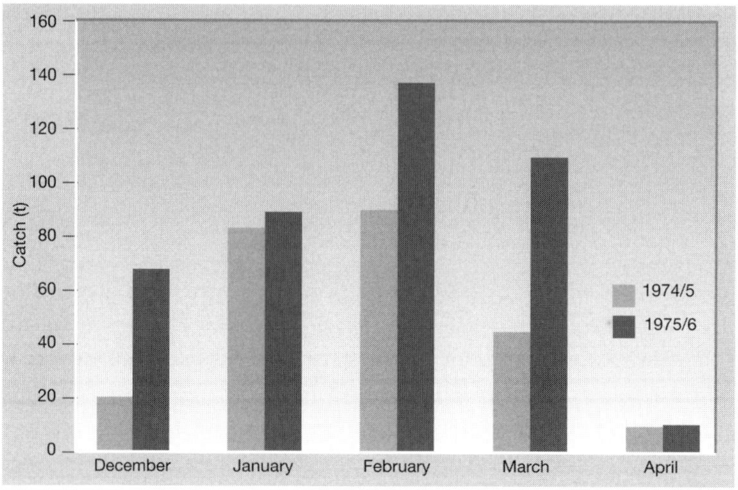

Fig. 3.4 Monthly distribution of glass eel catches in the Loire river in 1974/5 and 1975/6 (from data of Elie, 1979)

3.1.2.3 Light

Light influences crucially glass eel migration into fresh water. At this developmental stage, a strong negative phototaxis remains from the larval stage (Tesch, 1980; Tesch *et al.*, 1986). This is indicated by dipnet catches at the Den Oever locks, where reportedly a sudden transition to fresh water takes place (Fig. 3.5). There, the glass eels can be caught only when they are in open water, that is, when they may develop a certain activity. Accordingly, the activity was greatest during darkness. It began at sunset, increased until midnight, and decreased at dawn's onset. Similar tendencies were established for glass-eel catch of the two New Zealand species, *A. australis* and *A. dieffenbachii* (Jellyman, 1977). Van Heusden (1943, in Deelder, 1970) confirmed experimentally the apparent darkness-dependent activity. Negative phototaxis still could be observed about 70 km upstream from the freshwater boundary of Ems near Herbrum (Meyer and Kühl, 1952/3), where, in connection with a weir, the glass eels' freshwater entry was commercially exploited to obtain eels for stocking. Based on numerous commercial fishery catches in western France's Sèvre Niortaise, Gascuel (1987) could establish that under good fishing conditions the catches decreased at daybreak; the glass eels then stayed closer to the bottom, as was shown by setting nets at different depths. However, with progressing season (April) and rising temperatures and in fresh water, a change in the light sensitivity of the glass eels began. They migrated closer to the surface (Section 3.1.2.5).

3.1.2.4 Tidal current

To what extent does the eels' tidal rhythm observed in the Waddensea continue into the inner area of estuaries? The French conducted intensive studies of glass eels' freshwater entry and of glass-eel catch, especially in connection with tidal current (Elie, 1979; Cantrelle, 1981; Gascuel, 1987). The results of stocking colour-marked glass eels in the lower Sèvre Niortaise showed (Gascuel, 1986): 'The distances that the coloured eels

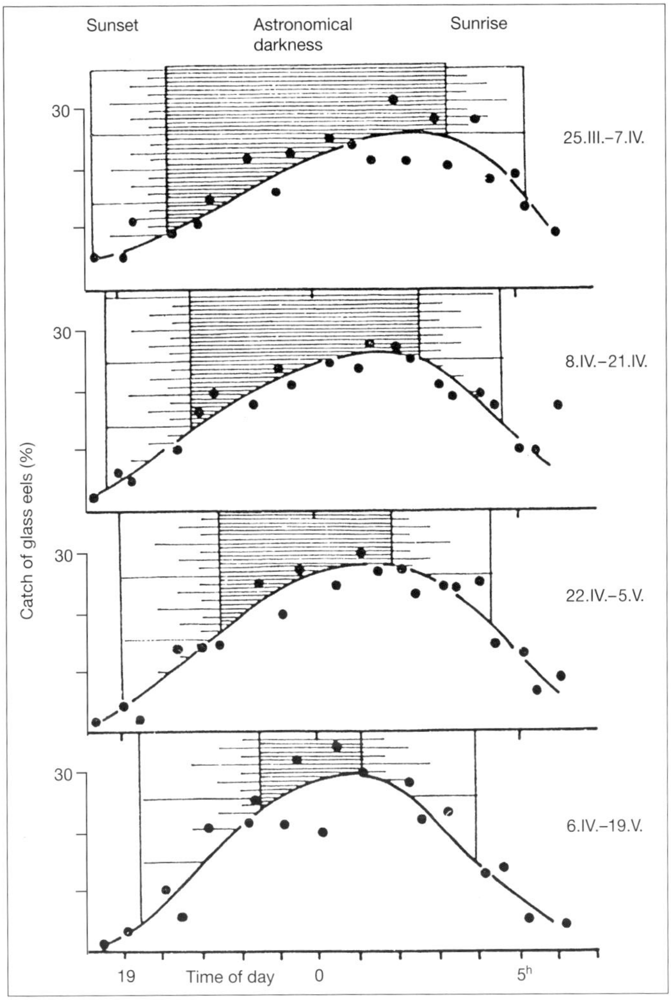

Fig. 3.5 Glass eel catches by vertical hauls (as a percentage of total catch) at various times during 1938–49 at the Den Oever sluice on the closing dyke of IJsselmeer (after Deelder, 1952, 1970)

moved after stocking during flood tide were less or equal to those of the water masses in which they existed. Only a few individuals appeared to move faster; but their swimming speed was very low. On the other hand, most of the animals did not use the full capacity of the current to make progress. This could be related to the fact that they only occasionally moved upward in the water column as the flood tide began', which was particularly observable during daylight. This supplemented the assessment made about the negative phototaxis at this stage. Many years of experience about the tidal current's significance for glass eel entry into fresh water were obtained in the Ems situated in Central Europe. There, in connection with a weir, the glass eels' freshwater entry was commercially exploited to get eels for stocking. Much as in the French rivers, it was confirmed here in addition that in fresh water, too, eels still use the flood tide to proceed upstream. Recording the hourly catch verified that during May the flood tide migration begins around 21.00 h, peaks at high tide, and ends 1–2 h after that (Kühl, 1955). The largest catches are to be expected if high tide happens around midnight. If ebb tide starts before and ends after midnight, for example lasts from 21:00 until 03:00, then no in-migrating glass eels are caught. Thus, at neap tide, the flood tide has to stop substantially further downstream, as it does in West France's Sèvre Niortaise ((Gascuel, 1986), see above).

A 14-day rhythm occurs through the relationship to darkness and current, as is observed elsewhere, as well, for example in Italy's relatively tide-weak Tiber (Ciccotti *et al.*, 1995), in Northern Ireland's Bann (Menzies, 1936; Lowe, 1951, 1952), or in England's rivers Severn and Leven (Lowe, 1951; Lübbert, 1910). Effects of the impulse imparted by the flood tide are still found further upstream later during the ebb tide and beyond the tidal area. Temporal shifts related to the time of peak high tide and the day–night boundary are to be expected then. An example of the 14-day tidal rhythm for other species of glass eels is provided by the two New Zealand species in a large river there (Fig. 3.6), in that the largest catches were also registered during daytime and evening 3 days after full or new moon (Jellyman, 1979a). Here, the moon has effect, of course, only indirectly *via* the tides. Likewise, the Japanese eel showed 14-day density maxima in coastal rivers of Taiwan (Tzeng, 1985).

3.1.2.5 Onset of the active migration

In his studies of the flood tide's transporting effect in west France's Sèvre Niortaise, Gascuel (1986) found that if water temperatures rose to 12–14°C, the glass eels became more active. They rose in the water column not only into the mid-depths, but for the most part reached the surface, as also observed in the rivers Loire (Elie, 1979) and Gironde (Cantrelle, 1981). From this moment onward, they all swam actively upstream, which also put them in position to pass eel ladders. At this time, the day-and-night upstream migration of the 'glass eels' starts, taking the form of a visible ribbon. This surface migration was observed at many places particularly near the streambanks (Tesch, 1965). It is known not only in the rivers of the coasts of the Atlantic Ocean, the North Sea, and the Mediter-

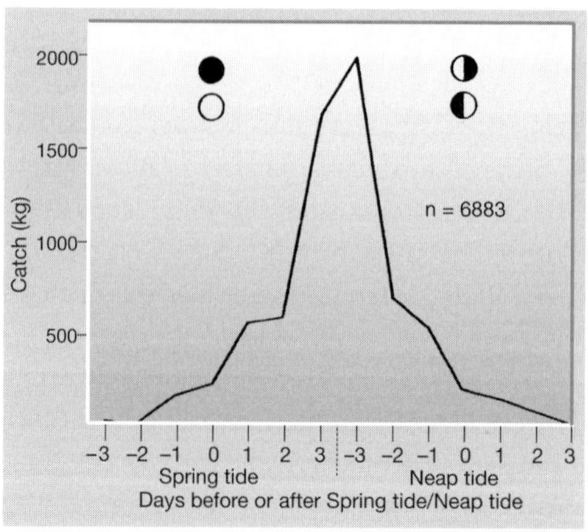

Fig. 3.6 Daily glass eel catches for the two New Zealand species in the Waikato river before and after full and new moon, as well as before and after the first and last quarters of the moon (after Jellyman, 1979a)

ranean, but also in the Baltic Sea (Herold, 1933). The ribbon of glass eels can be as much as 4 m wide in the French Atlantic tributaries, where it can also extend as much as 5-m deep (Wurtz-Arlet, 1961). But in German rivers the ribbon is usually only a few centimetres or decimetres wide and extends only about 5 cm deep (Leich, 1929; Löwe, 1930). The New Zealand species show a similar behaviour. In a large New Zealand river, Cairns (1942a) observed a 4.5-m wide, 2.5-m deep ribbon, which continued to go past the observation site for over 8 h. The passage supposedly lasted for several days in other cases (Cairns, 1942a). Deelder (1970) believed that contact with the streambank was not visual because such migrations also took place at night. Closeness to the bank is presumably detected rather more by the lateral line organ's sensitivity to pressure waves and current.

Surface migration has been observed far inland even 120 km in Holland during tidal influence that extended far upstream (Deelder, 1970). In the Elbe, it has reached as far as Lauenburg, 150 km above the estuary, but there the daytime upstream migration is restricted to 30–40 km, that is, the area above the upper limit of flood tide (Tesch, 1965). In the English rivers Severn and Leven, the eels are assumed to migrate upstream day and night (Lübbert, 1910; Lowe, 1951). Moreover, it is reported from south-western England's Severn that the eels there migrate only at the surface and near the bank during ebb tide (see also Fischer and Lübbert, 1908). Thus the surface migration of *A. anguilla* has been observed in many locations and under different light conditions. The glass eels' increasing tendency to migrate also by day at the surface indicates an at least transitory greater tolerance for light. Also during the upstream migration of the two New Zealand species, *A. australis* and *A. dieffenbachii*, it turned out that with decreasing duration of darkness in springtime, the migratory activity of the earlier glass eel developmental stages (primarily Vb) increasingly took place during daylight (Jellyman, 1977). More strongly pigmented glass eels entered fresh water primarily during daytime (Table 3.2), as also observed at another location (Jellyman, 1979a).

3.1.2.6 *Increase of pigmentation and change of body size*

The progress of development and migration can be measured by pigmentation, which starts as water temperatures rise and physiological changes take place (Strubberg, 1913; Section 1.3.3). Unpigmented eels have been found in front of the Dutch locks between the Wattenmeer and inland water bodies, as well as in the brackish water of the lower Elbe. These animals were principally night active and stayed in deeper places. In contrast, 100 km farther upstream, that is, above Hamburg, the present author (unpublished) found glass eels that were for the most part more strongly pigmented than in the lower Elbe. Of 16 specimens, five were at pigmentation stage VA_{II4}, the rest at VIA_{II4} to VIA_{IV2}, according to Strubberg's (1913) rating (see also Section 1.3.3). Eel pigmentation was studied to a greater extent in the estuaries of the French Atlantic tributaries, and seasonally increasing pigmentation was found. In November 1979, the proportion of almost unpigmented stage VB in the Gironde estuary approached 90%, but in May 1980, it was only about 25%. The more pigmented stage VIA_I composed 0–10% in November 1979, and 30–40% in April/May 1980 (Cantrelle, 1981). Concerning the Sèvre Niortaise, Gascuel (1987) wrote: 'From December onward, a detectable aging takes place; it increases beginning in March. In the uppermost study section, stage VIA_I dominates from then on.'

Besides the progressing pigmentation, the body length reduction of the glass eels' metamorphic stages continues to some extent (Fig. 3.2). In aquarium studies by the present author (unpublished), body length decreased from 7.5 to 7.0 mm in the period from 4 April to 31 July, during which the eels progressed from stage VB to, on average, stage VIA_{IV1}. Hardly any length decrease was noticed after the end of June at stage VIA_{III2}. The length decrease occurred in fed, as well as in unfed glass eels. Neither light nor darkness influenced length decrease and pigmentation decisively. The latter was somewhat prolonged by darkness or sand substrate, into which the eels could burrow. Water temperature determines the pigmentation and correspondingly also length decrease and later growth (Strubberg, 1913). Studies of the progress of body reduction and pigmentation in naturally developed eels confirmed the laboratory experiments, as was detected in the example of glass eel populations in the north African coastal area (Fig. 3.7) and in the Loire and Sèvre Niortaise estuaries (Gascuel, 1987). The mean length of glass eels reaching the Loire estuary decreased from 7.4 cm in January to 6.5 cm in April (Elie, 1979). Decrease in length and weight stopped during stage VIA_{III} on the North African and French coasts, much as in the laboratory experiments described. Juvenile eels resumed growth in the Sèvre-Niortaise estuary during March to May, but this was not so in the Loire estuary during mid- to late April (Gascuel, 1987). Glass eel increase and decrease (Fig. 3.7) give the impression that the changes in size and pigmentation proceed in parallel, but this could not be detected in the studies from the West French estuaries. Based on measurements of newly arrived glass eels (stage VB), Cantrelle (1981) found a decrease of length and weight from November to May. Furthermore, in seasonal comparisons of all stages she showed that the coefficient of condition decreased until May and the water content increased. Thus, apparently during the period of entry into fresh water there is a decrease in 'quality'. Likewise Santos and Weber (1992) furnished certain comments about that in their attempts to rear glass eels from the Rio Minho.

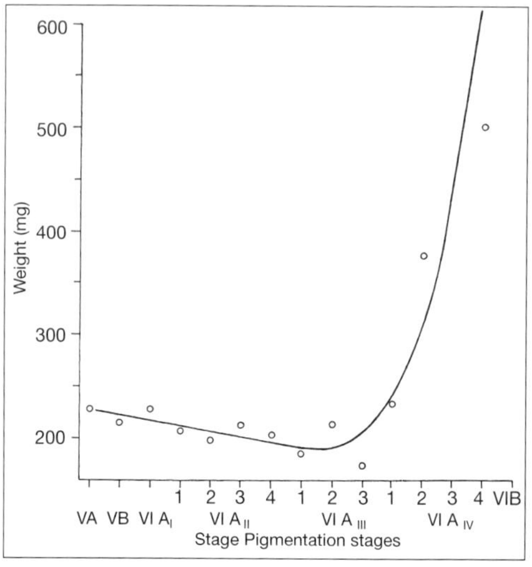

Fig. 3.7 Weight change in glass eels from Tunis Lake in relation to increasing pigmentation, according to Strubberg's (1913) pigmentation scheme; see also Table 3.2 and Fig.1.10 (modified after Heldt and Heldt, 1929a)

If one compares the mean lengths of glass eels from various regions (Fig. 3.2), the averages of the samples from the Atlantic (Severn, Bilbao) and North Sea obviously exceeded those from the Mediterranean Sea during almost all freshwater entry periods. Similar, even if not so very pronounced differences result from weight comparison (Heldt and Heldt, 1929b). In addition, length and weight differences can be recognised between the western and eastern parts of the Mediterranean Sea. The eels from the Balearic Islands, Italy and Tunisia are significantly longer and heavier than those of the Nile (Lübbert, 1930). Length differences similar to those of the European eel were also observed for the American eel (Vladykov, 1966). Mean length of glass eels increased by 6 mm from southern to northern North America. The decreases of body length and weight from north to south along the Atlantic coasts for both eel species are most likely attributable to the differing lengths of the eel larvae that are approaching the coasts. According to numerous studies and measurements, all leptocephalus and glass-eel stages in northern latitudes (e.g. the Bay of Biscay) are longer than in the south (e.g. in southern Portugal) (Kracht, 1982; Utrecht and Holleboom, 1985; Tesch *et al*., 1986; Bast and Strehlow, 1990; Antunes and Tesch, 1997a). According to Tesch and Niermann (1992), for instance, early oceanic glass-eel stages were on average 0.5–2.7 mm longer than in the south in 3 study years, and in 4 study years leptocephali had a north-to-south difference of 2.1–7.1 mm.

3.1.2.7 First food consumption

First intake of food happens considerably earlier than the end of length and weight reduction (up to stage VIA_{III2}) would lead one to expect. In the North Sea, glass eels ingested food when they had become 'light grey'. Before that, they had almost exclusively empty guts. A third of the 'light-grey' eels had eaten, and, with increased pigmentation, the majority of the examined animals had.

On the North and Baltic Sea coasts they ate primarily *Mysis*, *Idothea*, copepods and chironomid larvae, and to lesser extent polychaetes, oligochaetes, amphipods, *Asellus* and various insect larvae (Eichelbaum, 1924). Detritus also occurred in the gut in the first stages. Studies of upstream-migrating glass eels by the present author (unpublished) on 17 May 1966 at the Elbe river's Geesthacht Dam above Hamburg revealed that none had food in the gut until stage VIA_{III2} (light grey). Only at later stages was it evident that the animals had eaten something. Five eels of stage VIA_{IV1} all had some food in the gut, but only one specimen contained a significant amount.

In the Gironde, during the period of investigation from December to May, glass eels had eaten significantly more and earlier:

- two stage VB glass eels each had one mysid and one alga;
- one stage VIA_{I} had one fish egg;
- 17 stage $VIA_{II\text{-to-}III}$ had 15 chironomid larvae and two insect pupae in total;
- two stage VIA_{IV} had one insect larva, two adult dipterans, one spider and two masses of remains of polychaetes and plant matter;
- nine stage $VIA_{II\text{-to-}IV}$ had 13 mysids.

Half of the individuals studied from the Gironde contained detritus (Cantrelle, 1981). Similar to the Gironde results, aquarium studies using freshly hatched *Artemia salina* showed that feeding first occurred shortly after completion of stage VB (Table 3.1). However, in the initial stages (VIA_I) there was very little in the gut, and only with later development did filling increase. When presented with the aquarium's natural supply of organisms, glass eels ate planktonic *Artemia* and *Daphnia*, and benthic *Tubifex*. Regarding quantitative food demand, soon after first feeding they eat everything that they can manage from the standpoint of size.

3.1.3 Commercial use of glass eels

The mass immigration of glass eels into fresh water is commercially harvested at various locations in Europe and Japan. Originally, this harvest was not at all for stocking purposes. In particular, in the areas of great surplus on the Biscay coasts and on south-west England's river Severn, one did not at first know how to do anything with glass eels other than use them for human food and animal feed. Today, in Spain and Latin-American countries, they are served as a rather expensive specialty (e.g. in a soup strongly seasoned with garlic). They swirl in this like spaghetti, and it is doubtful whether they increase the culinary value of the dish very much. In French in fishing towns and villages, glass eels are prepared as a sort of biscuit. In the past, demand for them as food for humans was minor, though, and 'most were used as animal feed, mainly for swine' (Wurtz-Arlet, 1961).

In an 8-km stretch of the river Loire, farmers with their whole families caught 12–15 t of eel young daily in 1924. France's export to Spain amounted to 750 t. From another source it is evident that around 1930 the average annual glass-eel market in St Sebastian on the French border in Spain handled 237 t, of which 90% came from France (Gandolfi-Hornyold, 1936). The largest annual harvest was 494 t (1922/3), of which 250 t occurred in February alone. In Spain, glass eels are canned and used in glue production. Far more than 1455 t of glass eels were caught on France's Atlantic coast during the 1975/6 fishing season (Elie 1979), of which at most 800 t came from the Loire alone (Fig. 3.8A). In contrast, the amounts caught in the Ems (Fig. 3.8B) are modest: at best 6 t in the years 1960 and 1963, but very often being only around 1 t (Wiehr, 1967; see also Meyer, 1951; Schmeidler, 1957, 1963). River Ems harvests have recently decreased to negligible amounts.

The 'Deutscher Fischereiverband' (German Fishery Association), the traditional distributor of stocking eels, has been able to increase the imports from France to more than 10 t per annum. Lesser amounts came from Denmark and England to the Hamburg distribution centre. The annual yield of the pre-World-War-I German collection station for young eels in Epney on the Severn was about 2 t. In Northern Ireland's Bann from 1938 to 1943 about 10 t of glass eels were caught annually (Lowe, 1951). In 1965 and 1966, 13 t were exported from Hamburg to numerous European countries, the earlier German Democratic Republic, with its relatively large demand for lake stocking, receiving the main share. Other European importers, particularly Poland and the Soviet Union, were included. The demand from these countries, with their large freshwater surface areas, was especially high. Nevertheless, since 1956, glass-eel export from England into the Soviet Union has been comparatively low: 2.3 t/year (Orlov, 1966) and before 1939 it was only 100 kg/year (Kokhnenko, 1958). Under China's huge aquacultural demand for supplying the Japanese market, its 1966 importation of European glass eels rose to 200–300 t.

Table 3.1 Stages of glass eels (after Strubberg, 1913) that ate food during an aquarium experiment in 1966 and 1967. Feed: newly hatched *Artemia* larvae; temperature: 15 °C; for each date, left column: number of eels studied, right column: number with gut contents (after Tesch, unpublished).

	Date 1967																					Date 1966		
Stage	10.4.		14.4.		21.4.		2.5.		17.5.		29.5.		19.6.		13.7.		31.7.		Totals		%	1.4.–8.6.1966		%
VB	1	0																	1	0	0	4	0	
0																								
VIA_{I}	27	6	10	3	3	2	1	0	1	1	1	0	1	1	2	0	1	0	47	13	28	11	5	45
VIA_{II1}	3	1	4	1	4	4	1	0	1	1									13	7	54	18	5	83
VIA_{II2}			2	0	7	1	5	2	1	1	2	0							17	4	24	49	45	92
VIA_{II3}			2	0	3	2	3	2	2	2	2	0							12	6	50	53	52	100
VIA_{II4}							2	2	1	1			1	0			1	1	5	4	80	24	24	100
VIA_{III1}							2	0	6	6	3	1	3	1	1	1			15	8	53	29	29	100
VIA_{III2}									1	1	3	1	2	2	2	2	1	1	9	7	78	29	29	100
VIA_{III3}			2	0			2	0	3	3	3	0	7	5	2	0	1	1	19	9	47	9	9	100
VIA_{IV1}											1	0	1	1	4	2	11	7	17	10	59	17	17	100
VIA_{IV2}											1	0			1	0	15	7	17	7	41	8	8	100
VIA_{IV3}									1	1	1	0					2	1	4	2	50	3	3	100
VIA_{IV4}																						8	8	100
VIB													1	1	1	0	1	0	3	1	33	5	5	100
Total	31	7	19	4	17	9	16	6	17	17	17	2	16	11	13	4	33	18	179	78				
%	23		21		53		37		100		12		69		31		55		44					

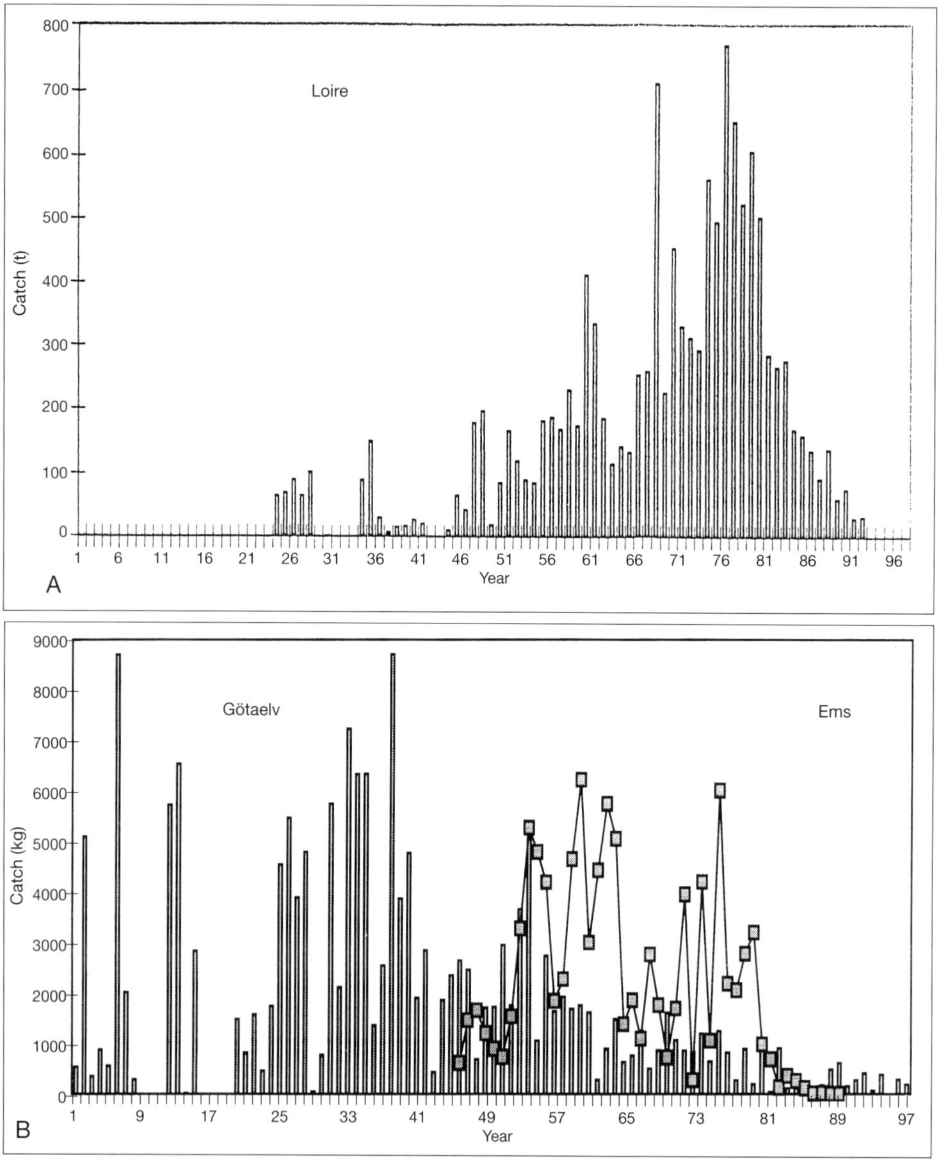

Fig. 3.8 Entry by glass eels into the Loire and Ems rivers (after Moriarty, 1990; Moriarty and Tesch, 1996) and combined entry of glass and juvenile eels into the River Götaelv (after Wickström, 1989, and pers. comm. Wickström)

A Loire (French west coast) 1924–94
B Ems near Herbrum 1946–90 (thereafter, no catch worth mentioning until 1999); Götaelv river 1901–97

Provision of stocking supply for the European eel population has suffered substantially under the consequent tremendously increased prices (Section 8.3).

Glass eels are caught from streambanks mostly with stow nets (Sections 5.4.3.2 and 5.4.3.3) or with scoop-like hand nets (Fig. 3.9A; Fischer and Lübbert, 1908; Wurtz-Arlet, 1961). In France, the streambank catch is often aided by light from flaming torches; each

Fig. 3.9 (a) Scoop net in operation in the river Severn, south-west England (photo: Wood). (b) Glass eel fishing operation with vessels in the river Vilaine, western France (photo: Hahlbeck)

fisher capturing about 20–30 kg per night. At present, glass-eel fishing in the Loire and other rivers is also conducted with nets towed by motorboats (Fig. 3.9B; Wurtz-Arlet, 1961), about 100 kg being caught per trip during the 1950s. Recently (2000/1), in the Vilaine (Fig. 3.9B), observations of the author indicated far lower catches, that is,

2–10 kg/trip. Glass eels in Northern Ireland's Bann are caught with automatic equipment, built much like a trap that was installed to catch elvers in the Elde (Elbe tributary) near Dörnitz (Gollub, 1959; see Fig. 3.14). A similar principle was applied in the Ems near Herbrum to catch glass eels migrating along the left bank. However, that upstream migration channel is not lined with straw and twiggy brushwood as is usual in German eel ladders, but is fitted with vertically oriented plastic brushes. A similar method is being used in some places in migration channels for elvers.

A Japanese method for catching glass eels and elvers resembled the twine fyke nets described in Section 5.4.2 for catching market-ready eels (Topp and Paulerson, 1973) and was tested, although with low success, for harvesting *A. rostrata* in Florida's St John river. A dip net proved to be better there. A trap for juvenile eels in the US state of Maine was likewise based on the fyke-net principle. The throats of this box-like trap net are composed of vertical walls and are open on the downstream side, like the aforementioned fyke-net-like gear. They are set on gently sloping stream banks, along which the upstream glass-eel migration passes (Sheldon, 1974). The author observed a fyke-net-like device also being used for catching Australian eels in a northern Tasmanian estuary in 1994 (Tesch, unpublished).

For transporting eels, gauze frames were already used in 1908 (Fischer and Lübbert, 1908), and this continued into the 1970s. Today light plastic containers are used, especially for air transport; with these, for example 292 t were already flown from France to Japan in 1973 (Forrest, 1974). Transport within Europe takes place primarily in special trucks having aerated fish tanks.

Misgivings have been expressed about shipping glass eels in February/March from France into then still wintry central and eastern Europe (Koops, 1968; Gedymin and Gottwald, 1968). It was felt that it should await more appropriate temperatures. However, with the present method of transporting in aerated containers, such injuries as found by Bogdan and Waluga (1980) are no longer to be expected. Anaesthetic calming of glass eels tended to cause higher transport losses (Gritzke, 1980). Studies by Peters *et al.* (1980) showed that poor holding conditions resulted in the glass eels deteriorating into a kind of stress and becoming sick.

In Japan problems in procuring glass eels are worse than in Europe. The very high demand for glass eels to stock rearing ponds leads to supply becoming the limiting factor in Japanese eel culture (Section 6.2.1, Fig. 6.3). More eels are produced for human consumption in Japan than in Europe (Tables 4.1 and 6.1). This explains the enormous demand for stocking material, which, in turn, is removed from an eel population that has a significantly smaller geographic distribution than does the European eel and hence the total population of *A. japonica* is significantly less. Japanese glass eel prices are correspondingly high. Whereas, in Germany 1 kg of glass eels cost 30 DM in 1972 and 600 DM in 1996 (*Die Aalpost*, 1997, page 1), the Japanese paid the equivalent of about 60 DM in 1973 (Forrest, 1974) and even 3000 or far more in 1996 (Section 8.3). Therefore, anticipation among Japanese people is great when glass eels come upstream, and harvest is intensive. The Far Eastern demand caused the enormous price rise for stocking material (*A. anguilla*) in Europe in 1996/7.

Human activity also harms upstream migration of glass eels. Germany's Weser with its many dams, which hamper migration in the tidal area, provides a cautionary example – just one among many on other European rivers. Wurtz-Arlet (1961) reports from France that since the diking of the Seine, few glass eels still migrate upstream there. In the Somme, which yielded 2 million juvenile eels in 1914, and still 500,000 in 1926, the fishery for them has been non-existent since the mid-1950s. The strong decline in upstream migration of glass eels on the French west coast and in other areas of Europe (Fig. 3.8; Moriarty and Tesch, 1996), as well as at the German glass-eel sampling site near Herbrum on the Ems since the early 1980s is so serious that anthropogenic influence on natural reproduction *via* shortage of spawners can no longer be ruled out (Dekker, 2001; Tesch, 2001). Besides blockage by dams, water pollution hampers upstream migration of glass eels, but this might not take effect until pollutants have reached high concentration; for example in the Baltic Sea off the Schwentine river mouth, pollution presented no barrier to advance of glass eels into that river (Neubaur, 1933).

In discussing the glass eel decrease, it is not often considered that a North European decline had started already in the 1940s (Fig. 3.8; Hagström and Wickström, 1990), which also showed up in the waning Baltic Sea catch of males (Section 4.2.2; Table 4.3). This aspect further complicates interpretation of the causes.

3.2 Migration of elvers

3.2.1 Dependence on environmental conditions

Far better known than the non- or little-pigmented glass eels are the elvers, which are already found in their completely pigmented state from 7 to 8 cm onward. This segment of life is attained upon reaching pigmentation stage VIB (see Section 3.1.2.6, Table 1.2).

As determined from questioning Elbe fishers 150 km above the estuary above Lauenburg, Germany, the surface migration described in the previous section is not seen that far upstream. The juvenile eels seldom appear at the water surface, many having already gone deeper while far downstream from this area. Of course, at appropriate temperature a glass eel continues to develop even if coastal area obstacles delay its migration (Strubberg, 1913). The rheotactic migration tendency is still maintained for a longer time. According to results of age–growth studies on such eels, the migration can continue for several years under some circumstances.

Catches in a freshwater entry area in the rivers Severn and Avon (south west England, not far from the oceanic area where larvae metamorphose) revealed ascending eels as long as 30 cm entering fresh water, but mainly specimens of 6–13 cm, within a population that averaged 7–8 cm (Churchward, 1996).

Length measurements of more than 8000 eels (Fig. 3.12; Peňáz and Tesch, 1970) at the Geesthacht Dam, which is still within the Elbe's tidal area, revealed maxima of 25–30 cm for eels at freshwater entry (Mann, 1961, 1963; Tesch, 1967d). Individual specimens had body lengths of as much as 40 cm. In the Weser, according to studies at four dams in river sections extending to almost 200 km above the estuary, up-migrating juvenile eels were >35 cm only in exceptional cases (Tesch, 1966). Opuszynski (1965) arrived at similar results in studies in the central Baltic catchment basin in Mazury, Poland. There, upstream migrating eels also were at most 35 cm long but usually not >25 cm. Similarly, Gemzøe (1906), investigating length-at-age of juvenile eels at various places in Denmark, found primarily only lengths of <25 cm. Larsen (1972) captured juvenile eels in two streams on the Danish island of Zealand by electrofishing and determined that >95% were 9–34 cm long, and 78–86% were between 9 and 24 cm.

Observations in the Elbe have shown that the upstream distance that the juvenile eel is capable of migrating before full pigmentation is at least 150 km (Tesch, 1965). If completely pigmented, it is able to cover much greater distances yet. Observations of *A. nebulosa* (Ibrahim, 1961) in India's river Godavari lead one to expect that, after the glass eel stage, the juvenile eels can migrate 650 km in 1 year. To swim such distances is a remarkable feat, especially when one considers that they grow from 5 to 15 cm in body length during this time. One is even less able to disregard the migration distance in view of obstacles, such as weirs, that lack adequate upstream passage facilities. For example, proceeding upstream through Germany's river's Weser dams, at the fifth, nearly 200 km above the mouth, the smallest eels caught were 15 cm (Tesch, 1966). In contrast, at the lowest dam, 80 km above the mouth, glass eels are also found, but here more large juvenile eels are found immediately above the dam than below it. Thus, based on the time needed to grow to 15 cm (Peňáz and Tesch, 1970), the eels take at least 2 years to reach the fifth dam. Therefore, they travel little more than 100 km during this time. It is also possible though that only the larger juvenile eels can negotiate the dam-cluttered section of river and then possibly cover the same stretch sometimes significantly faster.

In contrast to other kinds of fish, upstream migration of juvenile eels presumably occurs at relatively low speed, that is, although substantial upstream migration distances are covered, it takes a long time. This is evident from the participation of individuals of 2 freshwater years or older in the migration in lower river regions. Thus, juvenile eels do not

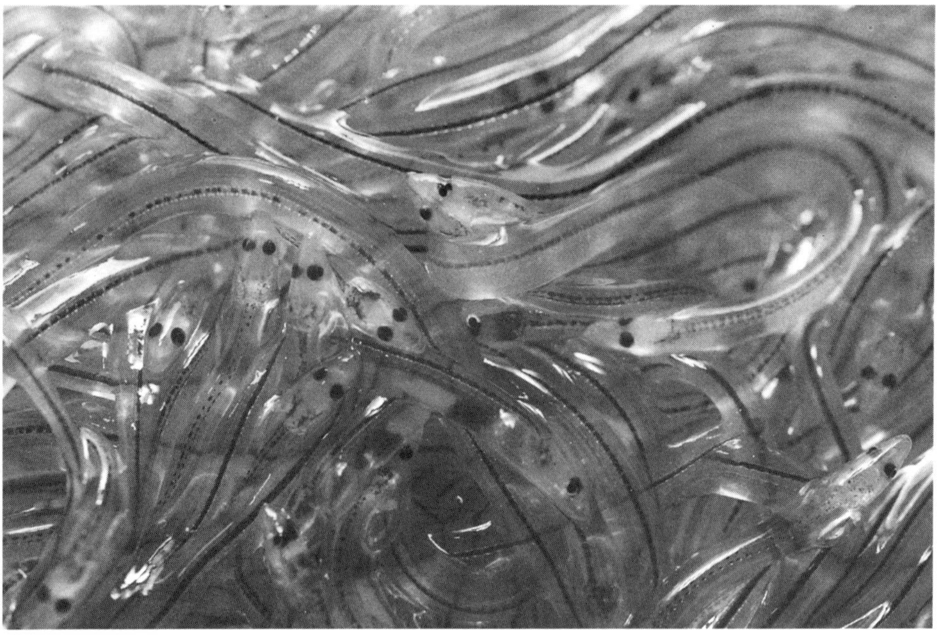

Fig. 3.10 Glass eels caught in the river Vilaine, western France (photo: Hahlbeck)

migrate so strongly and determinedly as, for example, salmon, lampreys and many other fishes that swim upstream to spawn.

An experiment enabled a comparison of the migratory behaviour of juvenile eels with that of the lampreys that were swimming upstream intensively (Fig. 3.11). An eel pass with trap, built right next to one weir, monitored upstream migration. If the weir section next to the eel pass was shut, the strong current adjacent to the eel ladder stopped, and water flowed out of the eel ladder only. The result was that almost no lamprey ascended anymore, but eels kept doing so. This result must be interpreted as follows: lamprey, like many other fishes (Fries and Tesch, 1965b), migrate upstream at the edge of the main current and orient to it. Therefore, they reach upper river sections without delay or detour. The elvers, on the other hand, do not use the immediate edge of the main current. Thus,

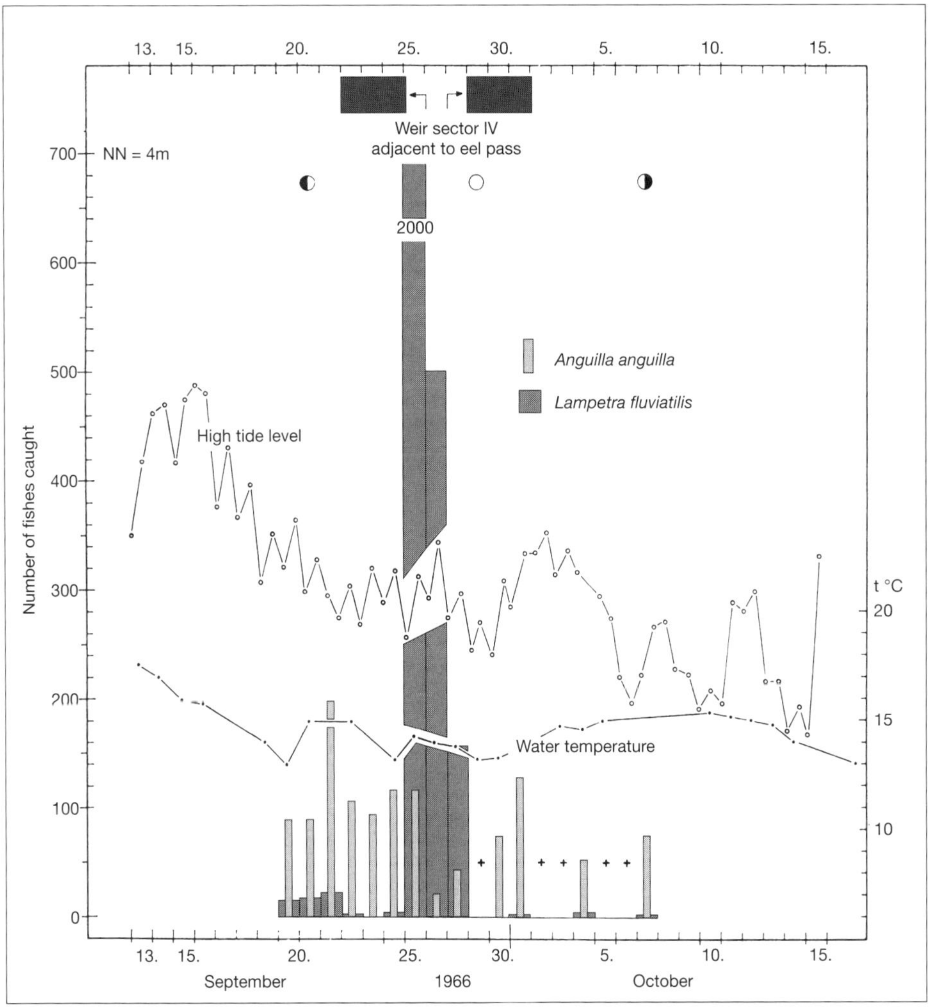

Fig. 3.11 Daily figures for entry of pigmented juvenile eels and lamprey (*Lampetra fluviatilis*) into the eel pass of the river Elbe dam at Geesthacht upstream from Hamburg, Germany, with data on water temperature, high tide levels, lunar phases, and adjustment of the dam's outflow. Crosses mark days when no observations were made in the eel pass (after Tesch, 1967d).

they migrate continuously, even if the current is weak or stops completely. In the process, they often wander into dead-end situations, finding the main current and, therefore, the way upstream again, only after some delay.

Schiemenz (1950, 1952), though, found strong rheotactic reaction in the glass eels he studied. He wrote: 'If there is an ongoing decrease in current velocity, then alignment [of the glass eel] to the direction of current decreases to a certain extent, and upon further slowing or in quiet water, it swims around aimlessly until it gets into the current, which induces it to swim up the thread of the current. In a current that is too weak, it searches for a current that is adequate to its migratory urge.' The glass eel has a migratory speed of about 5 cm/s (but according Deelder, 1970: 10–15 cm/s). In any case, the Elbe's midstream current, which flows at 100 cm/s and more at some places, is too strong for glass eels and even for somewhat older stages. The same surely applies to the midstream current in other large rivers. To proceed upstream, the juvenile eels must, therefore, swim in quieter areas, whereby they often get into dead ends. Then the current of an eel pass with its small amounts of water is detectable for juvenile eels over a fairly wide vicinity and stimulates their rheotactic behaviour. But the elver orients itself to the shoreline not so directly as do the younger or not-yet-pigmented stages, which then form a bright ribbon along the bank. It migrates somewhat deeper and only seldom appears at the surface; the somewhat larger juvenile eels also prefer the night, as is well known, and as the eel pass studies on the Elbe reveal (Fig. 3.12). The elvers' upstream migratory behaviour represents to some extent a compromise between migration in a fast current and that involving orientation to the bank, where often no current is observable.

The migratory activity of elvers also depends very much on water temperature (Mann, 1963; Tesch, 1967d; Larsen, 1972; White and Knights, 1994). It starts at 8–9°C, as in glass eels, but is presumably much more sensitive to thermal fluctuation than is the migratory activity of glass eels, which, when they have begun their ascent, are hampered only by temperature plunges below the minimum tolerable level. The thermal dependency of juvenile eel activity compared with that of the lamprey, measured by the numbers of individuals that surmount the Elbe dam at Geesthacht is shown in Fig. 3.11 (see last paragraph but one). Although the image of a temporally narrow, lunar-periodic or tide-dependent upstream migration can be recognised in the lamprey during spawning migration, the upstream migration of eels increases and decreases irregularly during the 18-day period (Mann, 1961, 1963). A dependence on tide stages, which in this section of the Elbe are strongly overlain by streamflow discharge, is not recognisable. A certain dependence on streamflow discharge is possible (Mann, 1961). With falling temperatures in autumn, upstream migration of eels decreases, and below 10°C halts almost completely. The strong thermal dependence of juvenile eels' migratory activity was also determined by experiment; simultaneous raising of the water temperature from 20 °C up to 25 °C, and of the current velocity from 0 up to 6.5 cm/s increased the activity 13-fold (Wehrmann, 1968; Lillelund, 1967).

With respect to diel periodicity, the upstream migration takes place mainly at night according to studies by Mann (1961) and Rosengarten (1954), but the difference between day and night migration is not so substantial as Table 3.2 shows. Moreover, upstream migration during daytime can even predominate, if the proportion of glass eels that has immigrated in the same year is large. When temperature plunges at night, upstream migratory activity likewise can be lower than on the previous day, as the author observed

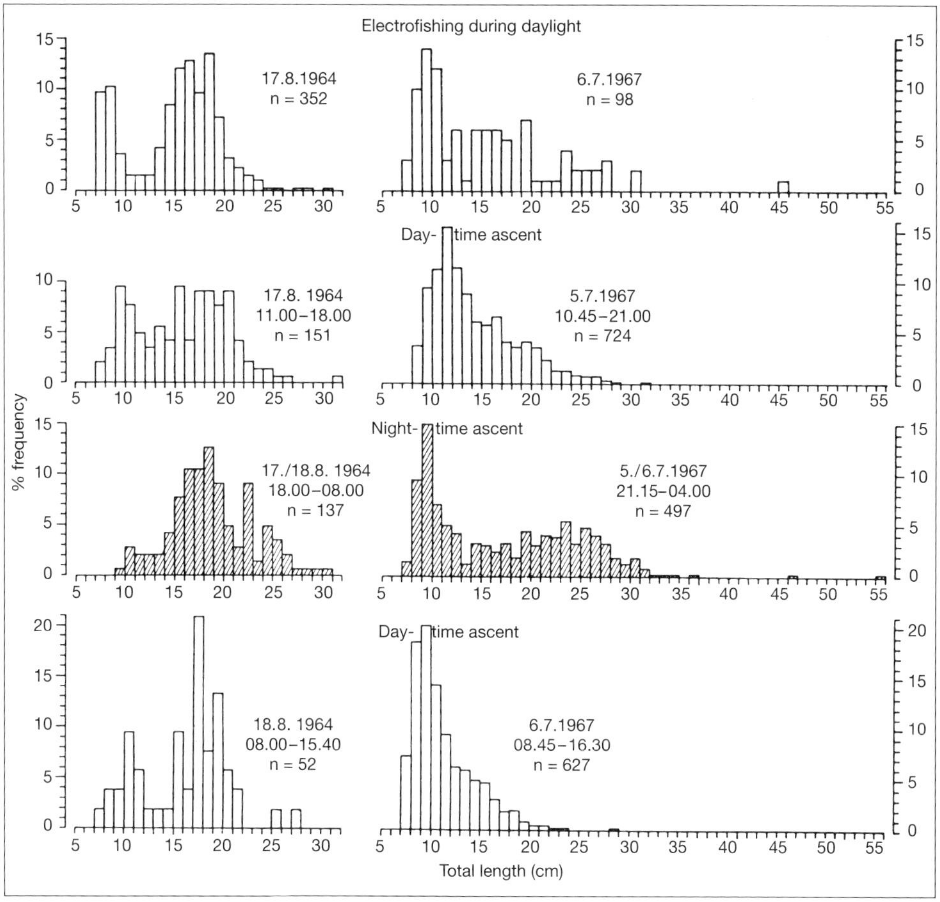

Fig. 3.12 Length frequency distribution of juvenile eels during day and night in the eel pass of the river Elbe dam at Geesthacht upstream from Hamburg, compared with the length frequency distribution from electrofishing catches below the dam (after Tesch, unpubl.)

in the Elbe at the beginning of July and in mid-August 1964. It follows from Fig. 3.12 that the proportion of large juvenile eels is relatively higher at night than in daytime (Tesch, unpublished). The studies lead one to surmise that, with increasing body length, juvenile eels change from initial 24-h to night-time activity. This also has seasonal effects. As studies on length frequency of eel in the Elbe's Hamburg Harbour area (Ladiges, 1936) showed, small juvenile eels of 11–20 cm predominated over those of 21–30 cm by 2 : 1 to 3 : 1 during summer, but in September the two groups were caught in equal numbers. Relative increase of larger juvenile eels in autumn was also found in later studies in the lower Elbe (Peňáz and Tesch, 1970). Moreover, the same was found further upstream in the Elbe near Dömitz (Figs 3.13 and 3.14; Larsen, 1972). Thus, the opinion that the significantly smaller juvenile eels have already disappeared upstream in autumn cannot be held valid; indeed, they would then have to show up in the upstream river zones. Rather, it is to be assumed that the activity of the smaller eels in autumn (in falling temperatures and decreasing day length) declines more than does that of the larger eels. A similar tendency

Table 3.2 The number (per hour) of juvenile eels ascending the eel pass on the right bank of the Elbe at Geesthacht and proportion that this number represents of the number of glass eels estimated to have immigrated into the Elbe during the same year (after Tesch, unpublished). In brackets: number of older juveniles, that is, not glass eels.

		Number of eels per hour			
Date	Hour	Mainly daytime operation	Mainly night-time operation	% of the number of immigrating glass eels 1964	Comments
31.5.–2.6.1964	22.00–10.30		13	0	
	10.30–14.30	2		0	
	15.00–19.15	3		0	
	19.15–20.50	7		0	
	20.50–08.15		20	0	
	08.15–10.30	1		0	
	10.30–12.00	2		0	
	12.00–10.35	1		0	
14.6.–16.6.1964	22.00–15.00		74 (11)	84	
	15.00–20.30	161 (14)		92	
	20.30–08.30		170 (15)	91	
25.6.–26.6.1964	08.30–18.30	15 (1)		96	
	18.30–08.30		3 (2)	17	
6.7.–8.7.1964	19.00–09.00		25	25	
	16.00–17.45	5		0	
	10.00–09.00	1	1	0	
	09.10–14.00			0	
17.8–18.8.1964	11.00–18.00	22 (18)		17	
	18.00–08.00		10 (10)	1	
	08.00–15.40	3 (2)		8	
26.8.–28.8.1964	06.00–20.00	18 (17)		3	
	20.00–08.00		22 (22)	2	
	12.00–15.00	9 (7)		23	
	15.00–18.30	11 (9)		15	
	18.30–08.00		29 (29)	2	
	09.00–12.00	13 (12)		13	
	02.00–15.15	3 (2)		3	
21.9–22.9.1966	10.30–14.30	0			The proportion of the immigrating number of glass eels (1966) was about 6%
	14.30–18.00	1			
	18.00–08.30		10		
	08.30–12.00	<1			
4.7.–6.7.1967	22.00–10.30		512		The proportion of the immigrating number of glass eels (1967) was about 50%, as estimated by electrofishing
	10.45–14.45	275			
	15.00–21.00	1067			
	21.15–04.00		185		
	08.45–11.15	765			
	11.30–16.30	868			

is possibly also evident in spring: in most years up-migrating juvenile eels were larger on average in May and the beginning of June than in summer (Peňáz and Tesch, 1970). To confirm the observations made in nature experimentally it is important to study differently pigmented glass eels, as well as elvers of various ages.

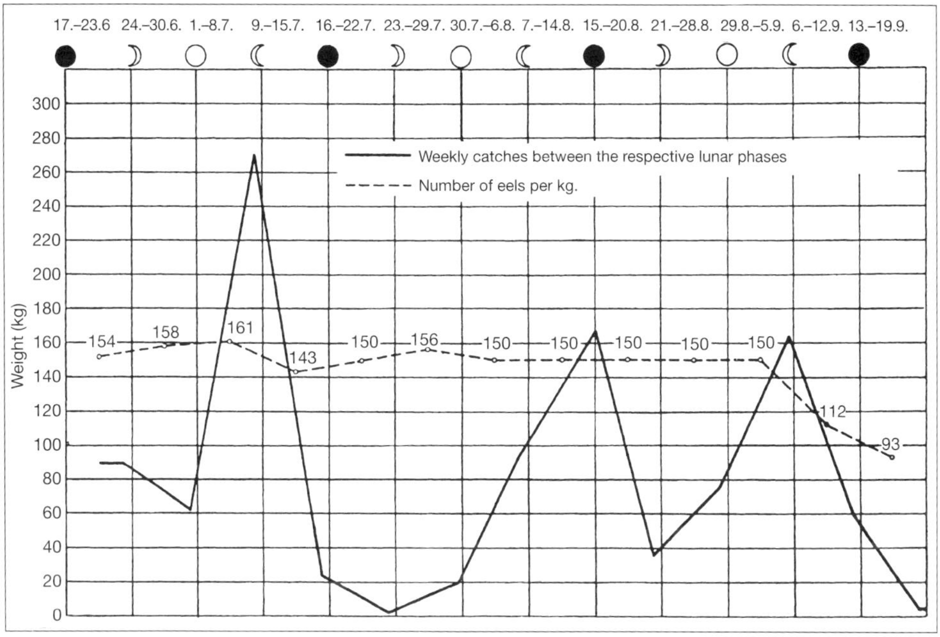

Fig. 3.13 Biomass of upstream-migrant juvenile eels in a trap in the river Elde near Dömitz/Elbe, Germany, in relation to lunar phases (after Gollub, 1959)

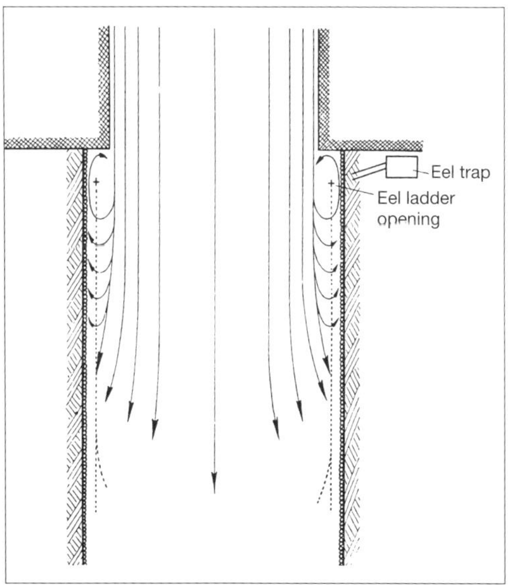

– Solid-lined arrows: current direction
– Dotted lines: the glass eels' favoured migration paths

Fig. 3.14 Location of an eel ladder with a trap box for stocked eels in the Elde river locks near Dömitz/Elbe, Germany (after Gollub, 1959)

It was established by experiment that 'Aal-brut' (eel brood) began to become active when the 120 lx light intensity that was normally offered was reduced to 9 lx; when light intensity was again raised from 9 up to 19 lx, the activity stopped (Wehrmann, 1968). Similar results were obtained with somewhat larger individuals of *A. rostrata* (Bohun and Winn, 1966). The light energy of wave length 548 nm that 'small' eels need in order to alter their movement activity, for example, to stop their activity amounts to 0.006 $\mu W/cm^2$ (Van Veen and Andersson, 1982).

Salinity is a further environmental factor that can influence the activity of juvenile eels in the lower reaches of rivers. As studies on the preference of young yellow eels for various salinities have shown, experimental animals from the lower Elbe area prefer brackish water of 1.8%, when they are also offered 3.6% salt water and fresh water from the tap (Schulz, 1975). Thus, conceivably, juvenile eels that have fully reached the yellow eel stage but not yet passed completely through the brackish zone develop no further impetus to enter fresh water. This could be a reason for the very dense population in the extended brackish region of large rivers.

Besides increased negative phototaxis, elvers presumably have yet another trait that causes a further activity rhythm. Studies by Gollub (1959) at an eel trap at the Elde navigation lock near Dömitz/Elbe showed an activity rhythm that obviously ran parallel to moon phase. During the 3-month studies, peak catches occurred in the last quarter of the moon or at new moon (Fig. 3.13). Heightened activity at this time was also observed in field experiments (Mohr, 1971).

Based on his collection from the Geesthacht Dam on the Elbe 80 km below Dömitz, Mann (1963) could not see such a rhythm. However he wrote: 'The only conspicuous thing is that the maximum of all catches (28–30 August) declined with the waning moon. It is to be inferred from this that the stimulating effect of the temperature rise at this moment was strengthened.' His further statistics on up-migrating juvenile eels (Mann, 1961, 1963), as well as some studies of the author's (Tesch, unpublished) at the same catch location likewise permit no sure interpretation in this regard.

For New Zealand juvenile eels (mainly 'glass eels'), as well, no lunar-periodic maximum could be determined by means of catch statistics (Jellyman, 1977), even though a clear lunar periodicity was evident from studies in a different location (Gibson and Boubée, 1992). But the negative results cannot disprove the findings described above (Fig. 3.13), for experimental study results (Wehrmann, 1968; Tesch *et al.*, 1992) support a lunar rhythm of movement, which also, by the way, like the tides, can be recorded in terms of 14-day peaks of activity and can be magnetically influenced. As an activity parallel to the moon phases was detected for silver eels (Section 3.4.4), a lunar rhythm of juvenile eel activity would not be surprising. Therefore, the question arises whether, already in earlier developmental stages, there is not only a tidal, but also a lunar dependency (Section 3.1.2.4; Kracht, 1982).

3.2.2 Economic use and furtherance of elvers

Similar to the upstream migration of glass eels (Section 3.1.3), that of elvers is also commercially exploited to obtain animals for stocking inland waters. Fisheries for stocking eels are known in the lower Elbe, Denmark, and southern France, the latter to supply Italian aquaculture.

The lower Elbe's the so-called 'Satzaale' (stocking eels) are fished for with fyke nets and stow nets, the same gear with which eels are caught for human consumption (Section 5). The length groups of the stocking eels caught there are distributed much as shown in Fig. 3.12, but the lower length groups are absent because eels <15 cm long pass through the meshes of the fishing gear. Thus, stocking eel shipment is limited to lengths of 15–30 cm.

Much discussed regarding these eels as stocking material have been their sexual development and potential growth (Schnackenbeck 1954; Wundsch 1954; Rahn 1955b, 1957a; Section 1.7.2). According to studies by Peňáz and Tesch (1970) on eels of the lower Elbe area it must be assumed that the fears that have existed until now mainly about prematurely emigrating male eels can no longer be shared (Section 1.7.2). Furthermore, it was indicated above that eels of this size are still rheotactically inclined. Thus, when stocked in waters further upstream, they have a strong tendency to migrate further upstream, perhaps as almost 'silver eels' or as stationary eels, instead of returning to the estuary by means of their homing ability (Section 3.3.3.4).

Stocking eels can also be harvested in eel traps in hydro-engineering facilities. Eel traps are often incorporated in eel ladders as at the outfall of the Elde locks near Dömitz/Elbe (Fig. 3.14). It seems important that the exit of the eel ladder leading to the trap should have its mouth in water that does not flow too fast (Section 3.2.1).

One can also improve eel populations by easing the way for natural upstream migration of juveniles, which is significantly hampered by water projects, such as weirs and dams, and to some extent by navigation locks. Thus, it is a responsibility of the water engineering authorities to install upstream passage facilities for eels. Only continual maintenance and, if needed, reconstruction for proper functioning ensures the success of such facilities. Beyond that, it should be kept in mind that within a series of dams or full-spanning weirs, each must have an eel ladder. Any single barrier renders all eel ladders above it pointless.

Eel ladders utilise the eel's capacity not only for swimming, but also for climbing. Other fish species having snake-like bodies, for example lampreys, also climb. This climbing, that is, upward wriggling, is enabled when the animal is supported by stable objects, and not only the caudal end is put to use, but also the rest of the body, including the anterior. If necessary, the eel even wraps itself around solid objects. Juvenile eels, with their long and relatively light bodies, can even climb almost vertical walls that are not too smooth. The surface only needs to be a bit moist. An example of this is that they can surmount the Rhine Falls at Schaffhausen, Switzerland (Meister, 1970). Nevertheless, the climbing of such obstacles should be facilitated for eels. Woods (1964) measured *A. dieffenbachii* and *A. australis* specimens before, during, and after they climbed a hydroelectric dam's face and determined that those of 8.7 cm body length and 0.53 g can get over best.

Based on the characteristics described, facilities for upstream passage of juvenile eels can be made easily and at much less expense than most proper fish pass constructions. Many fish passes are even unsuitable for juvenile eels because water flows too fast in them. So in the river Mosel, supplementary eel ladders were often installed next to the fish passes. The same was done at the former fish pass at the Elbe's Geesthacht Dam, even though, because of its moderate current, a supplementary eel pass would not have been required right next to it. This is also the case at the newly built fish pass there (Fig. 3.15) because it was fitted later in the form of a mountain stream that offers quiet water or much reduced current in many places behind rocks. The first fishing trials with the traps that are visible at the lower edge of the weir in the picture seemed to confirm proper functioning (Schubert, 1999).

Eels of 7–10 cm body length swam 0.6–0.9 m/s, those of 10–15 cm and larger swam 1.5 m/s (Sorensen, 1951). They maintained this speed for a distance of 120 cm. At the critical point of the Mosel fish pass near Koblenz, a current velocity of 2 m/s was measured (Rosengarten, 1954), and eels swam through this pass. But the juvenile eels swam up on the sidewalls, thereby utilising the places of lowest current. Besides that, the maximum current occupied only 10 cm of the pass. Juvenile eels took twice as long to migrate up through this fish pass as did the cyprinids. The behaviours of the juvenile eels coincide with those described above in the Elbe. Observations in the Finnish lake region (Järvi, 1909) give an idea of how capable juvenile eels are, despite their low swimming speed. There,

Fig. 3.15 This channel and flow-control structure are the main elements of the newly built fish pass (mainly for eels and salmon) at the Geesthacht dam on river Elbe dam upstream from Hamburg. The dam extends far to the left, and the impoundment is seen in the background. The channel was designed to simulate a natural stream, detouring around the dam, instead of the former, more traditional fish ladder (shown in this book's previous editions, Tesch, 1973, 1983) (photo: Tesch).

over a distance of at least 300 km, eels had to surmount 13 waterfalls and rapids that were 1.5–9.5 m high, and at another place even as high as 22 m.

About 100 eels per hour ascended the Mosel fish pass near Koblenz at night, a remarkable number at this far upstream compared with the amounts at the Elbe's Geesthacht eel pass (Table 3.2). Much closer to the sea at Geesthacht, 500 juvenile eels migrated upstream per hour, averaged over the whole day under optimal conditions, the maximum being 1000/h. Nevertheless, these rates were not often reached, and even 170/h represented quite a good value.

Even though normal fish passes promote upstream migration of juvenile eels, ladders especially designed for upstream passage of eels should be installed. They are significantly more economical than fish passes, so no barrier in the upstream migratory area of eels should be without an eel ladder. If a fish pass already exists in a large river, one or several should be installed nevertheless, at other points of the same obstruction. This possibility for improving yield is not even used sufficiently by commercial fishers; Schiemenz (1940) wrote that of 154 eel traps at dams or involving weirs in Germany's Hannover Province, only 18 were fitted with an eel ladder.

Regarding construction of eel ladders, Jens (1982), Dahl (1991), and Schiemenz (1940) give numerous suggestions. Figure 3.16 shows construction of eel ladders, as often installed today. The eel pass design depends largely on the construction of the dam or weir and the local characteristics. The principle is to connect the upstream and downstream areas such that water flows slowly within the connection. This is achieved by means of brushwork, straw, plastic bristles, or mat material (Köthke, 1964b; Schiemenz and Kühne, 1964; Dahl, 1991). The eels can then swim against the reduced water velocity or wriggle up the rough material (Fig. 3.17). The advantage of plastic material is that it does not rot as fast as woody brushwork, fascines, or heath plants. According to Jens (1969), the numbers that ascend bristle-brushes are even several times higher than for fascines.

In Germany's Mecklenburg province, many smaller eel ladders were installed (Gollub, 1955, 1959), which, as monitored by trapping, had good success. For example, after an eel ladder was installed leading to 180 ha Rudower Lake in the Elbe/Löcknitz catchment basin, enough eels immigrated within a week to equal what would have had to be stocked in the lake to provide for a yield of 16 kg/ha. In Dobbertin Lake, which has a lengthy outlet to the Baltic Sea, 4000 juvenile eels immigrate during the stocking season. This would have sufficed to provide 66% of the 6 kg/ha annual eel yield in this 290 ha lake. Each year, 75,000–225,000 juvenile eels ascend through the already mentioned Elde locks near Dömitz (Gollub, 1963). This would have covered about a quarter of the stocking requirement for the waters that lie upstream. Since then, the numbers migrating upstream have probably declined, possibly due to the Elbe's Geesthacht Dam that was built in the meantime. At another location where juvenile eels have already migrated a considerable distance inland from the Baltic Sea, the Mildenitz river near Sternberg (Mecklenburg, Germany), 16,000–80,000 of them were found annually.

Dahl (1991) and Legault (1991a, b) described how the upstream migration of eels is promoted in Denmark and France. Tube structures provide for migration over substantial dam heights in New Zealand (Gibson and Boubée, 1992). A 146 m-long tube-trough structure in the Matahina Dam's wall lets eels surmount a height of 63 m. From January to

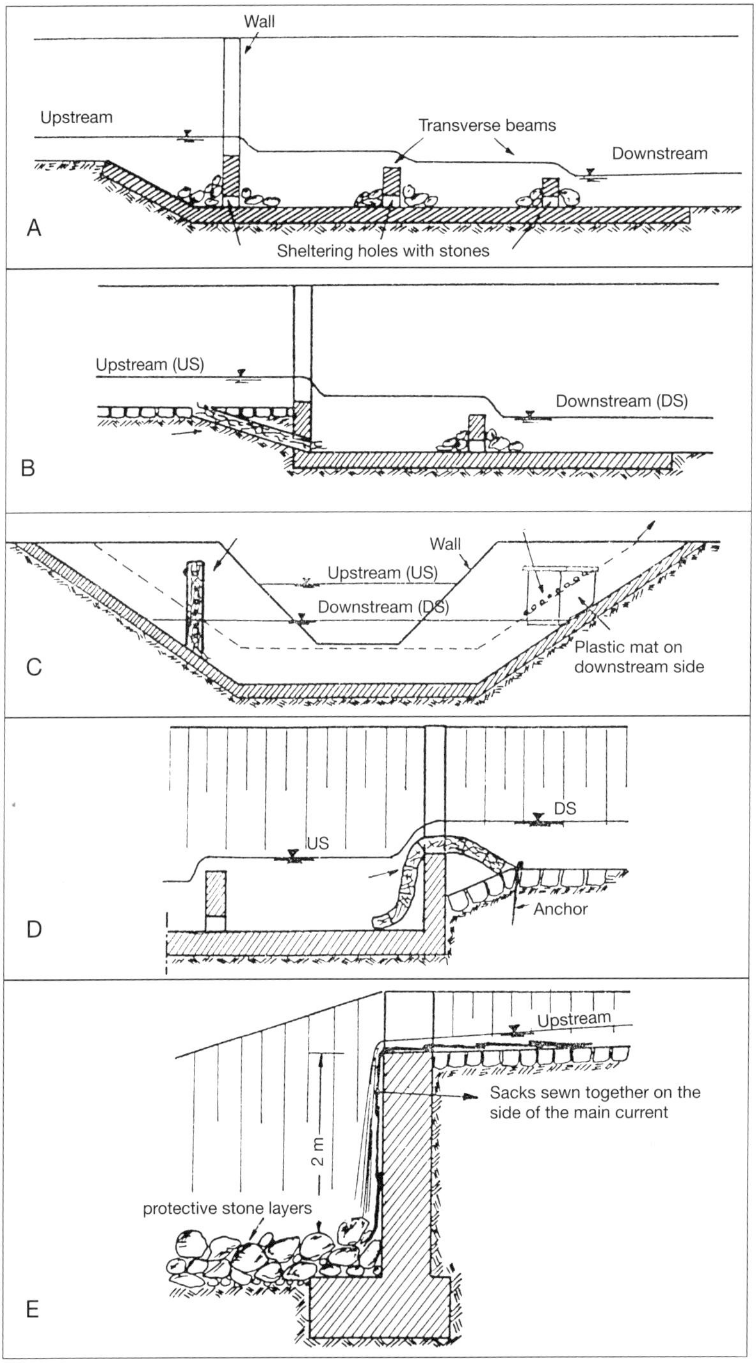

Fig. 3.16 Examples of eel ladder installations. (A) Passage holes at the base of a vertical wall; rocks and brushwork within and on each side of the openings reduce water velocity. (B) Tubing at least 30 cm in diameter, leading diagonally between the upstream and downstream areas and lined with wood fibres and brush. Better yet is a lining of removable bristles, as shown in Fig. 3.17. (C) Right: holes 2 cm in diameter in the dam, arranged at 5–10 cm intervals close above the embankment of the upstream area, the series of holes extending below water level even during low-flow conditions. Matting is hung in front of these holes on the downstream side. Left: the same holes as at right but aligned vertically and covered by a three-sided

[*cont'd opposite*]

Fig. 3.17 (a) Material for eel-pass tubes: bristle linings that were installed in one of the tubes in the eel ladder at the river Elbe dam at Geesthacht upstream from Hamburg (photo: Tesch). (b) Material for eel-pass tubes: Plastic tubes filled with plastic matting made of Enkamat – a trademark (after Dahl, 1991) .

March, an automatic counter recorded 15,000 juvenile eels, primarily of glass-eel size, 5000 in one night.

It is difficult to calculate the number of eels that migrate upstream past the Geesthacht Dam on the Elbe. According to studies from 1964 and 1966/7, over 100 juvenile eels pass there hourly on average, thus about 2500 per day (Table 3.2). That would amount to 375,000 during the annual 5-month period of upstream migration. It is to be assumed that similar amounts migrated up through the fish-and-eel pass on that river's opposite bank. In addition, the diverging lockway canal must be taken into account as a migration route. There, on mornings in good years, swarms of glass eels were observed and passed through. Thus, it is to be reckoned that the same numbers migrated up the lockway as in the eel pass of the Elbe's right bank. Therefore, a total annual migration of 1 million glass eels and

wooden box filled with brush and wood fibres. (D) If no holes through the dam are available, then brush matting laid over it and anchored in the upstream area will suffice; this consists of two layers of plastic-coated chickenwire filled with brush and wood fibres. (E) In an emergency, sacks can be sewn together and hung over the edge of the dam at one side of the main current (Wilke, 1975).

elvers could be expected. According to earlier observations 5–10 million used to ascend Germany's large North Sea tributaries annually (Schiemenz, 1930), a little more than 5 million the Ems alone in good years (German Fishery Association statistics; Wiehr, 1967). Thus, a considerable reduction in upstream migration seems to have occurred at Geesthacht or even in the lower Elbe area (Hamburg Harbour). The 1 million eels presumably passing the dam in those days could have supported an average yield of 5 kg/ha on a 30,000 ha area of waters (Section 4.4.3), but the catchment basin to be supplied contains many times that water surface area. Thus, meeting such a harvest goal would not have been possible without stocking in those days, still less so under today's severely reduced migration from the Atlantic.

Damming of the St Lawrence river (Canada) by the Moses Saunders Hydroelectric Plant in 1958 must have impeded eel migration into the catchment basin above, much as the Elbe dam did. First and foremost in this regard, Lake Ontario is included in the St Lawrence basin. However, yield statistics for the Canadian fishery 1959–70 did not show decline (Hurley, 1973). Also, an over-ageing of the population due to decline in later generations did not happen. On the contrary, to the disadvantage of older age groups, young became more numerous in the catches. While, for example, among all 10 cm length groups, the 90–100 cm group made up 17.6% of the catch in 1964, it declined to 2.6% in 1971, the 60–70 cm group increased from 14.8% to 38.2% in the same period (Section 4.2.5). As at the Geesthacht Dam, navigation locks are a possible way for juvenile eels to circumvent the Moses Saunders Dam. The plasticity in migratory behaviour of juvenile eels thus sufficed in the foregoing cases to withstand detours and substantial hydrographic alterations. Later, the Moses Saunders Dam was fitted with a permanent eel ladder, the prototype of which had already been successfully operated since 1974 (Whitefield and Kolenosky, 1978).

3.3 The yellow eel to the silver eel stage

In general, yellow eels are all those that have reached a certain size. Apart from their size, they are hardly distinguishable externally from migrating elvers (Section 3.2). This section will deal primarily with the developmental condition in which the juvenile eels' migratory stage concludes and they attain body lengths >30 cm. Eels that have reached this part of life are also at the feeding stage. Further migrations are undertaken only when weather and water conditions make it necessary to change home territory. But such changes in location are narrowly limited and hardly greater than those of territorial fishes. Nevertheless, many younger age groups must be included here, especially when discussion of feeding and growth is involved.

3.3.1 Feeding

3.3.1.1 Activity and food consumption

Just as the activity of eels and, therefore, of amounts harvested diminish in winter, a decline in feeding is also to be expected at this season (Thurow, 1959; Wenner and Musick,

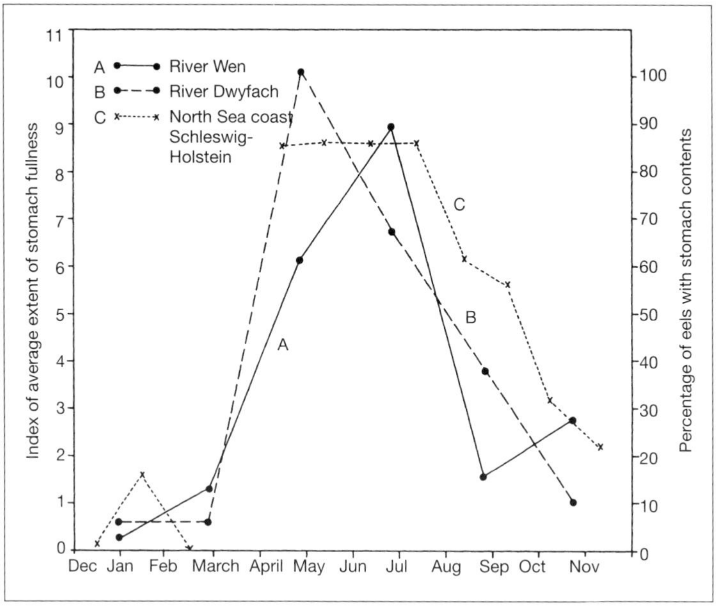

Fig. 3.18 Seasonal differences in diet. (A, B) Mean fullness of eel stomachs from two streams in Wales, the Wen and Dwyfach rivers (after Sinha and Jones, 1967c). (C) Percentage of eels having food in the gut in brackish areas on the west coast of Schleswig-Holstein (after Daniel, 1968).

1975). Figure 3.18 gives an idea of how much the amount of food consumed by eels rises during March to May. In summer, they obviously eat many times more than in winter, whether in the coastal area or in streams. For the eurythermal and almost subtropical eel, the greater food consumption in summer is significantly more pronounced than in other eurythermal species of temperate latitudes. In contrast to other warmth-loving species, the eel's increase in nutritional demand is particularly great, even if for no other reason than that no springtime spawning follows. The high springtime yields (Section 4.3) associated with the feeding migration and consumption are thereby explainable. However, the food consumption, measured according to stomach fullness, decreases from May to September (de Nie, 1987) (Fig. 3.18).

Besides the seasonal differences, diel periodic differences are to be expected also. The author's feeding trials in aquaria showed that eels ate substantially more at night than during daytime. Other authors also got this result. Eels over 25 cm long from a shallow lake in Holland had fuller stomachs just before morning than at the rest of the time (de Nie, 1987). A high percentage of eels in early evening catches in New Zealand had empty digestive tracts, but gear set over night caught animals with food in their guts (Cairns, 1942b). Likewise, baited fyke nets were effective in catching *A. dieffenbachii* in New Zealand only at night (Burnet, 1952b). Much the same is known from catching European eels with baited eel baskets and by angling (Sections 5.3 and 5.4). In a study of *A. nebulosa nebulosa*, all sampling for stomach analysis was done in early morning because they fed at night (Pantulu, 1956).

But in natural daylight, the eels are not completely inactive. Also in aquaria, they consume food during the day, especially if adapted to a particular feeding time. Anglers and commercial fishers also will attest to having some success in bait fishing for eels during daytime. Activity studies on individual 20–30cm American eels have shown that the daily activity rhythm is brought about exclusively by the light–dark cycle (Bohun and Winn, 1966). If the dark period shifts, the movement activity changes accordingly. Yellow eel activity may be even more closely associated with darkness than that of juvenile eels (Sections 3.1 and 3.2), which showed decreasing activity in light as they grew.

The question now arises whether translocational activity diminishes as feeding activity decreases and *vice versa*. According to the reported studies, this seems to be the case. But Polish investigations showed that a certain reservation must be made (Opuszynski and Leszczynski, 1967). Feeding studies on 10–40 cm juvenile eels in four lakes and a river that connected the lakes had a different result. Although, on average, 40–80% of eels from the lakes had food in their stomachs, and the stomachs had a correspondingly high degree of fullness, in repeated sampling of the river only about 10% of the eels there had fed, and these also had relatively little in the stomachs. The authors' explanation for this was that the eels in the river had to swim against the current, and thereby developed rheotactic activity that apparently reduced food consumption. Only easily obtainable prey (such as ephemeropterans and trichopterans) was eaten, although the posterior part of the intestine held chironomids, which obviously came from the lake that the eels had previously departed.

3.3.1.2 Food composition and season

As a primarily bottom-dwelling fish, the eels rely in their feeding on the prey population that is present there. Depending on the type of water body, these animals vary in abundance during the annual cycle and, therefore, also in the eel diet. In the English streams studied, greater amounts of fish were eaten in springtime and the beginning of summer than in later summer, autumn and winter (Table 3.3). This is associated with the fact that the primary prey were glass eels, in total 126 of them. The rest of the fish were 15 lampreys, 10 salmonids, four sticklebacks, and four flatfish. In this and other streams that were studied, the eels ate only a few salmonids. All salmonids found in the stomachs came from eels sampled in summer.

Mussels and snails were eaten only in one of the two streams, and that was almost entirely in summer. Also the ephemeropteran nymphs that were among the most important food organisms, were consumed mainly in summer, as were the trichopteran larvae, worms, and animal species of the water surface. Nevertheless, insect larvae were significant prey throughout the year. Dipteran larvae were especially important in winter. But it is very difficult to determine a preference during winter because the stomachs were usually empty then (Fig. 3.18).

In the Hamburg Harbour area of the Elbe, seasonally stratified studies revealed no particular differences (Ladiges, 1936). The main prey here was the bloodworm (*Tubifex*), which normally shows no seasonal fluctuation. Also for the next most important prey, particularly the cladocerans, no differences could be found. Mitten crabs (*Eriocheir sinensis*) were likewise present in the diet in almost all months. One important prey animal, primarily in springtime, is the earthworm, which inhabits newly flooded areas. Micheler (1967) made similar observations on eels of the Chiemsee (lake). They had also eaten land snails

Table 3.3 Percentage of eels that had fed on various kinds of prey at various seasons in two streams in Wales (modified after Sinha and Jones, 1967b).

	December/ January	February/ March	April/ May	June/ July	August/ September	October/ November
River Wen						
Fishes	–	5	10	17	–	2
Plecopteran nymphs	2	2	7	7	–	–
Ephemeropteran nymphs	2	7	17	23	7	8
Trichopteran larvae	3	2	10	30	12	2
Dipteran larvae	5	10	12	22	7	15
Coleopteran adults and larvae	–	–	–	10	–	–
Crustaceans	–	–	–	2	–	3
Annelids	2	–	2	5	2	10
Surface prey	–	–	–	7	–	–
River Dwyfach						
Fishes	3	–	–	17	5	3
Molluscs	3	–	–	13	5	–
Plecopteran nymphs	–	–	53	–	2	3
Ephemeropteran nymphs	–	2	60	36	9	3
Trichopteran larvae	–	2	47	40	9	–
Dipteran larvae	–	2	40	9	7	3
Coleopteran adults and larvae	–	–	–	5	–	–
Crustaceans	3			2	–	–
Annelids	–	–	13	4	5	–
Surface prey	–	–	10	4	2	–

during springtime flooding. In contrast, fishes were less frequent in the springtime diet there. *Rutilus rutilus* and *Perca fluviatilis* rose to 96% in August. The eels of Lake Balaton showed definite seasonal differences (Biró, 1974). There, *Asellus aquaticus* strongly dominated from March until June, after which a mud-dwelling amphipod (*Corophium curvispinum*), as well as scuds (Amphipoda) or *Limnomysis benedi* became the most frequent prey.

The seasonal diet pattern of tropical eels looks significantly different from that of temperate zone eels. The winter pause or restriction of the diet does not apply for them, but substantial temporal differences can also occur in the tropics, as was found for the Indian eel (Pantulu, 1956). Especially noticeable in this case was the strong domination of crab larvae in August, which composed 79% of all prey consumed. At that time, these organisms were present in enormous amounts. Even fishes (*Hilsa ilisha*, *Septipinna phasa* and *Pama pama*) that otherwise composed the main diet retreated completely into the background then. Prawns had a similar maximum (42–84%) in the eel stomachs from May until July, so that also, especially in May, fishes and other organisms declined greatly. Adult crabs, which likewise are represented very often, formed several maxima during the year and, primarily in January/February made the fishes seem unimportant in the diet. It must be emphasised that in the water bodies of their origin glass or juvenile eels were present year-round, especially in January/February. In contrast to the eels of English streams (see above), glass eels were eaten in only two instances.

3.3.1.3 Food selection and body size

The eel prefers prey animals that are suited to its body size and can be caught. Its jaws are not very big, so it can ingest only medium-sized chunks. If it can get a prey item piece by piece, it prefers that way of eating. This can be observed in the aquarium if one gives the eel meat or fish that has been minced. The eel bites into this and makes rapid backward or sideward jerking movements until a mouth-sized piece is torn loose. To rip pieces of food loose, it twists itself several times, if necessary, about its body axis with great nimbleness and rapidity. It does this in the wild when attacking crustaceans, for example. Elbe fishers often found mitten crabs attached in their basket traps with the dorsal shells opened up from the rear and the innards eaten away (Ladiges, 1936). The stomachs of the eels caught with them contained crab legs that were torn loose with the attached gill bundles. This indicates that the eel attacks the mitten crab actively rather than simply scavenging limbs that have been shed.

However, the eel is only able to indulge in its predilection for crabs (Decapoda) when it has attained a proper size. Eels >40 cm that were caught in water at least 3 m deep in the vicinity of Heligoland contained a tremendous share of crabs (Table 3.4). Bristle worms predominated in the diet of study eels <35 cm long. A year later, these studies on Heligoland eels by the author found many empty stomachs (Tesch, unpublished). Of the minor amount of stomach contents found, decapod crabs composed over 50% in eels of all sizes. But the crab remnants stemmed mainly from rock crabs (*Cancer pagurus*) that had been cut up to bait the basket traps. Thus, the smaller eels would have had no difficulty in consuming the otherwise hard-to-capture morsels. Among other prey that are relatively hard to capture, fishes increased in amount for larger eels, though not to a very convincing extent (Table 3.4). The fishes found were mostly gobies (Gobiidae), which at <5 cm, were also available to the smaller eels.

A study of several South African eel species showed a similar picture (Fig. 3.19). Whereas 10–20 cm eels had eaten insect larvae exclusively, those >20 cm contained crustaceans too, especially freshwater crabs of the genus *Potamon*, as well as fishes. The proportion of fishes and crabs rose for 50–60 cm eels. For those of 60–70 cm, crabs composed the highest percentage.

Table 3.4 Percentages of various sized eels that had fed on various kinds of prey from September to October 1965 in the lower tidal zone around Heligoland. Sampling gear: fyke nets (after Tesch unpublished).

Length range of the eels	30–35 cm	35–40 cm	40–50 cm
Number of eels sampled	19	24	27
Prey:			
Fishes (mainly Gobiidae)	11	13	19
Mussels and snails (mainly *Littorina littorea* and *Hydrobia* spp.)	21	13	15
Decapoda (mainly *Carcinus maenas*)	26	50	82
Other Crustaceans (Amphipoda, Isopoda, Mysidacea)	32	42	22
Hydrozoans (mainly *Sertularia* spp.)		4	11
Bryozoans (*Flustra foliacea*)			4
Polychaetes (mainly *Nereis* spp.)	68	63	7
Algae	11	4	

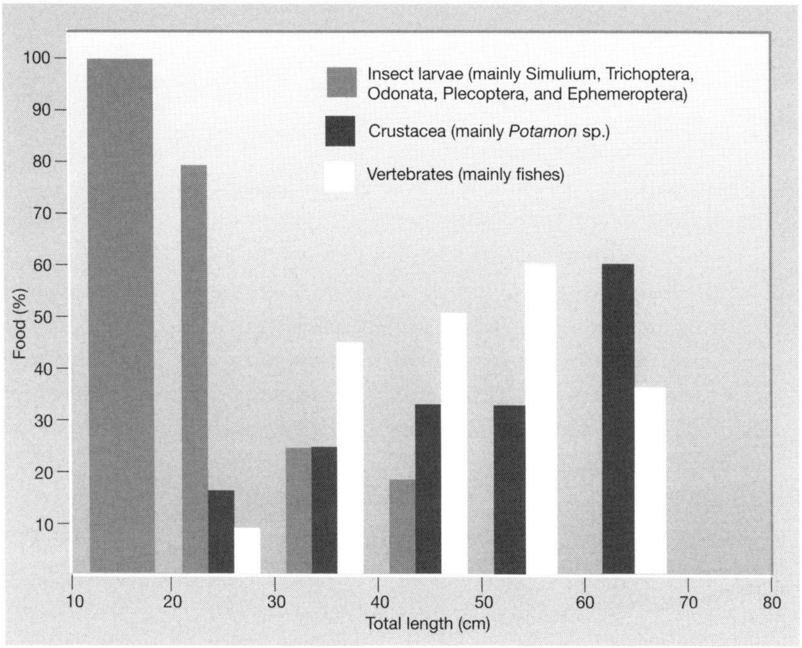

Fig. 3.19 Percentages of various prey animals found in South African eels of various sizes (after Jubb, 1961)

In the aforementioned English streams, the eels that had fishes in the stomach likewise seemed to be the larger ones (Sinha and Jones, 1967c). They had a mean length of 31 cm. Eels having insect larvae and worms in the digestive tract averaged only 21–30 cm. Only those groups of eels having molluscs and to some extent also crustaceans (*Gammarus* sp. and *Asellus* sp.) had more or less greater mean lengths.

Studies by Moriarty (1973b) on the Northern Irish Lough Gill showed the following percentages of eels in the various length groups that ate fishes:

- 30–40 cm: 7%;
- 40–50 cm: 19%;
- 50–72 cm: 71%.

Regarding caddis (Trichoptera) larvae, the proportions were as follows:

- 30–40 cm: 63%;
- 40–50 cm: 44%;
- 50–72 cm: 14%.

There was a similarly rising proportion of fish in the stomach with increasing eel length in the southern Irish lakes of the Corrib system (Moriarty, 1972). Prey that decreased with increasing eel length were trichopterans and, as in the aforementioned studies, *Gammarus* sp., as well as chironomids, ephemeropterans, and *Asellus* sp. Gastropods were preferred by the medium-sized eels.

Eels (*A. rostrata*) from the Bermuda Islands that ate fish had a mean length of 42 cm. In contrast, those having mosquito larvae in the stomach were only 23-cm long (Boetius and

Boetius, 1967b). Also American eels from the continent showed a pronounced relationship of body length to food choice (Ogden, 1970). Eels <40 cm caught by electrofishing in streams had eaten primarily mayfly larvae, and caddisflies, whereas larger eels had fed mainly on fish and crayfish.

De Nie (1987) showed that the size of fish in stomachs of eels increased as body length of the latter increased. He also found that as smelt size increased in the course of the summer, the size of smelt in eel stomachs rose correspondingly until, finally at summer's end the sizes in stomachs (50 mm) became less than those in the lake (55 mm).

Eels of substantially larger than normal size could be studied in New Zealand because the kind occurring there, *A. dieffenbachii*, usually reaches greater mean lengths than most other eels. Consideration of the amount of trout eaten by this New Zealand species showed that on reaching 1 m in length, it was converting to predatory feeding (Table 3.5; Burnet, 1952b), partially already at lower lengths. Further studies by Burnet (1952b) and earlier ones by Cairns (1942b) showed that, especially when at smaller body size, this species and the other New Zealand eel, *A. australis*, which does not become as big, are also able to sustain themselves on lesser animals of the benthos (in this regard, see also Table 3.5). Scattered stomach analyses were also undertaken on larger eels (50–80 cm) from the edge of the European geographic distribution (Volf and Smísek, 1955; Drapkin, 1964; Sedlár and Krcmárik, 1967; Morovic, 1970). These eels had likewise shown the feeding modes of typical predatory fish.

Besides fish and crabs, certain snail species are also eaten only by larger eels.

According to Polish studies in a lake, eels <20 cm ate no snails; midge larvae (*Chironomidae* sp.) composed the greatest portion, 46%, of stomach content in this length range (Opuszynski and Leszczynski, 1967). Molluscs made up an insignificant 14% of stomach content in eels of 20–30 cm, but 64% in eels of 30–40 cm, and 82% in those >40 cm. The

Table 3.5 Percentages of the New Zealand eel species *A. dieffenbachii* and *A. australis* of various size groups feeding on various kinds of prey (calculated from data of Cairns, 1942b).

	<40 cm	40–75 cm		>75 cm	
	A. dieff.	*A. dieff.*	*A.austr.*	*A. dieff*	*A.austr.*
Number of eels having stomach contents	145	2178	750	426	78
Prey					
Fishes		13	32	42	–
Mollusca	10	23	48	8	41
Crustacea	38	11	24	12	19
Ephemeroptera	11	39	–	7	–
Trichoptera	12	3	22	6	47
Orthoptera		3	7	–	–
Hemiptera		2	7	–	–
Diptera		11	27	–	–
Coleoptera	15	6	15	6	–
Odonata		–	1	–	–
Arachnida		–	<1	–	–
Various caterpillars		–	1	–	–
Oligochaeta	21	17	37	4	21
Various organisms		5	–	3	13

snail *Bithynia tentaculata*, which reaches a height of 12 mm, was the most numerous mollusc in eels of 20–40 cm. Those >40 cm preferred larger snails, above all *Radix limnosa*. In a shallow Dutch lake, more of the eels >25 cm ate mussels (*Anodonta* spp. and *Dreissena* sp.) than, on average, did all eels (de Nie, 1987). All told, molluscs trailed insect larvae in importance as eel food.

3.3.1.4 Food choice and head width

Information about the influence of head shape on food choice can be obtained only by exact studies involving measurements of body proportions. Studies of broad- and narrow-headed eels yielded the following differences (Törlitz, 1922):

- Distance between the eyes in relation to total length: 20% and 12%, respectively;
- Head width in relation to total length: 22–10%.

Of 16 eels defined as broad headed from inland water bodies, 13 had fish in the stomach and intestine, but among 19 narrow-headed eels, only three contained fish. This difference sounds convincing, if one regards as insignificant the somewhat greater 70 cm total length of the broad-headed eels, as opposed to 64 cm for the narrow-headed ones.

In Dutch lakes, consumers of chironomid larvae had narrower heads than those that ate fish (Lammens and Visser, 1989). In another Dutch lake, the broad-headed eels likewise presumably tended more to be piscivorous than consumers of small animals (de Nie, 1987). In lakes of Mecklenburg/Brandenburg, narrow-headed eels predominated where benthic invertebrate populations were dense, and broad-headed eels where benthic faunal density was less (Anwand and Valentin, 1981a). Micheler (1967) found relatively small differences regarding the intake of fish by narrow- and broad-headed eels: the former contained 42%, the latter up to 58% fish as food. Also in coastal water bodies (Table 3.6) broad-headed eels ate significantly more fish than narrow-headed eels did. Worms, in contrast, turned up more in narrow-headed eels' stomachs than in broad-headeds'. Thus, differences resembling those between smaller and larger eels became apparent (Table 3.4). On the other hand, no differences in preference for crustaceans were found between the various head shapes.

Table 3.6 Percentages of broad- and narrow-headed eels that ate various kinds of prey in Norwegian coastal waters (after Sivertsen, 1938).

	Broad-headed	Transitional forms	Narrow-headed
Number of eels examined	76	93	94
Fishes (Gobiidae, *Pholis, Gasterosteus, Aphya, Zoarces, Ctenolabrus, Labrus*)	32	16	11
Snails and mussels (*Rissoa, Nassa, Littorina, Mytilus, Modiolus*)	6	12	9
Crustaceans (*Carcinus, Portunus*, amphipods, isopods)	55	63	56
Worms (*Arenicola, Aphrodite*, Maldanidae, Lycoridae, Priapulidae)	6	9	22

In the previously discussed study results, the eels considered were primarily >50 cm. But because smaller eels can already be broad or narrow headed, the matter of whether dietary differences also exist at smaller sizes is of special interest. Poles studied such for small eels from various lakes (Opuszynski and Leszczynski, 1967). A comparison of dietary composition of the eels in Gardno Lake showed that broad-headed eels ate more snails by far than did narrow-headed ones. Micheler (1967) made similar observations on larger eels as well. Perhaps the high proportion of chironomid larvae, which indeed are already coped with by glass eels, is an indicator of narrow headedness in smaller individuals. Just as was evident in the discussion of food selection in different-sized eels, the proportion of molluscs in the diet in Gardno Lake rises as eel size increases. This can serve as confirmation that ingestion of snails and mussels requires a more capable jaw mechanism, which is made possible by larger body size, as well as by a broader head and correspondingly stronger head musclature (Törlitz, 1922).

Stronger head musclature is needed to break the snail shells open. After that, the shell remnants are for the most part spat out (Dröscher, 1897). If beyond that, one compares food organisms and head shape between eels that occur in fresh and sea water, the following picture emerges. Whereas in fresh water narrow-headed eels consume mainly insect larvae, in sea water they eat primarily worms (polychaetes). However, certain worms (*Tubifex*) are also eaten in fresh water if wastewater inflow makes insect larvae unavailable (Ladiges, 1936).

In a study on the development of broad- and narrow-headed eels, Lammens and Visser (1989) favoured the conclusion that width of mouth did not influence the choice of food primarily, but rather that the food that was available stimulated the development of mouth width. But a change in supply of small or large food organisms could result primarily in a decrease in coefficient of condition rather than in an alteration of mouth width (de Nie, 1987).

3.3.1.5 Diet and food availability

As the statements of the previous sections have shown, there are few benthic taxa that eels do not eat. The dietary spectrum of this fish species stretches from fish through mussels, crustaceans, insect larvae, adult insects of the water surface, and worms, on to plants. But plant remnants in the digestive tract of eels represent coincidentally ingested parts, on which the desired prey happened to be. The same applies to hydrozoans and bryozoans. That the eel does not limit its food choice to the aquatic fauna, but rather also takes anything nutritious that falls into or is inundated by the water, for example airborne insects and terrestrial worms, is a widespread phenomenon in fresh, as well as sea water (Tables 3.3 and 3.5; Heincke, 1894; Cairns, 1942b; Sinha and Jones, 1967c; Daniel, 1968; Ziepke, 1974).

A further question is whether animal corpses, other than bait organisms, are eaten. Most authors deny that. As already indicated, this is confirmed when eels are artificially fed. No eel touches food remains that have lain in water overnight. Eel bait on hooks and in traps must be replaced daily, otherwise the catch rate of eels declines markedly (Section 5.3; Schiemenz, 1910).

The enormous dietary breadth of eels reflects their great adapability with respect to nearly all conditions of water bodies. This characteristic is especially striking when two adjacent coastal areas that have different benthic conditions are compared in two German Baltic Sea bays, namely the Wismar Bay (which in its innermost part had lush plant growth at the time of the study) and the Wohlenburg Wiek (which for the most part had a bed of unvegetated mud at the eel fishing sites) (Table 3.7; Dröscher, 1897). Of 20 eels sampled in Wismar Bay, 15 had eaten fish, especially sticklebacks and sea gudgeon. Four eels had ingested isopods. In the plant-poor Wohlenburg Wiek, on the other hand, the food of the 42 eels examined consisted for the most part of mussels, and 88% of them contained *Macoma baltica*, which 60% of the eels had eaten exclusively. Further prey were likewise mussels for example, *Scrobicularia* sp., *Mytilus edulis* and *Mya arenaria*. Finally, some eels contained prawn (*Crangon crangon*). Snails and fish were completely absent.

Quite striking differences in the freshwater diet of eels resulted when the Polish authors Opuszynski and Leszczynski (1967) made a comparison in flowing and standing waters. The eels in the stream ate mostly trichopterans and ephemeropterans, whereas the eels in the lake preferred snails and, above all consumed larvae of midges. Similar characteristic differences were found in stomach analyses from a Northern Irish system of water bodies (Moriarty, 1973b); eels living in the lake had chironomids, and eels occurring in the river had trichopterans along with fish and isopods (*Asellus* sp.). Predominance of trichopterans and ephemeropterans was also observed in English streams. Besides that, these English eels contained largely dipteran larvae (Table 3.3), among which chironomid larvae and *Simulium* were strongly represented. Also the two New Zealand eel species, samples of which were available mainly from streams and rivers, insofar as they had not eaten crustaceans and fish, fed a great deal on trichopterans and ephemeropterans (Table 3.5; Burnet, 1952b). Snails, especially *Potamopyrgus* sp., also made up a significant part of the diet. In contrast, a lake study revealed large amounts of mosquito larvae, as is found for *A. anguilla*, and in flooded areas worms (Cairns, 1942b).

Also, different habitats within a lake result in different stomach contents of the eels there. In Lake Balaton, Hungary, eels of the littoral zone ate mainly *Asellus aquaticus* or other crustaceans like *Corophium curvispinum* and other scuds (Amphipoda). In the benthos of open water, the red midge larva, *Chironomus plumosus*, dominated numerically, and fishes, such as various cyprinids and percids, composed most of the volume (Biró, 1974).

Certain dietary differences were found between the two New Zealand eel species. Whether these differences are attributable to food selectivity or to habitat is questionable. Cairns (1942b) wrote that despite occurrence of *A. australis* in the upper regions of streams, ephemeropterans were not found in the stomachs, even though these areas no doubt harbour mayflies. The additional differences that Table 3.5 reveals between the two species can just as well be attributed to differential preference for certain types of water bodies. This applies primarily to *A. australis*, which occurs mainly in the coastal area or in lakes and spring streams (Burnet, 1968). Perhaps the strong proportion of molluscs and crustaceans results from that. A very large share, especially of short-finned eels >75 cm, had caddisfly larvae (Trichoptera), which, as already established, are strongly represented in all eels from streams.

Table 3.7 Eel diets in different environments (unless otherwise indicated, the species is *A. anguilla*) The various organisms are listed in order of frequency. Where specifically indicated, the percentage of the total wet weight or volume of all the prey is given, otherwise the percentage of eels which had fed on the organism in question is presented. Other localities have been described in Tables 3.3–3.6 and Fig. 3.19.

Sea water

Norwegian coast (Sivertsen, 1938)	(%)	North Sea, rocky littoral at Heligoland (Tesch, unpublished)	Wet weight (%)
Crabs (*Carcinus, Portunus*)	38	Crabs (*Carcinus*)	48
Fishes (*Gobius, Pholis, Gasterosteus, Aphya, Zoarces, Ctenolabrus, Labrus*)	21	Bristle worms (*Nereis* spec.)	35
Crustacea (*Gammarus, Corophium*)	19	Sandhoppers (*Jassa pulchella*)	8
Worms (*Arenicola, Aphrodite*, Maldanidae, Lycoridae, Priapulidae)	15	Fishes (Gobiidae, 1 eel)	8
		Bryozoans	<1
		Algae	<1
		Mussels and snails	<1

Brackish water

Baltic Sea coast, open sea bay Little vegetation (Dröscher, 1897)	Baltic Sea coast, open sea bay Much vegetation (Dröscher, 1897)	East coast, USA, Chesapeake Bay (*A. rostrata*) (Wenner and Musick, 1975)
Mussels (*Macoma baltica*)	Fishes (*Pungitius pungitius, Pungitius aculeatus, Gobius niger*)	Crabs (*Callinectes sapidus*)
Mussels (*Scrobicularia, Mytilus, Mya*)	Sea lice (Isopoda)	Mussels (*Mya arenaria, Mulinia lateralis, Macoma* spec.)
Midge larvae (*Chironomus*)	Mussels (*Cardium, Theodoxus*)	Bristle worms (Polychaetes) Small crustaceans (Amphipoda, Isopoda)
Shrimps (*Crangon crangon*)	Frogs	Fishes (*Alosa pseudoharengus*)
		Insects

Fresh water

Lake in Mecklenburg (Dröscher, 1987)	Lake in England (Frost, 1946)	Eutrophic lake in Holland (De Nie, 1987)	Lake in Ireland (Moriarty, 1969)
Snails (*Limnaea, Theodoxus*)	Snails and mussels (Sphaeridae, *Valvata, Limnaea*)	Chironomidae (*Einfeldia carbonaria, Chironomus plumosus, Glyptotendipes pallens, Polypedilum nubeculosum*)	Midge larvae and pupae
Caddis fly larvae (Trichoptera)	Insect larvae (Trichoptera, *Sialis*, Ephemeroptera, Chironomidae)	Fishes (*Osmerus eperlanus, Perca fluviatilis, Stizostedion lucioperca*)	Other insect larvae
Fishes (*Pygosteus pungitius*)	Scuds (Gammaridae)	Scuds (*Gammarus tigrinus*)	Aquatic sow bugs (*Asellus* spec.)
Midge larvae (*Chrionomus*)	Worms (Oligochaeta)	Molluscs (*Anodonta, Dreissena, Polymorpha, Sphaerium* spec.)	Fishes (*Perca fluviatilis*)
Dragonfly larvae	Fishes (*Perca*)	Various other organisms:	
		Oligochaetes, Hirudinea, Hydracarina	
		Copepoda, Malacostraca, Ephemeroptera, Trichoptera, Coleoptera	

River mouths, partial brackish water influence			
Elbe tidal reaches Hamburg harbour (Ladiges, 1936)		River at Calcutta (Pantulu, 1956)	
	Volume (%)		Volume (%)
Tubifex	55	Fish (*Hilsa, Septipinna, Pama,* few glass eels)	40
Cladocera	20	Crabs	26
Chironomidae	4	Shrimps (*Peneidae*)	20
Mysis	2	Insects	4
Mittencrabs	1	Crab megalopa larvae (*Varuna* spp.)	3
Earthworms	1	Plants (micro- and macrophytes)	2
Mussels and snails	1	Worms (Annelidae)	2
Leeches	1		
Gammarus	1		
Leander			

North Sea coast, diked waters (Daniel, 1968)		Bermuda (*A. rostrata*) (Boëtius and Boëtius, 1967b)	Kattegat, lagoon-like waters (Muus, 1967)
	Wet weight (%)		
Shrimp (*Crangon crangon*)	26	Fishes (*Fundulus bermudae*)	Fishes (*Gasterosteus* and *Pungitius*)
Fish spawn (*Gasterosteus aculeatus*)	19	Midge pupae (Chironomini)	Crabs (small *Carcinus maenas*)
Crabs (*Carcinus maenas*)	12	Snails (*Physa*)	Mussel fragments (*Mya*)
Earthworms	12		
Bristle worms (*Nereis*)	9		
Fishes (*Gasterosteus aculeatus*)	4		
Snails (*Radix ovata*)	2		
Midge larvae (Chironomidae)	1		
Fishes (*Pungitius pungitius*)	1		

Diked waters, North sea coast (Daniel, 1968)		Stream in England (Thomas, 1962)		Streams in North America (*A. rostrata*) (Godfrey, 1957)	Lake in southern Germany (Micheler, 1967)	
	Wet weight %		Wet weight %			Wet weight %
Earthworms	40	Mayflies (Ephemeroptera)	70	Ephemeroptera	Bony fishes	43
Fishes (*Pungitius, pungitinus*)	37	Caddisflies (Trichoptera)	45	Other insect larvae (Plecoptera, Chironomidae, Simulidae, Odonata, etc.)	Plant fragments	28
Fish eggs (Gasterosteidae)	12	Midge larvae (Diptera)	35	Fisches (*Rhinichthys* sp., Gasterosteidae, Cottidae, Salmon fry)	Dipterans	23
Insect larvae	3	Stonefly larvae (Plecoptera)	30		Gastropods	23
Leeches	2	Fishes (*Phoxinus, Lampetra*)	25		Annelids	14
Scuds (*Gammarus*)	1	Leeches (Hirudinea)	25		Algae	14
Snails (*Radix ovata*)	1	Mussels, Snails (Mollusca)	20		Detritus	11
Plant fragments	1	Aquatic sow bugs, Scuds (*Asellus, Gammarus*)	20		Trichopterans	10
Beetles (*Agabus lybius*)	<1	Alderflies (*Sialis*)	5		Tabanid larvae	7
Water spiders (*Argyroneta*)	<1	Beetles (Coleoptera)	5			

3.3.1.6 Prey selection

Finally, the question arises whether the eel eats indiscriminately whatever is available within reach, provided that it suits the eel's size, head shape or species. Regarding this, a special study from the brackish water zone is available (Daniel, 1968). It was observed that many eels in the area of the Eider estuary, Germany had each eaten a large number of individuals of the same species (monophagy), which was observed similarly in fresh water by other authors, as well (Dröscher, 1897; Ladiges, 1936; Cairns, 1942b; Pantulu, 1956). Of 145 'medium' eels in the Eider, nearly 60% were monophagous. Out of 96 'small' eels, 82% had this characteristic. In New Zealand studies of *A. dieffenbachii*, it was found that if a stomach held up to 10 animals at once, there was a great tendency for them to be of several species (polyphagy). However, beyond 10 to 30 prey in the stomach, monophagy increased, such that at about 1000 animals per stomach only a single species had been eaten (Burnet, 1952b). In the present author's studies of Heligoland eels from the rocky littoral, this was not confirmed (Table 3.4). The only organism that was eaten there in large number was the amphipod *Jassa pulchella*, but even when over 400 of them occurred in one eel's stomach, two further prey species were found. Of nine eels that had over ten *Jassa pulchella*, only three had eaten this organism exclusively. Most of the rest of the eels had at least two kinds of prey, seven having more than four species in the gut. Thus, monophagy is not always strongly pronounced, especially not where such diverse food is available as at Heligoland.

The problem of dietary uniformity or diversity plays a certain role with regard to biological production, and naturally this is primarily so when food competitors are present (see below). Whether eels use benthic food in proportion to its quantitative occurrence was studied in New Zealand streams (Cairns, 1942b; Burnet, 1952b). It was evident that a certain parallelism between the quantity of benthic animals and composition of the stomach contents exists. This applied to a special degree to the ephemeropterans. Other animals, for example trichopterans and crustaceans were preferred, that is, their proportion in eel stomachs was greater than in the benthos. In contrast, beetles (coleopterans) were spurned, which can also be observed in other fishes. American eels in streams also showed a certain parallelism between abundance of bottom fauna and stomach contents (Ogden, 1970). Stomach contents of eels from a lake in southern Ireland likewise showed compositions resembling those of the bottom fauna (Moriarty, 1972), for example chironomids, ephemeropterans, corixids, *Gammarus* sp., *Bithynia* sp.; in contrast, the stomachs contained isopods (*Asellus* sp.) in far higher amount than expected from the bottom samples. No eel ate the backswimmer, *Notonecta* sp. That the snail, *Bithynia* sp., was most extensively spurned contrasts with other studies (Table 3.7; Dröscher, 1897; Frost, 1946).

From Table 3.7 and with inclusion of the results cited in the present section, the following organisms are significant for the eel, although position in the list does not represent rank:

- Fishes, especially small, bottom-dwelling kinds, such as sticklebacks (in nine cases), gobies (Gobiidae; in three cases), sculpins (Cottidae; Frost, 1946), gudgeon (*Gobio gobio*), loaches (Cobitidae), conspecific eels and lampreys (Frost, 1946; Perlmutter, 1951; Thomas, 1962). Salmonids appeared disproportionately often, as they were a special research objective (Cairns, 1942b; Frost, 1952; Burnet, 1952b; Godfrey, 1957; Thomas, 1962; Sinha and Jones, 1967c; Köthke, 1968).

- Snails and mussels (Dröscher, 1897; de Nie, 1987).
- Crustaceans (Decapoda), such as crabs, prawns and such edible crustaceans as the brook crayfish *Astacus astacus* (Walter, 1910; Svärdson, 1969; *Fisch und Fang* **7**, 264, 1964; Runge, 1965) and the mitten crab (Jacob, 1970);
- Small crustaceans, especially scuds (Amphipoda; regarding *Gammarus tigrinus*, see also Tesch and Fries, 1963; Fries and Tesch, 1965a), isopods (Clarke *et al.*, 1993), bait prawns (Mysidacae), water fleas (*Daphnia*; Ladiges, 1936), the megalopa stage of crabs (Pantulu, 1956);
- Worms (Opuszynski and Leszczynski, 1967), especially tubificids, earthworms and in a special case the spiny-headed worm, *Echiurus echiurus* (Kühl, 1965);
- Insect larvae (Lammens and Visser, 1989).

3.3.2 Age and growth

3.3.2.1 Methodological problems of age determination

The following section does not present the methods for determining eel age. For that there are special publications that deal with techniques of such fish studies and, in general, that apply to the eel (e.g. Chugunova, 1963; Bagenal and Tesch, 1978). But the genus *Anguilla* poses a few special problems for the investigator, which must be mentioned.

It is unusual that the eel's life cycle has two distinct phases:

- the oceanic larval period;
- the continental period, the latter often incorrectly termed the 'freshwater phase'.

The duration of the larval period varies among eel species. For example, until recently it was assumed at 2.5 years for *A. anguilla*, and at 0.5 years for *A. rostrata* (Section 2.2). For the New Zealand species, *A. dieffenbachii*, otolith studies showed 2 years (Cairns, 1942a). For the Indian eel, *A. nebulosa nebulosa*, the central part of the otolith, formed during oceanic life, showed an apparent period of 1.5 years (Pantulu and Singh, 1962). However, determining annuli or daily circuli is difficult or even impossible for the European eel, which may have the longest larval life of any eel species (Antunes and Tesch, 1997b), but possible for the Japanese eel (Tzeng, 1990), and conceivable for the American eel, as is evident by comparison with other methods (Tesch, 1998). In age determination, 'seawater rings' could be included in the count only if they were interpreted as annuli. There is no evidence that this is valid. Furthermore, differing length of larval life among the various species hampers comparison, so it is advisable to count the years of life starting from the glass eel stage, as most authors do.

Counting the years from the glass eel stage onward was also not done uniformly (Thurow, 1959; Sinha and Jones, 1967c). In many cases, the age-group data from various authors are not strictly comparable. In addition, there are variations caused by differing interpretation of annuli (Holmgren, 1996a). Grinding and illuminating the otoliths does not always improve age determination. Burning the otoliths can lead to better identification of annuli (Möller-Christensen, 1964; Moriarty, 1973a; Jellyman, 1979b). This method was even given certain preference in an international working group (Vøllestad *et al.*, 1988). The method of measuring light transmissivity of otolith slices cut as thin as 0.2 mm, and of graphically displaying the values from centre to edge of the otolith met with little approval. Although it yielded well recognisable marks for the possible annuli (Deelder, 1978), confirming these values by measuring growth of known-aged eels is advisable. Likewise, contour analyses of eel otoliths could not yet gain acceptance, despite having the advantage over other methods in terms of low time consumption (Doering and Ludwig, 1990). Of course, other hard parts of the body, such as vertebrae are also usable for age determination (Ehrenbaum and Marukawa, 1914). However, nothing else reflects the early stages, including the larval period, as completely as otoliths do. Circulus structure on otoliths of tropical species, for example *A. nebulosa* in India, is also to be judged as a periodically conditioned marking. The extent to which it harmonises with the timing of annulus formation in temperate zone eel species is not known. Analysing eel scales for age and growth studies is difficult and inadvisable because the scales are formed first at a relatively large body size (Section 1.3.2). In that respect, still other differences exist between species (Pantulu and Singh, 1962):

- For Middle-European eels, various authors determined in most cases a deficit of 3 years in the annuli count (Ehrenbaum and Marukawa, 1914; Nordquist and Alm, 1920; Rasmussen, 1952; Rahn, 1955a);
- 2 years in Denmark (Gemzøe, 1906);
- 2–3 years in Masury, Poland (Opuszynski, 1965);
- 4 years in Germany and England (Marcus, 1919);
- 3 years for *A. rostrata* in Canadian waters (Smith and Saunders, 1955);
- 5 years for the New Zealand species (Cairns, 1942a);
- 0.5 years for the Indian eel, *A. bicolor* (Pantulu, 1956).

Finally, as Rasmussen (1952) explained on the basis of several authors' data, the difference between the number of otolith and scale annuli becomes greater as age increases, reaching >5 years at age 8. Moreover, the annuli count varies by up to ±2 years among different scales within the same area of an eel's body. For Indian eels (*A. bioclor*) that were raised from a length of 10 cm in aquaria, the more growth in length they underwent in a given period, the more annuli formed on the scales (Nair and Dorairaj, 1975). All these problems make age determination seem usable only in exceptional cases, especially because back-calculation of length for early years of life is also impossible.

For the visual identification of annuli on otoliths there are validations. Determinations made on known-aged eels from ponds, the otoliths of which were studied by examining surface structure, were quite consistent with actuality (Smith, 1968). Otolith annuli of American glass eels stocked in ponds showed good agreement with the actually determined growth of the fish after as much as 3 years' growth (Liew, 1974). After recapture, rather durably marked eels from the North Sea (Fig. 3.20) confirmed relatively slow growth that otolith studies had determined for the German part of that coast (Peñaz and Tesch, 1970; Burnet, 1969b). According to the otolith studies, individuals of about 40–50 cm grow 2–3 cm annually. On recapture, the marked eels evidenced the comparable value of barely 2 cm growth in length (Fig. 3.20). Likewise, eels from Lake Ontario showed 3.1 cm annual growth, according to otolith determinations at lengths between 78 and

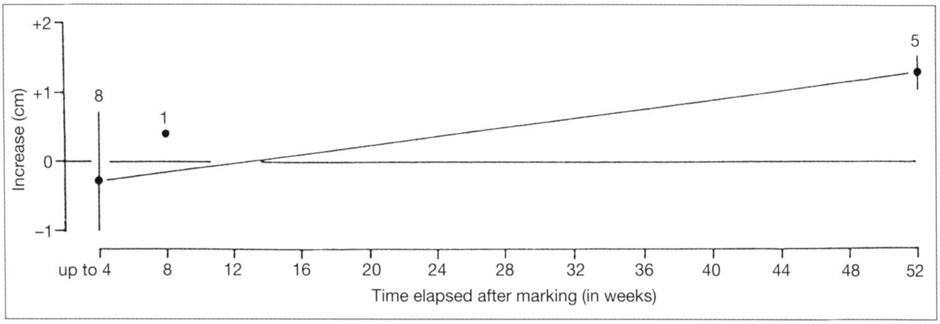

Fig. 3.20 A year's growth in length for Heligoland eels having internal stainless steel tags (after Tesch, unpubl., and, involving similar methods, after Lindroth, 1953)

- Vertical bars: range as determined by post-recapture measurements in the laboratory
- Numbers next to points or vertical bars: number of eels measured. They had been released at lengths of 32–45 cm. Some were recaptured 1–8 weeks later, and the rest, recaptured after 52 weeks, ranged from 47–51 cm.

91 cm. Markings of 73–87 cm eels revealed a very similar annual length increment of 3.3 cm (Hurley, 1972). Chisnall and Kalish (1993) achieved multiple validation of the annuli on otoliths of both New Zealand eel species by means of internal marks, external colour marking and fluorescent marking of the otoliths. *A. dieffenbachii*, the larger of the two species, grew 65 mm per annum, while *A. australis* grew 29 mm.

Back-calculations on otoliths are, contrary to other opinions (Rahn, 1955a), definitely practicable, as biometric studies have shown (Matsui, 1952; Pantulu and Singh, 1962; Peňáz and Tesch, 1970; Löwenberg, 1979; Lee, 1979; Nagieç and Bahsawy, 1990). However, the reservation must be made here that under certain conditions growth of body length and otoliths does not proceed proportionally, which can lead to mistakes in back-calculation (Holmgren, 1996a; Holmgren and Wickström, 1996). Under normal conditions the deviation from true growth does not exceed ±15%.

3.3.2.2 Growth differences between males and females

As was discussed in the section on sexual development, eels that came from overpopulated habitats and had slow growth during the first years of their continental residence tended to become males. It could, therefore, be expected that also, within one water body, the males grow slower than the females. A number of study results seem to confirm this (Ehrenbaum and Marukawa, 1914; Gandolfi-Hornyold, 1921, 1930; Marcus, 1919; Tesch, 1928; Matsui, 1952; Thurow, 1959; Pantulu and Singh, 1962; Sinha and Jones, 1967c; Peňáz and Tesch, 1970; Ask *et al.*, 1971). At the places studied in Europe and Japan, the females were as much as 10 cm longer than the males from the 4th year of life onwards. However, there are also artificial rearing results and age studies according to which no differential growth or even faster growth of males was found (Cairns, 1942a; Egusa and Hirose, 1973; Meske and Cellarius, 1973; Kuhlmann, 1975; Section 1.7.2). In addition, Holmgren and Mosegaard (1996a, b) found in comprehensive rearing studies that at a weight of 50–60 g males grew faster than females. In contrast, at 80–100 g, they gained weight slower than the females and almost stopped growing as they approached 150 g.

Table 3.8 Lengths and ages of male and female silver eels (*A. anguilla*, if not otherwise noted).

		♂♂				♀♀					
Water body	Year	Mean length (cm)	Length range (cm)	Number	Age	Mean length (cm)	Length range (cm)	Number	Age	Comments	Authors
Lake Esrum, DK	1937–8	–	–	–	–	57.8	49–67	85	11.5	–	Rasmussen, 1952
Lake Esrum, DK	1940–2		–	–	–	56.7	52–70	91	11.4	–	Rasmussen, 1952
Lake Esrum, DK	1943–5		–	–	–	56.5	46–79	82	11.1	–	Rasmussen, 1952
Lake Esrum, DK	1946–8	–	–	–	–	55.3	49–73	85	11.8	–	Rasmussen, 1952
Lake Esrum, DK	1949–50	–	–	–	–	55.7	49–79	83	11.4	–	Rasmussen, 1952
Windermere, E	1940–50	40.0	–	9	–	60.8	47–97	240	12.5	–	Frost, 1945, 1961
Cunsey Beck, E	1939–44	*c.* 40.0	37–46	–	–	*c.* 57.0	46–88	–	11	Number:* 1122	Frost, 1945, 1961
River Bann, NI	1944	38.5	34–44	159	7.3	–		–	–	–	Frost, 1950
River Bann, NI	1965	41.3	35–45	126	–	54.4	45–79	1531	–	–	Ministry Agr. NI, 1966
River Bann, NI	1966	41.0	33–46	157	–	55.3	46–78	813	–	–	Ministry Agr. NI, 1966
IJsselmeer, NL	1920	35.0	31–38	60	–	–	–	–	–	–	Tesch, J.J., 1928
IJsselmeer, NL	1956	36.0	29–43	590	9	–	–	–	–	–	Deelder, 1957b
River Dieze, NL	1920	–	–	86	8	–	–	–	–	–	Tesch, J.J., 1928
Danish coast	1906	*c.* 41.0	36–48	51	7	56.0	45–85	111	8	1 year older?	Gemzøe, 1906
LakeVrana, Y		37.2	33–39	30	6.7	–	–	–	–	♀ not representative	Haempel and Neresheimer, 1914
Comacchio, I[†]		37.2	31–44	190	<6?	–	–	–	–	Age 4 (Scales)	Gandolfi-Hornyold, 1934
North Adria, I[†]	*c.* 1973	43.2	36–54	198	4.6	59.6	39–100	1747	6.5	see also 1934	Rossi and Colombo, 1979

Valencia Lagoon S		*c.* 37.0	31–46	90	6.5	*c.* 55.0	46–100	44	9	–	Gandolfi-Hornyold, 1921
Arcachon Reservoir, F	1977/8	37.5	31–43	1089	3.9	49.8	43–83	1249	5.9		Lee, 1979
Tunis Lake	1930	36.0	33–44	347	7	–	38–101	249	–	♀ not representative	Gandolfi-Hornyold, 1930
Fresh and brackish water NF	1967/8	–	–	–	–	69.4	54–93	ca 300	12.3	*A. rostrata*	Gray and Andrews, 1971
River Makara, NZ	1971–3	46.7	38–60	65	14.2	73.7	56–93	34	19.4	*A. australis*	Todd, 1980
River Makara, NZ	1971–3	62.3	–	162	23.2	106.3	78–133	22	34.3	*A. dieffenbachii*	Todd, 1980
Ellesmere,[†] NZ	1974–80	43.2	34–55	12020	14.4	60.9	48–102	609	23.6	*A. australis*	Todd, 1980
Ellesmere,[†] NZ	1974–80	66.6	56–74	12	–	115.6	74–156	176	49	*A. dieffenbachii*	Todd, 1980

DK Denmark
E England
F France
I Italy
Y Jugoslavia
NF Newfoundland
NI Northern Ireland
NL Netherlands
NZ New Zealand
S Spain
* Total of males and females, combined
† Lagoon
♂ Male
♀ Female

In a 339 ha lake on the Isle of Gotland, artificially reared glass eels were monitored until their emigration as silver eels (Holmgren *et al.*, 1997). On average, the males grew 73.6 mm/a, the females 54.0 mm/a, even though the mean length of the females, because they emigrated at a greater age, was more than that of the males, and for both sexes annual growth rate decreased with increasing age. The change in growth rate just mentioned also led to crowding of otolith annuli together, especially near the edge for females that were becoming older. This phenomenon surely led, also in other studies, to more underestimation of age for females than for males. More recent results, as well as experiences from aquaculture (Section 6) also led to speculation that sex differentiation of the eel (Section 1.7.2) is a matter of early growth rate, not of socio-ecologic conditions: rapid growth after the glass eel stage, not overpopulation in the estuary, leads to development of the male sex.

3.3.2.3 Size and age at emigration

The mean total lengths of male European silver eels from different regions and water bodies lies within relatively narrow limits, namely 35–46 cm, with extreme values of 29 and 54 cm (Table 3.8). The mean lengths of the European eel do not deviate substantially from mean lengths, for example, of the short-finned New Zealand eel (*A. australis*) (Table 3.8; Burnet, 1969a), even though the males there reach a maximum length of 60 cm (Todd, 1980). Also Japanese silver eel males have an insignificantly greater mean length of probably 42 cm (Matsui, 1952) and a maximum length of 57 cm. Male *A. dieffenbachii* occupy a special position; they reach a mean length of 66.7 cm (Todd, 1980). The lengths of males appear to have higher values in northern than in central and southern Europe.

For female silver eels in European waters mean values of 50–61 cm were encountered (Apollova, 1969). The mean value for *A. australis* is only a few centimetres above that (Burnet, 1969b). *A. japonica* seems to resemble the European eel with regard to the size advantage of females over the males (Matsui, 1952). American eels averaged 69 cm during spawning migration (Table 3.8). The extreme values of *A. anguilla* females vary downwards from that to 38 cm; above that, lengths can reach as much as 100 cm (Gandolfi-Hornyold, 1930). Eels as long as 130 cm were caught in a Swiss lake. Female European eels of up to 150 cm are supposed to have occurred (Bauch, 1954). Considering all data, *A. dieffenbachii* lies far above that (Table 3.8; Burnet, 1969a). Its mean lengths equal or exceed the maximum lengths of most other species. The maximum lengths amount to about 180 cm at a body weight of 24 kg (Todd, *Freshwater Catch* Autumn **81**, page 5). But also for the South African eel species, the females of which, according to believable statements, can become 18–23 kg, the maximum lengths presumably lie at 180 cm and above (Jubb, 1961).

The mean lengths of a water body's silver eels can vary somewhat from year to year. A declining tendency of the mean total lengths was observed, for instance, for females in Denmark's Esrum Lake. Rasmussen (1952) attributed this to heavy stocking with juvenile eels, which must have raised the population density substantially, and therefore the food competition. The proportion of male eels is believed to have increased at the same time in this lake, but this was not based on any exact knowledge about the eels' sex ratio in this lake, and should not contribute to confirmation of the opinions about the dependency of sex differentiation on population density (Section 1.7.2).

The age difference between *A. anguilla* males, which averaged 6 (2–15) years old, and the females, which averaged 8.7 (4–20) years old, varied between 1 and 6 years (Table 3.8; Vøllestad, 1992). Probably, the females' greater body length at onset of their oceanic migration happens not by faster growth but rather by remaining about 3 years longer on the continent. In populations having smaller emigrating males, the emigrating female silver eels are also smaller; a similar correlation also applies to age of the two sexes. The mean lengths of silver eels do not correlate with age for both sexes, though. The age difference is presumably highest for slowly growing populations, and varies more than does emigration length. In northern and central European waters the males are older than in southern Europe by about 1.5 years. For females the difference is presumably greater yet, if Table 3.8 permits such a comparison. Also, mean length of silver eels increases as their distance from the Sargasso Sea spawning grounds. Presumably, it is not the eel's size but its fat content that determines the time of emigration from the continent (Larsson *et al.*, 1990) because the required energy reserves must first be available.

According to Svedäng *et al.* (1996), the reproductive strategy of the females is based on becoming sexually mature as early as possible. Exprience of Jellyman (2001) with a single population of *A. australis* showed that the onset of maturity of females was linked more closely with length than age and he presented the hypothesis that females adopt a size-maximising growth strategy to ensure maximum fecundity.

Mistakes in age determination of migrating eels or of those in advanced maturity seem to occur through studies of otoliths or scales of those individuals that are delayed in migrating from a water body because outflow is lacking. For that reason, eels stocked in Switzerland's Lake Como 45 years previously had at most 10-scale annuli (Gandolfi-Hornyold, 1932a). Eels that had lived for over 30 years in a clay pit near Brandenburg, and that showed advanced maturity externally, had formed not more than 6–8 annuli (Wundsch, 1953). That individuals of extraordinary age were involved in the Lake Como eel population appeared to be indicated by the fact that, besides a sometimes very low weight, they had extremely irregular scales. In all these cases, an exceptional condition was involved. The eels had become 'silver' and then had no opportunity to emigrate. Thus, further growth presumably could not take place, so that further formation of material on otoliths and scales also was not to be expected.

In addition to delayed formation of the first-scale annulus (Section 3.3.2.1), annulus formation in overaged eels presumably stops sooner on the scales than on the otoliths. This corresponds to an experience with other fish species (e.g. Campana and Neilson, 1985). That the eel can become very old when prevented from emigrating is seen in data published by Walter (1910). One specimen was held in a spring in Denmark for 55 years. It is also noteworthy that an eel was believed to have been raised by a French family in very small containers over a period of 37 years, becoming 1.3–1.4 m long in the process.

3.3.2.4 Head width and growth

As follows from the data on eel diet, broad-headed individuals ingest different prey from that of narrow-headed ones. Broad-headed eels generally eat larger, harder-to-handle animals. Their stronger and larger jaws enable them to do this. It follows then that the two varieties also grow at different rates. According to Table 3.9, the broad-headed eels grow

Table 3.9 Average lengths (cm) of narrow- and broad-headed female eels from Lake Sakrow near Potsdam in various age groups (after Rahn, 1955a).

Age group	V	VI	VII	VIII	IX	X
Narrow-headed	43.7	45.9	50.0	53.0	59.0	61.0
Number	6	26	35	20	10	6
Broad-headed	–	53.0	57.8	62.3	69.9	70.1
Number	–	8	33	19	7	8

significantly faster than the narrow-headed ones (Rahn 1955a). Their advantage increases from 5 cm at 6 years of continental life to about 10 cm in 8–10-year-old individuals. Someone else, who is not inexperienced in age determination, gives an even faster advantage of the broad-headed eel: 2–4 cm per 10 cm of length increase (Bauch, 1954). Other authors likewise give broad-headed eels the better growth potential (Wundsch, 1916; Ehrenbaum, 1930; Micheler, 1967). However, it must also be emphasised at this point that it depends on whether one calculates, for example, mean growth or specific growth during particular periods of life. In the latter case, the males definitely can also grow faster (Section 3.3.2.2). It should also be taken into account that eels presumably adapt individually to their head shape (Lammens and Visser, 1989), so recording head width and from that inferring corresponding body growth of the whole population is of dubious value.

Therefore, such differences should not be generalised, for they take place within a population. The Heligoland eels proved to be extremely broad headed (Tesch, unpublished), which is understandable on the basis of the abundant crustacean diet (Table 3.4). The head width amounted to, for example 36% of head length and had a ratio to total length of 1:21. For narrow-headed eels the ratio is about 1 : 40 (Walter, 1910; Törlitz, 1922). Nevertheless, Heligoland eels grow relatively well only until they reach about 30 cm long, then grow slower than other populations on the coast of the North Sea (Peňáz and Tesch, 1970).

3.3.2.5 Environmental factors and growth

Major factors that influence the growth rate of all individuals in a population are:

- population density;
- water temperature;
- amount of food that is available per unit surface area.

The available amount of food is in turn influenced by:

- age structure of the eel population;
- sex ratio;
- the different diet of different-sized eels.

Unfortunately, growth studies often do not take all of these factors into account. Two Welsh waters, the Ffraw and Rhyd-hir streams presumably offer an example of population density's influence on growth (Fig. 3.21). In the former, the females especially grew significantly faster than the eels of the latter until the 8th year of continental life. The length difference at that age amounted to as much as 4 cm, but, from the 9th year of life and 45 cm length onwards, diminished to almost zero. Population density in the stream that had poor

- River Ffraw: Population density about 1 eel/m^2, proportion of males 50%
- River Rhyd-hir: Population density about 4 eels/m^2, proportion of males 90%

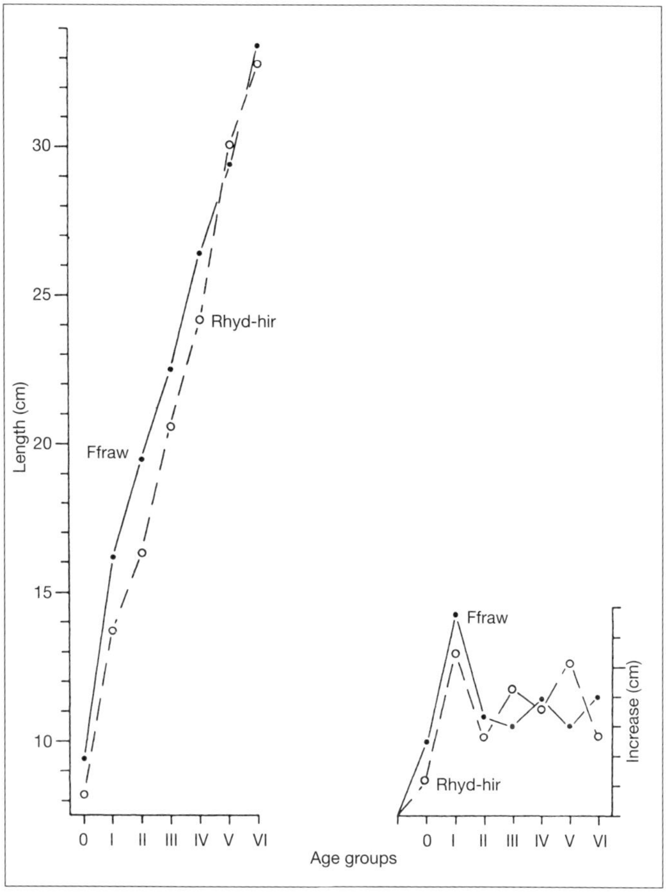

Fig. 3.21 Total length and length increase of male eels in two streams in Wales (data from Sinha and Jones, 1967c)

growth was four eels/m^2 (as converted from the data of Sinha and Jones, 1967c) and was one eel/m^2 in the one that had faster growth. In addition, the faster-growing eels lived in slow-flowing water, which may have given an advantage over the slower growing eels, which lived in faster flowing and, persumably also, colder water. For the larger eels, growth may not be so negatively influenced at higher population density because they feed differently, and in length groups >40 cm the males emigrate, so competition declines sharply from this length onwards. In addition, the proportion of females also declines at higher population density.

In contrast, Rossi *et al.* (1987/8) found no change in growth rate after the supply of eel offspring had declined, but did find a shift in sex ratio. The study did not involve a stream but rather the Lagoon of Comacchio. However, later studies on the same water body yielded a greater mean length for both sexes during emigration (de Leo and Gatto, 1996). As would be expected, a smaller proportion of females was found in the large Northern Irish Lough Neagh after heavier restocking (Parson *et al.*, 1977). A larger proportion of females and a greater mean length of both females and males were observed after eel production declined in the Irish Burrishoole river system (Poole *et al.*, 1990).

Moriarty (1973b) found differences in growth of eels in various waters. Slow growth and small mean sizes were detected in the Northern Irish Loughs Gill and Conn; at age 10 the animals were about 40 cm long. In the Erne river system, in contrast, the eels of this age were 50 cm long. The average size of all eels caught in the Erne area was likewise over 10 cm greater. Moriarty (1973b) could not find differences in the population density based on catch per unit effort. He attributed the small mean lengths in Loughs Gill and Conn to overfishing. It is conspicuous, however, that these two lakes lie just 10 and 25 km from the ocean, respectively. The waters of the Erne river system are over 65 km, and for the most part significantly over 100 km, from the ocean. Therefore, one should assume that a denser population in waters near the ocean is the reason for the poorer growth of the eels caught there. In a comparative study of eel populations in Ireland, Moriarty (1974) emphasises the food relationships of the eels (Section 3.3.1). Lake-dwelling eels that had fish in the stomach grew poorly, but those that had eaten few fish and, therefore, primarily chironomids grew well. This information enables speculation about the possible causes. Lakes that have an abundant population of benthic midge larvae (e.g. *Chironomus plumosus*), are known to be very productive. Thus, the eels have these and other invertebrates abundantly available. If this is not the case, then they switch to piscivory, that is, in infertile waters or under intense food competition. That they grow poorly in this situation is understandable. The studies by Lammens and Visser (1989) indirectly support the validity of this assumption; although they did not study growth rate, they found that eels that switched to eating smelt when midge larvae populations diminished, had lower body weight. But the problem of slowly growing piscivores in the Irish waters mentioned is not to be confused with the matter of narrow-headed eels growing slower than broad-headed eels in the same water body (Section 3.3.2.4; Table 3.9).

Determinations of growth rate of eels in a water body show that the slow growth of eels in natural waters results mainly from competition in a population that is too densely populated. Glass eels stocked in the eel-free Lake Wörther, Austria, weighed 500–800 g, 4 years later (Einsele, 1961); this corresponds to a length of 65–75 cm. No female eel in any natural water body containing a more or less natural eel population reaches a body length of >47 cm after completion of the 4th year of life (Table 3.10).

Glass eels newly introduced into a lake of the Upper Volga river area reached 13.9 cm after the 1st year and 25 cm after the 2nd year (Nikanorov and Smotryaev, 1962). These values show such growth as is in the range indicated in Table 3.10, but not too fast. It is to be considered that this initial stocking with 70–75 glass eels/ha was relatively intensive (Section 4.4.3), so a certain amount of competition existed right from the start. Besides that, the lake lies so far north that, based on low water temperatures alone, one would not expect particularly fast growth.

Newly introduced glass eels in Russian reservoirs grew to 13.1 cm in the 1st year after stocking and to 19.2 cm in the 2nd year (Kokhnenko and Borovik, 1958) which is less than in the previous case, but they were stocked at a density of 200/ ha. At the same time, Kokhnenko and Borovik stocked ponds in the same area with 500 glass eels that reached 22.2 cm in the 1st year and 39.9 cm in the 2nd year, thus grew significantly faster. Certainly, ponds that are warmer in summer and lack competition from other fish species offer better conditions for growth than the relatively deep reservoirs. Kostyuchenko and Prishchepov (1972) confirmed the growth for the later years of life in these lakes, which

Table 3.10 Growth of various non-European eel species compared with that of the European eel. (M: male, F: female, ?: sex unknown).

Species	Water body name and location	Sex	0	I	II	III	IV	V	VI	VII	VIII	IX	X	XI	XII	Author
A. rostrata	Altamaha river, Georgia, USA	M+F				22.7	28.6	32.9	37.5	46.0	54.6	52.5	60.2	70.5		Helfmann *et al.*, 1984
A. rostrata	Ottawa river, Canada	M+F				19.7	21.2	25.5	29.0	27.6	33.8	38.6	33.0			Hurley, 1972
A. rostrata	Lake Ontario	F					31.0		50.9	49.5	53.2	53.2	65.5	78.5	78.7	Hurley, 1972
A. rostrata	St. Lawrence river	F								58.8	61.9	64.8	68.6	69.8	74.6	Larouche *et al.*, 1974
A. rostrata	New Jersey Streams, USA	M+F					ca 17	19	28	34	37	45	45	50	55	Ogden, 1970
A. rostrata	Bermuda	F		ca 23	35	43	47	50								Boëtius and Boëtius, 1967b
A. japonica	Japan	F				27.9	31.6	40.3	45.2	52.9	56.1	65.8				Matsui, 1952
A. australis	New Zealand	?+F	5.7	7.8	9.5	17.5	24.5	31.2	39.5	47.8	ca 57	64	73	84		Cairns, 1942a
A. dieffenbachii	New Zealand	?+F	5.8	8.7	10.8	16.1	20.4	25.2	31.5	39.0	47	54	63	73	82	Cairns, 1942a
A. dieffenbachii	New Zealand	?+F	6.0	8.5	11.0	16.0	19.0	24.0								McFarlane, after Cairns, 1942a
A. nebulosa nebulosa	Calcutta, India	M	5.1	13.0	21.4	26.2	32.6	37.4	43.5							Pantulu and Singh, 1962
A. anguilla	Average for Various waters	?+M+F		9.4	15.3	23.2	26.4	30.0	33.7	38.6	43.9	52.8	54.8	64.0	70.7	Peñáz and Tesch, 1970

was nevertheless fast compared with values in Table 3.10. Based on a large amount of data, they showed that the eels attained the following lengths from the 5th to the 15th freshwater year: 53, (?), 59, 62, 67, (?), 72, (?), 77, (?), and 83 cm.

It is reported from Lake Balaton, Hungary, that eels stocked at a length of about 20 cm, weighing 7–14 g each, were 47–50 cm long after a year and 90 cm after 4 years (Koops, 1967a). The stocking density amounted to nearly one eel/ha, so intraspecific competition was very low. The relative warmth of this water body may have contributed to the unusually fast growth. Thus, extremely favourable conditions for growth existed at first.

It is generally hard to detect from ecological studies whether water temperature decisively influences growth. Methodological uncertainties are also involved. If length-at-age values from the Atlantic catchment basin are compared with those from the Mediterranean catchment (Tesch, 1983), one notices that in the north all female eels at ages V and VI were <40 cm long, but in the Rhône river estuary age V were 42 cm, and in the Lagoon of Commacchio age VI were 42 cm (Haempel and Neresheimer, 1914). Finally, for age group VII, southern values are available from Valencia, where females were 47 cm long, and from Comacchio, where they were 46 cm. In contrast, all age-VII females from the north reached lengths of <45 cm. According to Bellini, Italian female eels of age IV were 55–66.5 cm long (after Haempel and Neresheimer, 1914). Czechoslovakian data are available (Sedlar and Dobrota 1966), from which one could infer relatively fast growth in the rather warm, sparsely populated waters there. At ages V–VII the eels there were 48, 57 and 63 cm long, respectively.

It is confirmed experimentally that eels under favourable thermal conditions in aquaria and aquaculture (Chapter 6) are capable of growing significantly faster than in nature. Glass eels raised in the aquarium at 23°C and fed sea fish offal reached lengths of 35.8 cm and more after an April–November growth period (Meske, 1968, 1969). A glass eel reared by an aquarist in a 60-l container at 23°C attained a length of 37 cm in 14 months (Dreist, 1968). Another aquarist, who had also put an eel in an aquarium, reported that, at a feeding rate of 90% with the artificial feed Tetramin, it grew to 40 cm within a little less than a year (*Fisch und Fang* **10**, 167, 1969). Thus, under the thermal conditions of north and central European waters, the eel's growth potential is far from being fulfilled.

Obviously, the amount of food available likewise plays a crucial role in growth. Except under intensive aquacultural feeding, food availability varies with population density and space, therefore depends directly on it. One can observe competitive struggles among several individuals together in an aquarium. Eventually, one usually dominates and then grows much faster than the others. The smallest ones in the hierarchy wither and die. This severe deviation in growth is one reason why frequent sorting of the eels by size (grading) must be done in aquaculture (Section 6). According to intensive studies of rearing and sorting under aquarium conditions, Wickins (1987) concluded that dominance is more genetically determined and depends less on physiological condition (stress; Peters *et al.*, 1980).

3.3.2.6 Interspecific differences in growth

The great differences in growth rate between different populations of one species of eel make it hard to draw comparisons with other species. To establish definite differences, comparative rearing studies would have to be done under controlled conditions. Growth studies have been done on few species of eel other than *A. anguilla*. Table 3.10 gives an overview. *A. rostrata* appears to show no basic differences in comparison with *A. anguilla*. It is also evident that substanial differences exist within *A. rostrata*, as well. The fast growth of Bermudan eels is especially conspicuous. The relatively high temperatures there, compared with other places studied, may play a role in this. Growth in New Jersey (USA) streams proved to be especially low, also in comparison with the European eel. Growth of the two New Zealand species appears to be similarly low in the streams studied there. A difference in growth between the larger *A. dieffenbachii* and *A. australis*, which matures in the same area, cannot be determined from the available data, even though such assumptions have been reported in earlier publications (Cairns, 1942a; Burnet, 1969b). Jellyman and Coates (1976) note that growth rate of the two species was similar in fish culture on the South Island. However, each species attains a different body weight (Section 3.3.2.8). In general, the different eel species exhibit a very similar growth pattern, and even the tropical species, *A. nebulosa nebulosa*, fits in with this (Table 3.10).

3.3.2.7 Theoretical pattern of growth in body length

The best-known growth model is represented in the von Bertalanffy formula, which is much used for fishes, in particular for the eel. It is as follows:

$$L_{t_0} = L_\infty(1 - e^{-K(t-t_0)})$$

where:

L_t = length at age t
L_∞ = ultimate (asymptotic) length
K = the growth coefficient
t = age in years
t_0 = the pre-hatching time at which the growth curve begins.

Although the formula is much used for the eel (Table 3.11), it does not always match the actually measured situation very well. Therefore, Berg (1987b) applied special adaptations. It must be emphasised that there are no uniform values for the eel. For example, the growth patterns of males and females differ from each other. Also, very large, fast-growing individuals show different values from 'normal' eels. There are differences between different water bodies and populations, which above all influence ultimate length (L_∞) and the values for K and t_0.

Length and weight

Weight is especially variable for so elongated a fish as the eel. Good feeding conditions or a well-filled gut can cause an eel's weight to be twice that of another of equal length. This and other factors, for example gender, age, head width, season, and living space, make it necessary to have a great deal of comparative data available and take all variables into account in order to assess relative weight in eels. Growth comparisons are, therefore, almost always based on change in length. Nevertheless, comparing weights is interesting,

Table 3.11 von Bertalanffy formula values for eels (*A. anguilla*, unless otherwise shown) from various waters.

Water body	Explanation	L_∞	K	t_0	Author
Burrishoole System	Estuary, M, F	93.7	0.034	–3.02	Poole, 1994
Burrishoole System	silver eels, M, 1987	70.0	0.031	–3.71	Poole, 1994
Burrishoole System	silver eels, F, 1987	143.3	0.013	–4.26	Poole, 1994
Lagoons of Arcachon	normal size M	44.84	0.2857	0.5264	Lee, 1979
Lagoons of Arcachon	normal size F	68.79	0.1410	0.6888	Lee, 1979
Lagoons of Arcachon	giant eels	485	0.0088	8.93	Lee, 1979
Hooghly river, Calcutta	*A. nebulosa*, M, F	68.3	0.1415	0.3214	Pantulu and Singh, 1962

L_∞ ultimate length, growth curve asymptote
K growth constant
t_0 prenatal age and growth curve origin.

and it is important to have them as a basis for conversion factors in matters of biological production, fishery management, or catch technique.

Table 3.12 conveys data on average weight of individual length groups from measurements of European eels originating from different times and places. Further weight data are available for *A. rostrata*, based on length–weight regression curves as follows:

- Eels from Lake Ontario (Hurley, 1972):
  - 48–79 cm specimens: $\log_{10} w = 2.98 \log_{10} L - 5.59$;
  - 83–108 cm specimens: $\log_{10} w = 3.77 \log_{10} L - 7.88$.
  - Eels from two Newfoundland waters (Gray and Andrews, 1971):
  - 16–84 cm specimens: $\log w = 3.1797 + 3.2706 \log L$;
  - 29–57 cm specimens: $\log w = -2.8955 + 3.01812 \log L$.
- Eels from the St Lawrence river (Larouche *et al.*, 1974):
  - 56–110 cm specimens: $w = 4.461 \times 10^{-5} \times L^{3.01}$.

According to studies of the New Zealand eel species, the length–weight relationship is indicated *via* the so-called condition factor F:

$$F = 106w/L^3 \quad \text{or} \quad k = w100/L^3$$

where:

w = weight in g
L = length in mm.

Species differences

The European eel seems relatively light, based on Figs 3.22 and 3.23. *A. marmorata* clearly shows the greatest body weight at length. Interspecific differences are clearer when the species in question occur in the same area or especially in the same water body. For the two species native to New Zealand, all studies to date show that *A. dieffenbachii* is heavier than *A. australis* of equal length (Cairns, 1942a; Woods, 1964; Chisnall 1989; Chisnall and Hayes, 1991; Chisnall and Hicks, 1993).

Also compared with other eel species, *A. dieffenbachii* appears to lie substantially above the average. Quite considerable differences were found between three South African

Table 3.12 Summary of mean weights for individual length groups of various populations of the European eel[1]).

L_T (cm)	Weight (g)	Weight range (g)	N	L_T (cm)	Weight (g)	Weight range (g)	N
6.5	0.5		1	42.5	130.1	106–153	8
7.5	0.5		1	43.5	147.6	116–169	5
8.5	0.8	0.7–0.9	2	44.5	142.7	106–176	7
9.5	1.1	0.8–1.5	3	45.5	164.3	141–194	10
10.5	1.2	1.2–1.2	2	46.5	174.2	149–201	9
11.5	1.7	1.6–1.7	2	47.5	173.0	113–217	8
12.5	2.3	2.1–2.5	2	48.5	195.8	175–225	8
13.5	3.2	2.7–3.7	4	49.5	202.6	182–244	7
14.5	4.0	3.1–4.1	4	50.5	227.6	205–266	5
15.5	4.6	4.3–5.5	4	51.5	229.8	176–269	6
16.5	7.4	5.8–7.8	8	52.5	257.6	235–268	5
17.5	7.7	6.1–9.1	5	53.5	266.3	235–307	6
18.5	9.0	5.0–10.6	5	54.5	296.8	260–337	8
19.5	10.2	8.3–11.2	6	55.5	305.2	280–348	8
20.5	13.1	9.3–17.2	7	56.5	324.7	294–359	6
21.5	15.0	11.8–21	7	57.5	365.6	310–379	5
22.5	16.3	11.3–20.5	6	58.5	353.2	317–386	5
23.5	18.2	15.5–24.0	8	59.5	373.7	330–435	6
24.5	22.7	17.4–27.8	12	60.5	399.2	360–429	7
25.5	24.9	20.1–30.7	7	61.5	399.0	360–441	5
26.5	25.8	21.8–37.7	13	62.5	415.8	365–452	5
27.5	32.4	24.4–46.0	8	63.5	504.8	385–735	9
28.5	36.3	27.8–50.7	8	64.5	448	448	1
29.5	43.1	30.1–56.3	9	65.5	517.5	456–579	2
30.5	45.0	34.9–60.8	10	66.5	590		1
31.5	49.0	37.5–63.9	9	67.5	481		1
32.5	59.5	42.9–68.4	10	68.5	465		1
33.5	59.6	36.8–88	12	69.5	665		1
34.5	61.2	48–85.8	10	70.5	729		1
35.5	71.2	51.5–95	8	75.5	771	644–887	3
36.5	78.8	62.1–101	5	80.5	1046		1
37.5	82.1	67.4–104	7	85.5	1208	982–1434	2
38.5	97.6	76.5–117	5	90.5	1563		1
39.5	96.2	60.5–124	7	95.5	1800		1
40.5	115.0	90.2–137	5	100.5	1622	1360–1884	2
41.5	126.3	101–147	4				

N Number of eel populations studied.
L_T Total length.
[1] Data for calculation of the means were taken from Table 19 of this book's second edition and came from the following sources: Deelder and De Veen, 1958 (IJsselmeer); Gaygalas, 1969 (Kurisches Haff); Gandolfi-Hornyold, 1930, 1935 (Carmargue); Marcus, 1919 (diverse waters, primarily coastal); Min. Agricult. Belfast, 1966 (Lough Neagh); Peňáz and Tesch, 1970 (Elbe river, North Sea and adjacent waters); Sinha and Jones, 1967c (flowing waters in Wales); Sedlar and Dobrota, 1966 (Slovakian streams); Tesch, unpublished; Thurow, 1959 (Baltic Sea).

species (Fig. 3.22), but it should be noted that the weight for *A. nebulosa labiata* was determined by another author (Frost, 1954). The two species studied by Jubb (1961) appear to be rather heavier than *A. labiata* and *A. anguilla*. The difference between *A. marmorata* and *A. labiata* amounts to 2–4 kg for eels >120 cm, thus this difference of 30–40%, therefore, could be species specific. Also the two species *A. marmorata* and *A. mossambica*, studied by the same author, differ with respect to weight in the upper length groups, such that here a differential, species-specific weight gain would be an issue. In contrast to *A.*

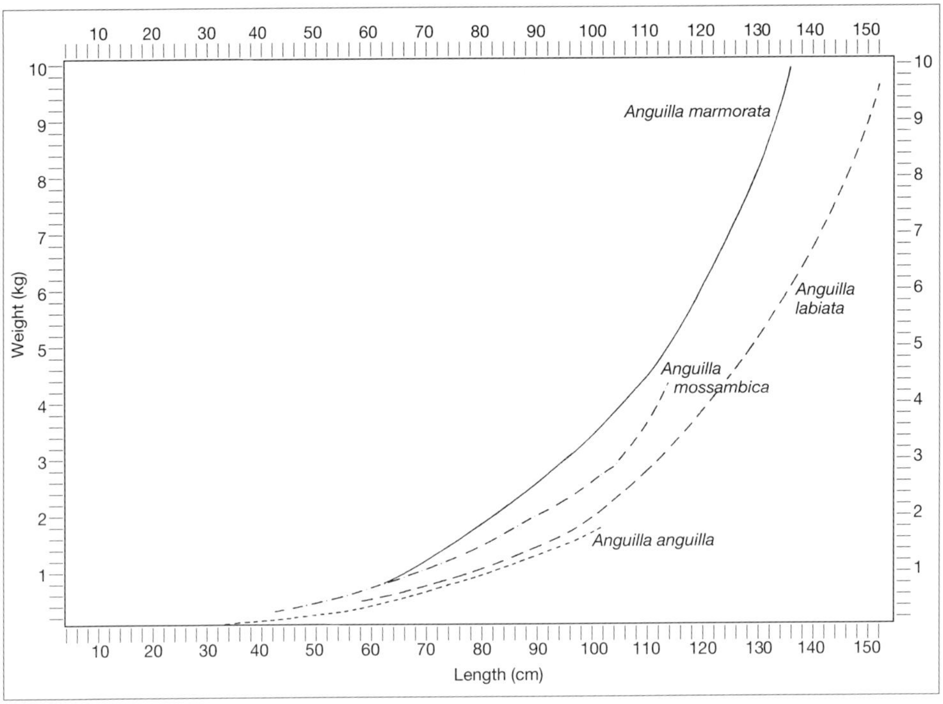

Fig. 3.22 Length–weight relationships of three African eel species (after Jubb, 1961) and the European eel, the latter according to data from Table 3.12

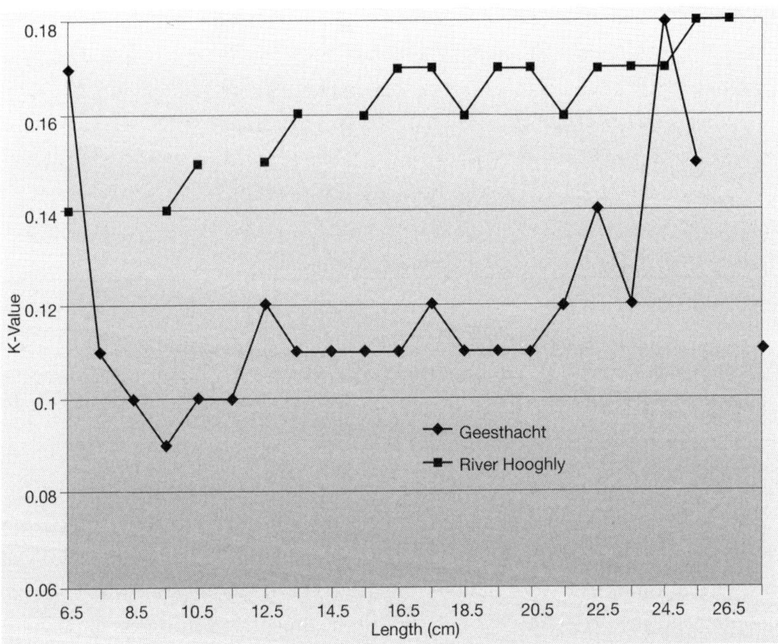

Fig. 3.23 Coefficient of condition (k = weight × 100/length^3) of *A. anguilla* ascending the eel pass at the river Elbe Geesthacht dam, Germany, in August 1964 (Tesch, unpubl.) and of *A. nebulosa nebulosa* from the river Hooghly near Calcutta, India (Pantulu, 1956)

nebulosa labiata, which has a relatively low weight (Fig. 3.22), *A. nebulosa nebulosa* seems to be significantly above average (Pantulu, 1956). At a length of 70 cm, this eel would weigh 1748 g, that is twice as much as, for example, *A. anguilla* at the same length. Although comparative studies under controlled conditions still remain to be done, the publications just referred to show interspecific weight differences, especially concerning the two New Zealand species.

Plumpness and body size

Figure 3.23 shows weight gain of two different species from the glass eel until yellow eel stage up to a body size at which sex differentiation is supposed to end gradually. During juvenile growth, relative weight increases considerably with increasing length. This was detected for the European eel, as well as for the two Indian species. Later, the relative weights also rose, which was attributed to increasing age (Table 3.13). In this case, though, head width can also have played a role.

Differences between males and females

There are always differences between males and females with respect to both length and weight, so it is not valid to compare weights between populations of water bodies that contain mainly females and those that contain mainly males. Remarkable differences between the sexes were observed in the summarised data from primarily north German waters (Peñáz and Tesch, 1970). There and in two-thirds of the English streams studied (Sinha and Jones, 1967c), the males were heavier than the females primarily in the upper length groups. The observed differences are explainable, if it is considered that the males start to become sexually mature at a length of about 30 cm, so, with maximal storage of body reserves for migration, they should be heavier than the females. This relative weight increase also applies to the females when they have reached the migration stage. Figure 3.22 shows, among other things, the length–weight relationship for *A. mossambica*. That curve rises more steeply after a length of 100 cm. The females of this species first attain the silver eel stage at 100 cm (Jubb, 1961), in which they become relatively heavier.

Annual and seasonal differences

Substantial differences in eel weight can occur from year to year or from one season to another. Thus, different populations should always be compared at the same season, regardless of the fact that time of day and other factors (e.g. age, head width) also can influence the weight. Season-to-season differences were detectable on the Stör river (Mar-

Table 3.13 Body weight comparisons for eels of equal length but different ages, head form not considered (after Rahn, 1955a).

	Age group						
Length (cm)	V	VI	VII	VIII	IX	X	XI
53	220	240	235	260	270	325	–
58	–	320	335	320	370	–	–
61	–	325	365	375	400	415	410

cus, 1919). Eels caught in April were lighter than those in the previous September. Weight loss took place during unfavourable conditions in winter. By the next September, this was regained by good diet. Nevertheless, compared with eels in the lower parts of other rivers like the neighbouring Elbe and Severn, those in the Stör river were below average in weight.

Tropical eels can also exhibit a certain annual cycle in their weight. *A. nebulosa*, for example, had condition factors (k) of 0.13 and 0.14 in May and October, respectively, but in the months before and after that the values were around 0.01–0.02 higher. The seasonal differences here are presumably caused by effect of the monsoons.

The eels from the North Irish Lough Neagh exhibited weights that varied remarkably from year to year (Min. Agricult. Belfast, 1966). They were significantly heavier in 1965 than in 1966.

Head width, weight and age

The influence of head width on body weight must not go unnoticed. Studies of the eels of Sakrow Lake near Potsdam, Germany (Rahn, 1955a) enable a comparison. Contrary to previous assumptions, it turned out there that the broad-headed ones were lighter than the narrow-headed ones (see also Törlitz, 1922). This difference is explained by the fact that relative growth in eel diameter 'increases with increasing age, independent of the length that the individual eels reach. From that, one must conclude that a slowly growing old eel is heavier than a rapidly growing younger eel of the same length, and this can explain the weight differences between narrow- and broad-headed eels of equal length' (Rahn, 1955a). As became apparent, broad-headed eels grow faster than narrow-headed ones (Section 3.3.2.4). But Micheler (1967) could not find a correlation between weight differences and head shape. That older eels must be relatively heavier than younger ones also became evident from the fact that older ones have temporally approached the silver eel stage closer than younger ones have, and silver eels are known to be heavier than yellow eels.

Salinity

Eels from the coastal area are usually lighter than those from inland waters. High mean weights were evident for eels from English streams (Sinha and Jones, 1967c), from the North Irish Lough Neagh (Min. Agricult. Belfast, 1966), and from an aquarium (Meske, 1969). In contrast, below average weights were found in all waters near the coast, for example the North Sea at the Isle of Föhr (Peňáz and Tesch, 1970); the Baltic Sea (Thurow, 1959); the Kurisches Haff (Kurskiy Zaliv) (Gaygalas, 1969); predominantly coastal waters (Marcus, 1919); and perhaps also the Camargue (Gandolfi-Hornyold, 1930a, b, 1935). Likewise, a comparative study in Dutch waters showed that eels from fresh water were heavier than those from brackish water (Heermans and Willigen, 1979). It is hard to determine in this regard the extent to which other factors besides salinity have played roles, for example head form (Peňáz and Tesch, 1970: the relatively light North Sea eels from Heilgoland are extremely broad headed). The question exists whether, as a result of osmotic conditions, salt- or brackish-water eels could be lighter (Sections 1.4, 1.5 and 1.7). That seems to be contradicted by the results of Löwenberg (1979), whose eels, caught with trawl nets in the German Bight's sea water, were heavier than average. But here only individuals of >40 cm were involved, which were not considered much in the previous discussion. In addition, eels collected by trawls are at a low stage of digestion of their stomach contents compared with, for example, fyke-net eels.

3.3.3 Habitat and behaviour

This section deals primarily with research results and isolated observations that characterise especially well the eel's special differences from other fishes, and partially its known species-specific forms of behaviour. As many very small individual findings exist in this subject area, only a representative selection can be dealt with here.

3.3.3.1 Occupation of tubes or cavities and thigmotaxis

The yellow eel's most prominent characteristic is its effort to stay negatively phototactic and in mechanical contact with its surroundings (Section 1.9.2). Light avoidance becomes more and more pronounced during the transition from glass eel to yellow eel. If tubular structures are made available to yellow eels in an aquarium during daytime, they are certain to occupy them. At night they leave the tubes, especially at the beginning of darkness, and swim around vigorously (Mohr, 1970). The mechanical-tactile behaviour is emphasised in that during daytime several yellow eels inhabit a tube at the same time, as observations of the author have shown also (Tesch, unpublished). Yellow eels also exhibit this behaviour when several tubes are offered. Thus, it can happen that numerous tubes stand empty, and several tightly packed eels peer out from a single tube, an expression of the so-called thigmotaxis.

The behaviour is similar in the wild; the yellow eel swims about at night and, by day hides in the available substratum. Divers in lakes of Schleswig-Holstein, Germany, even reported that eels aligned themselves facing obliquely upward at an angle of about 45°, sometimes in the metre-thick mud. At its entrance, the burrow resembles that of a fox or rabbit. There, about two-thirds of the eel head protrudes (Krause, 1961). Whether constructed burrows are involved here is questionable. Possibly, head movement formed a funnel-like expansion that could just as well collapse when the eel departs. Figure 3.24 shows how the eels stick their heads out of the substratum and lie together in groups.

Nevertheless, other observations seem to confirm that the eel can build burrows. During dike work along ditches of the lower Ems area that were affected by a 1.5 m change of tidal level, a middle-sized eel was sighted in a hole. It was caught, and, upon subsequent excavation, a metre-long burrow was found that extended obliquely downwards and ended in a spherical cavity. Still more eels were there, and they could not escape. There were several such burrows, and a maximum of 15 more eels found in them. Other kinds of animals that build such burrows and could have vacated them should not have occurred in the area (Rudsinke, 1960; Schnare, 1960).

That eels crawl deep into cavities and possibly contribute to excavation is also explained by the following seemingly fantastic report of Benecke (in Walter, 1910): 'In the Brake stream from 1846 to 1847 near Mühlhof above Rittel a high dam was built in order to irrigate a large meadow complex by impounding the stream. Downstream of the dam, a

Fig. 3.24 Eels in a Baltic Sea mussel bed (photo: Krause)

suitable surface of planks was installed, which is about 100 paces long and is supposed to prevent erosion of the bed and banks by water that plunges down when the sluice is opened. This board floor consisted of two layers, a lower one of 2-inch and an upper one of 3-inch planks. The substantial height of the dam (13 m) has cut off entry for eel young ascending into the upper reach of the Brake and its connected lakes, and the formerly very considerable number of eels caught above the dam has gradually declined to almost zero. In 1847 construction of the dam and the suitable [downstream] surface was completed. In 1852 the plank flooring heaved up in various places in a very irregular way, such that it had to be torn out for extensive repair. Upon doing that, the cause of the heaving was immediately discovered. Thousands of finger-thick eels, colour-faded from lack of light and for the most part pressed more or less flat, filled the space between the two layers of boards, and their combined pressure must have made the flooring give way. In any event, these eels had wedged between the two layers of flooring, had found enough food there, and had grown until their increased volume had sprung their prison's ceiling'.

Walter (1910) relates further observations that speak to the fact that the eel likes to hide in bottom materials. For a picture of the way it inhabits the tidal zone of the coast, it is especially important that the beach mud at the North Sea Isle of Sylt was dug up by many people 'and in it were captured many quite lively eels only a foot deep and in contact with firm sandy earth'. Moreover, eels burrow into the mud when ponds are fished out, which complicates raising eels in ponds (Section 6). Regarding the American eel, it is reported in connection with the study of the overwintering of lobsters in a river estuary that it hides in the mud (Thomas, 1968). Presumably, this is to be interpreted not as fleeing from the pond drawdown, but rather as avoiding the effects of winter.

The burrowing into the beds of inundated areas that are being laid dry is likewise a protective reaction. If the eel remains on the surface, then it is vulnerable to becoming dried out or to predation. Some of the eels use the outflowing water of the tidal current to avoid becoming trapped in dry areas, but many eels, especially the smaller ones, seem to stay behind (regarding choice of bed materials, see Section 3.3.3.2).

The eel's predilection for congregating in direct physical contact obviously is not restricted to its occurrence in burrows and cavities. It is reported for *A. rostrata* that in a lake's outflow in Nova Scotia, Canada, 15–30 specimens had formed a clump of half-metre diameter. They lay there at a depth of 1.5 m, staying motionless. When eel spears were used, they disbanded immediately. A similar observation was also made for drifting eels. The ball that they formed is thought to have had a diameter of up to 2 m. Upon being speared, they disbanded more slowly than the smaller ball (Medcof, 1966). Such associations are not limited to yellow eels; they were also observed for silver eels. A similar behaviour is also known for *A. anguilla* silver eels (Nilsson, 1860).

3.3.3.2 Depth zone and substratum

The question of which depth zone the eel prefers in fresh- and salt-water loses meaning in the eel's predilection for living in the boundary zone between water, earth and air. One encounters eels of all body sizes at every depth. Also experiments in which yellow eels were subjected to a pressure corresponding to a water depth of 1000 m showed that after 30 days the enzyme activity had adapted to the condition before the pressure was raised (Simon *et al.*, 1992). Therefore, the depth changes occurring in the continental area are physiologically meaningless for the eel.

If one electrofishes in small streams, it is no problem to catch eels at quite shallow depths. Nevertheless, it is to be assumed that in shallower water smaller specimens are more numerous and at greater depths more of the larger ones occur. This applies, of course, especially where water levels are constant.

In population studies in a coastal water on the Danish Isle of North Fyn, which is no longer influenced very much by tide, it was found that eels >20 cm long occurred primarily in the sea grass area at depths of 1–3 m. In contrast, eels of 10–20 cm were found more in shallower water (Muus, 1967). Saltwater sport fishers recommend fishing for eels at the edge of weed beds at depths of 1–3 m (Loebell, 1966). As anglers are interested in catching larger eels, these areas can be regarded as the preferred habitats of such animals. In the author's experience, electrofishing in shallow water close along lake shores where eels are abundant yields primarily small eels that have hardly reached migratory size (Tesch, unpublished).

The old notion that eels, especially in the sea, occur only in the coastal region and in relatively shallow depths (Walter, 1910) has not been borne out. People used to assume that eels occurred only on mud flats. On the North Sea Isle of Heligoland, too, eels were supposed to be restricted to the cliff-base bench. This idea was based on the fact that in the past eels were very seldom caught in trawl nets, and haddock lines baited with sand worms likewise never caught them in the open ocean (Walter, 1910). However, it can be established that favourable living conditions for the eel exist at depths of 20 m and more in the North and Baltic Seas. Results of trawling in deep parts of lakes are consistent with this (Section 5.8).

The ability of eels to live in these deep zones depends on having a certain dissolved oxygen concentration. This does not always occur in lakes. Nevertheless, eels venture temporarily into such areas. Dead eels in fyke nets that were set on the bottom of small lakes give proof of this. The deep region of lakes often has significantly lower temperatures than the upper areas. Thus, the depths that offer the eel refuge from light by no means always provide optimal habitat for this warmwater fish. This is no problem at depths of 20–30 m in the North Sea. The water masses there are almost continually mixed, so such thermal stratification as occurs in lakes seldom happens. This also applies to deep channels of rivers.

Almost all water bodies offer possibilities for the eel to find its way into the bed materials or, on submerged cliffs or stony bottoms, to occupy cavities. Therefore, there are hardly any waters within the normal geographic ranges of eels that are not populated with them if thermal and oxygen requirements are met and if barriers to entry do not exist. Small and large eels have different requirements in the bed materials though. It is known that *A. marmorata* mostly stays in the sand and under stones until it is 30 cm long. In contrast, larger specimens prefer cavities (Nishi *et al.*, 1969; Lecomte-Finiger and Prodon, 1979). Ideal habitat for eels consists of plant thickets, especially edges of weed beds, as well as all sorts of other hiding cover, such as tree stumps, root masses and sunken debris.

3.3.3.3 Survival in air

As is known from many physiological studies, neither fresh- nor salt-water limit eel distribution (Section 1.5.1). The eel can also live exceptionally long after leaving water; it need only be protected from drying (Sections 1.4 and 1.7.1). Nevertheless, survival in air is only temporary. Contradictory observations exist about that and the eel's ability to find its way back into water. That they find their way over beach walls at the beginning of the silver eel migration is well verified and believable. 'At the time of the autumn new moon in July to October, the outlets of beach lakes are often clogged with sand by strong storms from the west and north. This forms a sand wall. From time to time, a breaker from the still strongly surging sea overflows the sand barrier with salt water. At dusk, the eels that exist in the outlet throng toward this obstruction and, in the meagrely trickling sea water, slither over the barrier and into the sea beyond' (Buckow, 1956). Of course, eels are also harvested in this situation. Behaviour may have been similar for the lone silver eel that Bergmann (1970) once spotted about 3–4 m from the edge of a little German low mountain stream during midday at the end of July. This fish had surely escaped from a mill pond 70 m away at higher elevation; the millrace was not used anymore and was clogged, so that water could only seep out, and the eel had to use the overland route while it was moist. Two further cases are cited from the inland region.

Juvenile eels also basically undertake similarly purposeful 'overland migrations' while migrating up rivers. Nevertheless, this cannot be called an actual overland migration in either case. In such migrations on solid ground, enough moisture is always present to indicate to the animal that water is nearby, so venturing away from or finding the way back into inundated areas seems possible.

Further, it is conceivable that external circumstances force the eel to make its way in dry places at times. Its reaction to the sudden discharge of wastewater into streams is well known. It avoids the unbearable conditions by crawling onto the stream bank. In this way also, eels escape from ponds that have been fertilised with quicklime or too much liquid manure (Walter, 1910) and will leave aquaria if conditions such as water quality or competition from other eels prove unsuitable. Eels also escape from aquaria or other containers to follow the migratory urge (Boëtius, 1967) but often cannot find their way to their natural environment.

In aquaculture, the eel is lured onto dry areas for feeding (Section 6). Also in aquaria, it rises above the water surface if one puts food there. Should this ability of the eel to get food 'from land' not also play a role in nature? In this regard, Bergmann (1970) provides 11 reports of persons whom he considers credible eyewitnesses. Three involve the aforementioned emigration of silver eels from landlocked waters. The rest, in Bergmann's opinion, represent foraging for food. Although the fish were repeatedly seen in pea fields at night, Bergmann (1970) rejected the idea that peas were the objective of the foraging. Pea fields would be more attractive because they, like low meadows and other moist terrain, are favoured habitat for snails, worms, and many other food animals. Eels are supposed to be able to find the way back into water *via* their slime tracks. Besides these reports, there are numerous further statements of variable credibility (*Allgemeine Fischerei Zeitung* **27**, 341 (1972); **28**, 230 (1973); *Fisch und Fang* **14**, 694 (1973); **15**, 66 and 204–206 (1974)). On the question of whether eels had peas in their digestive tracts after having been captured in the pea field, Bergmann (personal communication) stated that this was not the case in a report that he reviewed.

However, the eel does not find its way into water bodies, if forcibly transported onto dry areas nearby. The following practical experiment illustrates this. In a Weser tributary, 12 eels were caught on baited hooks and then released 4 m away from the stream bank in the short grass of a meadow's slight depression. After half the night none of the eels had left that place. Had the place where the eels were deposited sloped toward the water and not been basin shaped, then most of them would undoubtedly have succeeded in getting back to the water. Of 10 eels placed on level meadow terrain, only two or three reached the water; they got there by chance, not by some 'water instinct' (Rudsinke, 1960).

3.3.3.4 Territoriality and homing ability

Burrows and cavities serve as dwelling and resting places for the eel. Now the question arises as to whether the eel occupies one such hideaway continually or changes after nightly excursions, and whether it generally remains a long time in one place or moves on frequently. Apparently, there are no specific study results on residential behaviour despite the fact that modern diving techniques would easily permit such observations. On the other hand, several studies of the eel's home range have been conducted. The following areal extents of the home range have been reported:

- Lake Champlain, Vermont, USA (LaBar and Facey, 1983): 28 ha;
- small lake, South Spain (LaBar *et al.*, 1987): 0.13–0.27 ha;
- streams, lower Mississippi river area, USA (Gunning and Shoop, 1962): <100 m;
- tidal creek, Georgia, USA (Bozeman *et al.*, 1985): 1.04 ha;
- estuarine creek, tributary to the Gulf of St Lawrence, Québec, Canada (Dutil *et al.*, 1988): 0.5–2.0 ha;

- salt marshes, Massachusetts, USA (Ford and Mercer, 1986): <100 m;
- river Elbe, lower reaches (Mann, 1965): 30 km;
- Ellesmere Lagoon, New Zealand (Jellyman and Glova, 1996): most of the eels within a few kilometres; some in the whole lagoon;
- Waikato river, New Zealand (Chisnall and Kalish, 1993) mainly <20 m.

Thus, depending on characteristics of the water body, the eel can inhabit an extraordinarily limited area, usually for more than a year, and the area tends to be more restricted in smaller water bodies.

As is known from further studies, eels tend to move upstream during high water and downstream during low flow (Koops, 1962; Tesch, 1966; Aker and Koops, 1973). During high water the stream becomes much wider, and eels find abundant food in the flooded areas. When the water recedes, they have to leave these areas to avoid drying. Sometimes, however, eels burrow into the moist ground (see below). The nature of such unstable aquatic areas thus brings about a certain directed movement in the eel population.

In contrast, when habitat and food conditions are rather constant, the eel maintains a very restricted home range, even if broad expanses of water are open in all directions, for example, in the North Sea. On the *c.* 3 × 4 km bench at the base of Heligoland's cliffs, 350 eels were marked internally in 1968; of the 17 recaptured, not a single one was outside the bench. Five of these recaptures were not until a year later. Such constancy was also observed at other places, for example, in the Elbe estuary near Cuxhaven and at a lock-passage through a dike on the North Friesian Isle of Föhr. Of the 73 eels recaptured in total at these three places, only one was outside its home area. Among them were 21 olfactorily injured animals. The home area off the Isle of Föhr covered 100×500 m at most, but the one near Cuxhaven (Elbe estuary) was relatively large, extending for some 30 km (Tesch, 1967a, 1970, and unpublished). In the inner part of the Penobscot river estuary in the north-eastern US, the eels' home range covered 6.7 km, as rather precisely determined by territorial observations of yellow eels tagged with ultrasonic transmitters (Parker, 1995).

But when eels do change their territories or even their home ranges, this tends to be especially during the transitional periods between summer and winter. In such cases, the switch from summer to winter habitat and *vice versa* is involved, as is also seen in many other species of fish (Schiemenz, 1960b). Ice cover, rapidly cooling water, and anchor ice are environmental conditions that fishes, especially eels, avoid. Regarding sensitivity of eels to extreme winter weather, see Section 1.4. In streams, in brackish water, and in tidal areas, eels seek out sufficient depth and quiet side channels and coves, or (in rivers) deep, mainstream pools, where thermal stratification maintains minimally suitable living conditions, at least near the bottom. In Germany's Hunte river, 'marked' eels were recaptured in summer exactly a year after initial capture at the same places in shallow side arms that were 1–1.5-m deep. In winter they were situated in the Hunte river in depths of 2–2.5 m, sometimes 5 km away from the location of summer capture (Lübben and Tesch, 1966). It is also reported for individuals of the two New Zealand eel species that they occupy deeper places in sand and mud in winter. In summer, on the other hand, they are not to be found in these places, and this was confirmed by tracking tagged *A. australis* (Cairns, 1942a, Koops, 1962; Tesch, 1966; Aker and Koops, 1973).

Extensive tagging studies in the German Bight and its estuaries confirm that eels undertake no movements whatsoever during summer. In September/October, however, 'stronger shifts of the populations' occurred. Eels captured on the coast near Büsum were encountered from September onward in the sea region of Heligoland or in mainland fresh waters. The shallow mudflat area is thus unsuitable habitat in winter. However, eels captured in the sea near Heligoland in summer retained their home range in winter (Aker and Koops, 1970). There, climatic changes did not influence the eels' home range as much.

According to fishers' observations, the American eel undertakes similar springtime and autumn migrations (Smith and Saunders, 1955; Medcof, 1969). Such a habitat switch was observed particularly in the immediate coastal zone but also on the mainland. In autumn, yellow eels migrate from the coast into fresh water as soon as the first frost occurs. The migratory movement can then be observed for about 3 weeks. It lasts about as long in spring, when the eels return into brackish water. Fishers of the Elbe estuary reported quite similar observations. Also, the eels used for homing experiments near the Isle of Föhr's town of Wyk were captured on the basis of such seaward habitat shifts in spring, that is, at the beginning of May (Tesch, 1967a, 1970). Eels supposedly become active in the same way in autumn, although from the ocean into fresh water. The autumn migration is not only from the sea into mainland waters; it can also proceed in the reverse direction. The same applies to the springtime migration. Aker and Koops (1973) experienced in their marking experiments that in the Eider, a North Sea coast river in Schleswig-Holstein, the autumn migration of yellow eels goes downstream in the river's middle reach and upstream in the coastal area. This can mean that both eel populations migrate to a common overwintering area. Likewise, the two New Zealand eel species increased their activity during the yellow eel stage, and in freshwater habitats in autumn as well as in spring (Fig. 3.26). As determined by tracking in a New Zealand lagoon, shortfinned eels migrated into the deeper areas and remained motionless at temperatures $<12°C$ there (Jellyman *et al.*, 1996). The question arises whether the increased activity of silver eels (Boëtius, 1967) in both autumn and spring also applies to yellow eels (Section 3.4.1).

Besides the summer or winter territoriality, yellow eels have a pronounced homing capacity. This enables a yellow eel, when displaced, to find its way back over great distances to its original territory. This tendency to return has prevailed in all displacement trials in rivers (Mann, 1965; Tesch, 1966), in lagoons (Jellyman *et al.*, 1996), on coasts and in the open ocean (Tesch, 1967a, 1968b, 1970; Deelder and Tesch, 1970). The trials shown in Fig. 3.25, as well as a series of further experiments showed that, up to distances of well over 100 km from the home location, most eels find their way back. At greater distances, the proportion of returned eels decreased markedly. Individual eels could, however, even find their way back at distances of over 200 km. Olfactory impairment presumably did not reduce homing performance.

On the other hand, odours could have played a role when eels homed after displacement from one estuary into another (Tesch, 1970; Vladykov, 1971). Capture-marking-and-recapture studies on eels at the outer coast of the Po river delta in the Adriatic Sea showed low homing performance despite a seaward displacement distance of 10 km which was much shorter than in the North Sea experiments (Rossi *et al.*, 1987). As established above, in what follows, and in many other studies, eels find their original home location rather

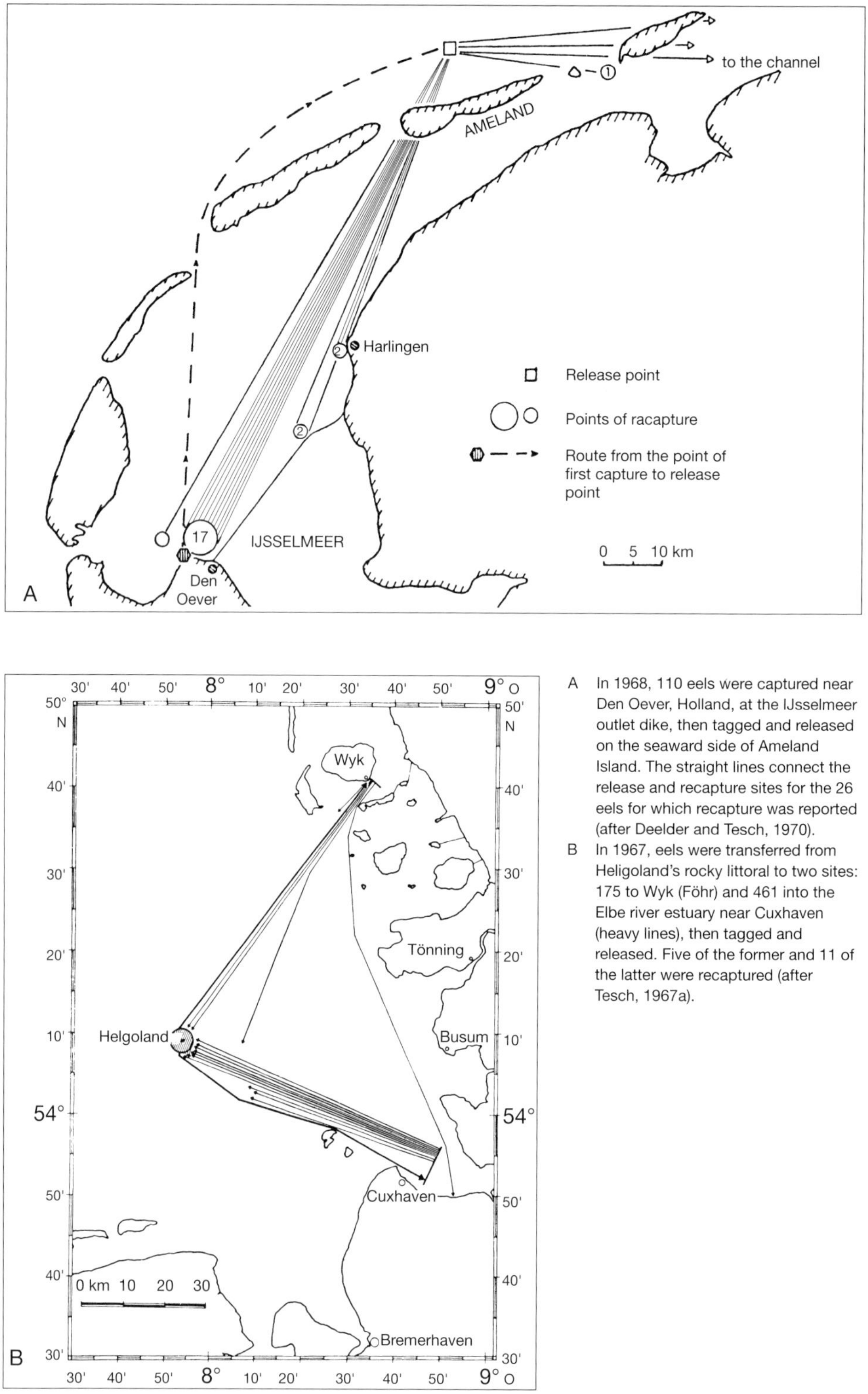

A In 1968, 110 eels were captured near Den Oever, Holland, at the IJsselmeer outlet dike, then tagged and released on the seaward side of Ameland Island. The straight lines connect the release and recapture sites for the 26 eels for which recapture was reported (after Deelder and Tesch, 1970).

B In 1967, eels were transferred from Heligoland's rocky littoral to two sites: 175 to Wyk (Föhr) and 461 into the Elbe river estuary near Cuxhaven (heavy lines), then tagged and released. Five of the former and 11 of the latter were recaptured (after Tesch, 1967a).

Fig. 3.25 Homing performance of eels transferred within the North Sea coastal region

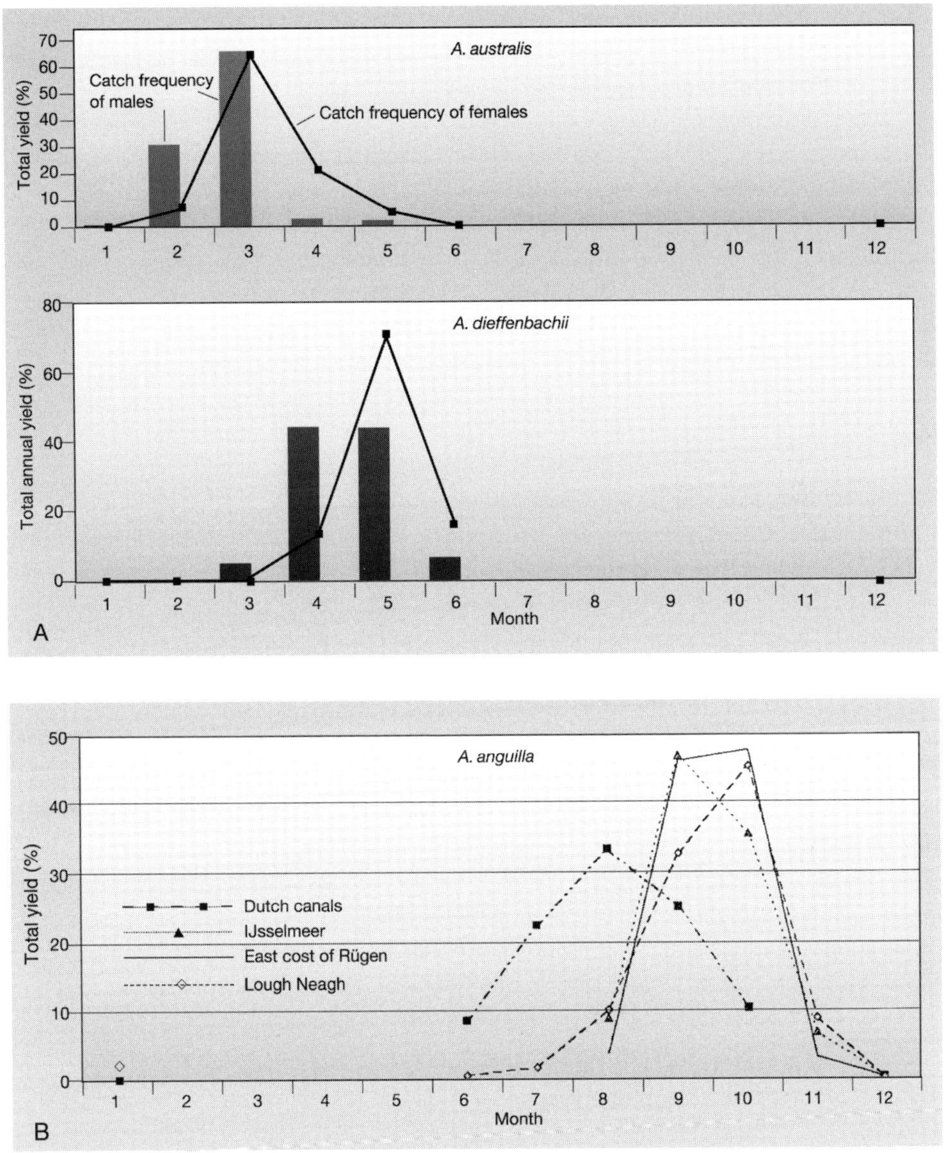

Fig. 3.26 Times when silver eels begin to migrate in the northern and southern hemispheres, based on commercial fishery catches in New Zealand and Europe

A New Zealand: fyke net catches of *A. australis* and *A. dieffenbachii* in Ellesmere Lagoon on the east coast of the South Island (Todd, 1981)

B Europe (males and females combined):

- Average fyke net catches in Dutch canals, 1947–52 (Deelder, 1954)
- Average fyke net catches in the IJsselmeer, 1950–1 (Deelder, 1970)
- Average pound net (Kummreusen) catches on the east coast of the Isle of Rügen, 1928–36 (Meyer, 1938)
- Seven-year average of stow net catches in the outlet (river Bann) of the Northern Irish Lough Neagh (Frost, 1950)

accurately, but the last mentioned authors' interpretations of this are not completely convincing. They say that salinity alone should cause the return to the coast, and that all other locally existing factors would have too little influence.

As the conventional marking methods could not solve the question of orientation during homing, Tesch (1974a, 1975a; Fig. 3.30) tracked displaced yellow eels tagged with ultrasonic transmitters. The eight experimental animals came from northern Denmark, from the Elbe estuary, or from the Baltic coast near Gdingen. All experimental animals, regardless of origin, swam south-eastward (mean direction: 126°) from their release sites in the south-eastern North Sea during the *c.* 17 h trackings. Thus, only some of them swam towards their home area. Laboratory tests showed that, when deprived of all hydrographic directional stimuli, both silver and yellow eels can maintain a particular compass bearing (Tesch and Lelek, 1973a, b; Tesch *et al.*, 1992). This choice of direction can be influenced by alteration of the magnetic field (Fig. 3.33; Tesch, 1974b; see also references in Tesch *et al.*, 1992). The orientation of the yellow eels to their home areas could then be explained as follows. Eels released in the south-eastern North Sea swim, under guidance by the Earth's magnetic field, along a south-east compass course and reach the coast. There they then orient themselves along the coastline if they do not immediately meet their home area. Danish eels would, for instance, have only a 50% chance of finding their way back because the other half, after reaching the coast, would keep swimming westward. Numerous recaptures of conventionally tagged eels after displacements of >200 km (Deelder and Tesch, 1970) south-westward from the release site can be explained in this way. Homing to a seaward home area, that is, to one lying to the north-west (e.g. Heligoland: Tesch, 1967a), also is explainable. The eels in a laboratory test showed a bimodal (i.e. north-western and south-eastern choice of direction). Thus, if they swam south-eastward in the sea, they would have to be able to orient themselves equally toward the north-west, as well. Thus, yellow eels from the English east coast, when released in the western North Sea, swam toward the English coast (Tesch, unpublished).

The migratory speeds and depths for some of the displaced yellow eels resembled those of the silver eels (Tesch, 1974a, 1975a); they also seemed to migrate pelagically (Section 3.4.5.5).

The homing behaviour of *A. rostrata* yellow eels was studied *via* tracking experiments also in the tidal zone but within a small river estuary (Parker, 1995; see also Dutil *et al.*, 1988 for behaviour of yellow eels in the St Lawrence river). The eels began homing (distance: 10–17 km) shortly after release and in the process used suitable tidal current for the return journey. They swam primarily at night but without the influence of moon phases; for this they needed 220 h on average (range: 17 h to 4 weeks).

3.4 Migration of silver eels

The migratory movements of silver eels are much more clearly understood than yellow eels' changes from one locality to another. Early on, an extensive sampling operation had already drawn attention to the fact that there must be a spawning migration (Section 5.42). Therefore, time and place of the migration from the interior to the coast and into the Baltic Sea are by and large well known. Meagre indications about the oceanic migration have arisen only recently (Fig. 3.15). The following sections will, therefore, mainly deal with migrations on the continent and the bordering shelf area, which are easier to study than those in the expanses of the world's oceans. But the two parts of the migration are not to be regarded as separable.

3.4.1 Seasonal pattern

The commercial fishing industry, by exploiting the silver eel migration for profitable harvest, has long known its timing. But most of the available statistics say nothing beyond how much of the catch consists of silver or yellow eels. Some information about the special conditions in the Elbe is provided by Fig. 4.7 (see also Fig. 5.24). However, monthly catch statistics give definite information about the months of the main silver eel migration (Figs 3.26 and 4.5).

Although the largest catches of migrating eels in Dutch canals occur in August, in a Northern Irish stream and on the central Baltic Sea coast they are most abundant in October and hardly less so September; in Holland's IJsselmeer they are more numerous in September. It would have been expected that the eels of the central Baltic Sea migrate sooner than those of Northern Ireland and the IJsselmeer; in Irish waters, though, they are at least 1,500 km closer to the spawning grounds than those near the Baltic Isle of Rügen (see also Walter, 1910). Thus, it is very problematic to draw definite conclusions about an earlier or later onset of migration from migration-time differences between waters that lie east and west or north and south (Section 4.3).

More definite information provides a comparison between seaward water bodies and those that lie more towards the interior. The eels in Dutch canals are earlier than those of the more seaward IJsselmeer (Fig. 3.26). In the coastal area of the Baltic Isle of Rügen, silver eel migration starts 2–3 weeks later at Sassnitz and 4 weeks later around Lohme, which is still nearer the central Baltic Sea, than in Greifswald Bay, which adjoins the mainland. Furthermore, the peak catches that indicate migration normally occur on the coast of Rügen in September/October but do not do so in The Sound (Øresund) until October/November (Meyer, 1938). This and the migratory direction are verified by tagging studies (Martinköwitz, 1960, 1961). According to this, there is likewise generally a shift of peak catches from September in Greifswald Bay in the south to October in Tromper Bay (Wiek) at Rügen's north end. A partial confirmation of the temporal shift in silver eel catches from inland to seaward locations is provided by Fig. 4.5 (discussed in Section 4.3), though inclusion of yellow eels confounds the results.

The experimentally determined peak of springtime migratory activity is also problematic (Boëtius, 1967). There would be a natural explanation for this in yellow eels, as discussed above (Section 3.3.1.1). For silver eels, few of which commercial fishers catch in spring, various interpretations are possible:

1. The autumn peak of activity marks the start of migration, and so to a certain extent, departure of the eels from their home waters. The springtime peak might be related to spawning activity, which happens about March in the Sargasso Sea. But why are laboratory eels inactive between the autumn and spring, despite the fact that eels in the wild migrate during this time?

2. Silver eels that are artificially or naturally hindered from emigrating in autumn deteriorate into an inactive state during winter, in order to become active again for the resumption of their migration in spring (Frost, 1950).

As the spring migration includes large female silver eels in the Elbe, it seems reasonable to suspect that these were latecomers for the autumn migration. Likewise, among the last eels caught in autumn, there were always particularly large ones (Table 3.14).

Therefore, differences in the main migratory period exist between males and females. This situation first came to attention when a shift in the size distribution of migrating eels was perceived during the autumn migratory period (Fig. 3.27). According to this, fyke netting at the beginning of the migratory period yielded mainly small eels, but the proportion of larger eels increased in subsequent months, and in November reached 90% (Meyer, 1938; Nolte, 1938). In the fishers' opinion, the large, late arrivals came from far inland and by chance had to migrate past this location. It is well known that under normal recruitment conditions, the eels occurring in the inland waters are primarily large female eels, and on the coast mainly small males. Thus, the males do not need to start migrating sooner, but rather they have a shorter distance to cover before reaching the catching gear.

That eel size increases from beginning to end of the migratory period at particular locations was also observed in Dutch waters (Deelder, 1970). The proportion of large eels was zero in August, 11% in September, 26% in October, 47% in November, and 60% in December. Much the same is seen for the two New Zealand eels (Fig. 5.24). Finally, the

Table 3.14 Proportions (%) of silver and half-silver eels of various lengths caught in the middle Elbe river at various time of the year 1958 (after Lühmann and Mann, 1961).

	Date caught						
Length group (cm)	21.5	30.5	20.6	4.7	23.7	29.8	24.9
40–50	4.8	7.3	10.6	16.9	2.0	10.2	0
51–60	4.8	41.9	59.6	45.8	33.3	38.6	10.5
61–70	23.8	37.1	29.8	27.7	35.3	40.9	50.0
71–80	42.9	10.5	0	6.0	19.6	10.2	29.2
>81	23.8	5.2	0	3.6	9.8	0	8.3

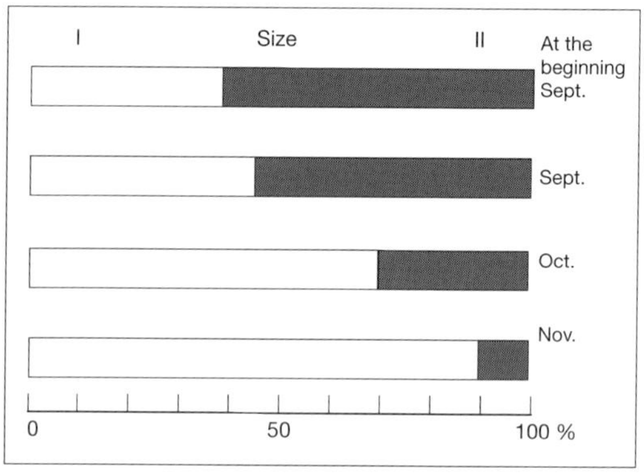

Fig. 3.27 Proportions of large (I) and small (II) silver eels delivered at the harbour of Sassnitz, Isle of Rügen (after Nolte, 1938)

percentage of small eels in the Elbe decreases steadily from summer to autumn, so that by the end of September no more males can be present at all (Table 3.14).

The New Zealand yellow eels show quite similar, seasonally restricted migrations (Fig. 5.24). The migration to spawning sites takes place in late summer and autumn in the southern hemisphere. According to statistics for both of the species that occur there, males appear on the monitoring sites first and then females, much as in Europe. *A. dieffenbachii* migrates first and then *A. australis* (Fig. 5.24). However, previous studies found a reversed sequence (Hobbs, 1949). The differences between the two studies may be ecologically based.

In contrast to the predominance of males at the onset and females at the end of migration, mean body length of males declines as the season advances during the migration in Dutch waters (Deelder, 1970):

- July 36.0 cm;
- August 35.5 cm;
- September 34.0 cm;
- October 33.0 cm.

This may indicate that the smaller males migrate slower because they are smaller and, therefore, less capable and so pass particular sites later than the larger males.

3.4.2 Diel periodicity and the influence of light at the onset of migration

Silver (migratory) eels are caught mainly at night. They supposedly put in their appearance from an hour after sunset until an hour and a half before sunrise (Walter, 1910). It was observed on the southern Baltic Sea coast that under unfavourable conditions, migration was obvious only in the evening after sundown until about midnight. At the height of the harvest season, most of the eels that were in motion were even encountered during a still shorter period, that is, between sunset and moonrise. As peak harvest occurs only during certain lunar phases (see below), this period was limited to 1 or 2 h (Bräutigam, 1961b; see also Frost, 1950; Lowe, 1952). Much the same was observed for *A. rostrata* (Medcof, 1966; Winn *et al.*, 1975; Hain, 1975).

Thus, these studies show that the appearance of moonlight may also play a role in the relationship between activity periods and lunar phase. The question exists whether eels prefer deeper water on brighter nights, if it is available. They also avoid artificial light on the banks of water bodies (Rumphorst, 1930), so lights are used to guide eels into fyke nets (Section 5.9). Furthermore, it was observed on the Isle of Rügen's coast 'that in pound traps having several chambers, the seaward-most chambers in deeper water caught more eels during full moon than those in shallower water, whereas, the reverse is often true at the dark of the moon. However, the wind is substantially involved in this ...' (Nolte, 1938). Under these conditions, the increased silver eel catch observed when water is very turbid (Määr, 1947; Frost, 1950) seems to result more from low light intensity in the water than from other influences of the turbidity-causing substances. Edel (1975b) conducted studies to analyse silver eel activity in relation to light and cover. He found that mean total activity of the animals increased if the cover was removed, so that the activity that normally took place at night hardly occurred anymore.

Table 3.15 Silver eel catches in various ocean areas.
a *A. anguilla*, r *A. rostrata*, j *A. japonica*, d *A. dieffenbachii*.

Ocean area	Place caught	Depth (m)	Date	Species	Length (cm)	Weight (g)	Ovary breadth (cm)	Oocyte diameter (mm)	Comments	Source
North Sea	NW Dogger Bank	–	11.10.24	a		100	–	–	sex organs poorly developed	*Fischerbote* **17**, 120–121 *Fishing News,* **13,** 12–24
	S to W Skagen	180	18.11.24	a	50	*c.* 500	–	–		
	?		–	a	70	–	–	–		
	Noss Head	–	Dec.23	a	75	–	1.2	0.13		
	Tod Head S-Aberdeen	–	Nov.24	a	67	–	1.2	0.23		
	Langeoog	10	Apr.24	a	50.5	197	–	–	pectoral fin length 3 cm	Schnakenbeck, 1932 (see also Schnakenbeck, 1934)
	North grounds W-Doggerbank	75	Oct.24	a	46.5	170	0.9	–	pectoral fin length 1.9 cm	
	N-Doggerbank	66	Oct.32	a	65	510	1.9	–	pectoral fin length 3.1 cm	
	NW-Doggerbank	60	Oct.32	a	61	410	1.7	–	pectoral fin length 2.4 cm	
	SE Devils Hole	50	Oct.32	a	61	365	1.3	–	pectoral fin length 2.7 cm	
	Doggerbank	–	autumn 58	a		–	1.3	–		Lühmann & Mann 1958
Kattegat	SSE of Skagen	–	18.11.24	a	51	500	–	–		*Eh. Fischerbote* **17**, 120–121, 1924
	Swedish west coast near Vinga	–	Dec.24	a	51	230	1.1	–		
	Swedish west coast near Vinga	–	Dec.23	a	50	250	–	–	sex organs poorly developed	
North Atlantic	Rosemary-Bank	725?	22.1.64	a		228		0.15	in stomach of *Mora mora*	Reinsch, 1968
	Rosemary-Bank	725?	22.1.64	a		210		0.12	in stomach of *Aphanopus carbo*	
	Färöe Islands	325	26.11.74	a		66	450	0.3–0.4	gonad weight 13 g	Ernst, 1975

Table 3.15 *(Cont'd)*

Ocean area	Place caught	Depth (m)	Date	Species	Length (cm)	Weight (g)	Ovary breadth (cm)	Oocyte diameter (mm)	Comments	Source
Western North Atlantic	Block Island, USA	–		r		–	–	–	5 specimens	*Maritimes*, 3–5 (winter) 1969
	SE Chesepeake Bay	15–20	5.10.67	r	51–58	–	–	0.28	3 specimens	
	SE Cape Cod	68	7.11.69	r	64	–	–	0.17		Wenner, 1973
	SE Cape Cod	45	7.11.69	r	37	–	–	–	males	
	Assateague Island	–	–	r	61–66	–	–	0.36	6 specimens, gonad weight 30–37 g	
Japanese Sea	Awashima Island		20.1.64	j	64	540	–	0.30	gonad weight 39 g	Honma, 1966
	Near Niigata	0–120	8.4.55	j	69	338	–	0.28	gonad weight 13 g	
	NE of Sado Island	0–350	30.11.55	j	73	760	–	0.29	gonad weight 22 g males	
	N of Awashima Island	0–250	5.10.55	j	40	74	–	–	gonad weight 0.48 g	
	Off Agona Estuary	0–15	16.10.67	j	84	1140	–	0.20	gonad weight 31 g	
	Strait of Sado	0–290	4.1.61	j	65	–	–	0.33	gonad weight 34 g	
	Strait of Sado	0–210	20 10.61	j	71	–	–	0.29	gonad weight 19 g	
	Strait of Sado	Shallow	4.4.62	j	65	360	–	0.28		
	Awashima Island	–	24.1.62	j	58	575	–	–		
	Awashima Island	0–210	29.12.62	j	85	938	–	0.29	gonad weight 51 g	
SW-Pacific	New Zealand, E of Cape Farewell	113	10.5.71	d	100	2700	5.7	>0.5	gonad weight 302 g, pectoral fin length 6.6 cm	Todd, 1973

3.4.3 Eel migration in subterranean waters

A further capability of the eel clearly shows that they can do something that is only encountered in other fishes in exceptional cases. Eels can migrate long distances in underground waters at the juvenile and yellow stages, as well as during the spawning migration. Evidence of this comes from the North Adriatic karst region, where many streams flow underground for long stretches before entering the Adriatic Sea. Among these is the Timavo; eels are caught in its estuary at the little town of the same name. Here, 200 yellow and 300 silver eels were marked by finclipping, then released at three sites 28–42 km upstream. The 29 reported recaptures occurred 40–188 days post-release. All were in the migratory stage and had returned to the sites of their first capture after passing the Timavo that flowed underground in 200–300-m stretches with only short open reaches.

The eels in underground stream sections are pale coloured, a well-known attribute of cave animals (Sella, 1929). But this does not apply to silver eels caught during migration. The observations of eels from areas having underground streams are not surprising in that eels also take shelter in cavities when in normal surface waters (Section 3.3.3.1).

American eels also travel long distances underground. In the Santa Fe river of Florida, sampling around a 3–4 km long 'natural bridge' captured 12 eels downstream and one eel upstream from it (Hellier, 1967). The inference from this is that, although eels can also migrate upstream underground, such stream reaches present a certain hindrance.

Correlation with lunar phases

It is known that eels are caught in greatest abundance during the last quarter of the moon, that is, at waning half-moon (Fig. 3.28; Todd, 1981a). The previous section indicated that moonlight probably influences the daily activity cycle. Statistical analyses of eel catches on the Baltic Isle of Rügen's coasts raised doubt about whether light intensity alone causes the lunar-monthly rhythm in eel migration (Meyer, 1938). In the English Lake District, silver eels showed activity phases even in enclosed tanks (Lowe, 1952). Later studies on escape behaviour of eels in enclosed test facilities showed that 61% of the silver eels tried to escape during the critical period of the waning half-moon. In contrast, at new, waxing, and full moon, the values were 9%, 13% and 18%, respectively (Boëtius, 1967). During complete darkness in circular tanks, silver eel activity increased significantly at new moon, and that of yellow eels up to 3-fold. Regarding the earth's magnetic field, strongly reduced or directionally altered magnetic conditions resulted in temporal compensation or shift in the activity maxima (Tesch *et al.*, 1992). The catch curves shown for the Upper Rhine (Fig. 3.29) can be divided into lunar phases having much and little cloud cover (Jens, 1953). When it was very cloudy, there was no reduction at all in the peak catches observed during the waning half-moon. In addition, Hain (1975), in his laboratory experiments, could not rule out that other diel-periodic and lunar-monthly factors also play roles, besides light intensity.

Further analysis of the lunar-monthly catch curve showed that it can be regarded as an ellipse (Jens, 1953), perhaps indicating planetary nature of the intensity curve. An interesting correlation is revealed by comparing the catch curve from the Isle of Rügen (Fig. 3.28) with the number of meteorites detected by radar in the Ottawa area, Canada (Lang, 1972). Although the two curves almost match and do correspond to the known peak catch during the moon's last quarter, no causal relationship can exist here. According to the

A IJsselmeer and Dutch canals (after Deelder, 1954, 1970)
B River Rhine (after Jens, 1953)
C The Isle of Rügen's coast, calculated by Jens (1953) from data of Meyer (1938)

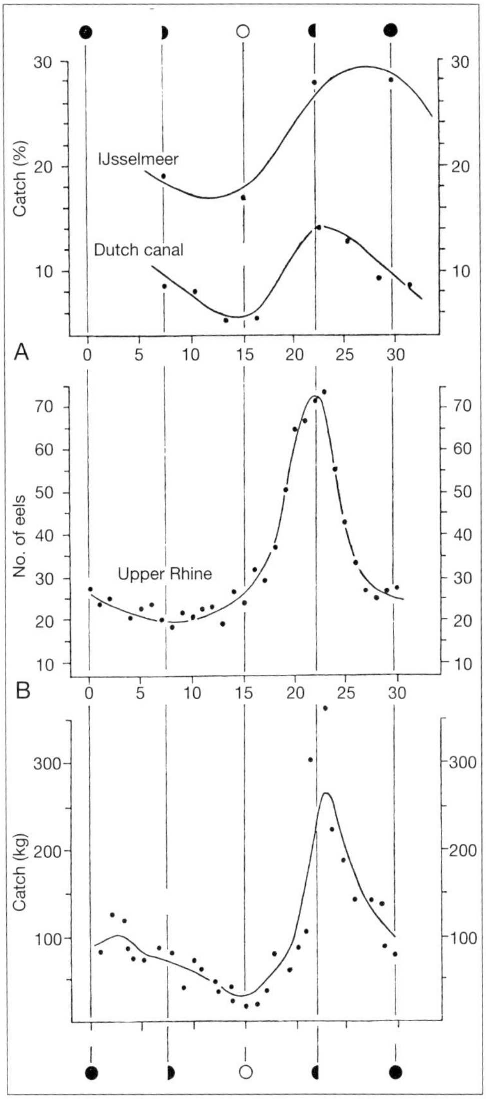

Fig. 3.28 Indices of frequency for catches of silver eels during lunar months in various European waters

cited publications, it can hardly be inferred that moonlight alone causes the lunar-periodic activity of silver eels. In tagging tests on the Swedish east coast, silver eels travelled shorter distances during the period before full moon than after it (Lindroth, 1979). Therefore in the sea, the lunar cycle continues to cause a corresponding activity rhythm.

3.4.5 Effects of weather and hydrography

3.4.5.1 Low pressure areas (cyclones)

A weather phenomenon has become known that could affect the eel somewhat indirectly, though also not *via* hydrographic conditions. This involves microseismic oscillations that

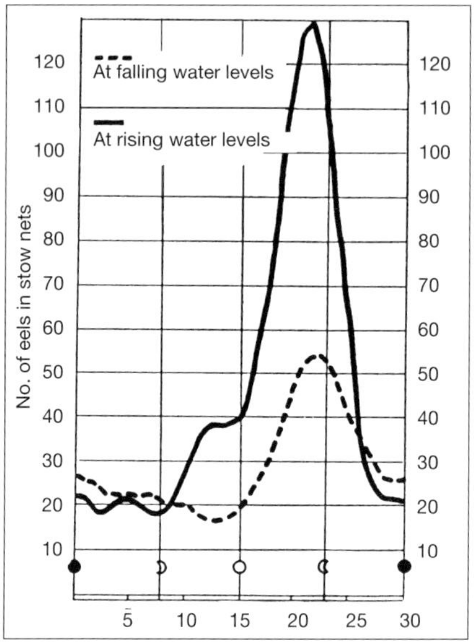

Fig. 3.29 Comparison of daily stow net catches in the upper Rhine during rising and falling water levels according to the lunar month model (after Jens, 1953)

are caused by low pressure areas over the ocean and transmitted long distances in the sea bed (Deelder, 1954; see also Määr, 1947). Analysis of 1000 fishing days showed that microseismic oscillations (3 s period) preceded major increases in catch. No other weather phenomenon, such as air pressure, wind direction and strength, temperature, precipitation, and cloud cover, was as well correlated with catch increases. Studies showed that microseismic oscillations possibly also influence New Zealand eels (see below).

3.4.5.2 Water level and current in streams

Current is the dominant aspect of streams, and it is strongly influenced by weather. It transports eels seaward and affects them in many other ways, as well. If there is more precipitation or more ice and snow thaws, streamflow discharge increases, water level rises, and current becomes stronger. It is well known that more downstream-migrating eels are caught during high than low flow. Studies were made on this in English and Irish lake outlets (Frost, 1950; Lowe, 1952).

In the previously mentioned Bann, the greatest amounts of eels emigrated during highwater periods. In the Elbe as well, catch increased considerably when stream discharge was greater. Although daily catch was about 10 kg per stow net, averaged over the whole year (Table 3.16), 40–50 kg/day were landed in individual cases during high water, and in low-flow years, in contrast, mean catch was barely 7 kg/day (Lühmann and Mann, 1961).

However, yellow eels were mixed in with the Elbe harvest. In fact, other factors often confound the situation so much that conclusions can be reached only by special analysis of the data. This was done with stow net catches in the Upper Rhine (Fig. 3.29) by separat-

Table 3.16 Otter-board stow net catches (Kilograms of eel per net and day) in the Elbe river downstream from Schnakenburg (after Lühmann and Mann, 1961).

Streamflow discharge (m^3/s)	Year	Catch rate, Fishery A (kg/net-day)	Catch rate, Fishery B (kg/net-day)
877	1956	15.22	11.95
904	1957	10.18	5.60
920	1958	18.12	8.88
571	1959	7.24	3.85
499	1960	5.98	4.58

ing out the catches made during rising or falling water stage within each lunar month. In both cases, the lunar effect dominated, but catches during the rising stage were so much greater that stimulation of eel migration by strong discharge cannot be doubted.

Also, for the two New Zealand species, strong rainfall also brought about higher silver eel catches (Burnet, 1969a). The observations were made in streams, so more rain can be equated here with increased streamflow discharge. But in large rivers a marked temporal lag between rainfall and stream flow would occur. The author is not sure whether the increased catches occurred because of the greater stream discharge or as a direct result of low atmospheric pressure passing through. Thus, the question arises here, too, about the influence of microseismic oscillations (see above).

The various authors generally characterise the increased catches as depending on higher water levels. Surely, this is not accurate. Direct influences of water levels on the number of emigrating eels are hard to imagine except in special cases. In Dutch polders, which as a rule are kept at constant water level, silver eel catches varied anyway; they increased according to amount of water released through the sluices (Deelder, 1970).

Does the increased volume of water or the stronger current cause the increased abundance of eels caught? It was determined in the Elbe that during high water, eel abundance can rise 5-fold (see above). This no doubt corresponds to the increased volume of water. Similar relations are seen in the Bann. However, the current becomes essentially stronger only in limited parts of the river, and then only by a factor of one and a half or two. Thus, in rivers, one should attribute the essential effect to water volume.

However, water level, too, cannot be completely without influence. Under certain circumstances, it even has decisive effect. If, for example, increased precipitation causes ponds or larger standing waters to overflow, this gives eels a chance to exit. This situation may have played a larger role in the past than now, when streams were not yet so channellised and diked as they are today.

3.4.5.3 Tidal currents

Quite apart from the fact that the current can only be of subordinate influence on the number of downstream-migrating silver eels passing a certain point, it does have critical effect on speed and direction of migration. Under certain preconditions, this effect continues when the eel reaches the sea. In the North Sea estuaries, migrating eels are carried passively with the ebb tide, so they (as well as the larval stage) are called 'drifters.' Their behaviours presumably do not change from those of river migration. Only during flood tide do they no longer drift. The fishers' stow nets in the North Sea estuaries catch the drifting eels exclusively during ebb flow. In contrast, yellow eels, depending on the season, are caught in both directions (Section 3.3.3.4). North American tracking experiments confirmed that silver eels can use the ebb tide in lower river reaches to reach the sea faster (Parker and McCleave, 1997).

Tracking animals tagged with ultrasonic transmitters gave an idea of how silver eels move in tidal currents (Tesch, 1972, 1974a, 1992). Figure 3.30 summarises the results and compares yellow eels that were displaced from the eastern North Sea and from the Baltic Sea. The silver eels moved mostly to the north or north-west, but the courses were much contorted. If the flood tide runs opposite to the swimming direction, the fish appears to move only a little; in some cases, it even goes in the opposite direction. However, if tidal current is accounted for, the direction of migration appears to be more constant. In general, silver eels swim without tidal current or time of day causing change in their direction of movement and speed. Thus, they do not use tidal current in the open sea, which is the case, for example, in the Baltic Sea and in the Mediterranean. Also, the observations described by Parker and McCleave (1997) were not made in large European and North American estuaries (Tesch, 1994; Stasko and Rommel, 1974).

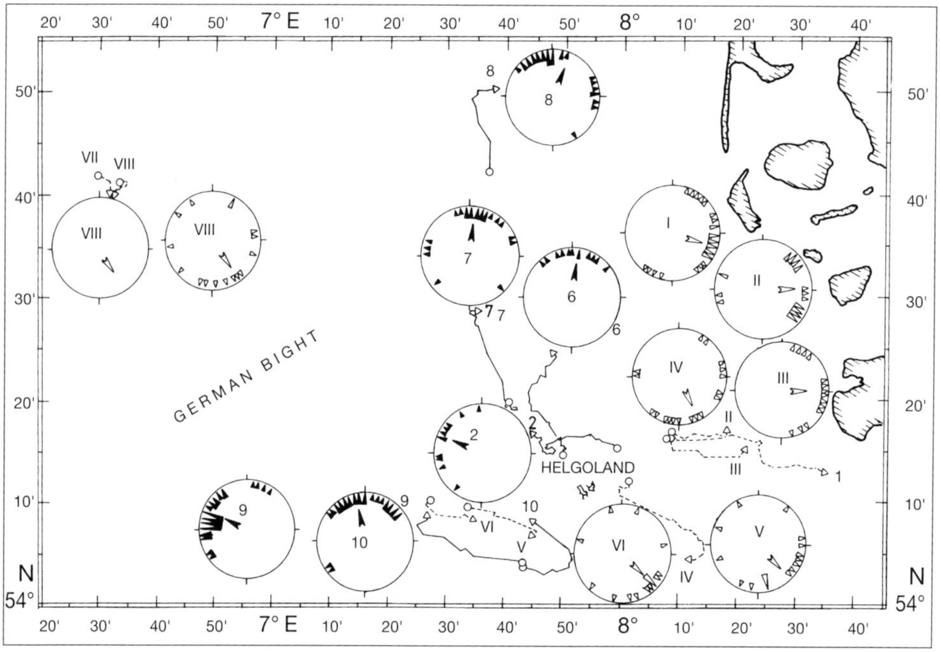

Fig. 3.30 Migration routes of silver eels and transplanted yellow eels in the southern North Sea (from data of Tesch, 1974a, c, 1975a)

- Solid lines: silver eel tracks
- Dashed lines: tracks of transplanted yellow eels
- Triangles in circles indicate swimming direction recorded every quarter hour (hourly in two cases) during tracking; arrowheads in circles show the resultant mean direction

3.4.5.4 Wind and currents

Eels already encounter strong currents in freshwater streams and rivers before reaching the tidal zone. Downstream migration with the current is used by the stow-net fishery, which catches eels not only on the bottom, but also in upper layers of the water (Section 5.4.3). The more water flows, causing current velocity to increase, the more eels are caught (Table 3.16). Tests in a freshwater current showed that silver eels stayed in the thread of the current, and this also applied to the tidally influenced area (Tesch, 1978c, 1991, 1994). In the middle reaches of the Weser, three ultrasonic-tagged eels swam 3.3 km/h on average. In stronger current (5 km/h) they were slower than the river; in weaker current (2.3 km/h) they were faster than the current.

Currents other than tidal flow affect the eel's route and must be taken into account. In the Baltic Sea, tidal current has almost no effect, but wind-dependent currents have much. In agreement with Nordquist, Rumphorst (1930) reported: 'On-shore winds are usually favourable; they bring the eel closer to land, hence into the traps. Off-shore winds are usually unfavourable; they keep the eel away from the coast.' On the Isle of Rügen's east coast, for instance, catches increased when the wind blew from the east (Meyer, 1938; Nolte, 1938). On Rügen's north coast, catches rose in northerly to easterly winds. Likewise, in Greifswald Bay, which is open toward the east, this occurs especially often during east wind. A west wind is unfavourable at most of the places mentioned

One should assume that a west wind could increase the catches on the west coast, for example, at Hiddensee Island (which parallels that coast). However, this is not so (Rumphorst, 1930): 'An offshore wind is a very favourable eel wind for Hiddensee, especially for its northern half and indeed an east and a north-east wind which blows for a long time, maybe also a north wind. At such times, the strong coastal current takes some of the eels along Rügen's north coast to the Cape of Arkona, the northernmost point, where they turn south and thereby wind up in the Hiddensee Island traps.'

Observations confirming that relatively many silver eels show up in fishing gear during on-shore winds are available from the North Irish Lough Neagh, from the English Lake Windermere (Frost, 1950), and from the Dutch IJsselmeer (Deelder, 1970). In the latter, for instance, 83.5% of the entire eel catch was landed during a strong west wind, but during moderate east wind only 50%. Also, in fyke-netting Berlin's Müggelsee (lake), the author caught eels in especially great number on the north shore when a south or southwest wind blew (Tesch, unpublished).

Thus, with the exception of the Hiddensee Island example, all observations seem to show that certain wind relationships bring eels closer to the coast. In trying to explain the phenomenon for the Isle of Rügen's east coast Nolte (1938) speculated that the eels, provided they migrate on the sea bed or at least in lower layers of the water, were induced to swim toward the shore against a deep counter-current created by an on-shore wind that pushed surface water shoreward. However, already at that time there was justifiable doubt about this. Indeed, several observations exist that silver eels also migrate near the surface (see below). Specifically for Rügen's coast, nets caught marked eels at all sorts of depths, including near the bottom as well as at 2 m beneath the water surface. Thus a surface current could carry at least some of eels to the coast, perhaps even most of them, for otherwise the greater catches during an on-shore wind would not be seen. Another explanation would be that, in the protection offered by greater turbidity during a strong on-shore wind, eels get farther into shallow water without their negative phototaxis forcing them to turn and swim at a normal distance from shore. However, on the west side of Hiddensee Island, they do not approach the coast during an on-shore wind because their westward (north-west to south-west) direction of swimming (Section 3.4.5.7) generally keeps them away from it. That eels choose direction according to current in the Baltic Sea (Westerberg, 1975) was not confirmed in later studies (Tesch *et al.*, 1991).

3.4.5.5 Temperature and depth

According to commercial fishers' observations in the Elbe estuary, when summer temperatures last into autumn, the silver eel migration is delayed or there is even a decrease in the numbers that emigrate. This suggests that certain minimum temperatures trigger the migratory behaviour. Of three silver eels tagged with ultrasonic transmitters and tracked in September in the Baltic Sea, one showed yellow eel behaviour (Fig. 3.30); it swam south-eastward (Tesch, 1979a). The surface temperature was then >16°C. The other two eels, tracked later at about 15°C, behaved 'normally', that is, swam west. In the process of studying thermal preferences of emigrating eels in a Norwegian stream, 9–18°C was the main activity range (Vøllestad *et al.*, 1986).

As the annual pattern of silver eel migration suggests, there may also be a certain lower thermal limit. Migration declines sharply in November. Only in exceptional cases can it last into January, for example, when seasonal rains are very late in the Bann (Frost, 1950). From there it is also reported that the onset of frost interrupts the migration. It can be surmised from silver eel tracking in November that migration takes place only above a certain minimal temperature. An eel released in the sea near Heligoland at 9°C water temperature migrated westward, that is, in the normal direction (Tesch, 1974a). Two days later, an eel tagged with an ultrasonic transmitter and released in 6°C brackish water of the Elbe estuary near Brunsbüttel let itself be drifted back and forth by ebb- and floodtide, thus may have been rendered inactive by temperature that (in brackish water) was too low (Tesch, 1974a). According to the aforementioned studies of Vøllestad *et al.* (1986), water temperature of 4°C is the lower limit for spawning-migration activity. However, a fisher of the Weser's mid-reaches reported to the author that some few eels still drift (passively) downstream even at water temperatures almost as low as 0°C. Maybe negative rheotaxis in fresh water is not much different from being inactive.

Perhaps it is low temperature that delays the start of silver eel migration in deeper, more gradually cooling inland waters. When eels that came from a canal were forced, at the beginning of their spawning migration, to pass through a lake deeper than 10 m, their migration was delayed. Below this lake, migration had to continue through a shallow canal, where the majority of the eels were found 2 months later rather than in the canal above the lake, which indicates a considerable delay (Deelder, 1957a).

Table 3.17 shows the water temperature and depth conditions as well as preferences of eels tracked from freshwater streams to the Sargasso Sea. It is clear from the Baltic Sea studies that eels avoid temperatures of 4°C at the bottom of water bodies (Tesch *et al.*, 1991), despite that fact that 6°C suffices at onset of migration in daylight to initiate migration at the bottom (Westerberg, 1979). After diving briefly into the 4°C zone, they immediately returned into warmer areas above the thermocline, where they continued their migration, which was somewhat deeper during the day than at night (Tesch *et al.*, 1991: in most cases not deeper than 4 m). Before that, during seaward migration in the streams, they had already switched from living on the bottom to swimming in the water's upper layers (Tesch, 1994). This behaviour is known from the stow-net fishery also (Köthke and Klust, 1956). For American eels, too, it was found that those emigrating in rivers rise from the bottom into upper layers of water during the ebb tide and in this way progress seaward faster (Parker and McCleave, 1997). This could not be confirmed for European eels in the lower Elbe (Tesch, 1994), perhaps because water temperatures at this place and for starting the spawning migration were too high (Table 3.17) or depth and breadth of that shipping channel were too great. Therefore, before active swimming began in water having higher salinity, only the residual current (stream flow ± tidal current) can have contributed to seaward movement (Section 3.4.5.6). Contrary to other ideas (Arnold and Cook, 1984), in the southern North Sea there were no preferred depths (Tesch, 1994; see also Tesch, 1974a). The northern exit of the 'North Sea shelf' has tidal currents that decrease in velocity and would not significantly aid the long migration to the Sargasso Sea.

Although eels take a rather long time in traversing the North Sea to reach the edge of the deepsea shelf, those from the Atlantic coasts and Mediterranean Sea usually reach it quicker. Once beyond the shelf edge, the eels go to much greater depths, especially during daytime (Table 3.17), which increases protection from enemies (Vaillant, 1896). As a consequence of its vertical migrations (Fig. 3.31) the eel is exposed to a wide range of water temperatures (13–22°C), which indicates great thermal tolerance during oceanic migration.

One can encounter an eel at strongly differing temperatures within just a few hours, changing from above to below the thermocline. For example, eels tracked with ultrasonic transmitters in the Bay of Biscay and west of Spain preferred depths of 40–250 m at night

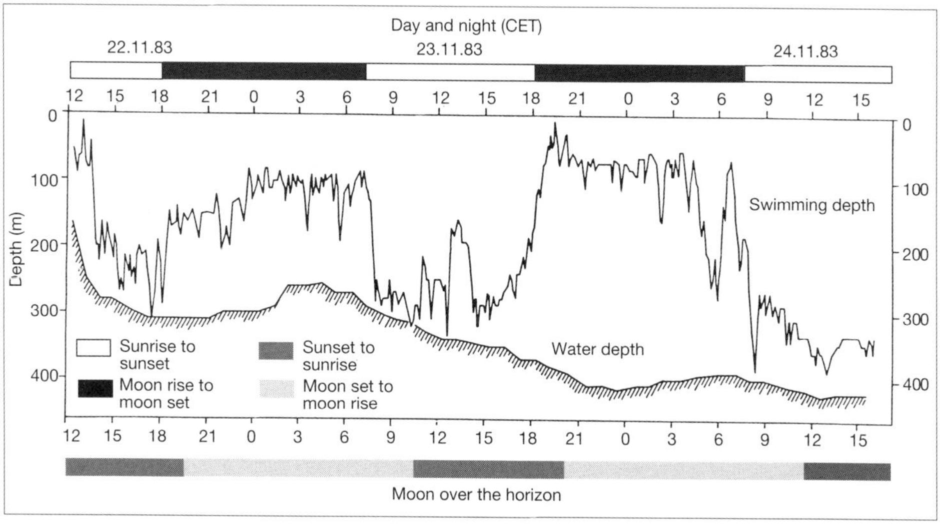

Fig. 3.31 Swimming depth of a silver eel 85 cm long that was released and tracked off the Moroccan coast of the Mediterranean Sea, 22 November 1983 (after Tesch, 1989)

Table 3.17 Depths and temperatures in which silver eels bearing ultrasonic transmitters were tracked during their spawning migration in fresh water, over continental shelves, in the deep sea region, and in the Sargasso Sea spawning area.

Water body	Date	Eel depth (m) night	Eel depth (m) day	Water depth (m)	Thermo-cline depth (m)	Water temperature (°C)	N	Remarks	Source
Weser river, mid-reach	29.XI.78	1 m over bottom		3.5–5.5	–	–	1	Eels in thalweg	Tesch, 1994
Weser river, mid-reach	XI.77	?		3.5–5.5	–	9	3	Eels in thalweg	Tesch, 1994
Elbe river, lower reach	IX,X.75/76	?		>15	–	14–18	10	Drifting with ebb tide	Tesch, 1994
Penobscot river	IX.X.91–93	0	bottom	10–15	–		6	SW-New Brunswick	Parker and McCleave, 1997
Ocean bay	X, XI.73	upper 5 m		10–80	–	7–11	5	On bottom 4 °C	Stasko and Rommel, 1974
Baltic Sea/Falster	23.IX.74	20	25	*c.* 50	25	Surface 16	1	on bottom 4 °C	Tesch, 1979a
Baltic Sea/Öland	X.85	13	16	*c.* 60	25	Surface 11	1	on bottom 3–4 °C	Tesch *et al.*, 1991
Baltic Sea/Öland	17.IX.85	11	8	20–60	16	Surface 11	1	on bottom 4 °C	Tesch *et al.*, 1991
Baltic Sea/Öland	IX.86	10	12	*c.* 60	17	Surface 11	5	on bottom 6 °C	Tesch *et al.*, 1991
Baltic Sea/ Bornholm	IX.85	1	15	*c.* 60	21	Surface 11	4		Tesch *et al.*, 1991
Baltic Sea/ Bornholm	IX.74	12–30	bottom	*c.* 40	18–30	Surface 15	1		Westerberg, 1979
North Sea/ sourth-east	2.XI.88	12	12	32–37	none	11.5	1		Tesch, 1992
North Sea/ south-east	4.XI.88	13–4	8–2	18–34	none	11	1		Tesch, 1992
Biscay	IX.75/76	150	>350	<2200	<100	Surface 13	3		Tesch, 1978b
Biscay	X.73	0–7		*c.* 50	>25	13.5	1	2 downward movements/h	Tesch, 1979a

Western Mediterranean Sea	8.XI.82	47	258	62–560	80–180	Surface 17	1	night above, day below the thermocline	Tesch, 1989
Western Mediterranean Sea	19.XI.83	*c.* 84	*c.* 420	106–600	90	13–19	1		Tesch, 1989
Western Mediterranean Sea	22.XI.83	*c.* 90	*c.* 300	165–420	*c.* 90	13–19	1	temperature 16°C night, 13°C day	Tesch, 1989
Western Mediterranean Sea	4.XII.84	*c.* 278	*c.* 405	100–920	*c.* 73	<14	1		Tesch, 1989
Sargasso Sea	III.79	100–670		*c.* 5000	*c.* 100	12–25	2	usually below thermocline	Tesch, 1989
Sargasso Sea	III.93	250		4–6000	*c.* 100?	18.8	2	usually below thermocline	Fricke and Käse, 1995

N Number of eels.

(Tesch, 1978a, b). Within 3–4 h after daybreak they swam 150–200 m deeper, that is, they then preferred depths of 300–500 m. In the process, they swam through the thermocline, which was at 75–100 m, and where the temperature fell 2–3°C. The depths occupied by an eel tracked in the Mediterranean Sea are presented in Table 3.17 and Fig. 3.31. Of the 11 eels that were successfully tracked there (Tesch, 1989), the 8 tracked long enough (in sufficiently deep water) had average depths of 84–278 m (mean: 167 m) at night and 212–623 m (mean: 347 m) during daytime. The average daytime swimming depths presumably would have been deeper yet, but ocean depth was not always sufficient to meet their requirements, as Fig. 3.31 indicates.

It was calculated on the basis of the vertical swimming speed that the limit of descent, perhaps also of pressure tolerance, lies at a depth of some 600 m (Tesch, 1995). This value is of the same order of magnitude as tracked eels showed on the east Atlantic continental slope, in the Mediterranean Sea, and shortly before they reach the Sargasso Sea spawning grounds (Table 3.17). According to tracking studies, two completely ripe eels that had reached the spawning grounds there appeared to prefer depths of only about 250 m and did not undertake vertical excursions anymore (Fricke and Käse, 1995). This depth is consistent with the occurrence of the smallest larvae (still yolk-sac stage) (L_T <5 mm), which is why the depth of spawn deposition is estimated at 200–300 m (Castonguay and McCleave, 1987; Tesch and Wegner, 1990). Hormone-treated Japanese silver eels, tracked by the same technique in the presumed proximity of the spawning sites but only for a few hours, swam only about 40–190 m deep in 2–3 h; an eel tracked for 7 h swam with increasing tendency in about 120–260 m, thus at about the same depth as European eels do before spawning (Aoyama *et al.*, 1999).

Besides different day and night choice of depth, the eel tracking often showed short vertical excursions, sometimes down to the bottom (Stasko and Rommel, 1974; Westerberg, 1975; Tesch, 1995). An eel tracked for almost 6 days in the Mediterranean Sea undertook vertical excursions of >100 m: 12 already on the 1st day, five on the 2nd day, and an average of only about three down- and upward journeys per day thereafter (Tesch, 1995). Therefore, the initially large number of vertical ventures is apparently an artefact of the experiment, but they do seem to decline in number. Perhaps the vertical excursions enable the eel to orient itself regarding depth conditions. Decreasing depth can be a sign of coast or shelf edge and calls for change in the direction of swimming. In addition, continual depth change possibly contributes to increased production of gonadotropic hormones, hence promotion of spawning maturity (Fontaine *et al.*, 1985).

3.4.5.6 Salinity

Concurrent with the eel's physiological adaptation for migration, its body functions change with respect to salinity (Sections 1.4 and 1.7.1). Quite possibly, it now not only tolerates increased salinity, as does the yellow eel, but even prefers it. Logically, it would have to exhibit behaviour opposite to the glass eel, which moves in the other direction. This was indeed detected for silver eels in a North American estuary (Parker and McCleave, 1997). Beyond that, commercial fishers confirm that eels swim against the seawater flood tide current (Deelder, 1970). However, the glass eel does not seek low salt concentrations but rather the odour of natural fresh water (Sections 1.4 and 1.7.1). Are there corresponding behavioural patterns for the silver eel regarding sea water, or does it orient itself directly to the osmotic gradient, as is suspected for salmon (McInerney, 1964)?

Experimental studies showed that silver eels became active (positively rheotactic) when presented with sea water; in fresh water, in contrast, they showed themselves to be negatively rheotactic (Hain, 1975). Thus, although they swim downstream in fresh water, they actively turn against the current in salt water, quite contrary to yellow eels, which remain inactive in sea water. Olfactorily impaired silver eels do not show this activity toward sea water. Thus, it is to be inferred that silver eels detect sea water with their olfactory organs, much as glass eels recognise fresh water by smell. That immediate recognition of sea water is helpful when emigrating from fresh water into the ocean is shown by a previously mentioned example (Section 3.3.3.3). Where sea water washes over sand bars that block coastal lakes, eels wriggle out of the lakes towards the inflowing sea water.

This activation of silver eels by sea water becomes especially significant, if it is considered that, just as they enter it, a certain migratory course tendency takes effect (Section 3.4.5.7; Tesch and Lelek, 1973a, b; Tesch, 1974a; Tesch *et al.*, 1992). This only makes sense when the eel has reached the estuary and the open sea. Before that, in fresh water, the tendency to swim north (see below) is not apparent, as laboratory experiments showed (Tesch, 1974b). Experiences by stow-net fishers in the lower Elbe show that a northward tendency becomes evident as soon as the silver eels have reached the brackish water of this river, which flows westward before emptying into the North Sea. Here, they are caught only on the north bank, that is, apparently at the start of their northward tendency in the increasing salinity.

3.4.5.7 Directional choice in the ocean

The Baltic Sea represents a particularly favourable study area for research on the silver eel's oceanic migration. An extensive silver eel fishery, perhaps as found nowhere else in the ocean, enabled recaptures along almost the entire Baltic Sea coast (Sections 4.2.2 and 5.4.2.3). Due to that, tagging operations to study the eel's spawning migration periods already began at the start of the 20th century (e.g. Nordquist, 1904; Trybom and Schneider, 1908; Määr, 1947; Martinköwitz, 1960, 1961; Karlsson, 1984). Ever since, most silver eels by far are recaptured on the Baltic north coast. This is so for eels released in the eastern part, that is, in the Finnish Gulf, and for those from the south coasts, as well as those from the north coast. Eels tagged on the coasts of Estonia, Latvia or Lithuania crossed the Baltic Sea and were recaptured on the Swedish coast. Those from the area of Rügen headed west or north-west to the nearest coast, that is, to Denmark or Sweden. Those released on the Swedish coast were recaptured apparently without exception on the same coast or by Danish fishers on The Sound (Øresund), the Baltic Sea's northernmost exit (Berntsson *et al.*, 1974; Ask and Erichsen, 1976). The general course of all eels marked in the Baltic Sea, from release site to recapture site, is, of course, south-west (Karlsson, 1984) because the Baltic Sea runs from north-east to south-west. This complicates interpretation of a free directional preference of eels in the Baltic Sea, as set forth below.

The only two eels recaptured after conventional marking in the North Sea region travelled from the river Elbe into the north-western coastal area of Denmark, thus must have swum north (Lühmann and Mann, 1958). In addition silver eels in the south-western North Sea swim north-westward (Fig. 3.32; Tesch, 1974a, 1992). The constant north-westward course makes sense for finding the North Sea's northern exit. These tracking results together with those from other areas are summarised in Figure 3.32. Among these are also earlier initial studies of this kind in the south-western Baltic Sea. This eel's mean course lay to the north–north-west, in so far as it was not blocked by land or shallows. Therefore,

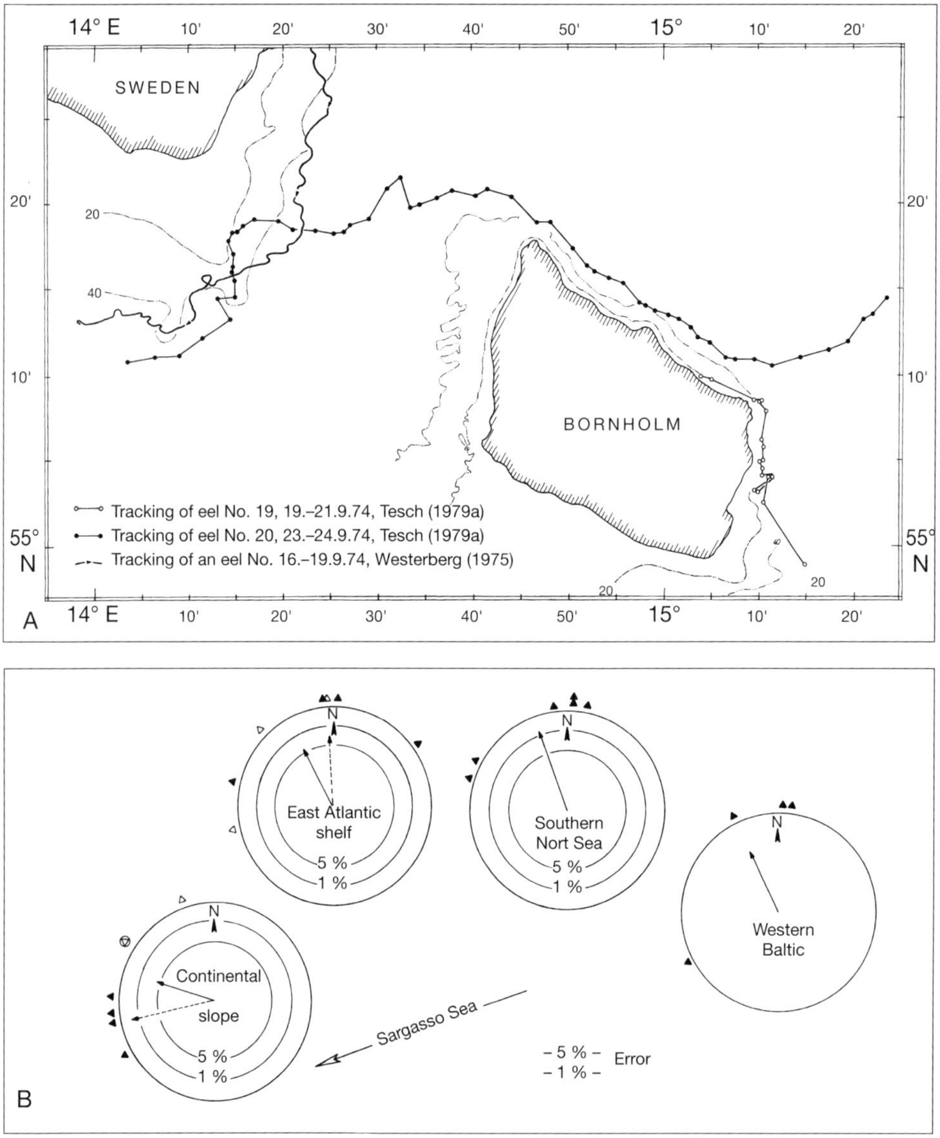

Fig. 3.32 Silver eel migration in the Atlantic and in the Baltic Sea. (A) Tracking courses for three eels between Bornholm and the Swedish coast (from data by Tesch, 1979 and modified after Westberg, 1979a). (B) Distribution of compass directions of individual mean bearings (triangles) for silver eel tracking courses in various ocean regions (continental slope and east Atlantic shelf after Tesch, 1979; southern North Sea after Tesch, 1974a; western Baltic Sea after Tesch, 1979). Arrows: mean direction within the particular region, with length indicating probability of error (scaled on interior circles) – but this does not apply to the dashed arrows, which represent directions exclusive of doubtful bearings (open triangles). Two further eels, having courses of 348° and 294° in the southern North Sea, were consistent with these results (Tesch, 1992).

it corresponded to the directional choice also found for eels in the North Sea, and beyond that, to the course found by tracking on the Atlantic shelf. Figure 3.32 also shows a selection of eel migratory courses near Bornholm, which demonstrate the influence of coastal shape on the compass migration.

The experimental animals had a north- to westward tendency on the Atlantic shelf, in the North Sea, and in the south-western Baltic Sea immediately after release, but this could not be detected by tracking in the central Baltic Sea (Tesch *et al.*, 1992). There, almost all eels swam south-westward (mean: 209°) after release, but later, during as much as 2 days of tracking, they altered their direction clockwise to the right, so that they finally did show preference for the north-west (mean: 294°). This could indicate that in northern latitudes, for example, in the central Baltic Sea, the south-west direction is preferred. With decreasing latitude and/or longer time swum, the eels alter course, usually passing through west to north. Another explanation for the change to northward course in the south-western Baltic could be the increasing salinity there (Section 3.4.5.6). Measurements of current were also made in the Baltic Sea studies in order to investigate Westerberg's (1979) speculation that the eels could be orienting themselves against the current. But this could not be confirmed in any of the 16 healthy eels that were tracked (Tesch *et al.*, 1992).

Then, as long as they are not diverted by shallows, the migrating eels' north-west course takes them to somewhere on the shelf edge which slopes down to the deep sea. Results of tracking at these places suggest it is there that they first take the south-west-to-westward course required for reaching the Sargasso Sea (Fig. 3.32).

It is unknown what orientation mechanism is available for maintaining the compass course across the shelf or the deep sea. For a fish that migrates mainly at night and often at great depth (see below), optical features of the heavens (sun, stars) are excluded. Also, the eel's extremely sensitive nose (Section 1.9.3) is out of the question for orientation to a compass course or to guide the eel across greater distances (Tesch, 1970). Laboratory experiments showed that the earth's magnetism enabled the necessary course-keeping for the eel, and even changes in strength and inclination of the magnetic field could affect the direction of swimming (Fig. 3.33; Tesch, 1974b; Souza *et al.*, 1988; Tesch *et al.*, 1992). Such an orientation is also a possibility for the eel larvae that migrate from the Sargasso Sea toward Europe or America (Section 2.2.1). Orientation aided by the earth's magnetic field is detected in the animal kingdom, especially in birds (Ossenkopp and Barbeito, 1979; Wiltschko and Wiltschko, 1996), in which the field's inclination is significant. Similar sensitivity to the magnetic field's inclination, as appears possible from tests with eels (Tesch *et al.*, 1992), was detected in the sea turtle, *Caretta caretta* (Lohmann and Lohmann, 1996), which also has to cross the Atlantic Ocean.

3.4.5.8 Behaviour and speed

In rivers, silver eels probably drift in mid to deep layers of the water (Section 3.4.5.5). If, for example, stow nets are set right on the river bed, they catch fewer eels than if in mid-water. Silver eels do not necessarily migrate singly. Supposedly, they also drift in groups like yellow eels (Section 3.3.3.2).

Various people have observed that, at least at night, eels can be encountered immediately beneath the surface. It was found in spotlighting the sea surface in the Great Belt

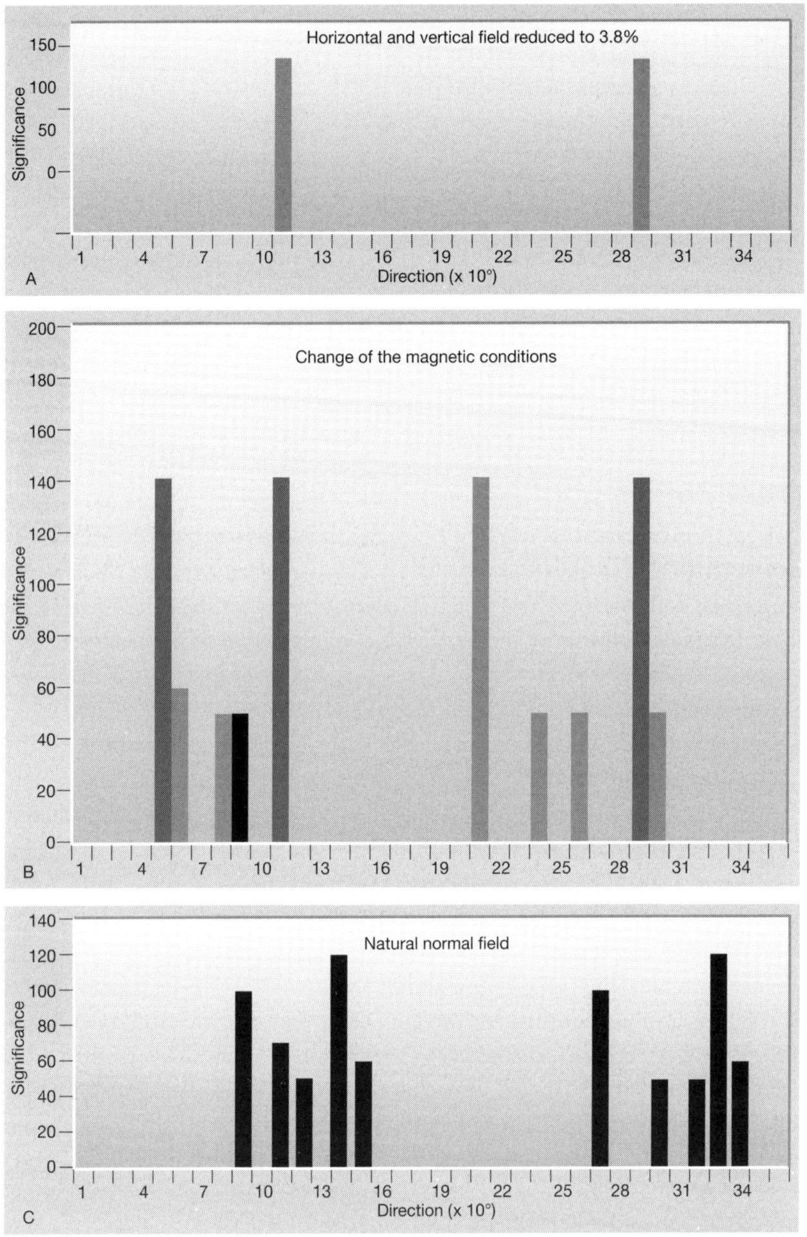

Fig. 3.33 Directional preferences of yellow and silver eels related to the Earth's magnetism, as studied in various laboratory experiments involving different techniques. (A) Compensation for Earth's magnetism in a residual field that was 3.8% of the normal field strength. (B) Artificially opposed horizontal field and artificially altered inclination, direction, and strength of the magnetic field; the various bar shades signifying different conditions. (C) Normal field conditions of the Earth's local magnetism in Hamburg

Scale of significance on the ordinate:

50 = 0.10 or less

100 = 0.05

140 = 0.03

In various paired-comparison tests, the difference from normal choice of direction at altered inclination, direction, and strength proved to be significant.

Table 3.18 Daily distance travelled by silver eels according to results of tagging and recapture and of tracking (mainly headway through the water) by means of ultrasonic tags.

Ocean area	Distance travelled from release site to recapture site (km)	Travel speed (km/d)	Remarks	Source
Baltic Sea	>1000	13	–	Trybom and Schneider, 1908
Baltic Sea	>230	16	–	Trybom and Schneider, 1908
Baltic Sea	–	55–63	maximum speed	Määr, 1947
Baltic Sea	–	16	mean speed	Määr, 1947
Baltic Sea	Rügen→Swedish coast (95 km)	15	resolute swimming, October	Martinköwitz, 1960, 1961
Baltic Sea	Rügen→Danish coast	10	resolute swimming, October	Martinköwitz, 1960, 1961
Baltic Sea	Rügen→Swedish coast (125 km)	5	less resolute, September	Martinköwitz, 1960, 1961
Baltic Sea	Rügen→Danish coast (125 km)	8	September	Martinköwitz, 1960, 1961
Central Baltic Sea	8 eels tracked for 1–3 d	25	middle to end of September	Tesch *et al.* 1991
Southern North Sea	1 eel tracked for 34 h	36	beginning of November, Heligoland	Tesch, 1992
Eastern North Sea		25	–	Lühmann and Mann, 1958
Mediterranean Sea	6 eels tracked for 1–6 d	27	Oct./Nov. east of Gibraltar	Tesch, 1989

(one of the straits connecting the Baltic Sea and the Kattegat) that some of the silver eels migrated just under the surface and with slow body movements. This surface migration ended at moonrise (*Fischerbote* **21**, 316–318, 1929). Similar behaviours are known for *A. rostrata* (Maritimes 1969, Winter: 3-5). A female eel was sighted near the surface in the US state of Florida's coastal area at new moon in November. Further, many eels were caught by night-time spotlighting 5–10 nm off the coast of North Carolina (USA); the same for one individual over the 180 m contour off Rhode Island (USA). Other eel-like fishes behave similarly during their spawning migration, for example the tropical, shallow-water worm eel, *Ahlia egmontis* (Cohen and Dean, 1970).

Regarding distances traversed, quite considerable daily performances have been reported. This does not involve just individual performance, but rather the combination of swimming speed and current. In the Elbe, tagged silver eels could cover at least 30 km/day (Lühmann and Mann, 1958). In the ocean, speed was less in most cases (Table 3.18). For the Baltic Sea, it should be kept in mind that augmentary currents are seldom an issue. There, progress of 15 km/day can be reckoned with during the period of most active migration. At the beginning of the migratory period, the eels were apparently less energetic, and only 5–8 km/day were covered. In the North Sea, 25 km/day can be expected, according to tagging and recapture. The figures for the North Sea are supported by observations of yellow eels, which likewise homed at 25 km/day under favourable conditions. Moreover, daily performances 3-fold greater appeared possible in exceptional cases (Tesch, 1967a; Deelder and Tesch, 1970), which also proved reachable in individual cases for silver eels in the Baltic Sea. Silver eel tracking in the North and Baltic Seas enabled direct investigation of migration speed, which, of course, had to turn out higher because all small detours like, for example, zig-zag swimming could not be taken into account (Fig. 3.30; Tesch, 1972, 1974a; Tesch *et al.*, 1992). The performance of relatively large female eels amounted to 48 km/day in the North Sea. Two eels of about 90 cm long maintained speeds of 42 and 44 km/day, respectively, for 3 days. In the Baltic Sea, 13 experimental animals covered about 0.6 body lengths per second. This would suffice for reaching the Sargasso Sea (Section 2.2.1) by the spawning season in March, if the established compass course were maintained.

4
Harvest and environmental relationships

4.1 Development of eel fishing

Eels were fished already in pre-historic times. From numerous archaeological discoveries, we know the diet of our ancestors included fish at all times. Eels should not have been any exception. The extent to which they were harvested can be inferred from publications or archives. Publicly available statistics already existed relatively early from Denmark. The amounts of eels landed from a major part of Danish waters are known since 1885. As the reported amounts caught at that time were little less than present-day average harvests, it can be assumed that considerable amounts must have been landed and sold long before 1885. Thus, the origin of fishing techniques and to a lesser extent also the beginning of statistical registration must date from earlier still. The most important Danish fyke-net fishery originated in the Great and Little Belts (straits connecting the Baltic Sea and the Kattegat). It took 400–1400 t of eels annually between 1902 and 1916. The same fishery, including also hook-and-line fishing, already reported 333–585 t/year during 1885–95 (Jacobsen and Johansen, 1922). For 1940–59, the fishery of the Belts, still based on fyke netting for silver eels, had not risen above the level of that at the turn of the century: it yielded 500–1000 t annually (Jensen, 1961). Thus, the harvest levels have stayed essentially constant since 1885.

The most important Italian eel fishery, the catch in the north Adriatic lagoons, can be traced even much further back (Bellini, in Walter, 1910). In the Lagoons of Comacchio, 962 t were caught annually on average during 1798–1924. Yield declined in 1825–70 to 493 t/year, then rose to an average of 858 t/year for 10 years, only to finally diminish again at the beginning of the 20th century to 391 t/year. The catches achieved at the end of the 18th century in Comacchio amounted to a third of today's total Italian eel landings. But it is not just since 1800 that the lagoon fishery of Comacchio has been very productive: it was already that way in the 14th century. At that time, the leasing of fishing rights alone took in 266,600 marks annually from this fishery (Walter, 1910), so its yield could hardly have been below that of the 19th century. Eel fishing is thus an old branch of the economy in Europe.

Less information is comprehensively available about eel harvests and their development up to the present time on other continents, to say nothing of worldwide. Surely, eel fishing has also been significant at various places outside Europe since long ago. The statistics of the International Council for Marine Research and of the United Nations Food and Agriculture Organisation (FAO) first give a certain, even if very incomplete, overview. From these, a certain increase in the world catch seems to occur from about 1960 onward, that is, after the consequences of the Second World War had been overcome (Fig. 4.1). Then, world catch went into a decline, and this included the European and East Asian (Japanese) catches, as is also to be inferred from comparison of the years 1963 and

1993 in Table 4.1. The production in Japan's natural waters amounted to about 3200 t in 1938 but had shrunk to 1300 t in 1996 (Fig. 4.1). Nevertheless, the total amount available for consumption rose tremendously. That is attributable to the development of eel aquaculture in Japan and China (Fig. 6.1, Table 6.1). The aquaculture began at the end of the 19th century and in the 1980s reached a production of almost 40,000 t/year in Japan alone.

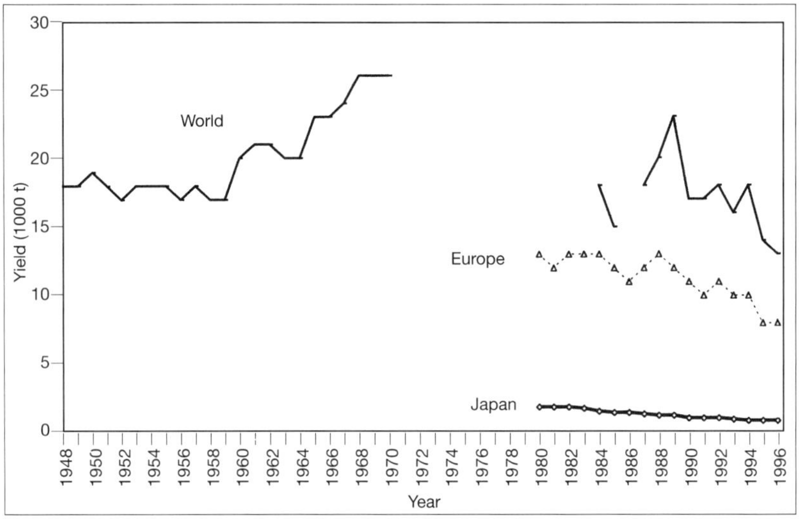

Fig. 4.1 Eel catch for the world, Europe, and, as an Asian example, Japan without aquacultural production (in part estimated, based on data from the *FAO Yearbook Fish. Statist.* and *Bull. statist. Cons. perm. Explor. Mer.*)

Table 4.1 Eel catch by weight (aquacultural production not included) in various European and North African countries in 1963 (Tesch, 1983) and 1993 (after *Yearbook Fish. Statist.* FAO).

	Thousands of tons	
Country	1963	1993
Italy	3.5	3.5
Denmark	4	1.8
Germany	2.1	1
Sweden	1.9	1.3
Holland	1.9	0.4
Spain	1.9	0.2
France	1.4	1.7
Poland	1	1.1
N. Ireland	0.7	0.6
Portugal		0.5
Turkey		0.3
Tunisia		0.4
Norway		0.3
Greece		0.4
Hungary		0.3
Ireland		0.2
Morocco		0.2
Albania		0.2
Belgium		0.1
Others		0.3

Including figures from aquaculture in China and Taiwan, worldwide eel production climbed from 38,000 t in 1938 to 205,000 t in 1995.

It should be noted that the FAO statistics traced here are incomplete with regard to yields and to some extent do not describe the real values. Therefore, not all of the data that are available are represented in the following discussions. One example: Taiwan, which was not included in the FAO statistics, produces yields in its aquaculture facilities, which nearly equal those of Japan (Table 6.1). Its annual export just to Japan is said to have amounted to 23,000 t at times (*Inform. Fischw. Ausland* **30** (6), 1970). Furthermore, it was difficult to separate aquacultural production from open water catch in other countries. Commercial fishery catch is underestimated in many cases because part of it does not get recorded in the statistics. For example, Ireland's landings, rather than being the 150 t given in FAO statistics, are estimated at 250 t (Moriarty, 1996).

4.2 Regional yearly fluctuations in harvest

4.2.1 Fluctuations in Europe and the North Sea

As natural stock density of a given year class within eel populations of, for example, all Europe or parts of it, can presumably depend on the results of a single spawning run, it can be assumed that fluctuations in reproductive success influence the eel yields in all European countries, or at least part of them (depending on the genetic relationships of the European eel population, Section 2.5.3), in the same way. Furthermore, climatic fluctuations could raise and lower the eel harvest equally throughout large areas. When, for

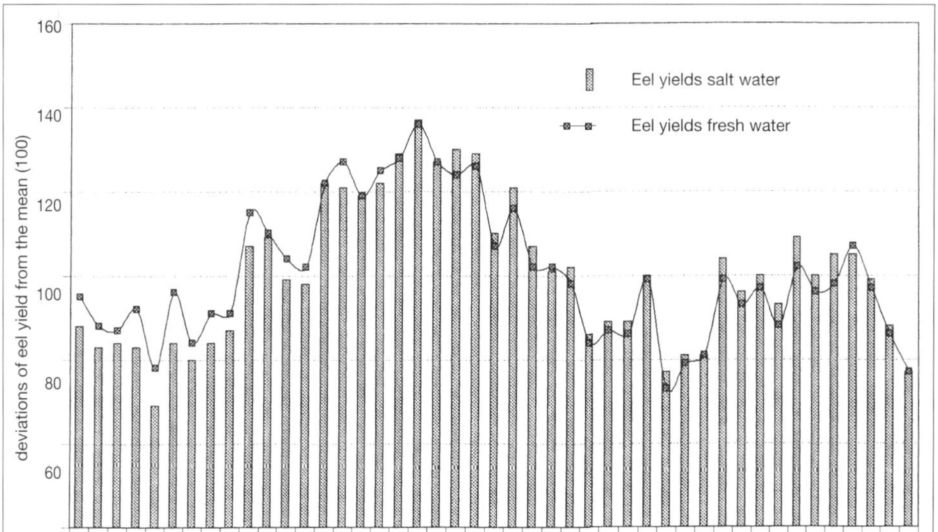

Fig. 4.2 Deviation from the long-term mean for fresh-water and saltwater eel yields, calculated by Kuhlmann (1998) from data for Germany, Denmark, France, Ireland, Italy, The Netherlands, Northern Ireland, Norway, Poland, Sweden, and Spain

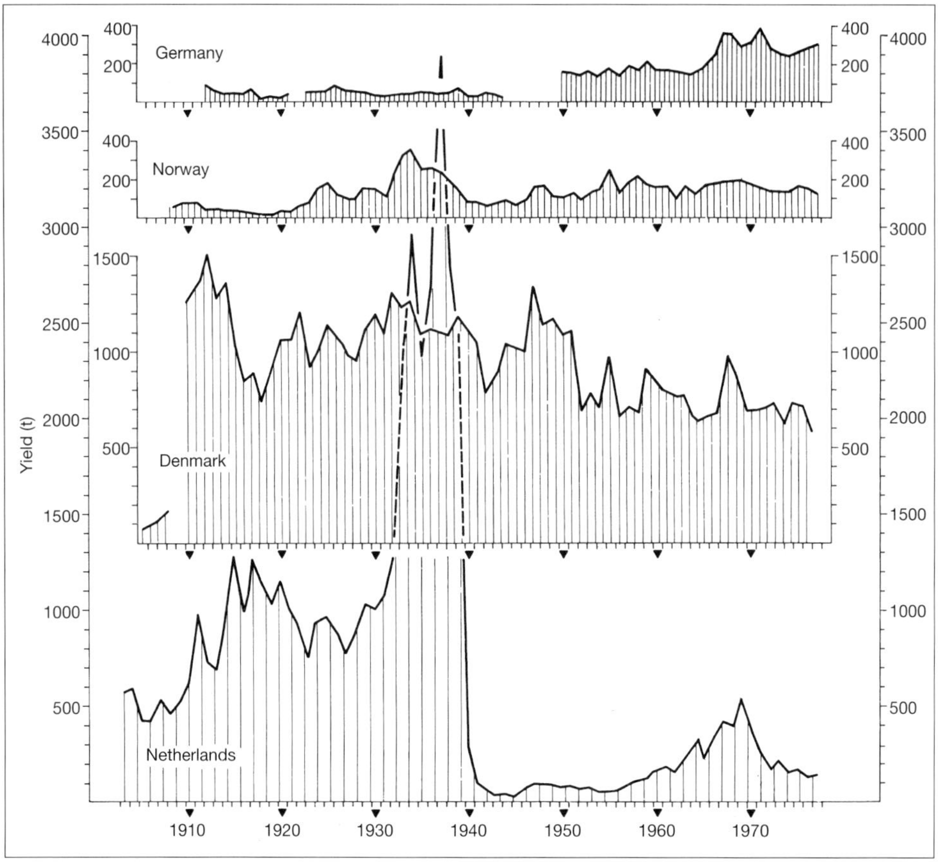

Fig. 4.3 Eel yields from the North Sea coasts of countries adjoining the North Sea (data from *Bull. statist. Cons. perm. Explor. Mer.*)

several European countries combined, the salt- and freshwater eel catches are compared, both show astoundingly similar fluctuations year to year, as well as in long-term periodicity (Fig. 4.2). This is particularly remarkable for the North Sea, whose individual bordering countries had rather similar fluctuations in yield (Fig. 4.3). In this example, if each year's yield level relative to that of the previous year is recorded as positive or negative, the result is that in 39 of the 49 comparable years, yield either rose or sank at the same time in at least three of the four countries involved; in 14 years it even showed the same trend in all four countries. As the economic conditions in these countries do not always follow the same trend, economic influence is not usually the cause, when their eel harvests fluctuate in the same direction.

4.2.2 Fluctuations in the Baltic Sea

In the past, smaller fluctuations of eel landings in the Baltic Sea, as well as differing proportions of biomass of males and females were recorded often and from several fishing areas simultaneously (Tesch, 1973). As was evident in former years, above and beyond this, a general decline in yield dominates these smaller fluctuations in the whole Baltic Sea.

This was especially obvious on the Swedish southeast and south coast (pers. comm., Svärdson, and according to statistics from Wickström). Clearer yet is the development from the silver eel catches of various trap net sites in Denmark (Table 4.2). The catches of small, primarily male eels declined from 600–2000 kg to 37–100 kg per trap at all three sites in the course of the last 20–30 years. The catch of the large and almost exclusively female eels likewise decreased, but only by about half. Svärdson (1976) reported that the proportion of females at various sites on Sweden's southwest coast was 44%, 19%, 75% and 65% in 1890, that in 1910 it was 7% and 59%, and that in 1970, at seven sites in southeast Sweden it was 96%, 40%, 60%, 87%, 97%, 92% and 98%, thus a considerable increase. The decline in male eels corresponds to the well-known tendency of lower stock densities to have lower percentages of males (Sections 1.7.2 and 3.3.2.2). Observations by divers along the southern part of the Baltic Sea's east coast showed about 10–15 eels grouped together in 1955, but only 2–3 eels at most in 1998 (pers. comm., Breitling).

Table 4.2 Mean annual catches (kg) of small and large eels per pound net in the Danish fishery districts Køge (south of Copenhagen), Faxe (Southeast Sjælland) and Rødby (South Lolland) (after Hoffmann *et al.*, 1979).

	Køge			Faxe			Rødby		
		Catch (kg)			Catch (kg)			Catch (kg)	
Year	Number of nets	Small eels	Large eels	Number of nets	Small eels	Large eels	Number of nets	Small eels	Large eels
1949	25	1191	1903						
1950	24	1101	2056						
1951	27	1029	1243				3	2053	1629
1952	25	1276	1544				3	1300	1829
1953	21	1384	1309				6	1016	1158
1954	18	1075	1176				7	645	937
1955	24	976	996				7	1070	887
1956	22	826	1011				6	816	876
1957	20	864	940				6	404	1291
1958	17	610	1094	11	640	842	6	1052	696
1959	18	836	1256	11	681	898	6	850	1718
1960	14	711	2157	11	934	1230	6	575	721
1961	16	590	1577	11	840	1058	7	444	834
1962	16	505	1840	11	703	927	7	464	1280
1963	17	543	2464	13	370	883	6	311	850
1964	21	450	2317	13	492	760	9	259	449
1965	30	306	1916	13	278	747	9	311	528
1966	34	290	1589	13	340	922	10	198	478
1967	32	196	1493	13	259	413	10	232	536
1968	32	226	1546	14	236	944	11	298	907
1969	32	153	1253	14	267	718	11	127	954
1970	32	101	971	14	168	599	11	197	1263
1971	36	93	918	14	183	604	12	177	1071
1972	36	56	1267	14	147	605	11	57	990
1973	36	117	1513	14	122	577	14	72	946
1974	36	62	994	14	143	605	14	82	799
1975	36	87	1026	14	140	566	14	78	736
1976	34	100	1498	14	130	670	14	37	408
1977	30	47	1244	15	84	515			

The reasons for the reduced eel population in the Baltic Sea are unclear. Although less influx of juvenile eels was detected at various sites on the Swedish coast (Fig. 3.8; Svärdson, 1976), reduced reproduction in the Atlantic Ocean cannot be concluded from that. Up to that time, other European areas did not show any correspondingly extensive decline. The juvenile eel immigration could also have had a regional character and such causes as the west-wind tendencies described later (Section 4.4). According to the fact that panmixia of the European eel population probably does not exist (Section 2.5.3) the regional character of the eel's population decline in the Baltic needs increased consideration.

Hoffmann *et al.* (1979) reported increased fishing for yellow eels in southeastern Denmark and on the Swedish south coast. The same can surely be reported as well from other countries bordering the Baltic Sea. This sort of intervention into the eel population, of course, also reduces the number of emigrating silver eels, but surely not to such an extent as is expressed for the long term in Fig. 4.4.

4.2.3 Correlation of west wind and glass-eel invasion

Svärdson (1976) reported that in the 1960s a minimum of westerly winds was recorded in the immigation months of December and January, which could have had the effect of reducing drift of glass eels into the North Sea and the Skagerrak. A cause of higher glass-eel invasion on to the European coasts is, according to Danish opinion, presumably the occurrence of stronger west-wind drifts in springtime (e.g. Jensen, 1961). They give rise to increased surface currents and thereby promote the washing of glass eels on to the coasts. Therefore, the silver eel yields in areas, where primarily males occur were compared with the air pressure difference between the Danish North Sea island Fanø, off southwest Jutland and Skagen, a town at Jutland's northern tip. The air currents at the Earth's surface are known to flow in about the same direction as the isobars. If a major difference in air pressure occurs between north-lying Skagen and Fanø to the south, then the lines of equal

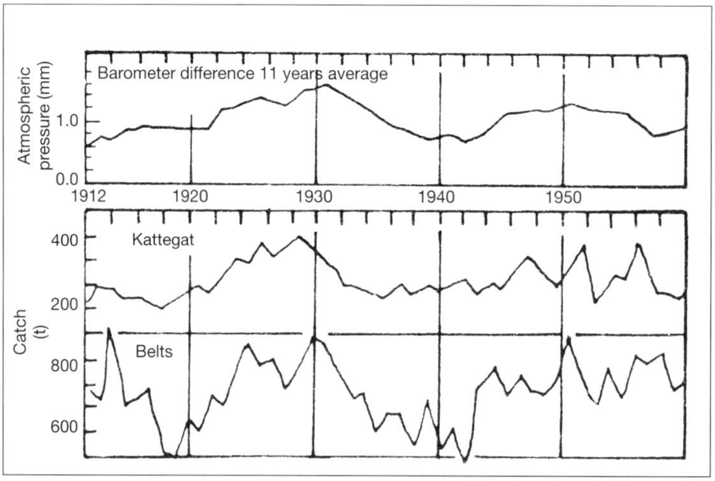

Fig. 4.4 Landings of the Danish silver eel fishery in the Kattegat and the Belt straits, 1912–59, and the differences in barometric pressure between Fanö and Skagen 7 years previously (after Jensen, 1961)

air pressure will very probably run east-west at both points, and west winds will occur depending on strength of the pressure difference. This barometric difference was, therefore, applied here as an index of west-wind drift strength, and parallel trends in catch and air pressure difference resulted. As glass eels are also caught in Denmark (Section 3.1.2), it was, in addition, investigated whether the glass-eel catches 7–10 years previously correlated with the silver eel harvests. Agreement occurred here, as well. To be able to make predictions under the present conditions of severely diminished yield, further long-term comparisons would have to be conducted.

4.2.4 Influence of seasonal temperature

The possibility suggests itself of correlating eel catch, particularly that of the yellow eel fishery, with summer temperatures and this has been done by Danish scientists since 1922 (Jacobsen and Johansen, in Jensen, 1961). Although a certain correlation does exist here, it is likewise thought possible to see a relationship with winter temperatures, namely, that in cold winters, the eels suffer substantial losses (Johansen, in Jensen, 1961). Commercial fishers from Heligoland confirmed this to the author by pointing out that, for example, after the cold winter of 1962–3, particularly severe losses among blinded eels occurred (Section 1.4). Eel mortalities were observed during the cold winters of 1962–3 and 1969–70 in inland waters, as well (Rahn, 1963; Tesch, 1964; Mattern, 1971).

The extremely cold winter of 1962–3 also caused higher mortality in other aquatic organisms (*Helgoländer Wiss. Meeresunters.* **10** (14), 1964). This would have had to affect eel yields in central and northern Europe. Quite to the contrary, however, especially many eels were caught on the southern coast of the Baltic Sea in 1963 (Falk and Lauterbach, 1965; Gaygalas, 1966). In German inland waters, the relationships were different. On the German North Sea coast (Fig. 4.3), the 1963 and 1964 harvests, as in East German inland waters, were less than average. However, the North Sea fisheries of Denmark, Holland and Norway, in contrast to Germany, showed increased yields in 1963. For these reasons it is questionable whether cold winters can significantly influence eel populations and thereby yields, especially when temperatures are not extremely low.

4.2.5 Fluctuations in Canada and North Atlantic aspects

Substantial fluctuations in yield were reported for a 45-year period of Canada's eel fishery (Eales, 1968). The maximum, 1200 t, was landed in 1933; the minimum, 350 t, in 1948. Since 1948, the catches rose again and reached 800 t in 1965. The causes of the strong fluctuations are not completely apparent, but are presumably related to the economy and to fishing technique. The eel became more significant in Lake Ontario's fishery after more valuable kinds of fish became rarer on the market (Hurley, 1973). In the 1960s and 1970s, eel landings in Lake Ontario attained an 8% share of the Canadian fish harvest there and a 14% share in monetary value. From 1959 to 1970 the total annual Canadian eel harvests varied between 49 t (1962) and 109 t (1964). When the Moses-Saunders hydroelectric dam impounded the St Lawrence river in 1958, a drop in yields would have been expected because this severely hindered upstream migration of eels, but this was not apparent in the yields of 1959–70. However, there was a decline in mean body length, which was attributable to intense fishing, not to any poor recruitment caused by reduced upstream migration.

From the middle of the 1980s onward, a major decline in upstream migration of juvenile eels could be seen at the Moses-Saunders dam (Castonguay *et al.*, 1994a). Parallelling this, the population density of yellow eels in St Lawrence river tributaries diminished. Various reasons for this were considered, among them also effects of waste-water and hindrances of upstream migration. In addition, there was discussion of a possible influence of decreasing transport by the Gulf Stream since the 1980s, which could have reduced the immigration of eel larvae toward Canada. Finally, it was not to be overlooked that the Canadian eel decline showed parallels to that in European eels (Castonguay *et al.*, 1994b). This pointed to influences of the North Atlantic circulation system, of which the Gulf Stream is part. A more direct effect of the Gulf Stream on the migration of *A. anguilla* toward Europe is questionable (Section 2.2), so other influences of the system may be involved, for example, the Azores Current or the central eddy, which affects the spawning area of both species.

To summarise matters of the regional fluctuations in yield, the eel population fluctuations covering extensive areas of Europe depend primarily on strength of the glass-eel populations. Direct observations of harvest and analyses of meteorological conditions during migration have yielded clues about this. Hardly comprehended with regard to the eel are the super-regional influences, which presumably dominate. Yield fluctuations are brought about less by population size than by movement activity, the conditions of fishing technology and differences in temperature. From a regional standpoint, these fluctuations are usually more limited. The sex ratio of eels must be considered.

4.3 Variation in yield throughout the year

There is great seasonal variation in quantity of eels caught. In addition, differences exist between the various water bodies. One reason is that age groups have different activity peaks. Age group strength, in turn, varies among water bodies. The seasonally-dependent activity rhythm and thereby the monthly distribution of catch is in many ways determined by the yellow eels, that is, the eels that live for about 7–10 years in the continental waters. In contrast, waters that the eels pass through in the seaward spawning migration show another picture in this regard. However, many water bodies that have yellow eel populations are influenced by the peak migration of silver eels to a greater or lesser extent.

Yellow eel activity depends primarily on water temperature. This is shown by comparing the monthly catch distribution of the German fishery in the western Baltic Sea with water temperatures in 1953 and 1955 (Thurow, 1959). The largest catches coincide with the highest temperatures and, therefore, occur in mid-summer. Yields decline almost to zero by December and do not start rising considerably until April. This pattern of monthly catch distribution and that for the North Sea are similar (Fig. 4.5). Variation in the percentage of silver eels shifts the peak more or less strongly toward autumn or creates a second peak then.

The catch distribution in rivers resembles that in coastal waters that have few migratory eels (Fig. 4.6). In the rivers Oder and Rhine, the highest catches occur during summer. In

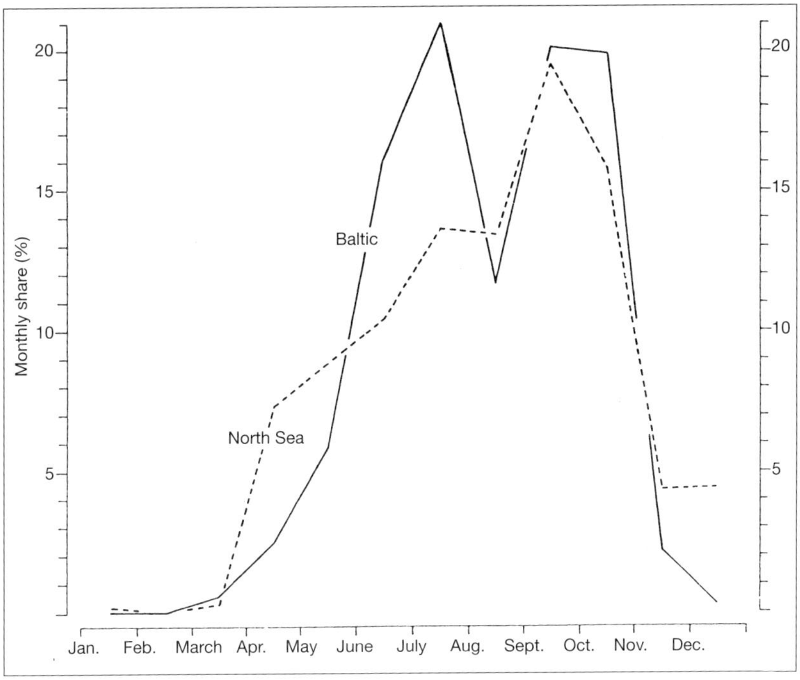

Fig. 4.5 Mean monthly eel catch expressed as percentage of mean annual eel catch in the North and Baltic Seas, 1977–9 (based on data in *Land- und Forstwirtschaft, Fischerei, Fangergebnisse der Hochsee- und Küstenfischerei*)

Lake Constance, the peak is significantly shifted into springtime. This water body is located in the Rhine's upper catchment area, where silver eels would have to start their migration significantly earlier in order to reach the river mouth by autumn. Thus, perhaps this early peak harvest can be attributed to earlier onset of migration in the upper river. Other interpretations are hardly possible, because the slower springtime warming of Lake Constance, an Alpine foothills lake, would be unlikely to stimulate increased activity.

The catch distribution in the middle and lower sections of the Oder indicate a similar tendency. A relatively early peak in June is noted in the middle but not in the lower section.

Monthly recording of eel yields in lakes of the German area of Mecklenburg, back to Dröscher (1897) showed peak catches in spring and autumn (Table 4.3). Likewise, Russian studies in Lithuania indicated that a silver eel migration is already observable there in springtime (Matschenis, 1965). However, only studies having clear separation of silver and yellow eels permit such definite conclusions as are known from the investigations in the lower Elbe (Fig. 4.7; Section 3.4.1).

Increased total harvests in springtime are a well-known phenomenon on the coast (Section 3.3.3.4). However, for the most part they may not be attributable to increased silver eel migration but rather to the movement of yellow eels in leaving winter habitat. This is matched by an autumnal movement in the opposite direction, when retiring to winter habitat.

Table 4.3 Monthly catches for a stationary eel trap in the Weser area and of another in the lake region of of the Mecklenburg Baltic Sea coast (after Walter, 1910), as well as of one in the Wadang river, Masury (after Seligo, 1926).[1]

Year	Feb.	Mar.	Apr.	May	Jun.	Jul.	Aug.	Sept.	Oct.	Nov.	Total
				Weser area catch (kg)							
about 1900	–	–	1	2	9	12	15	38	28	–	–
				Mecklenburg lake region catch (kg)							
1883–97 15-year mean	–	–	274	1119	727	886	884	687	477	69	4998
				River Wadang catch (number of eels)							
1896–1903 8-year mean	9	74	276	358	286	247	269	167	110	27	1831

[1]For catches in the individual years and months see Tesch (1973, 1983).

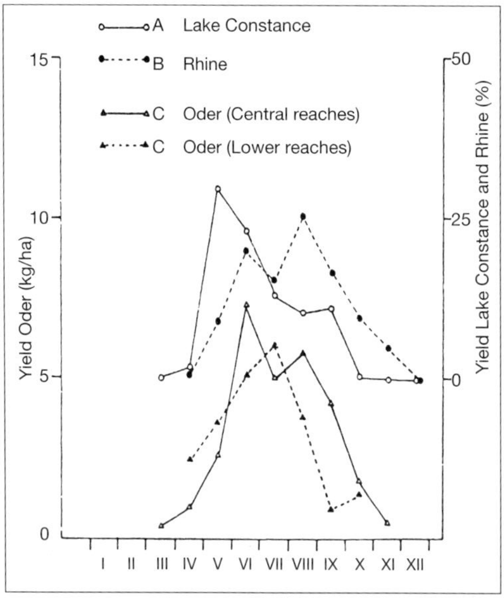

Fig. 4.6 Monthly distribution of eel catches. (A) Monthly distribution of eel catches in Lake Constance (Bodensee), as percent of total annual eel catch, 1964–6 (based on data in *Land- und Forstwirtschaft, Fischerei, Fangergebnisse der Bodenseefischerei*). (B) Monthly distribution of Rhine eel catches by vessel-mounted stow nets, as a percentage of total annual eel catch there by the same gear (after Hass, 1964). (C) Monthly yields (kg/ha) in the Oder's middle section (near Seelow) and lower section (near Gartz), 1958–63 (after Albrecht, 1964).

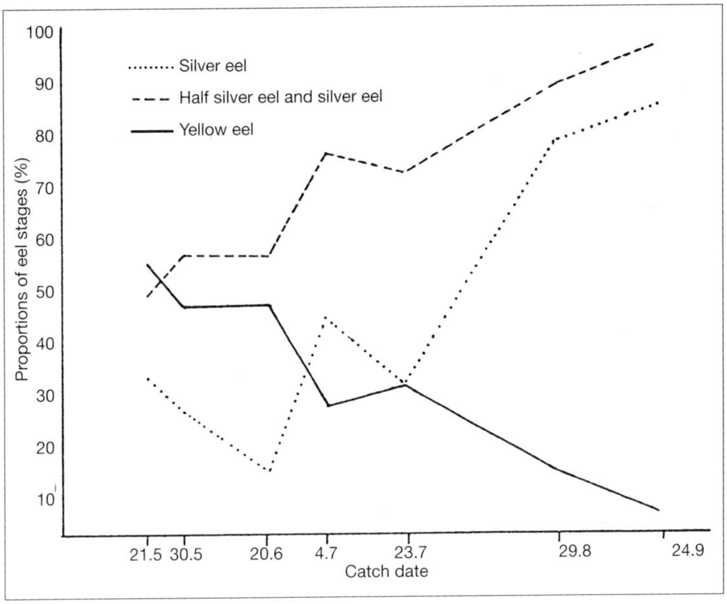

Fig. 4.7 Proportions of yellow and silver eels in a fishing operation in the Elbe upstream from Hamburg, 1958 (after Lühmann and Mann, 1961). Further observations in 1967 had a similar result (Mann, 1968).

4.4 Population density and catch per unit area from an economic and ecological standpoint

4.4.1 Regional differences

It is evident from an overview of the eel's geographic distribution that its population density under natural conditions in continental waters is mainly a function of juvenile invasion, which, in turn, depends on the length of waterway that connects with the sea.

This is one of the most pronounced characteristics of the yield and population data presented here (Tables 4.3–4.7). No direct correlation exists though, because local conditions are too variable. For example, the large West German coastal rivers and their catchment basins cannot be compared with the waters of Nova Scotia and New Brunswick. Although, given the dimensions of the German rivers, up-migration of eels can be traced for hundreds of kilometres, only about 10 km are involved in each of the rivers and streams described in Canada (Smith and Saunders, 1955). There do not seem to be any visible differences among the small areas that New Zealand eels occupy (Burnet, 1952a). Anthropogenic effects on the environment, of course, overlie the natural conditions. For example, today a lake near the sea may have relatively low eel populations because people built dikes and weirs.

Furthermore, it is conspicuous that biomass and yield in New Zealand's streams far exceed the values for streams and rivers of other continents. Whilst the streams studied in New Zealand held eel standing crops of 66–530 kg/ha (Table 4.4), only 1.5–100 kg/ha were reported for streams in Lower Saxony (Germany), among which, 43 kg/ha represents an

Table 4.4 Eel biomass in various waters.

Water body	Geographic location	Surface area studied (ha)	Year	Eel biomass kg/ha	Total fish biomass kg/ha	Other fishes	Water body characteristics/ remarks	Sources
Horokiwi Stream	New Zealand, Wellington	0.3	before 1952	530	–	*S. trutta*	stream 4 m wide	Burnet, 1952a
Wainui-o-mata	New Zealand, Wellington	0.9	before 1952	320	–	*S. trutta*?	stream 8m wide	
Doylestone Stream	New Zealand	*c.* 0.1	1954	130	263	*S. trutta* 133 kg/ha	stream 3 m wide	Burnet, 1959
Hamer Road	New Zealand	*c.* 0.3	1956	66	120	*S. trutta* 54 kg/ha	stream 3 m wide	
Hinau stream	New Zealand/ North Island	0.2?	1968/9	188	218	*S. trutta*, *P. breviceps*	small stream	Hopkins, 1971
Hinaki stream	New Zealand/ North Island	0.2?	1968/9	809	881	*S. trutta*, *P. breviceps*	small stream	
Ten Mile Creek	Narrowsburg, New York	–	1968/9	69	173	*S. corp.*, *C. commers.*	small stream	Engstrom-Heg and Loeb, 1971
Bills Lake	New Brunswick, near Passamaquoddy Bay	4.2	before 1955	79	80	*Couesius plumbeus*, *Salvelinus fontinalis*	T: ø 3, D:1.6	Smith and Saunders, 1955
Streams	Denmark	–	1971–88	163	–	–	21–1300 eels/100 m^2	Pedersen, 1997
Brand-strup stream	Jütland/river, Gudena	–	VI+XI 1950	75	246	Remainder: *S. salar* and *S. trutta*	stream 1 m wide	Larsen, 1955
Nivelle	Pyrenees, Atlantic side	–	–	108	–			Neveu, 1981
Breagagh	South Ireland, tributary to Nore river	0.77	1988	52	(70?)	–	4–22 m wide, D: >35	Moriarty and Nixon, 1990
Tweed tributaries	England			36–328				Hussein, 1981
Coleback Lake	Maine, NE coast of USA	8.0	1958	47.5	117	*C. commers.*, smelt	high primary production	Rupp and Deroche, 1965

Potter's Lake	New Brunswick, near Passamaquoddy Bay	44.8	before 1955	9	23	black bass, *Esox niger*	T: 2, D: 6	Smith and Saunders, 1955
Trefrys Lake	Nova Scotia, SW coast	21.6	before 1955	3	20	*Morone americana*	T: 1.1, D:1.6	
Tedfort Lake	Nova Scotia, SW coast	21.0	before 1955	2.6	41	*M. amer.*, *F. diaph.*	T: ø2.1, D:5	
Pools at Gundsmagle	Værebotal, Sjælland, Denmark	1.0	1959	1.9	482	roach, perch, pike	occasional connection with Værebo river	Larsen, 1961a
Loch Kinord	Scotland, watershed, of river Dee North Sea	82	1992–93	20–38		*S. trutta*, *Perca fluviatilis*, *E. lucius*	T: 1.5; D: 56	Carss *et al.*, 1996
Loch Davan	Scotland	42	1990–96	20–30			T: 1.2; D: 56	
Jesse Lake	Nova Scotia, SW coast	18.2	before 1955	0.8	22.2	*M. amer.*, Catfish	T: ø 2.4, D: 25	Smith and Saunders, 1955
	Nova Scotia, SW coast	22.6	before 1955	0.2	19.3	*C.commersonii*	T: ø 2.7, D: 48	
Naere Stand	North Fünen, Denmark	430	1958	150	–	–	lagoon, S = 2–20‰	Muus, 1967
North Sea	German Bight	128,500	1978	2.9	–	–	T: 20–50	Löwenberg, 1979
7 lagoons	North Adria	14,860	1971–8	76		mullet	S = 1.5–2.8‰	Rossi and Colombo, 1976
Mediterranean lagoons	France	–	–	70–<20		–	reduced quality of water and habitat	Fontenelle *et al.*, 1997

T depth D distance to ocean (Atlantic) S Salinity
The following fish species are shown in abbreviated form: *Salmo trutta*, *S. salar*, *Philypnodon breviceps*, *Semotilus corporalis*, *Catostomus commersonii*, *Morone americana*, *Fundulus diaphanus*, *Esox lucius*.

Table 4.5 Yields of eels in flowing waters.

River	Catchment basin	Surface area covered (ha)	Year	Eel yield kg/ha	Total fish yield kg/ha	Other significant fishes, catch method	Water body characteristics/ remarks	Sources
Oder	central Baltic Sea	3198 ha[1]	1961–3	32–60	68–100	25% bream, 10% pike	bream zone	Albrecht, 1964
Beste, Barnitz	Trave, W Baltic Sea	3 ha, Oldesloe	1958–64	11–38	44–83	hand line	lowland streams, 2–5 m wide	Herrmann, 1967
Trave	W Baltic Sea	12 ha above Oldesloe	1958–64	11–35	77–174	hand line	lowland stream, 20 m wide	
Trave	W Baltic Sea	18 ha below Oldesloe	1958–64	8–25	55–129	hand line	lowland stream, 20 m wide	
Ems	S North Sea	638 ha above estuary	1954–5	2.8	26	pike, coarse fishes	bream zone	Tesch, 1967b
Weser	S North Sea	190 ha above Nienburg	1961–3	12	27	bream, pike	bream zone	
Weser	S North Sea	297 ha below Nienburg	1961–2	7	13	bream, pike	bream zone	
Elbe	S North Sea	Werben above Havel	1896–1928	2–35	34–127	pike, 11% bream, roach	bream zone	Pape, 1952
Elbe	S North Sea	Werben above Havel	1927–37	0.4–6.2	12–110	pike, 12% bream, roach	bream zone	
Elbe	S North Sea	below Schnackenburg	1956–63	25–50	110	pike, burbot	bream zone, tidal zone	Mann, 1964
Lahn and tributaries	Rhine, North Sea	104 ha in Hessen	1960	9	64		barbel zone to salmonid zone	Buhse, 1967
Werra and tributaries	Weser, North Sea	281 ha in Hessen	1960	46	64	trout	barbel zone to salmonid zone	
Fulda and tributaries	Weser, North Sea	1652 ha inHessen	1960	8	82	trout	barbel zone to salmonid zone	
Diemel and tributaries	Weser, North Sea	230 ha in Hessen	1960	8	68	trout	barbel zone to salmonid zone	
Mosel	Rhine, North Sea	Trier to Koblenz	1951–61	7		cyprinids, pike	stow net catch	Jens, 1967
Rhine	North Sea	3500 ha below Mannheim	1956–61	4		pike, cyprinids		Hass, 1964; Rameil, 1967
Imsa	Norway		before 1985	1.9				Hvidsten, 1985
Burrishoole	NW Rep. of Ireland	near ocean estuary	1959–88	1.1		silver eel trap	lake outlet	Poole *et al.*, 1990

[1]Frankfurt/Oder to Gartz.

above-average value (Tesch, 1967b). In Denmark, one stream held 75 kg/ha and another had 163 kg/ha. Fyke netting for eels in New Zealand yielded between 33 and 1520 kg/ha, and the average amounted to >100 kg/ha (Burnet, 1952b). Corresponding values for Germany were 3–50 kg/ha with an average of little over 10 kg/ha.

In New Zealand the eel populations seemed to be higher than in the temperate zone of northern continents. The main reason is apparently not higher production but rather the relatively low-fishing pressure in those waters at that time. This applied not just to flowing waters but also to standing waters. The lagoon-like waters of the South Island's Lake Ellesmere may be shown as an example. Relatively intense fishing yielded the following catches (kg/ha) for the respective individual years from 1972 to 1991: 12, 13, <28, 42, 22, 26, 18, 15, 10, 5, 5, 4, 5, 4, 5, 5, 5, 4, 6 and 8 (Jellyman, 1993). Thus, after several years of fishing, yields were reached which, in European standing waters, are equalled and sometimes substanially exceeded (Table 4.4). Perhaps, higher temperatures in the rivers studied on the North Island played a role.

In winter the water temperatures hardly sink lower than 6°C, for example, in the Horokiwi Stream of the northern island (Allen, 1951). It is also to be considered that competition from other fishes, aside from introduced salmonids, is low in New Zealand because species diversity is poor there. New Zealand's high mean temperature situation has parallels in Europe, of course, although New Zealand is especially favoured by its marine climate. Accordingly, large populations or yields of eels would have to be expected in southern Europe's near coast streams. This is certainly true. An angler told the author that eels are very numerous and of large size in Spanish streams (see also *Fisch und Fang* 1966, p. 276), which indicates an extensive fishery. Eel population studies in a mountain stream that flows into the Atlantic Ocean (Esva, Cantabria) show, in the course of the seasons, mean standing crops of 107, 69 and 60 kg/ha in each of 3 years and extremely high values up to 254 kg/ha (Lobon-Cervia *et al.*, 1995).

Also in European standing waters, the highest eel populations and yields are found in the south (Tables 4.4 and 4.7). One northern European water body that is used as intensively as, for example, the Comacchio Lagoon, is Lough Neagh in Northern Ireland (Table 4.6), a lake of about the same size, which is intensively stocked with eels, from which the out-migrating silver eels are almost entirely caught out, and in which yellow eels are intensively fished for, too. In this lake, however, the 20 kg/ha yield was not so high as in the northern Italian lagoons, which, before Europe's present decline in juvenile recruitment, amounted to 30–40 kg/ha and even in some cases to as much as 90 kg/ha (D'Ancona, 1961). Not really comparable is Conventer Lake in northern Germany (Table 4.7), which likewise is routinely cited as a classic example of high eel yield. Its 36 kg/ha undoubtedly represents optimal commercial use of an eel population. The high yield is attributable to abundant natural immigration of juvenile eels from the nearby Baltic Sea. The fishery may be favoured by both a rather high productive capacity of the water and the small area of the lake, which facilitates fishing. Also, peak yields do not far exceed 20 kg/ha in lakes of Germany's Schleswig-Holstein area, which are intensively stocked with eels and sometimes abundantly supplied by natural immigration. However, Herrmann (1967) states that as much as 40 kg/ha are attained in exceptional cases. Here, small lakes are involved, which should be ranked with the aforementioned Conventer Lake with respect to yield.

Table 4.6 Yields of eels in inland lakes.

Lake	Geographic location	Surface area (ha)	Year	Yield of eels (kg/ha)	Total fish yield (kg/ha)	Other significant fishes	Water body characteristics/remarks	Sources
Lough Neagh	Northern Ireland	39,000	before 1966	about 20		Salmonids, Coregonids		Frost, 1950; Tesch, 1967c Rep.Foyle Fish. Comm.,
Lough Neagh	Northern Ireland	39,000	before 1996	17				Rosell, 1997
Lough Derg	Rep. Ireland		before1996	<2				Moriarty and Reynolds, 1997
Pike and tench Lakes	Schleswig-Holstein	*c.* 50	1954–64	9–20	21–51		average of several lakes	Herrmann, 1967
Bream lakes	Schleswig-Holstein	50–100	1959–64	7–16	24–127		average of several lakes	
Bream lakes	Schleswig-Holstein	*c.* 50	1954–64	5–14	43–71		average of several lakes	
Bream lakes	Schleswig-Holstein	100–200	1954–64	3–9	17–61		average of several lakes	
Roach lakes	Schleswig-Holstein	1000	1949–64	4–8	20–46		average of several lakes	
Coregonid lakes	Schleswig-Holstein	1000	1949–64	2–6	13–26		average of several lakes	
Rögglin Lake	Mecklenburg, Schwerin	200	1957–61	2–15			the eels were all stocked	Gollub, 1963
Steinhuder Meer (lake)	Lower Saxony	3000	1959–62	5	20	bream, pike perch	pike-perch lake	Tesch, 1967b
Lake Constance		53,900	before 1988	3–6		Coregonid lake	eels were stocked	Berg, 1987
18 lakes	Sweden			0.1–3.4			average of several lakes	Wickström and Hamrin, 1997
Large Masurian lakes	Poland		before 1986	2.6			eels were stocked	Leopold, 1986

86 lakes	Poland		Before 1990	5.2			eels were stocked	Moriarty *et al.*, 1990
Crecy Lake	New Brunswick	20	1949–63	1.5–5.1		Salmonids	Bay of Fundy, low productivity	Smith and Saunders, 1955
Gibson Lake	New Brunswick,	24	1945–51	1.1			Bay of Fundy, 4 m deep	
Mecklenburg waters	District of Schwerin	25,700	1954–61	1.9–4.5	24–28		mainly lakes	Gollub, 1963
Dümmer Lake	Lower Saxony	1500	1960–2	2.8	11		pike and tench lakes	Tesch, 1967b
Storkow Lake	S of Berlin	385	1937–49	0.3–6.1	1.5–27	bream 18%, pike 12%	bream lake	Müller, 1952
Bederkesa Lake	N Lower Saxony	232	1957–8	2.0	10	bream, pike-perch		Tesch, 1967b
Eder Reservoir	SW of Kassel	800	1937–58	0.03–0.9	15–99	roach, pike-perch, perch	in low mountain region, commercial and sport yield	Buhse, 1967
Eder Reservoir	SW of Kassel	800	1959–65	2.2–4.1	59–81			
Sakrow Lake	SE of Berlin	100	1949–56	0.6–3.7	7–38	bream 14%, roach 14%, pike 12%	bream lake	Rahn, 1957b
West Berlin waters		2510	1960–3	1.5–2.0	14	bream, pike, pike-perch	commercial harvest of River Havel lakes	Buchin and Müller, 1967
Oder Reservoir	southern Harz Mtns.	136	1958	0.3	4.9	trout, roach	sportfishing yield, low mountain region	Tesch, 1967

[1]Frankfurt/Oder.

Table 4.7 Eel yields in coastal waters and coastal lakes.

Area	Geographic location	Surface area (ha)	Year	Biomass (kg/ha) Eel	Fish total	Other fishes	Characteristics of the water body and remarks	Sources
Lagoon	Commachio Italy	50,000	Before 1980	29		Mullets		Rossi *et al.*, 1987
Lagoon	Commachio Italy		after 1980	5–7				Ciccotti, 1997
7 Lagoons	North-Adria Italy		before 1976	19			Decreasing growth	Rossi and Colombo, 1976
Tortoli Lagoon	Italy		1957–64	120–130				Ciccotti, 1997
Tortoli Lagoon	Italy		1965–78	40				
2 Lagoons	Mediterranean France	90,000	1985–96	99–>45				Fontenelle *et al.*, 1997
Conventer Lake	Bad Doberan/ Mecklen burg	65	1954–61	29–45	59–123	Pike 13%, Tench 4%	Lake near the coast, 300m to the Baltic	Gollub, 1963
IJsselmeer	Holland	220,000	1954–62	*c* 10	–	Pikeperch, Smelt	separated from the sea by a dike	
Strelasund	Southern Rügen western Baltic	24,300	1963–5	7	15	Roach 30% Perch 20%	Connects 2 marine Bays	*Fisch. Forsch.* Rostock, 1963–5
The Small Oderhaff	Estuary of river Oder	29,000	1963–5	6.5	42	Roach 27%, Bream 12.8%	Salinity low	
Wismar Bay	South coast of western Baltic	12,200	1963–5	4.3	8	Marine fish prevailing	Open marine bay	
Bodden Chain Darss	South coast of western Baltic	20,000	1963–5	3.1	22	Roach 33%, Bream 20%	Elongated lagoon, salinity low	
Rügenbodden Chain	South coast of western Baltic	15,500	1963–5	3	15	Roach 30%, Bream 17%	Elongated lagoon, salinity low	
Peene Stream	Estuary of river Oder	17,700	1963–5	2.3	39	Roach 41%, Bream 24%	Estuary, low salinity	
Greifswalder Bodden	Estuary of river Oder	53,400	1963–5	2.8	7.5	Herring, Perch, Roach	Shallow marine bay	
Frisches Haff	South coast of central Baltic	86,100	1936	4.2	13	Bream, pikeperch, Pope	Lagoon	Gaygalas, 1966
Kurisches Haff (Kurskiy Zaliv)	South coast of central Baltic	161,300	1927–38	0.8	119	Smelt 29, Pope 13 kg/ha	Lagoon	Neuhaus, 1967
Kurisches Haff (Kurskiy Zaliv)	South coast of central Baltic	161,300	1947–64	1.2		Bream 12, Pike-Perch about 3 kg/ha	Lagoon	Maniukas, 1961

4.4.2 Eel populations and interspecific competition

4.4.2.1 Coastal waters

Extensive coastal waters generally yield lower eel harvest per unit area than do small coastal lakes (Table 4.7). It should be kept in mind that these waters serve to some extent as migratory passage for silver eels, therefore, the harvests do not stem completely from their own production. Moreover, their productivity is above average. For the Baltic Sea's coastal waters yield in Table 4.7, only the catches of freshwater fish are shown, but many of these waters have saltwater fishes, too. Greifswalder Bodden (Bay), for example, actually shows significantly greater yields because they include saltwater fishes, which tend to be rather transient. In 1952, the harvest of freshwater fishes, including eels, was 11 kg/ha, but in total, 76 kg/ha were caught (Subklew, 1955). Maybe these waters had only a relatively low yield of eels because the many other species there competed with them.

4.4.2.2 Salmonids

Also with regard to streams of the coastal region, the question exists as to whether larger eel populations could exist if other, competing fish species did not occur with them. This question applies particularly to such trout streams as are known in Denmark (Larsen, 1955, 1961b). Within a total standing crop of 246 kg/ha, 75 kg/ha eel is not very much (Table 4.4). The same must be said for bottomland tributary streams of the Elbe river estuary (Tesch, 1967b). Although the total standing crop of fish there, 286 kg/ha, was the highest found in streams studied in Lower Saxony (German province), eels made up only 35%. Also individual data reported for 20 Danish streams look similar (Larsen, 1961b): on average, there were 5100 trout/ha, but only 850 eels.

In many cases the competing fish were trout, which feed to a substantial extent on the same organisms as the eels do (Sinha and Jones, 1967c). Larsen (1961b) drew the conclusion from his population studies in trout streams that trout dominate in small streams, but only if the environmental conditions are favourable for trout. Otherwise, the eel dominates, especially in larger streams, where various unsuitable conditions for trout reproduction prevail. There however, as previously said, cyprinids are abundant and for the most part compete just as much with the eels for food. The reason that eels do not reach the expected share of the standing crop in typical trout streams may be because, contrary to trout, the warmwater eel species do not attain optimal metabolic rate there.

A further criterion for eel productivity in flowing waters, given optimal juvenile immigration, is stream morphology. Eels and trout have very similar requirements in this respect. As Burnet (1952a) found, streams that have large amounts of cover provide the greatest eel yields. Particularly included are aquatic plants, undercut stream banks and bank vegetation – including importantly trees and brush, that has fallen into the water or drapes into it. Such habitat features are also in keeping with the eel's negative phototaxis, and this applies to trout, as well. Thus, competition between eels and trout also exists for structural habitat.

The results of studies on eel diet lead one to expect that eels and trout are food competitors (Section 3.3.1). The literature of English-speaking countries has given this matter much attention. As both the market and sport-fishing values of eels there are far lower than those of salmon and trout, the main objective is to protect the salmonids from the eels. However, this bias is decreasing these days in view of concerns for maintaining ecological balance in waters (Tesch, 1986). In Germany, as well, one should pay more attention to this aspect and foster the eel, for example in the Danube basin.

Based on the findings of many studies, there can be no doubt that eels and trout compete intensely, especially in summer (Sinha and Jones 1967c and references therein). This problem was often put forward as a major argument against toleration of eels in salmonid waters. Whether eels could act as predators on trout spawn, fry and juveniles tended to be considered as secondary (Sinha, 1969). Nevertheless, studies of New Zealand (Cairns, 1942b) and American eels (Godfrey, 1957) have found substantial predation on salmonids. The worst losses of salmonid fry to eel predation occur in Canadian streams, apparently in spring and early summer, when the fry emerge from the streambed gravel. At that time, the eels, which have starved during winter, launch into their first and most intense feeding activity of the year (Section 3.3.1.1).

Intensive population studies on the influence of eels on trout in New Zealand (Burnet, 1959, 1968) indicated that it is not quite right to say that eels affect trout negatively. In a 3–4 m wide stream, where fish population inventory by electrofishing revealed standing crops of 56 kg/ha for trout (*Salmo trutta*) and 68 kg/ha for eels (numerically: 64% *A. australis*; 36% *A. dieffenbachii*), hardly any trout population increase happened after the eel population was removed. However, in a 2–6 m wide stream which, due to good cover, had very abundant eels (numerically: 47% *A. australis*; 53% *A. dieffenbachii*), major reduction of the eel population had quite a considerable effect. The amount of eels removed in the first year was 400–1090 kg/ha in the various stream sections. In contrast, the trout standing crop was estimated at only 16–121 kg/ha. There was already a considerable trout population increase the first year after eel reduction. Four years later, trout standing crop had risen to 134–240 kg/ha. *Via* substantial decrease in mortality until age one, the number of age-2 trout increased 2.7–9.9-fold. As a result of this increase in trout biomass and numbers, body growth of the trout declined substantially. In one of the stream sections, the trout needed four instead of the previous 2 years to reach a length of 28 cm. Thus, the trout were less attractive for anglers, who fished this stream. In a control section of the stream the eel population was not reduced, and no trout population change occurred.

If the eel and trout populations of European streams (see above) are compared with those of New Zealand, in most cases a detectable impairment of the trout population seems to be out of the question. Such dense eel populations as are known from the past in New Zealand streams occur only seldom in European trout streams. In addition, if eels inhabit European streams very densely, then, as a rule, only the small-sized males compose the main population (Section 3.3.2.2). As food studies show, these are not so dangerous to juvenile trout (Section 3.3.1.3). It should be mentioned that the European trout (*S. trutta*) evolved with the European eel, whereas, they were just recently stocked among New Zealand eels, so probably are not well adapted to coping with them.

It is far more difficult to answer the question of how much a strong eel population affects interspecific competition in standing waters. A high population of other fishes affects eels positively to the extent that they serve as eel food, but negatively insofar as they prey on the eels or interfere with them *via* food competition. On the other hand, abundant eel predators exert positive regulation on eel populations, in which natural immigration is so high as to cause intense intraspecific food competition.

Just on the basis of greater species diversity, the role of the eel in fish community balance in standing waters is more complex than in streams. For many fish species, the stream poses an extreme environment, in which only salmonids and eels can occur as the primary components of the fish community. In standing waters, many other fish species find optimal living conditions. In addition, the plasticity of these species enables them to populate other biotopes as replacements for community members extirpated there.

It is highly probable that eels had a devastating effect on the brook charr (*Salvelinus fontinalis*) population of a 154 ha Canadian oligotrophic lake in the St Lawrence river basin (O'Connor and Power, 1971). Besides sticklebacks and smelt, this lake harboured a brook charr standing crop of 0.34 kg/ha and a substantially greater eel population. Limnologically similar lakes in the same area had no eels, but in one case, besides Arctic charr (*S. alpinus*), the brook charr. Standing crop was 3.1 kg/ha, and in two further cases, 1.6 and 4.3 kg/ha for brook charr.

In addition, in the lake that contained eels 7 of 22 examined eels had eaten brook charr – 56% by volume of the eels' total stomach contents. In this lake, brook charr mortality during the third year of life was twice as high as in any of the other lakes. Surely this example, like the one that follows, finds parallels only when intruders, such as eels, disturb the ecological balance in high elevation or otherwise isolated lakes containing unusual biotic communities.

4.4.2.3 Coregonids

Studies of interactions between eel abundance and the whole fish population also have management implications. Included are intensive stocking of certain kinds of fish and long-term observation of the yield of all fishes at a constant level of fishing effort. Various examples of this are known from lakes in the northern German area of Schleswig-Holstein (Herrmann and Marre, 1961; Herrmann, 1967). The data on yields of the large whitefish species, *Coregonus lavaretus*, and of eels in one whitefish lake are very convincing. Figure 4.8 is a compilation of those data from various sources. It shows that from 1898–1905, eels were caught in only small amounts compared with the whitefish. After eel stocking was begun about 1900, the whitefish yield declined sharply. Between 1919 and 1949, it was under 0.5 kg/ha. From this period, the years 1930–37 are represented in Fig. 4.8. 'Around the middle of the 1950s, under the influence of heavy stocking and intensive fishing with monofilament gill nets, the whitefish yield again showed a brief rising trend (up to 3 kg/ha), during which eel yields of about 3 kg/ha were likewise realised. Upon further intensification of eel stocking and increase of its yield to 4.5–6 kg/ha, the whitefish yield then declined so much that; from 1950 to 1964, it has become practically meaningless' (Herrmann, 1967).

The main cause is presumably the predator–prey relationship between eel and whitefish, that is, feeding of eels on whitefish spawn, as well as on the fish themselves. This was especially obvious in that whitefish caught in gill nets ('namely, when the nets are set near the bottom') had often been eaten by eels (Herrmann, 1967). Similar observations were made, according to information of the author, in coregonid lakes of the Alps region, as well as in Lake Ontario, Canada. There, eels ate whitefish in gill nets and tangled the nets (Hurley, 1973).

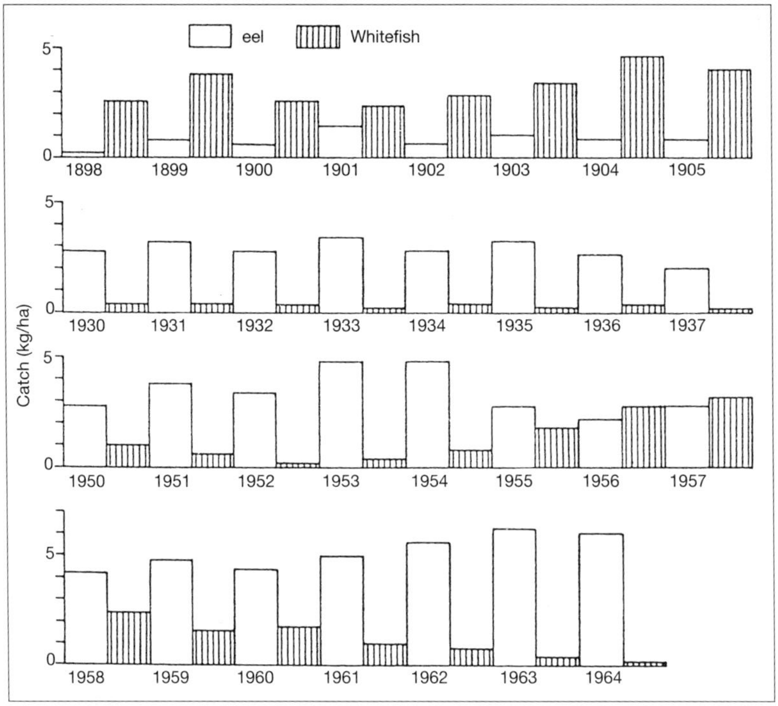

Fig. 4.8 Yields of eels and whitefish (*Coregonus*) from a whitefish lake in Schleswig-Holstein during various periods since 1898 (modified after Herrmann and Marre, 1961; Hermann, 1967)

4.4.2.4 Other fish species

Further examples of eel competition with other fishes in lakes of Schleswig-Holstein are to be found in Herrmann and Marre (1961) and Herrmann (1967). Competition between eels and carp is particularly understandable, as the latter also eat benthic fauna and these two species did not co-evolve. Therefore, intensive management for carp has been avoided in many cases. Similar competition between eels and tench (*Tinca tinca*) has been observed.

The bream (*Abramis brama*) must also be included among the eel's direct competitors, though controlling it so as to benefit eels means a positive management result for the fishery. For a further example involving competition with cyprinids, perch and pike, see Section 4.4.3.

Competition of eels with other predatory fishes is also thought possible. At Oneida Lake, a site of intensive fish ecology study in the US state of New York, anthropogenic influences caused both eels and pike to decrease after 1900. In contrast, walleye (*Stizostedion vitreum vitreum* – closely related to Zander) increased considerably. The walleye is very important in that lake's sport fishery (commercial fishing is not done there), as is the yellow perch (*Perca flavescens*), which dominated the fish community before and after the changes in the other species' populations, but few people, if any, fish for the eels (Forney, 1977). The competitive (predatory) relationships between eel and walleye could have consisted of feeding on walleye eggs by eels, as well as of both having the same prey, such as juvenile fish of various kinds. An influence of the eel on the perch (*Perca fluviatilis*) population can be recognised also in Lake Constance (Radtke and Eckmann, 1996). The eels devastate the youngest year class of perch. Presumably this is only significant in years, when the cannibalistic effect of predatory perch on its offspring is minor.

Finally, one might point out much the same sort of competition with another, very similar predatory fish, the conger (*Conger conger*). Both are found together on the west European coast. The conger is the stronger predator. Moriarty (1978) reported that in the outer areas of Irish estuaries he often found small specimens of *Anguilla* species in conger stomachs, and that larger eels could not be caught at places where conger are harvested. Further upstream, however, *Anguilla* could be caught in abundance, but no conger. Therefore, the author infers that during co-evolution of these two species, which have very similar habitat and food requirements, *Anguilla* had to avoid the conger by leaving sea water; only on northern coasts, where conger are rare are *Anguilla* encountered in major amounts.

4.4.2.5 Crustaceans

From the commercial fisher's viewpoint, eels compete undesirably with valuable crustaceans, such as the lobster (*Homarus vulgaris*) and the crayfish (*Astacus astacus*). However, there is evidence of this only for the crayfish (Svärdson, 1969, 1972). In southern Sweden, eels and crayfish occur in many waters, the eel more in the west, the crayfish more in the eastern part of the country. Both species inhabit many lakes together. Lakes that lack eels yield significantly higher annual crayfish harvests: 2.3 crayfish/ha *versus* 1.1 ha^{-1} in eel-less lakes (sample size: 159 cases). In addition, where mean yield of eels was about 0.22 kg/ha, there were hardly any crayfish. In contrast, mean eel yields of 0.11 kg/ha enabled crayfish to occur in the same lake. Svärdson (1972) provides several examples, according to which crayfish yield rose when the eel population declined, for whatever reason (e.g. blockage of natural upstream migration). Conversely, reduced crayfish yields coincided with rising eel harvests. The best-known example is from Lake Vänern, Sweden, which, when a canal connecting it to the sea was built, began getting eel immigration, and at that time lost its good crayfish population. It is also known from North America that eels there prefer to eat crayfish (*Oronectes* sp.) (Facey and LaBar, 1981). Suspicions that the increased occurrence of eels could have brought about the decrease of the lobster population in the German Isle of Heligoland's cliff littoral have no basis in fact. That trawl catches of eels have increased since 1963 in the German Bight (Section 5.8) is no indication that the eel population has risen because this fishery did not use small-mesh nets in the past. Also, signs of negative effects of eels on lobsters have not yet been noticed on the Swedish west coast, where eels and lobsters occur together (pers. comm., Svärdson).

4.4.3 Improving yield by stocking

The farther a water body lies from the coast, the less are likely to be its natural immigrations of eels and, therefore, its yields. Eel yields in Brandenburg and much of Mecklenburg (German states) have been significantly lower than those of Schleswig-Holstein (Table 4.6), where they averaged 6.5 kg/ha in the 1950s and 1960s (Herrmann, 1967).

Although they were 7 kg/ha in northern Mecklenburg, just 1 kg/ha was attained southeast of Berlin (Koops, 1967a). To compensate for such differences, artificial stocking projects were already begun before 1900 (Section 2.2.2). Lake Vilm in East Pomerania is a well-known example of successful eel stocking, which has been conducted since 1909; in 13 years, yield went from 0.7 kg/ha to a peak of 8 kg/ha (Fig. 4.9). On average, the managers had stocked 14 eels/ha annually (Strophal, 1930), from which, after the project became effective, an average annual yield of 4 kg/ha resulted, so net yield of at least 3 kg/ha/year can be credited to the stocking. The stocking consisted largely of young, stocking-sized eels (hereafter, 'stocking-elvers', that is, juveniles about 15–30 cm long) weighing, on average, 20 g. According to Table 3.21, this could involve eels of 24 cm length. An illustration from Robak (1994) shows how greatly so-called stocking eels can vary in size. For example, in August they were 11–24 cm long, in November 17–30 cm and in April 14–34 cm. This should be considered also with respect to all further examples. The manager himself felt that a 2–3-fold higher stocking rate would have been appropriate for reaching and sustaining maximum yield. In comparing the numbers stocked and harvested, Strophal (1930) found a 40% return. He assumed a return rate of 20% for those stocked as glass eels. The eel stocking coincided with declining yields of other fishes, such as pike, roach, perch and tench, but the extent to which these decreases were due to other changes in the lake is not apparent. Financially, those losses were more than offset by the eel harvest.

A further example is known from Lake Rögglin in the German area of northern Mecklenburg (Table 4.6; Fig. 4.9). Gollub (1963) wrote that until 1949 catching an eel in this lake was a great rarity, but in 1954 the first 500 g eels were caught. Harvests seem to have been economically important beginning in 1957 (Fig. 4.9), most being taken in stationary eel traps.

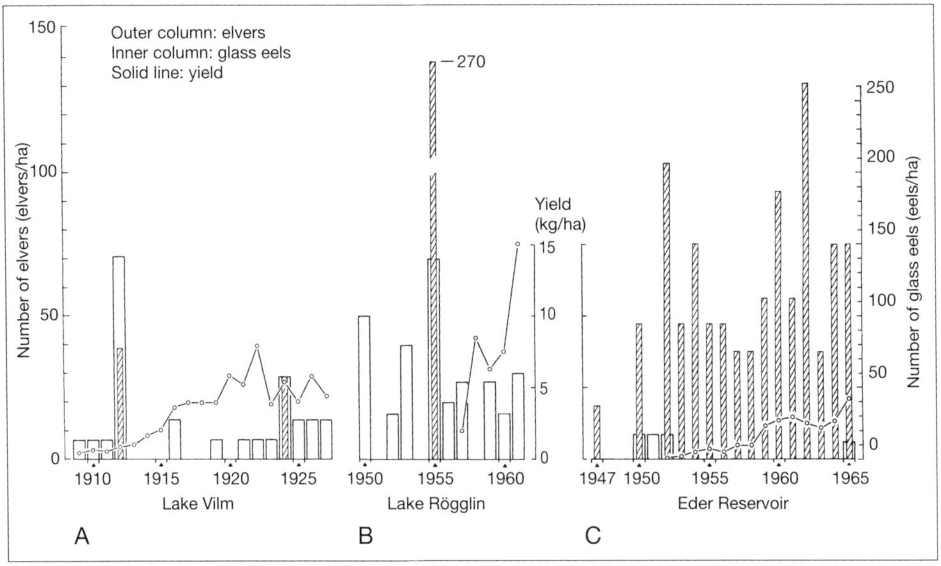

Fig. 4.9 Eel stocking and yields. (A) Lake Vilm, 1750 ha (based on data from Strophal *et al.*, 1930). (B) Lake Rögglin, 200 ha (based on data from Gollub, 1963; Koops, 1967b). (C) Eder Reservoir (based on data from Buhse, 1967)

In the warmer Hungarian Lake Balaton, initial results of stocking were already detected after 3 years (Koops, 1967a; Biró and László, 1970), and the strongest effect seemed to become evident after 6–7 years (Southern European conditions) – as opposed to 8–10 years in the Pommeranian Lake Vilm (Fig. 4.9). If the size of the stocked eels used is taken into account, an annual stocking rate of 30 young eels/ha is needed to provide for a yield of 8 kg/ha. This stocking rate and yield apply to Lake Vilm, which is only half as intensively managed as Lake Rögglin, where the results indicate that, for every increase of 4 stocking-elvers in the stocking rate, yield should rise by 1 kg.

A less definite result came from stocking (mainly) glass eels in the previously eel-free Eder reservoir (Table 4.6; Fig. 4.9). The stocking was augmented with stocking-elvers for only a short period. In preparing Fig. 4.9, the weight data of Buhse (1967) were converted to numbers of individuals (3000 glass eels/kilogram; 55 stocking-elvers/kilogram). The first heavy stocking, 1950, was followed by an initial maximum yield between 1959 and 1962 (Fig. 4.9) According to that, the benefit of glass-eel stocking is realised after about 10 years. This corresponds with the data of Jensen (1961), which indicate that female eels live in continental waters for 10 years. Judging from the above examples, the peak catch around 1961 can hardly be attributed to the planting of stocking-elvers in 1950–2, that was more likely to have brought about the minor peak of 0.7 kg/ha in 1955. Conclusions about the amount of stocking needed cannot be drawn from this example. The reservoir's upper end is too open for that; it is easy for the juvenile eels to escape into tributaries. Thus, the only way to evaluate the stocking would have been in terms of a basin-wide result, obtained by installing an eel trap at the outlet and capturing all emigrating silver eels. Nevertheless, this example shows that stocking barely 100 glass eels/ha and year has detectable success. The yield was around 3 kg/ha.

Schäperclaus (1949) indicated the relative effectiveness of using stocking-elvers or glass eels, when he said that one can expect losses of 40–60% for elvers and 80% for glass eels. Strophal (1930), as reported above, came to a similar conclusion. Anwand and Valentin (1981b) calculated losses of 93% in the former East Germany (GDR). However, their calculation did not include elvers, small amounts of which were stocked there. Therefore, the losses seem to be too low. There was a loss of about 96% from initial stocking of 23–300 glass eels/ha in 43 Byelorussian lakes, which totalled 48,000 ha and had free outflow (Kostyuchenko and Prishchepov, 1972). Polish analysis of glass-eel stocking in many lakes came to the conclusion that losses of 89–94% occurred in lakes that had little natural immigration of glass eels (Leopold, 1976). Thus, return to the fishery from glass-eel stocking is 10% at most. For Lake Constance, as well, a good 10% chance of survival is expected for glass eels, based on long-term observations of stocking results (Hahlbeck and Kuhlmann, 1997).

In order to get a yield increase of 1 kg/ha in Northern Ireland's 39,000 ha Lough Neagh, 45 glass eels/ha were needed (Koops, 1967a). However, the annual yields there were almost 20 kg/ha, so the stocking for this yield must have been relatively successful.

Albrecht (1975) reported on stocking of eels in originally thinly populated lakes near Berlin, Germany. There, annual yields of at most 40 kg/ha were managed, for which 750 glass eels/ha (i.e. 19 glass eels/kilogram harvested) had to be stocked. The result from Polish investigations of glass-eel stocking and yield in 559 lakes was that 21–40 glass eels were needed to achieve a 1 kg/ha increase in yield. In central Europe, less favourable stocking-to-yield relationships are to be expected from management that is even more intensive. For instance, if one compares the amounts of stocking-elvers against yields for various Schleswig-Holstein lakes (Herrmann, 1967), high stocking success is by no means as apparent as would be expected from the previous examples. In Schleswig-Holstein, stocking-elevers were stocked at an average of almost 1 kg/ha, but in a few places at 5 kg/ha on average, for example, in the approximately 50 ha pike and tench lakes. There, however, a yield increase of only 10 to, at most, 20 kg/ha resulted. Certainly entering into this was the fact that above-average populations of eels already existed here. In addition, as indicated above, because 20 kg/ha already lies at the upper limit of waters that have average productivity, the amounts stocked must have been abnormally high. Therefore, planting five stocking-elvers/ha to increase catch by 1 kg/ha will work only in eel-free or thinly populated waters, as was the case in the three examples above.

With regard to the initial effectiveness of using stocking-elvers, as in Schleswig-Holstein, it has been the experience in other waters that one must wait 3 or 4 years before noticing results in terms of increased yield: with glass-eel stocking, it takes 8–10 years. Jens (1967) found otherwise. He compared the yield curve against the glass-eel stocking of 2 and 5 years previously, and the curves matched in both cases. From this Jens (1967) draws the conclusion that the effectiveness of stocking can already become evident within 5 or even 2 years. This is improbable, according to the previously presented examples, as well as Danish studies (Jensen, 1961). In Scandinavian waters, the results of glass-eel stocking were first apparent after 7 years, the maximum occurred after 9–10 years, and effects were still detectable after 14–15 years (Eh, 1929). In Byelorussian lakes, newly stocked glass eels were caught for the first time after 5 years and mainly after 7–11 years (Kostyuchenko and Prishchepov, 1972); here, one stocked year class of glass eels improved the fishery for about 11–12 years. Stocking glass eels in Lake Constance first had noticeable results after 10 years (Deufel and Strubbelt, 1977).

Initial stocking in southern waters takes effect much faster than in central and northern Europe. As is known from initial stocking studies in the Hungarian Lake Balaton (70,000 ha), eels 47–50 cm long were already caught a year after stocking-elvers were released. However, there, as well, it takes several years before economical effects of stocking can be expected (Biró and László, 1970). But because the Lake Balaton fishery was certainly carried on in a way that was technically ineffective for eels, this may have contributed to the delay.

In summary, it can be concluded from the above results that success of stocking can vary greatly, depending on environmental conditions. Low success can be expected if a considerable eel population is already there (e.g. 5 kg/ha); if a substantial, species-rich fish population and/or an abundance of other competitors or predators (Figs 4.10 and 4.11; Carss *et al.*, 1998) exist; if cover is scarce; or if primary production is poor. Under favourable conditions, five stocking-elvers or 15 glass eels planted per hectare are needed to provide for a 1 kg/ha increase in yield. Results of stocking aquaculturally reared eels into the wild presumably are better than for glass eels, which are smaller in size, but poorer than for stocking-elvers that are caught in natural waters, which are more adapted to normal environmental conditions. In an eel working group of the European Inland Fishery Advisory Commission (EIFAC/FAO), a stocking rate of over 300 glass eels/hectare (0.1 kg/ha) was recommended for waters, where small populations of eel

Fig. 4.10 Cormorant (*Phalacrocorax carbo*) with eel (photo: van Daalen)

exist (Moriarty and Dekker, 1997). If continued annually toward the above-mentioned values, this could lead to a very high yield. These single-year stockings of a comparatively large number of young eels should be spread over several years if a rapid or high yield is not really called for, or if, for economic or technical reasons, heavy stocking is advisable.

4.4.4 Summary of yield considerations

The following classification of yield is based primarily on commercial fishing, which is really the only way that silver eels are caught. Theoretically, intensive sport fishing could also approach these yields, but in practice, sport fishers usually aim to catch large fish, and do not particularly try for great harvest in terms of total weight. All comparisons of water bodies with regard to eels were always based on yield and population (also converted from density or biomass). Of course, comparisons of production and mortality would have given more information about the ecological conditions underlying yield, but too little literature is available for that. However, the reader is referred to the following publications concerning studies and comments on these questions:

- Lagoons and other coastal waters: Rossi (1979); Rossi and Colombo (1976); Fontenelle *et al.* (1997); Wickström *et al.* (1996);
- Running waters: Mann and Penczak (1994); Aprahamian (1988); Mann and Blackburn (1991); Fontenelle *et al.* (1997);
- Tabular data for some water bodies from the cited literature: Moriarty and Dekker (1997).

Fig. 4.11 Osprey (*Pandian haliaetus*) with eel (photo: Smith)

The following yield classification constitutes comprehensive guidelines.

- Open ocean bays:
  - the amounts caught per unit area may be moderate due to competition from many other species of fish and to intense fishing. In central Europe, the range is <1 to 10 kg/ha.

- Lakes, including those near coasts:
  - high yields, especially in small coastal lakes: 10 to >100 kg/ha;
  - moderate yields: 2 to <10 kg/ha;
  - low yields: <2 kg/ha.

- Streams:
  - High yields: 20 to50 kg/ha;
  - Moderate yields: 5–<20 kg/ha;
  - Low yields: <5 kg/ha.

In New Zealand, the populations and yields are higher than in Europe and North America. This is presumably a matter of temperature. Southern Europe probably is also higher with respect to yield, as shown by comparison of the yields in Italian lagoons against those of the coastal areas of the North and Baltic Seas.

5
Fishing methods

Updated and revised by O. Gabriel and T. Wendt

5.1 Introduction

Hardly any other kind of fish is caught in so many different ways as is the eel. With few exceptions (e.g. purse seine and gill nets), almost all important fishing techniques known are used. Trapping devices (eel pots, fyke nets, etc.) and stow nets are set for eels during their migrations or other movements that are determined by season and time of day. When they forage for food, trapping devices, trawls and angling are employed. During the spawning migration, weirs, barrier (block) nets, fences and chains of lights in combination with stow and fyke nets are used. In fresh water, yellow eels are caught by electrofishing and with towed nets. Seines, spears, scoop nets, lift nets and cast nets are now used less often than in the past. For a whole series of special catching gear, one finds the target fish, eel, embodied in the names of implements: eel spear, eel rod, eel pot (eel basket), eel fyke net, eel stow net, eel trawl, eel seine, etc. With these, the eel's various needs and behaviours are exploited: its periodic migratory drive, its food seeking, its avoidance of light (negative phototaxis) and its affinity for physical contact (thigmotaxis), as well as the ways it occupies and moves between habitats. The eel is found on the bottom in shallow and deep water, or at other times in open water. Under some circumstances, it will even leave the water altogether. It drifts in tidal flows and in rivers and streams, or it swims against these currents (rheotaxis). It is found on and in sand, mud and stony beds in fresh and salt water. There is almost nowhere within the continental slope that it cannot be caught. Not until it reaches the ocean's depths does the eel become unavailable, and no fishing technique has yet been able to capture it there (Post and Tesch, 1982), aside from exceptions during surveys and larval research sampling. As the variety of fishing methods and their modifications to adapt to specific waters, personnel, vessels or other conditions is so great, it is hard to cover the subject exhaustively within the limitations of this book. Therefore, only the best-known and most typical gear and methods for eel fishing can be presented here. These, for the sake of better overall view, conform to the latest FAO classification (Nedelec and Prado, 1990) and are organised beginning with the simplest and progressing into the very effective technologies. The fishing methods discussed will be dealt with primarily as commercial activities, except that the section on angling will also include an emphasis on sport fishing. The last two sections are dedicated to use of lights and electricity in combination with various conventional fishing gears. For further fishing gear and details, special literature on fishing technology must be referred to, though this will not necessarily deal directly with eel fishing (e.g. Seligo, 1926; Peters, 1935; Schnakenbeck, 1940, 1942, 1953a; von Brandt and Rutkowski, 1956; Breitenstein, 1956, 1960; von Brandt, 1957, 1959, 1961, 1984; Jäger, 1960; various authors in *EIFAC Technical Paper* **14**, FAO, Rome, 1971).

As a result of a worldwide decline in fishing harvest and present-day demands for effectiveness, the methods presented here have changed in importance and in the geographic spread of their application, and the change continues. This trend has been included only to a limited extent in the information that follows. As ecological and animal protection issues enter increasingly into use of various fishing techniques, such as with injurious (e.g. impaling) gear or small-mesh trawl nets, unrestricted use of some methods described here is no longer legally permitted. This aspect is but briefly touched upon here because laws and their administration vary by region and country, and because continual changes in the situation occur.

5.2 Impaling and clamping gear

To this group of fishing equipment belong various spears, rakes, forks, clamps and tongs, with which the eel is captured by impaling or grasping.

The use of fish spears extends back into the middle stone age about 10,000 years ago, so they are among the oldest capturing implements known. At that time, worked bone, horn or wood were still being used, but later, metal (copper and bronze, finally iron) replaced them. The eel spear (eel stabber, eel iron) that evolved from fish spears was used in Germany into the 1960s, including almost 100 years as a commercial fishing implement. It consisted of a wooden shaft, whose 2–6-m length was suited to the water depth involved, with a forked, iron spearhead attached. On each side of the actual spear (an iron spike, with or without barbs) were sturdier prongs that guided the eel onto the spike and held it there (Fig. 5.1A). On another type of eel spear, from the Pomeranian coast and especially suited for use on soft bottom materials, these guide-prongs were replaced by a springy, bow-shaped, open and flared collar of iron that served the same purpose (Fig. 5.1B). Spears were used mainly in winter when the eels lie motionless on or in the bottom. This is a season when fishers might otherwise have to be idle. In especially severe winters with solid-enough ice cover, eel spearing could be done through the ice, but this was more difficult and less productive than spearing from a boat. A fisherman could cut three ice holes per day at most, and by doing this, take as many as 30 eels (mainly small ones) under good conditions (Kienast, 1961).

Operational depth for eel spearing extended to about 4 m. Whether done legally or illegally, this method was more widespread than commonly thought. On the Mecklenburg coast of the Baltic Sea (Germany), the catch per fisherman-day was between 4 and 11 kg and averaged 7 kg. This required considerable practical knowledge. Overcast weather was best for spearing: 90% were grade III eels (Section 8.3), that is, small ones. Over and above that, and this was the questionable aspect of the method, the eels caught were mostly unmarketable (Kienast, 1961).

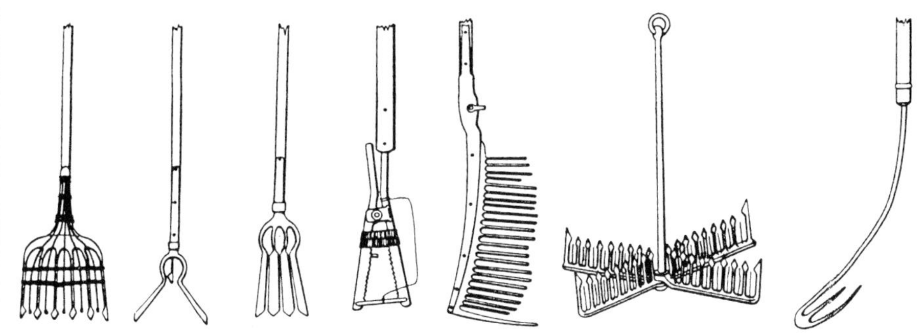

Fig. 5.1 Impaling and clamping gear

A North German fish spear or 'fish iron' with four tines
B Eel spear with a flexible, pronged, clamp or shackle for use on soft bottom
C Danish eel clamp with flexible prongs
D Mechanical eel clamp from Bornholm
E Eel comb (German Baltic coast [according to von Brandt, 1984])
F Eel anchor
G Japanese eel fork

In Canada, under favourable conditions, a skilful spear-fisher was said to catch 40–50 eels (200–1000 g) in 2 h. There, as in Europe, eel spearing was restricted mainly to coastal areas (Eales, 1968).

A special form of eel spearing was night-lighting (German Blüsen: literally, flushing), that is, aided by artificial lights at night. This was done from a rowed or drifting boat in shallow water when the weather was calm. For this, oval pans filled with blazing pinewood were floated on the water. Now and then, petroleum lanterns were used also. This lit a 1–2-m radius area of the lake or river bed, which was watched carefully. When disturbed by the light, the darkness-loving eels would crawl out of their burrows and could be speared selectively with practice. This night-lighting can be regarded as the most profitable spearing method, and most conducive to protecting small eels. During summer, such spearing by two fishers working together could take as much as 35 kg of eels per night (Kienast, 1961).

The eel clamp functions on another principle. It stemmed from Denmark and was introduced into Germany about the end of World War II. This has neither a spike nor large prongs, but rather four to seven very flexible, thin, flat, serrated, metal leaves, 1–2-cm wide and paralleling each other (Fig. 5.1C). A variation that had a back-angled barb on each leaf was used in America (Wight, personal communication). When this device hits an eel and presses it against the substrate, the pressure and the eel's struggles force the leaves apart such that the animal is clamped tight between the serrated edges. An advantage of this Danish clamp over the German spear is that the eel is not stabbed, but only receives skin wounds, which may or may not be severe. The eel clamp is especially suited for a soft, obstacle-free substrate. Disadvantages are that it soon becomes weed-clogged in dense aquatic vegetation, and that it also catches very small eels.

Eel tongs are also known from the Danish Isle of Bornholm (Fig. 5.1D). These are closed mechanically by a spring or rubber band when an eel gets between the jaws.

The eel comb or rake (Fig. 5.1E) is definitely a destructive implement and has long been banned. It usually consists of a long iron bar, its end slightly curved and bearing as many as 30 sharp, usually unbarbed tines of equal or varied length. Two comb heads fastened together crosswise and dragged like a grapple are called an eel anchor (Fig. 5.1F). Besides catching the intended eels, an eel comb having only about 0.5 cm spacing between its tines catches very small ones and wounds a disproportionate number of eels without catching them. From rowed boats, the comb was wielded like a scythe, or it could be towed from sailboats or rowboats.

Operated in much the same way as the eel comb is the one- or two-tined eel fork from such Asian lands as China, Japan, Korea and Thailand (Fig. 5.1G).

5.3 Angling

Classified as angling for eels are all methods involving bait that the eel bites. One can use bait on a gorge or hook, or simply, at the end of a line, attach a hookless batch of bait that the eel will bite and refuse to let go. In sport angling, a single line is usually fished, whereas commercial fishers tend to use several at a time. Sport fishing for eels is extensively described by Loebell (1965) and Schicker (1981), for example. For German sport fishers, the eel is among the most sought-after of fishes because it tastes so good, and because catching it keeps presenting the angler with all sorts of intriguing challenges.

The commercial fisher's hook-and-line gear for eels, as opposed to that of the sport fisher, is not tended and, therefore, is not put out as single lines but rather as an assembly of many lines.

As an expensive fish in Central and Northern Europe and East Asia, the eel is especially well suited for commercial angling, which is relatively labour-intensive and does not result in large catches.

The simplest and perhaps oldest form of fish hook is the gorge, essentially a large needle, pointed at each end, with the line tied to its middle. The gorge is completely covered by the bait, and, when swallowed by the eel, it lodges crosswise in the mouth, throat or stomach. The early gorges were made of bone, shell, wood or stone. Today, the gorge used for eels is made of steel, with an eye for the line at the middle, and with barbs at both ends (Fig. 5.2). It is still used for eel fishing in France (Gironde).

With regard to the bent hooks for eel fishing, either barbed or barbless hooks are used, as well as bends that are either straight or offset to the left or right (kirbed). The fishing line's leader is knotted to the eye of the hook or to a flat tab (Fig. 5.2C) in that position. In inland waters, eel anglers tend to prefer the so-called Kirby hook, a particular shape (Fig. 5.2A), in sizes 2/0, 1/0, 1 and 2, but at the coast, the straight, round-bend hook (Fig. 5.2B) with long shanks, which are especially suited to baiting with small fish (von Brandt, 1959). Bergmann (1962) recommended the Norwegian Mustad hook, size 1. Anglers like to use the relatively large hooks of size 1/0 so as to avoid hooking eels that are too small (Bergmann, 1962; Baumgärtner, 1968;). Barbless hooks (Fig. 5.2C) are still sometimes used, above all in Italy.

The eel angler's tackle consists essentially of rod, reel, line, lead weight and baited hook. Almost any kind of fishing rod can be used, but specialists recommend having one

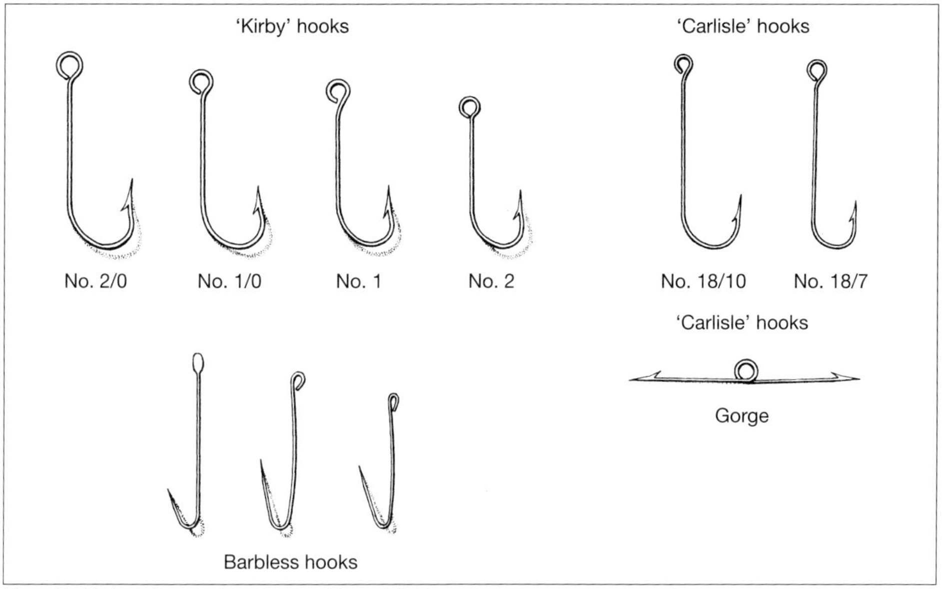

Fig.5.2 Various types of hooks used for eel fishing

with a sensitive tip and lots of backbone (Weiss, 1995). Rod length should be chosen so that the hooked eel can be manoeuvred around submerged obstacles while being played. As a rule, a 3 m rod does this.

Reels to be considered include small- to middle-sized stationary models of high-enough gear ratio to bring fish in fast. For this, anglers in the British Isles like Nottingham reels with large-diameter spools, which enable direct contact with the fish.

For fishing line, synthetic monofilament, as well as braided Dyneema lines are used. Although the latter are much more expensive, they are much stronger than monofilament of the same diameter. Furthermore, braided lines do not stretch much under tension, which has the advantage of letting the angler pull the eel away from submerged obstacles at greater distance.

Depending on local conditions, eel anglers use monofilament lines of 0.25–0.40 mm diameter or braided lines about half as thick (Eggers, 1996). The breaking strength of these lines is between 5 and 13 kg, strong enough for landing a heavy eel that could anchor itself to aquatic plants, rocks or other such objects.

To keep the hook and line on the bottom or in the path of the current, sinkers with holes bored through should be used. Such gliding and running lead weights are necessary. 'When the eel bites, the lead weight should not drag, otherwise the animal would become wary. The sinker's only purpose is to keep the bait on the bottom, especially in a strong current. I usually use a lead ball because it doesn't get hung up so easily on bottom rocks' (Pape, 1966). In standing waters, any light fishing sinker will do, but in the sea or in a strong current, pyramid-shaped or hexagonal sinkers weighing up to 250 g are the most advantageous due to their relatively large contact surface (Loebell, 1965). In a strong river current, Roth (1997) used special claw-like sinkers that are really intended for surf fishing.

'A swivel with spring clip is attached to the line in front of the sinker to prevent the eel's wriggling, twirling motions from twisting the line up. Besides that, with a spring clip, the leader can be changed quickly and easily, even in the dark' (Pape, 1966). Leader length should be about 30 cm (Loebell, 1965).

It can happen that sinker, leader, hook and bait sink into deep mud. To avoid this, a dropper line for the sinker is attached to the main line with a simple swivel. The main line runs through the loop of the dropper line, and a bead or split shot ahead of the main line's swivel can serve as a stopper. The bottom sinker is attached to the end of the dropper. The leader and bait can be held up in the water about a hand's width above the bottom mud by buoying the leader with a little piece of cork sized to the baited hook's weight (Castan, 1967).

Eel angling does not absolutely require a float (bobber), which is used mostly for fishing in strong current, from ocean beaches and from boats. But whenever it is difficult to keep the line taut, for example on long casts, in the presence of weed beds, or in particular currents, it is well to put a sliding float on the line (Loebell, 1965; Pape, 1966). To detect when an eel has bitten at night, the float is fitted with a snap light that shines brightly for about 8 h and can be seen from a great distance. Angling between the plants in water lily beds requires a float.

Baumgärtner (1968) vividly described details of how eel angling is done *via* the example of a night-time catch in southern Germany's Lake Chiemsee region: 'The little eel always swallows the bait right there, usually without making the rod twitch perceptibly. It senses the hook, pushes against the line with its tail, and, at the first tug, usually fastens itself tightly to the bottom in a flash. The eel hunter's task is now to land the eel by pulling it free of the bottom, little by little, without breaking the line and without having the eel anchor itself again. The predatory eel, in contrast to its little brethren, takes off with its prey, in which case, setting the rod hard is in order. The running trophy, once its speed is up, will always set the hook in the tip of its maw. Thus far, I have not yet had to cut an eel free. The predatory eel gives a considerable battle. It makes no long runs, but rather tries to shake itself free. If that does not work, it displays real proficiency in holding tight to sturdy objects on the bottom. It winds itself around tree limbs, boards and reed stalks with great strength, and it would be impossible to dislodge it by force. Either the eel is reeled in together with its anchoring material, or the angler has to wait until it relaxes and unwraps itself, which can take hours. A gentle tug on the line now and then discloses how far this disentwinement manoeuvre has progressed.'

Czechoslovakian anglers indexed the monthly probability of angling harvest size for various fish species, including the eel. If the eel catch for July is set equal to 100, then the proportional expectations of catches for the months from January to December turn out to be 0, 0, 0, 0, 0, 48, 100, 70, 18, 5, 0 and 0. There is no catch in April and May because that is the closed season. Otherwise, appreciable catches are expected only in the warm months (*Fisch und Fang* **5**, 316, 1964). In areas having denser eel populations (e.g. Hamburg Harbour, Germany), daily catches of 8–10 eels per angler were possible under favourable conditions in the 1960s; in one case, an angler took 46 eels in 15 h (*Fisch und Fang* **7**, 268, 1966). Today, good catches are still possible despite decreased populations. Success is being increased by feeding the eels artificially.

Undoubtedly, the eel's sense of smell guides it to the bait. Reportedly, this involves certain amino acids of animal matter which act attractively only in combination (Section 1.9.3).

These amino acids act in the same way when combined artificially, but only to lesser extent. Therefore, there is value in making a material that attracts eels without such behaviour being produced by habit or training. However, this bait would have to smell so strong as to overwhelm all other attractive smells in the water. There are several synthetic eel attractants on the market, but their effectiveness is controversial. In the practice of angling, several natural variants have proved effective when used with an appropriate technique. Therefore, some anglers attach little 'feed baskets' on the main line instead of the sliding sinker. These consist essentially of fine wire mesh and a lead weight. The little basket is filled with scraps of fish or dough made of breadcrumbs and worms or the like. These naturally attractive materials form, mainly in flowing waters, an odour trace that leads the eels to the main bait. In any event, the bait used should always also be the main ingredient of the basket's attraction feed (Weiss, 1995). Table 5.1 shows a compilation of the variety of baits used successfully in different places, the most frequent being the earthworm. The great advantage of this bait is presumably attributable in large part to the fact that it is easy to obtain almost everywhere. No less successful are lugworms (*Arenicola marina*), ragworms (*Nereis* spp.), and similar, related kinds of worms. Almost all kinds of fish are also good angling bait, among which smelt and sand eels take the forefront in commercial fisheries of Europe's coastal areas. In one case, fish that had begun to decay made good bait, despite the fact that using only fresh ones is otherwise recommended. In addition, eels reportedly refused to bite on ordinarily well-liked earthworms perhaps because a rotten odour had stayed on them from some dead worms in their container. Thus, only fresh bait should be used (Hardt, 1981). Crustaceans also are good bait, primarily shrimp (*Crangon*) and mitten crabs. Likewise, snails and mussels are usable if the shells are broken first. Leeches and frogs can be used, too. When no sort of aquatic animal is available, protein-rich foodstuffs will work, including pieces of meat, liver, lungs and sausage. Cheese can be an especially successful bait. In comparing the catch by various gear and baits, on the Baltic coast's Kurisches Haff (Kurskiy Zaliv), fishing with baitfish caught mainly eels between 54 and 59 cm long, and with earthworms primarily eels of 40–50 cm (Gaygalas, 1969).

A special Japanese variant of sport-angling is done in rivers, ponds and marshes, where, during daytime, eels occupy the dark shade, such as is found in rock hollows or cracks (Okuhara, 1958). The baited gorge or hook is thrust into the dark place with a stick, then carefully drawn out again, and the angler waits for the eel to bite.

There is also a sport angling method involving no hook but a rather special form of bait that has proved its worth in hook-and-line fishing: the earthworm – actually a bundle of them. This method is eel bobbing (called naring in Britain and Pöddern, Budden or Pören in Germany). To rig this bait, rather medium- to large-sized worms are strung lengthwise with a sturdy, blunt needle on to wool yarn, making a 1–3 m long 'worm chain'. At least 20 and as many as 80 worms are needed. The chain is then wound around the hand to make loops, which are tied tightly together at one point with the ends of the yarn, creating a bundle of perhaps 10 dangling loops (Fig. 5.3). This is fished with the strongest possible, 1–2 m rod; a broom handle will even do. The line should be *c.* 3 m long according to water depth. Suspended on the line's end is a lead weight of at least 100 g, under which the worm bundle is fastened, possibly by means of a spring clip. Then, from a boat or shore, this bait is hung in the water just a few centimetres above the bottom. A float is not needed.

Table 5.1 Eel baits that are successfully used or recommended for fishing with various angling tackle.

Type of bait	Tackle	Location and water body	References
Arenicola, mussels,shrimp (*Crangon*), sprat, sandeels (worms and mussels mainly for narrow-headed eel, sprat for broad-headed eel, shrimp for smaller eels)	eel line	coastal water	von Brandt, 1959
Pieces of herring	mackerel tackle	Baltic Sea, Lübeck Bay 15 m deep (pelagic)	*Fisch and Fang* **8**, 6, 1967
Nereis spp. in November, better than *Arenicola*	hand line	Flensburg Firth	*Fisch and Fang* **8**, 30, 1967
Stickleback, during its, spring spawning migration	eel line	River Hunte tidal area	*Fisch und Fang* **10** 10,165
Earthworm, small herring, cheese canned meat, sausage pieces	hand line	brackish water, Bremerhaven	*Fisch und Fang* **8**, 298, 1967
Bleak, gudgeon, roach, weather loach, pieces of larger fish (skinned) or shelled crabs; alternatively small frogs, shelled land and aquatic snails, leeches, meat pieces (in emergency), worm baits for small eels, fish baits for large or broad-headed eels	jugging	inland waters	Walter, 1910
Small fishes, e.g. surplus small perch,leeches, frogs, earthworms	eel line	inland waters	Seligo, 1926
Bleak, roach, stickleback, ruffe, smelt etc. or earthworms	eel line	inland waters	Jäger, 1960
Strong-smelling cheese and sausage heavily spiced with garlic	hand line	inland waters	*Fisch und Fang* **7**, 360, 1966
Pieces of coarse fishes (Cyprinids) cut into strips, tapered to a point	hand line	inland waters	*Fisch und Fang* **7**, 367, 1966
Preferred fishes: perch, ruffe, roach, white bream, rudd, bream, stickleback, gudgeon, smelt (spawners especially favourable); earthworms, caddisfly larvae, May bug grubs, Snails (crushed)	eel line	lakes	von Brandt, 1957
Bream and other coarse fishes, gudgeon, Snails *Tubifex*, shrimp, earthworms	eel line	lakes	Buchholz, 1948
Eel pieces, glass and juvenile eels up to 15 cm long	hand line	lake near Cuxhaven	*Fisch und Fang* **9**, 300, 1968
Whitefish, possibly cut in pieces	hand line	Plön lake	*Fisch und Fang*
Dead, rotten fish (!)	hand line	gravel pit	*Fisch und Fang* **9**, 142, 1968
Smelt cut into pieces in January during this fish's spawning migration	hand line (bottom tackle)	Weser, cooling water outlet	*Fisch und Fang* **7**, 279, 1966
Cheese	hand line	rivers in the North Sea tidal area and in S. Germany	*Fisch und Fang* **8**, 356, 1967
Earthworm, mitten crab, crustaceans (cut up, possibly boiled), for predatory and broad-headed eel, baitfish, cheese, fresh liver and lung, dead frogs, *Arenicola*	hand line	–	Castan, 1967
Shad, smelt, herring, various coarse fishes, mussels	eel line	Canadian waters	Eales, 1968

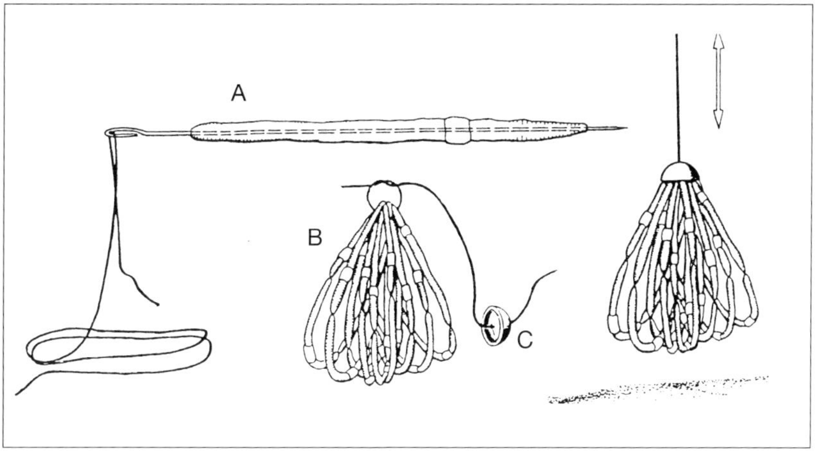

Fig. 5.3 Making an eel bobbing lure (after Schicker, 1981). (A) Using a baiting needle, large worms are threaded lengthwise on to wool yarn, and the resulting worm chain coiled into handwidth-sized loops. (B) The loops are tied together at one point and secured to the fishing line. (C) An appropriate lead sinker is threaded on to the line.

A gentle up-and-down motion (bobbing) can be advantageous. If an eel bites, a light, steady tug is generally felt, but often also jerking. One must start pulling the eel in almost immediately with a steady but not too rapid motion until after the eel is out of the water.

The eel will have bitten into the bundle's wool yarn and caught its teeth in the fibres – from which it can easily drop. To lessen the risk of losing the eel, a perforated live-box is often hung from the gunwale and floats in the water next to the fishing rig, so the eel can be quickly flipped into it before falling back into the water. An upside-down umbrella also works instead of the box. In Great Britain and France (Schalensky, 1975), eel bobbing or naring is done in slightly different form than in Germany. It also used to be done on the Hudson river and perhaps elsewhere in eastern North America (personal communication, October 2001, Robert A. Ragotzkie, University of Wisconsin, Madison; Paul Hamer, New Jersey). The method is most common on the North and Baltic Sea coasts, areas of high eel population density, where, with some degree of skill, it can yield considerable catches. Often several eels at a time cling to the wool threads. Reportedly, well over 100 eels were caught using one worm bundle in one morning on the lower Weser (Grunert, 1969). Reportedly, catches from the lower Ems catchment basin (Germany) can differ quite a bit between seasons, being good in May and October. Thus, one evening, an angler landed 6 kg of large eels suitable for smoking, as well as a bucket full of smaller ones. Several other anglers each caught over 25 kg of relatively large eels, apparently in just 14 days. These good results were achieved during west winds and high tide (*Fisch und Fang* **10**, 11, 1969). Similarly, successful eel bobbing was also reported from elsewhere, namely, 30 and more eels per hour is said to have been no rarity (Walter, 1910). On the whole, though, bobbing lands mainly small eels. On mud flats, eel bobbing is done in narrow channels and lower river reaches during ebb flow. The largest catches and biggest eels are taken at night, but during daytime acceptable catches are also made on the mud flats if one keeps small eels. Loebell (1965) even recommended going further upstream and bobbing for eels in the trout zone. In contrast to normal night-time hook-and-line fishing, the hookless worm bundle catches only eels, so one can severely reduce the eel population without endangering trout.

Jugging (Fig. 5.4) represents a step towards a more commercial type of fishing. The 'jug', which serves as a drifting float, can be an actual, corked jug with the line tied to its handle, or it can be made from a hollow tube (in German: Aalpuppe – eel doll), or occasionally from a bundle of rushes or other light, buoyant material, such as cork or tin cans. A line with a hook is wound around this object, the line being twice as long as the water is deep (von Brandt, 1957). The jug and its dangling line drift freely in the water. A light wind is good for moving it slowly along. As it keeps changing position, it has more chance of catching eels than does a setline (stationary line). However, this device is suited only for smaller water bodies or bays of larger ones, where the wind drifts it around. Richard (1938) reported that when the jug line is released in shallower places, the line should not be completely unwound so the bait will not get caught on bottom materials and stop the jug line from drifting. An added suggestion for shallow-water use is that, when set adrift, almost all the line should be left wound up, with the forward half metre or so (perhaps more in deeper water) dangling from a lightly tied knot, so the whole line can unroll as soon as an eel bites (von Brandt, 1957). This kind of unattended angling is well suited to the eel's previously described biting and escape behaviours.

A further unattended method, the roll line (German: Wickelangel – reel tackle), involves gear similar to that used in jugging but does not drift (von Brandt and Rutkowski, 1956; von Brandt, 1984). It is rigged as a setline on the bank or under a hole in the ice, with the line on a reel or wound on to some other object from which it can pay out when the eel takes the bait.

The aforementioned eel long line, a commercial fishing device, consists of a main line about 200 m long with many short snoods (also called hook lines, dropper lines, drop lines, droppers, leaders, gangions or gangins) suspended from it at intervals of about 2 or 3 m. In operation, several such main lines are tied end to end, creating a 3–5 km series having 2000–3000 hooks. This gear is then termed a row-line and can be regarded as a special form of long line. The snood should not be too short, ideally 1–1.5 m long (von Brandt,

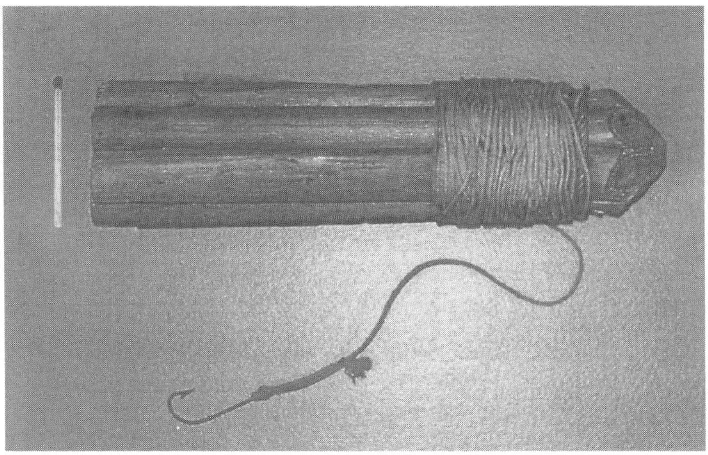

Fig. 5.4 Eel doll (photo: Gabriel)

1959), or, according to Lübben (1968), 60–70 cm, so that after biting, the eel feels resistance as late as possible. Another advantage of a longer snood is that it can be used for a long time, even if the tip is occasionally broken off and has to be replaced. To reduce tangling, the distance between snoods along the main line must be at least double that of their length. The main materials suitable for the main line are polyamide and polyester, which can be in braided or twisted form. The thinner the line, the better it will catch eels. The diameter is usually about 2 mm, and that of the snoods about 0.5 mm. Polyamide monofilament 0.4–0.6 mm thick is also used. Instead of knotting monofilament snoods to the main line, it is advantageous to fasten them on with clips having swivels.

Eel long lines can be stored in various ways. The main line is often kept in a box, the hooks arranged on top of one another in wooden clamps, three or four of which are fastened to the end of the crate, each holding up to 100 hooks. Although these clamps have to be rather exactly adapted to the size of the hooks, a metal magazine that has flexible lips offers the advantage of being able to accommodate hooks of variable size and shape (Gabriel, 1997). It also causes fewer hook problems and enables an arrangement with more storage capacity yet. Another form of storage is to arrange the hooks on a ladder-like frame, which is fastened vertically on the crate's end and has a cork layer to hold the hook points. Such a cork layer for storing the eel hooks can also be placed right on the edge of the crate. A further variation of hook storage consists of metal rods fastened horizontally in the box.

Operating an eel long line requires skill and experience (Buchholz, 1948), but often a single boat's crew (usually at least two persons) baits and sets out up to 3000 hooks. Experienced fishers can bait and set out over 1000 hooks/h – and can also retrieve them at about this speed.

If the snoods are supposed to lie on the bottom, they are weighted with sinkers. This is necessary during daytime fishing because eels stay on the bottom then, but long lines are generally first set towards evening and pulled the next morning. Night-set long lines can even have floats because eels swim up into higher layers of the water overnight. Floats may also be necessary to prevent the long line from sinking into soft mud.

The baitfish should be put on the hooks so that the point of the hook barely protrudes. For storage, the long lines are best kept in an eel hook holder or clamp so they can be cleaned or washed, which is conducive to good catches. Cleaning in a potassium permanganate bath may even be useful; a pinch of this chemical per bucket of water will do (von Brandt, 1957; Jäger, 1960). Special cleanliness is needed not only in handling the long lines, but also in baiting the hooks. The smell of petrol and oil from operating motors could frighten away the very odour-sensitive eels.

Besides the kind of bait used, the way the line is set (laid out) is important. On this, von Brandt (1957) wrote: 'First, a stone is attached to one end of the line. (And, so it will smell right, the stone should be from the lake.) If possible, the line should be moistened (so it will sink faster) and then paid out loosely into the water in a curving pattern – the slacker the better. This is done from a boat moving slowly forward. If the hooks are baited in the boat while the line is paid out, it is better to let a missed (mistakenly unbaited hook) go by than to hold onto the line such that boat's motion draws it taut. The slacker the line lies on the bottom, the less likely the eel will be to feel resistance, so the more likely it will be to swallow the hook. It is preferable to do the baiting in the boat; only if weather is windy should it be done beforehand on land. The baited hooks are packed in sand, either in a flat box or on a board, so that the line can pay out smoothly'.

The commercial long-line fishery for eels is concentrated in inland lakes and, above all, in the coastal region of the Baltic Sea, where it is conducted from about May to October. With occasional harvests of over 100 kg per long-line per day, it can be even more productive than trawling, especially in areas where the latter is not permitted (Gabriel, 1997).

Long-lining for eels is also done in other regions, for example, southern Europe (Rankovic, 1957), North America (Eales, 1968) and Japan (Okuhara, 1958). The best harvests are achieved in temperate climates when the water has first reached normal summer temperatures, that is, in May/June (Eales, 1968), but eels can also be caught later, well into autumn. In Lake Skutari on the border between Montenegro and Albania, where water temperatures seldom or only briefly fall below the minimum for eels, the highest catches occurred during March to May, and the low was reached in the warmest months, June to August.

As already mentioned, long-lining for eels is commercially more profitable than jugging because it involves a much greater number of hooks. However, as studies in the German Sakrowe Lake have revealed, the probability of catch *per hook* is greater for jugging (Rahn, 1957b). Averaged over a 5-year period, the jug fishery for eels caught 2.9 kg/100 hooks, long-lining over the same period caught only 0.7 kg/100 hooks.

The eel line (Jäger, 1960) functions much like the long line. It is 2–3 mm thick and is stretched across a river. Its snoods are checked morning and evening to remove hooked fish and rebait hooks.

5.4 Traps

Traps form eel fishing's most important and diverse group of catching gear. In this category are not only the most varied kinds of the fyke net, by far the best-known methods, but also quite simple hiding covert traps, stow nets and stationary traps. Eels enter traps either under their own power or by drifting with the current. They usually reach a terminal catching chamber, *via* one or more entrances, which they cannot exit until the fisher removes them. In the final analysis, all those eel catching devices can be counted as harvest traps with which people are not directly involved.

A harvest trap is usually not used alone but rather with supplementary gear or facilities.

The operational principle of all traps is much the same. The differences lie in the ways that fish are lured or guided into them. The guiding is typically done with various wing structures, that is, weirs or net walls, which block routes of migration or other movement and divert the eel's direction of travel. The eels are very commonly lured into the trap with bait. Furthermore, the eel's cover- or refuge-seeking behaviour, its negative phototaxis and its thigmotaxis are also exploited in various trapping methods.

5.4.1 Covert traps

As already mentioned, the eel has pronounced thigmotaxis (affinity for contact) and at the same time negative phototaxis (tendency to move away from light). These behaviours are purposely used to capture eels by offering places that they perceive as suitable refuge. One of the oldest kinds of covert-trap fishery in Europe and elsewhere is the so-called eel bundle fishery. For this, twigs or brushy materials are bound together in loose bundles, weighted with stones, sunk on to the bed of the lake or stream and marked with a float or flag. In Poland's lower Vistula river, eel bundles were also suspended from poles driven into the river bed at an angle, so that the bundles partially protrude from the water (Seligo, 1926). Eels like to hide in such bundles. A fine-meshed dip net is thrust under the bundle when it is lifted so that the eels cannot escape. Most eels that gather in such bundles are small ones.

Walter (1910) reported on a similar fishery on the North Friesian Wehlen (brackish lakes and ponds). When there was heavy ice and snow cover, and eels left their winter quarters, holes were cut in the ice, into which straw bundles were stuck, and eels that came to gulp air at the ice holes would crawl into the bundles.

In Japan, covert traps have been made from bamboo tubes (Okuhara, 1958), usually two bundled together, and several of these suspended in the water from a line (Fig. 5.5). Eels hide in these. To make the tubes more attractive, they can be baited. Of course, the tubes must be retrieved carefully to keep the eels from escaping. Sometimes divers are even supposed to go down and close the tubes before they are lifted. Like these bamboo tubes, tubes made from fired clay or other material are also used.

As eels were sometimes captured in animal corpses or were found in drowned persons' clothing, it was often erroneously said that they eat decaying meat. Actually, they were just trying to get away from light when they took refuge in such 'organic cover'.

5.4.2 Eel traps and derivatives (German: Reusen)

The category eel traps (Reusen), which evolved from cylindrical or conical devices made of reeds or wicker, includes a whole palette of devices differing greatly in design, material,

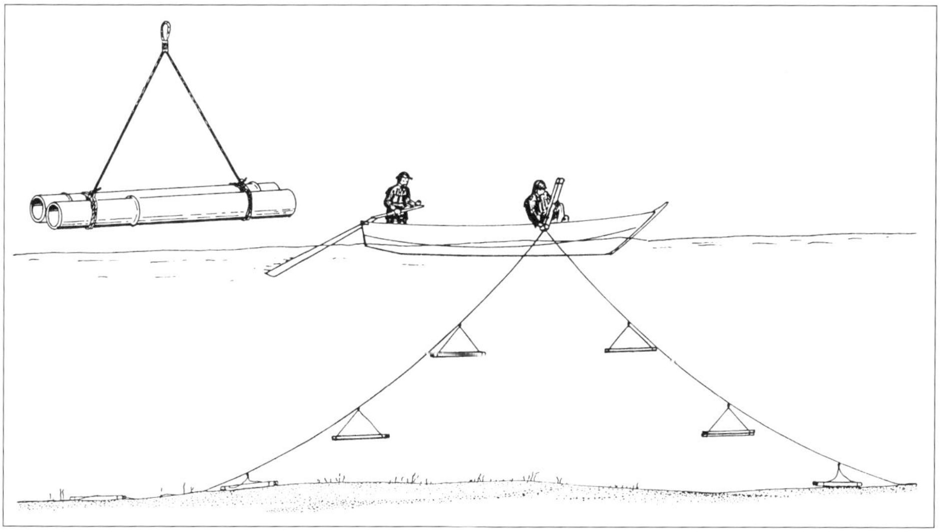

Fig. 5.5 Covert traps made of bamboo tubes, and setting them out (redrawn by Kuhlmann from photo by Gabriel)

size and the way they are deployed. Therefore, they do not necessarily resemble the original form of a tube or brush bundle any more. Classified today, in the broadest sense, as cylindrical eel traps (Reusen) all are stationary capturing devices (except for outright weirs, Section 5.4.4.1) made of wood, wicker, plastic, metal or netting, that have, at the structure's entrance and usually made of the same material, a funnel (throat) which lets eels enter the interior but blocks or conceals the way to get back out. Furthermore, such devices, usually (except for small pots or baskets) include accessory parts, such as leaders (weiring or netting), wing nets and hearts – and are not installed as a single unit but in series or combinations of several. Above all, devising such combinations of eel trapping gear requires specific knowledge of this animal's behaviour at the site in question. Enclosure gear is used for eel fishing in streams and lakes, as well as on the coasts of the Baltic and North Seas.

It is not simple to organise the great diversity of traps and other enclosure gear into a definite classification. Besides that, the classification has to accommodate differing viewpoints, colloquialisms and downright terminological contradictions that exist among persons involved in eel fisheries within any geographic region, even if they speak the same language, such as English, French or German. For example, to call a certain common little trap an 'eel basket', as is done in some places or languages (e.g. Aalkorb in German), is strictly valid only for a container woven from plant material, but the term is also sometimes applied to a widely used little hoop net made of cloth netting. Moreover, whereas the same device may in some English-speaking localities be called an eel basket, it is usually termed 'eel pot' in that tongue, even though it is a reticulation of wicker, wood, netting or other such materials (now including plastic) and not at all a ceramic object. Also, in German, the Bügelreuse (a large, hoop or fyke net) is sometimes called a Korbreuse (basket trap) to differentiate it from a Kammerreuse (pound net), and, on the other hand, the term Reuse (usually meaning trap, in general), is used exclusively for woven baskets in some areas. Due to the existence of such definitional problems, the extensive group of stationary traps is organised according to size and type of effect as follows:

- Small, simple traps (tubular traps and basket-like devices).
- Hoop or fyke nets (with extensions, such as wings and leader weirs, as well as systems of several such traps).
- Large traps (pound nets).

5.4.2.1 Small traps (tubular traps and pots)

The best-known small trap is the eel pot or eel basket. Its original form (Fig. 5.6), as mentioned above, consisted of plant material and undoubtedly was made easily by our ancestors long ago. Its characteristic of having two or more funnels is seen in all the more highly developed versions, whether made of natural or synthetic fibres, or of moulded plastic.

A somewhat more convenient type of eel pot is made from Spanish reed (peddig reed). It sinks immediately when set out, thus guarantees quicker results. In Japan, the device is made from similar material, which is also well known in North America (Eales, 1968), that is, it is used worldwide. Wooden barrels, slats and, as a further development, wire are used to form an outer casing. For centuries, fishers had to put up with the fact that eel pots

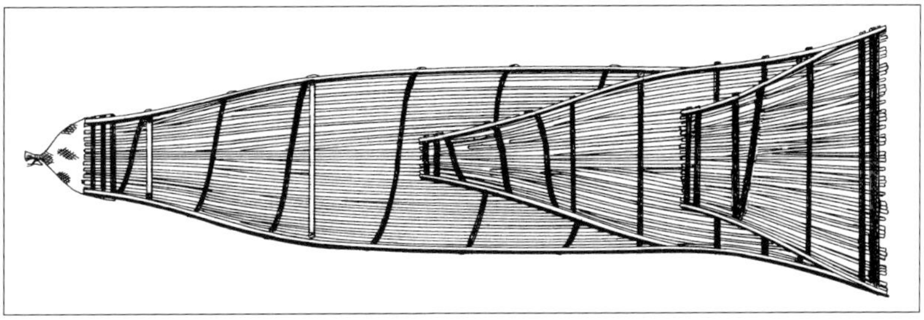

Fig. 5.6 Two-funnelled eel pot (schematic longitudinal section), as woven from reed or flexible twigs in many lands

made of natural materials rotted after about 2 years. That changed when plastic models made their appearance (Mohr, 1963a, b; Köthke, 1964a). The fishing effectiveness of these was like that of wicker pots, as long as tidal waters were not involved, which apparently require a pot of somewhat different construction. Originally, such a plastic device was twice as expensive as a wicker pot. Therefore, the plastic pot described by Mohr (1963b), which did not yet involve moulded parts but was a woven form, had to last at least twice as long as a wicker pot because of its higher price. The price ratio changed for the later plastic design that was composed of moulded parts because ever-increasing labour costs affected the wicker pots more, and the cost of manufacturing plastic parts declined. As Fig. 5.7 shows, such a plastic eel pot consists of a funnel-shaped front section with an oval opening, and the first funnel is fastened to the rear edge of that section. Following that is a cylindrical middle section, composed of two parts and holding a second funnel at its centre. A conical end piece forms a rear chamber (cod end, so to speak), which can be opened to remove the catch.

The eel pot is well regarded not just in commercial fisheries, but among all kinds of fishers. As it is so handy, the pot is easy for lay persons to use. It is popular with sport fishers and in fishery research and management is used for investigations and surveys. It is most commonly used in running and tidal waters that have dense eel populations.

Its small size and robust construction suit it to strong and changing current. In the Elbe's tidal zone, as many as 60 eel pots are set out on a long, anchored line at 5–6 m intervals, each one tethered by a short line. The main line is positioned obliquely to the current,

Fig. 5.7 Plastic eel pot (photo: Gabriel)

so that the individual baskets do not knock together and damage each other in the shifting current. Attached to the inside of the door on the pot's end piece is a bait to attract eels *via* odour.

How strongly the eel depends on the bait and its odour for finding the basket is shown by studies in which the eel's sense of smell was impaired (Tesch, 1970). Eels were caught, primarily in eel pots, in the Elbe estuary and were marked. Some had their sense of smell blocked (impaired), the others were left intact (unimpaired), and then they were all released. Only 12% of the impaired eels were recaptured in baited pots, but 22% of the unimpaired ones were. At another, nearby site, fished with unbaited eel traps made of netting, mainly fyke nets, recapture rates of the marked eels were 12% for the impaired and 11% for the unimpaired ones, an insignificant difference. Thus, normal eels entered baited traps more than did eels whose sense of smell was impaired, and both types of eel entered unbaited traps at the same low rate. Every commercial eel fisher, who has set baited and unbaited pots next to each other, can confirm this experience.

Success in catching eels also depends very substantially on the type of bait, as can be verified from the hook-and-line fishery (Section 5.3). At times, the eel reacts only to particular kinds of bait, for example, at certain times it can be caught only with smelt. People who feed eels in aquaria attest to this fastidiousness. It always takes a certain amount of time before the animals get used to a different kind of feed. Fishers of the lower Elbe have the following explanation for the eel's insistence on preferring smelt as bait only at certain times. As long the smelt are small, they cannot escape eels that chase them. During this period, the smelt is just big enough to pass through the mesh of wicker pots. When the smelt get bigger, they can swim fast enough to escape from eels, and the eels, having developed an appetite for the smelt, which just became unavailable in the open water, are then particularly attracted by the bait's odour. This is when the time for using smelt bait begins. Other suitable baits are shrimp (*Crangon* spp.) or prawn (*Neomysis vulgaris*), soft mitten crabs, earthworms (raised in wood shavings), crushed mussels or whelks (*Buccinum undatum*), fish spawn, shredded lamprey and even pieces of salt herring (Mohr, 1963a). According to fishers, eels take only fresh bait. Bait left in the water longer than 12 h loses effectiveness. It should be kept in mind that every kind of bait works size selectively. In a trap baited for New Zealand eels, the proportional distribution of the catch by length group was (Burnet, 1952b):

- 51 to <58 cm: 6%;
- 58 to <61 cm: 6%;
- 61 to <66 cm: 11%;
- 66 to <71 cm: 19%;
- 71 to <76 cm: 22%;
- 76 to <81 cm: 20%;
- 81 to <86 cm: 8%;
- 86 to <91 cm: 5%;
- 91 to 94 cm: 5%.

Unbaited pots can be placed in front of weirs and dams. If the upstream migration of eels is blocked, they immediately seek cover (negative phototaxis, thigmotaxis). For this reason, the middle part of plastic eel pots was roofed over with darkened material. Today, the use of such pots combined with woven material as leader weirs (Walter, 1910; Seligo, 1926) is less common and is surpassed by modern facilities.

Although the eel pot is typical gear for river and tidal areas, it can also be used in lakes under certain conditions (Brümmerstädt, 1955). This means combining the pots with construction of artificial spawning nests for cyprinids (*Leuciscus rutilus*, *Blicca björkna*, *Alburnus alburnus*), the nests being made of conifer (e.g. pine) boughs or else setting the pots in the natural spawning nests of these fishes. As a predator on spawn, the eel finds such places and then, at times, seeks refuge in the pots, which have to be close by. During the pike spawning season, eels are also supposed to be catchable on the shallow weed beds (spawning meadows) involved. There, eel pots are laid in dips, channels and ditches where the eels retreat as the water level drops when springtime high water recedes. Of course, not only eel pots are used on the spawning grounds (Feddersen, 1893; *Fisch und Fang* **10**, 160, 1969), but also small fyke nets of various forms, which will be discussed presently in more detail.

Baited wooden eel boxes, which became well known in Holland during the 1960s, also work much like the eel pots described above. These involve boxes (Fig. 5.8) made of water-resistant (marine) plywood, about 65 cm long and 10 × 12 cm in cross section, and fitted with funnels of fine-mesh netting at the ends. For baiting, removing the catch, and for thorough cleaning, the boxes have a sliding top cover, held shut by a little wooden wedge-peg. Each of the long sides has a circular opening to let undersized eels escape; this hole is supposed to be 14.5 mm, according to present regulations. To aid sinking and lend stability on the stream bed, the bottom of the box is weighted with a steel band. The boxes are hung vertically by lines about 1.2 m long, fastened with clips at intervals of about 40 m along a 4-mm thick main line (long-line), and this whole assembly is lowered on to the sea bed. Mechanised winches have made it possible for two fishers on a 17-m cutter to operate about 750 such boxes, thus landing an average of more than 100 kg of eels daily (Gabriel, 1998). The preferred bait in the IJsselmeer is smelt, caught with a little trawl just before the boxes are baited. The boxes are deployed at noon and retrieved the next morning. New boxes must sit several weeks in the water before they can catch eels effectively. The low cost of these boxes (they can be made by fishers themselves) and their simple

Fig. 5.8 Wooden eel pot or eel box from Holland (photo: Gabriel)

operation led to their almost complete replacement of the IJsselmeer's customary long-line fishery. The eel box fishery's catches climbed rapidly: 4855 kg (1966); 32,406 kg (1967); 227,340 kg (1968); 277,967 kg (1969); 449,617 kg (up to September 1970) (Deelder, 1971) and in 1997 an estimated 1000 t.

'Eel pots' made of netting (net traps) were constructed according to the design of the wicker eel pot and then further developed. Formerly, they were made of cotton and other natural fibre, but today from synthetic fibres and wire (on the great advantages of synthetic net materials, see von Brandt [1961] and Oray [1968]). They were, for example, already operated on the coast of Mecklenburg (Germany) around 1900.

The net pot, when deployed in its operating position, almost completely resembles an eel basket in size and shape. It is cylindrical or somewhat conical in shape and consists of netting stretched tight around hoops, which are held apart by 'spreading sticks' that are jammed between them or tied to them (Figs 5.9 and 5.10). A net pot with a funnelled opening at each end (Fig. 5.9; German: Trommelreuse [drum trap], Bunge or Bollreuse) is used only occasionally by central European commercial fishers, but more often by laypersons because it is easy to operate. It consists of three hoops, over which netting is stretched. The terminal openings of the two opposed funnels are slit shaped. The device is kept longitudinally taut by spreading sticks jammed between the outer hoops. The centre hoop must be a bit smaller so the mouths will contact the stream bed. A drawback is that eels escape from these devices relatively quickly.

More common in the commercial fishery is the little conical net pot (Fig. 5.10), open only at one end. It consists of five hoops or rings of progressively smaller diameter toward the rear and contains two funnels. Its length is about 1.3 m, and the mouth diameter is 35 cm. At least two spreading sticks tied to the hoops hold it taut. (In German, it is sometimes called a Wurfreuse, i.e. throw-trap because it is thrown out from a boat.) This little trap's effective fishing area is small, and it catches relatively little. Therefore, they are seldom set out individually but are assembled in series aligned with the current, as already mentioned for the wooden eel boxes from Holland and for the pot systems set obliquely in the lower Elbe. By having the front and rear ends of the series weighted with stones or held with small anchors, these conical net pots hold each other in fishing position by tension.

The so-called eel pot chain, used mainly for yellow eels on the Baltic Sea coast, can be included in the small trap category, even though these installations, arranged in series,

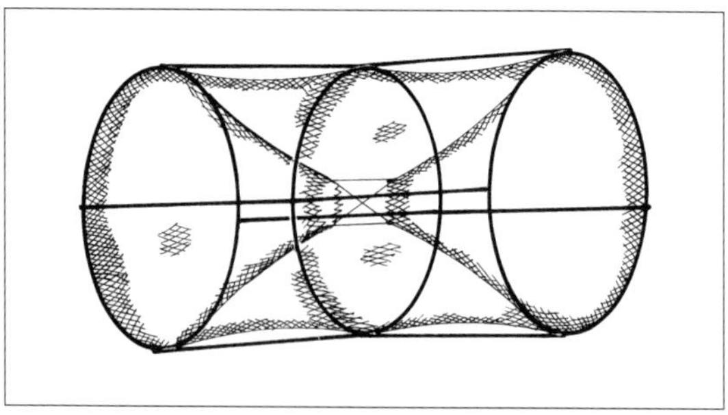

Fig. 5.9 Net pot with a funnelled opening at each end

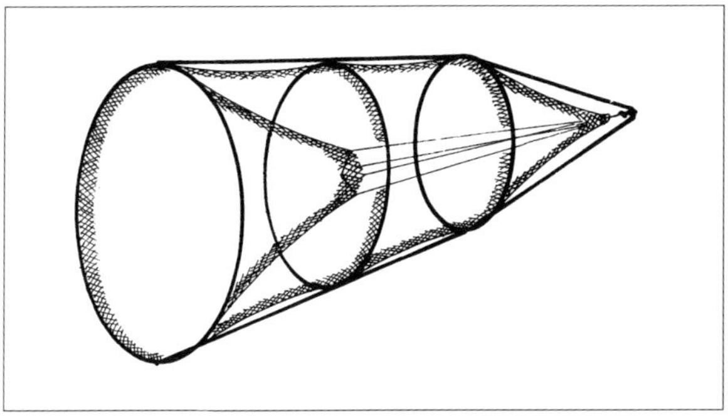

Fig. 5.10 Conical hoop net as a stretched bag

commonly reach lengths of 300–500 m, sometimes over 1 km. The basic element is a double net consisting of two small, conical net pots (each about 1.5–1.7 m long and with mouth diameter of about 35–45 cm), which face each other and are connected by a leader net that is about 3–3.6 m long. This connecting net and, to some extent, the first hoop of each pot are weighted with lead for good contact with the stream bed. Figure 5.11 shows the design of such a double net and its connection to the next one by means of a codend snood. A net pot chain from Germany's Wismar area, for example, consists of six central and two end pots, making a chain 42 m long. The central pots are each about 2.3 m long, which is a bit more than the end pots. These eight-pot chains are deployed, much as gill nets are, from open or covered boats in series of up to 20 chains. At the end of each series, the chain is anchored and marked with a buoy. Further marker buoys are used at intervals on long series of pots. Eel net pot chains are easy to handle and seaworthy, so they are even used in the strong tidal currents off the North Sea island of Heligoland. They are also encountered in rivers. Despite the small pot size, this fishing device's effectiveness can be confirmed as rather high.

5.4.2.2 Fyke nets

The stretched-bag devices and net pot chains described above are indeed fish traps stretched on hoops. However, true hoop nets, generally called fyke nets, are of larger dimension, have more extensive arrangements of accessory features, and may be part of whole integrated fishing systems.

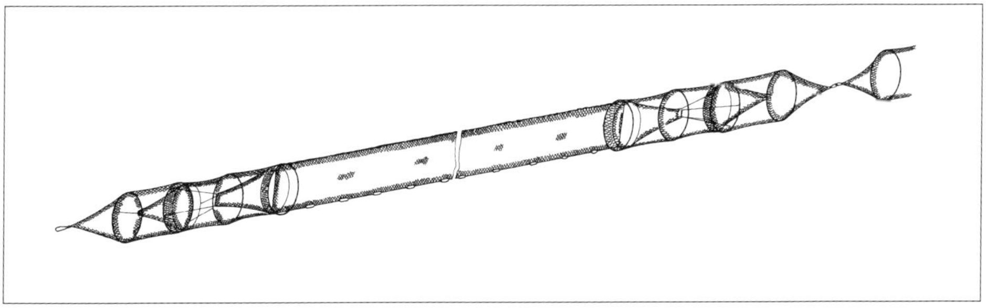

Fig. 5.11 Paired or double eel net pots connected with similar adjacent gear to form a net pot chain

The following material will deal exclusively with the true hoop or fyke nets. Figure 5.12 shows the fyke net, a very simple device that has long been known in the form that was, and to some extent still is, used in many inland regions or on the coast. Although in such fyke nets the front funnel's small end is relatively wide and is rectangular or U-shaped in cross-section, the second funnel's exit into the rear chamber (the codend) is a very narrow, slit-shaped opening (Breitenstein, 1956). Eels seldom find their way out of that. Instead of two funnels, three can be installed. The number of hoops usually varies between five and seven and, in the coastal fishery, can even reach 10. The diameter also differs a great deal, ranging for the larger fyke nets from 0.5 to about 2.5 m. The two wings can be lengthened as much as one wants. In addition to the wings that extend at angles, a longer net or weir, the leader, often also extends outward, usually aligned with the fyke net's axis. This is to intercept silver eels that pass by. In coastal fisheries, leaders can be several hundred metres long.

A large fyke net's substantial area of effectiveness can even be enlarged, especially in a lake, by combining it with several others in a system. Figure 5.13 shows a selection of possible deployment patterns and their designations. Above all, the 'Kreuzsack' (cross-net) or German 'Kossack' that originated in East Prussia is used very successfully for eel fishing in north German lakes. It consists of a lakeward-positioned main net and three smaller side nets connected by leader nets. Setting this device up is relatively simple; readying all four traps for fishing requires driving only four stakes. Anchors can augment the stakes when securing hoop nets if, for example, outer beach sites exposed to wind or current are involved.

As can be seen, fishers' enthusiasm for different designs knows almost no bounds. Practically every eel-trapping site has its own characteristics, including weather, current, water depth, bed slope, bottom conformation and extent of suitable bed, as well as considerations of hydroengineering structures and navigation. Therefore, installing these fishing devices requires training and experience, for although basic components like fyke nets, wings and leaders are available on the market, these are not always the right sizes, so the fisher has to take up the netting-needle for alteration and assembly. Special deep-water

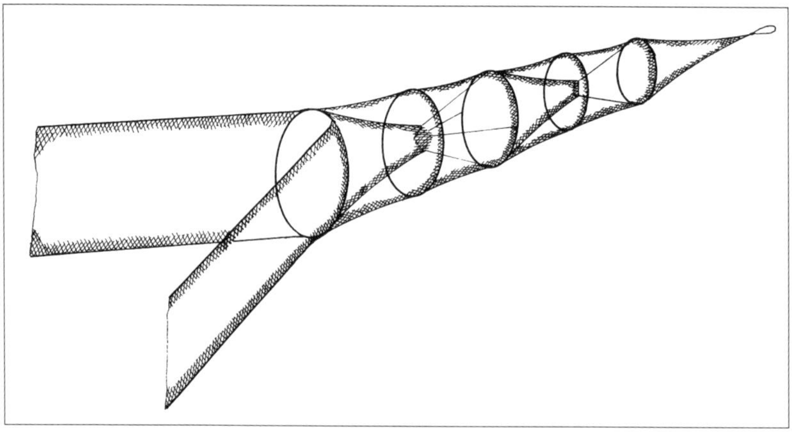

Fig. 5.12 Fyke net with wings

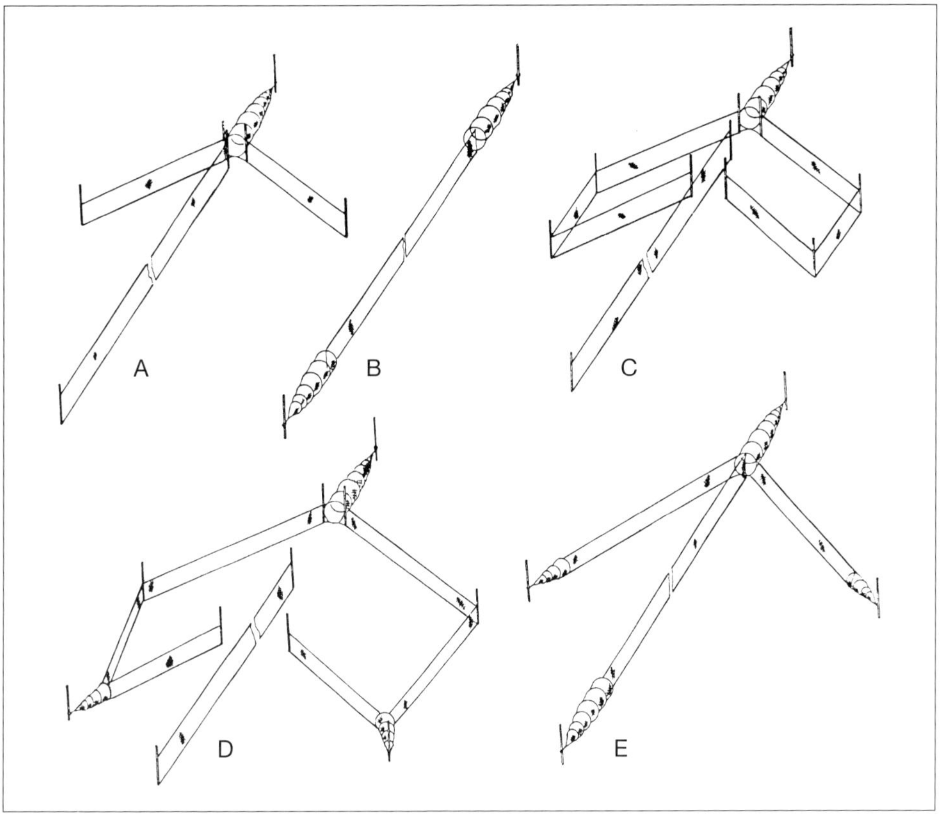

Fig. 5.13 Forms of fyke net installation

A Simple fyke net
B Paired or double fyke net
C Bocksack
D Triple bag
E Kreuzsack or Kossack

fyke nets were developed for the deep meander cut-offs of the upper Rhine (Kiekhäfer *et al.*, 1974) and for Lake Constance (Quoss, 1976), for example.

5.4.2.3 Large traps/Pound nets

Large traps, as the name says, have larger dimensions. They can be produced either by making porch-like additions to fyke nets or by building spacious catch chambers (cribs) out of netting and adding wings and possibly antechambers. They are used on the coast or in large lakes where it is necessary to cover an extensive area and catch large amounts of eels. Pound nets can be set up with stakes or else with anchors. In order to increase effectiveness of a trap, its construction can be altered by combining a three-dimensional trap (a completely closed one) with a two-dimensional one (open at the top). If these big traps are made of netting, they are called pound nets.

A classical region for eel pound nets is on the Baltic Sea coast, especially in the western and central parts, where rather large concentrations of emigrating eels are encountered. However, pound netting no longer attains the annual eel harvests of *c.* 1000 t from Sweden and 3000 t from Denmark that were taken in the early 1970s, and the over 2000 t on the German coast in 1939 (Neuhaus, 1967). Still today, though, eel catch takes first place in value within some fishery enterprises, as Schlieker already established in 1957.

Emigrating silver eels often encounter coasts, especially those that restrict them in the northern and southern Baltic Sea. Many coasts also are not as exposed to prevailing winds as are the eastern North Sea coasts, so weather and sea conditions present fewer problems, especially on coasts of Sweden, Denmark and Germany. However, it is crucial that the Baltic Sea has almost no tidal currents or water level changes. Therefore, pound netting developed into a branch of the coastal fishery that is significant in terms of fishing equipment use. This applies mainly to the silver eel fishery.

The eel pound net evolved from the herring pound net, which, with its large dimensions, showed that large catches of eels were possible, too, without the outer coast's rigorous conditions causing insurmountable difficulties. They were developed in the most important eel-fishing area, the coasts of the Isle of Rügen, in 1911 after the actual autumn silver eel fishery had begun in these waters in 1895 (Schlieker, 1957). Moreover, still today large herring pound nets are equipped with pairs of little fyke nets in the after chamber, so as to catch eels in the course of the springtime herring fishery.

With respect to the eel pound nets themselves, around Rügen, the catching chamber complex has a somewhat rounded appearance, so that the general type is referred to there as a chamber pound net. Two basic designs are used: combined chamber pound nets and double-chamber pound nets (Schlieker, 1957). The first is a combination of pound and fyke net. It is smaller than a double-chamber pound net, hence easier to set and to empty. Two or three fishers working together are usually enough to operate one.

A double-chamber pound net for eels set with two funnels and an outside and inside heart (Fig. 5.14) resembles, in exterior dimensions, a single-funnelled herring pound net, that is, it is about 6 m wide. The leader extends from the beach or a beach reef outward 160–1080 m into the sea (Nolte, 1938). This device can be set in water as deep as 12 m (Bobzin, 1965). Using very long leaders, two or three pound nets can be positioned in line. Operating such a pound net requires eight to ten men, generally individual fishers working as a cooperative, which may also be needed for pooling money to cover the pound net's great expense. The fishers usually reserve rights lasting several decades for a definite pound-netting site.

Stake weirs initially existed on the outer beaches but were increasingly replaced by anchored floating weirs, and such anchoring later proved its worth for pound nets, as well (Martinköwitz, 1961; Bobzin, 1965). In doing this, the anchors for the leader are attached to its lower edge, those for the catching chambers to a rope framework stretched around the chambers. The device is held at the water's surface with floats, sized so the pound net will start sinking when current velocity exceeds 0.3 m/s. The first catch chamber contains an upper 'apron', the second a 'roof' made of thin netting which is supposed to prevent eels from escaping in heavy sea and strong current. The elasticity of anchored pound nets makes them more durable than staked installations. Moreover, their setting and recovery require about half as much labour. Their catching effectiveness has proved equal to, if not greater than that of staked pound nets (Bobzin, 1965).

Since the 1950s, pound nets have also been set on the Baltic coast of Schleswig-Holstein (Germany) in much the same way as on the Isle of Rügen's coasts. However, they

Sc Stern chamber
Sf Stern funnel
Fc Forward chamber
Ff Forward funnel
Oh Outside heart
Lw Leader

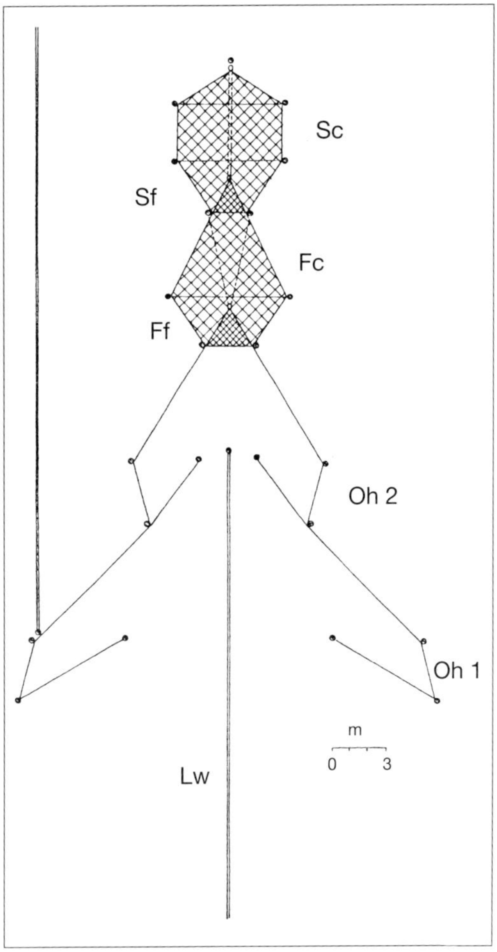

Fig. 5.14 Double chamber pound net (after Schlieker, 1957)

did not evolve from the Pomeranian chamber pound net, but rather from the Danish–Swedish bundgarn (Fig. 5.15) – literally, bottom net in Danish – which, with other design specifications, has long been well known for catching herring, cod, flounder and lumpfish. In 1964, there were already 111 bundgarns from Flensburger Förde to Neustädter Bay, which gave this area's eel harvest a substantial upswing and pushed other eel-fishing methods into the background. The bundgarn is the largest eel-fishing device on the Baltic Sea coast and, of course, catches other fishes at the same time. It is another combination of basic types (Klust, 1965), having, in plan view, the shape of a chamber pound net, but with vertical walled chambers open at the top, and with two fyke nets protruding from the main chamber as the final traps (Fig. 5.15). The fyke nets have closed, four to six hoops and two funnels, and they look tiny compared with the two big chambers. The rear chamber (crib) is more than 15 m in diameter (47 m or more in circumference) and is closed with netting at the bottom. The bundgarns in Schleswig-Holstein extend to depths of 6m and are installed with floating leaders 150–600 m in length (Henck, 1965).

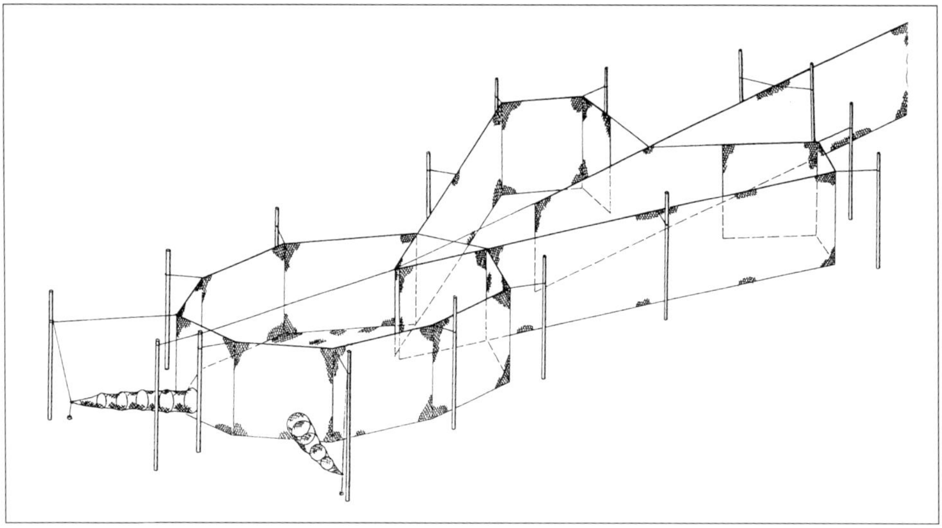

Fig. 5.15 Danish pound net

Henck (1965) described the pound net's catching process as follows: 'The eel that moves along the coast mainly in the east–west direction and later goes north–south, initially encounters the leader that was set at a 90° angle to the coastline and will try to find a way through it, probing around all the way to the water's surface. As the leader blocks the way, the eel takes the path into deeper water. Upon reaching the end of the leader, the eel, guided by the wings that were installed as fishing devices, enters the front chamber and later the main chamber. Often it then finds the opening of one of the fyke nets, which are fitted with two funnels. To remove the catch from the pound net, one needs only to lift the fyke nets and not, as was previously necessary, the whole crib'.

The annual silver eel harvest extends from the end of August to the beginning of December. The best catches are in October, as a rule. It is well known that small eels predominate in the catch early in the season, whereas, large eels are caught from the peak of the season until the end (Section 3.4.1).

5.4.3 Stow nets

Stow nets are typical eel-fishing gear in running and tidal waters. They are also known as coghills in Britain and Hamen in Germany. Included under this general name are a whole group of devices. Their common trait is a long, conical net bag having its mouth held open by a frame or other rigging.

The category of small stow nets used in eel fisheries includes such gear as the scoop net, the bottom stow net, and the scrape or skimming net.

One way to fish with the bottom stow net, is to pump a flushing board (a board on a handle long enough to reach the bottom) up and down to flush eels, mainly from heavily vegetated areas, driving them into the net. For this particular method, the stow net bag not only has a solid frame at the mouth but is also held open at its rear end by further framework. The netting of the bottom stow net can also be drawn over the

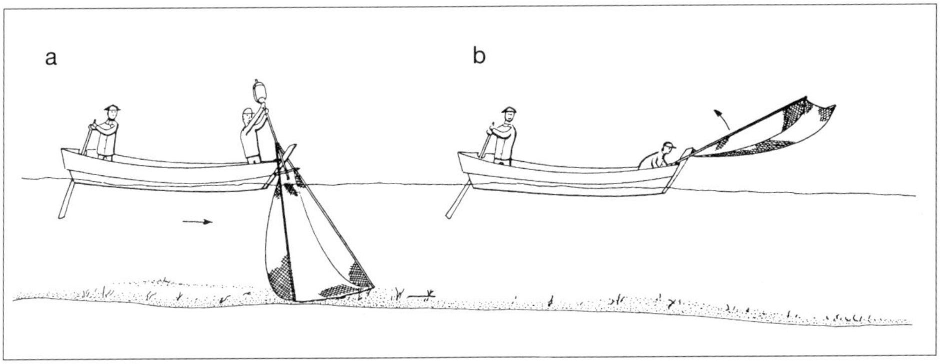

Fig. 5.16 Japanese skimming net of the type called a scrape net (after von Brandt, 1964), set in (a) and emptying in (b)

mouth's edge and extended funnel-like back inside the net bag to retard eels from swimming back out.

The scrape net consisted of poles, held scissors like, between which the net was stretched. It could be as much as 5 m long. Operated from a boat as a hand-held device, it was faced into the current and lifted about every 2 min. A similar Japanese device for river eel fishing is mounted on the bow of a boat, thus can be larger (Fig. 5.16).

So-called large stow nets have recently developed into important eel-fishing devices. Preference for stow nets in European river and coastal eel fisheries frequently came about because this fish was the only valuable species that was left, or because market demand for less valuable species no longer existed.

From the standpoint of construction, large stow nets are basically very simplified large fyke nets. Even a funnel, indeed the typical characteristic of a fyke net, can be found in stow nets. The essential difference between stow net and fyke net is that eels are carried more passively than actively into the stow net bag by the current, whereas, their movement into the fyke net is primarily active. In a certain sense, the large stow net's bag (sometimes called the codend) can be considered a predecessor of the trawl.

For the continental part of the eel's spawning migration, it uses the current. It does not move long the stream bed as the juvenile and non-migratory stages do, but rather in open water where the current is strongest. Thus, it wanders in the part of the channel where it is best, for hydrodynamic reasons, to put the stow net. This situation makes the stow net a valuable eel-fishing device, no matter what its structural design.

A technical problem in using stow nets is how to keep them open in the current. For that, there are three solutions:

- stakes;
- anchors;
- otter boards.

5.4.3.1 *Staked stow nets*

These consist of net bags, usually in a row abreast, held open and against the bottom by wooden stakes or metal pipes. Each net's rectangular opening (as much as 6×3 m) is

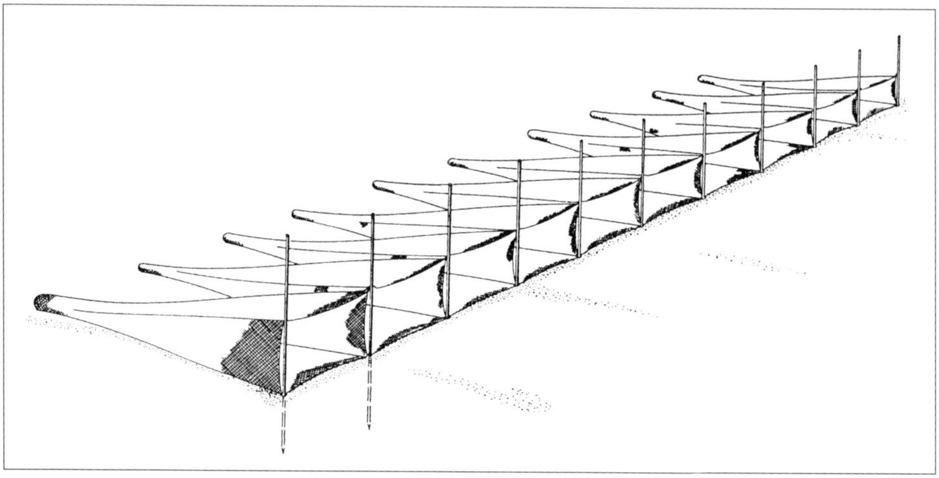

Fig. 5.17 Staked stow nets

stiffened with a rope frame, which has its four corners hung by loops or metal rings on the stakes. In Germany, such stow nets were common in the lower Weser and Ems river area above all. They existed in Holland, as well. On the coast, a set of them would consist of up to 50 next to each other in a row, having net bags as long as 25 cm, the openings facing flood or ebb tide (Fig. 5.17). They were, however, gradually forced out by navigation. In Germany, they were already restricted to the Ems and Stör rivers in the mid-1960s (Nolte, 1967; Pape, 1970).

5.4.3.2 Anchored stow nets

The original form of this type of stow net had its mouth held open by a rectangular or trapezoidal wooden frame (later steel, too), and wire cables led from the corners upstream to an anchor in the river bed. That made it possible, as a further development of the staked stow net, to set stow nets also in deeper water and on hard bottom. As their placement is stationary, anchored stow nets in this form also eventually became obstacles to navigation just as staked ones had, and could no longer be used. Therefore, moveable, boat-rigged stow nets were introduced that have two ways of holding the mouths open. The first corresponds to the scheme described above. This type is up to about 25 m long and has a mouth of about 9 m × 5 m.

The second variant, originated in the lower Rhine, has only upper and lower beams of 10–12 m long, the lower beam weighted with chains. The net bag is about 30m long and usually ends in a fyke net. This type of stow net has special names in German (Schokkerhamen) and Dutch (Ankerkuile) referring to the anchored vessel from which the nets are suspended (Fig. 5.18).

Formerly, the vessels used for fishing these stow nets did not have their own means of propulsion, but motor vessels towed them to and from their stations at the beginning and end of the operation. Their main function, as seen in Fig. 5.18, was to hold the harness and sometimes to serve as fishers' quarters for night-time fishing. Today these 12–17 m vessels are usually motorised (60–120 hp) in large rivers in order to change fishing sites. A cable

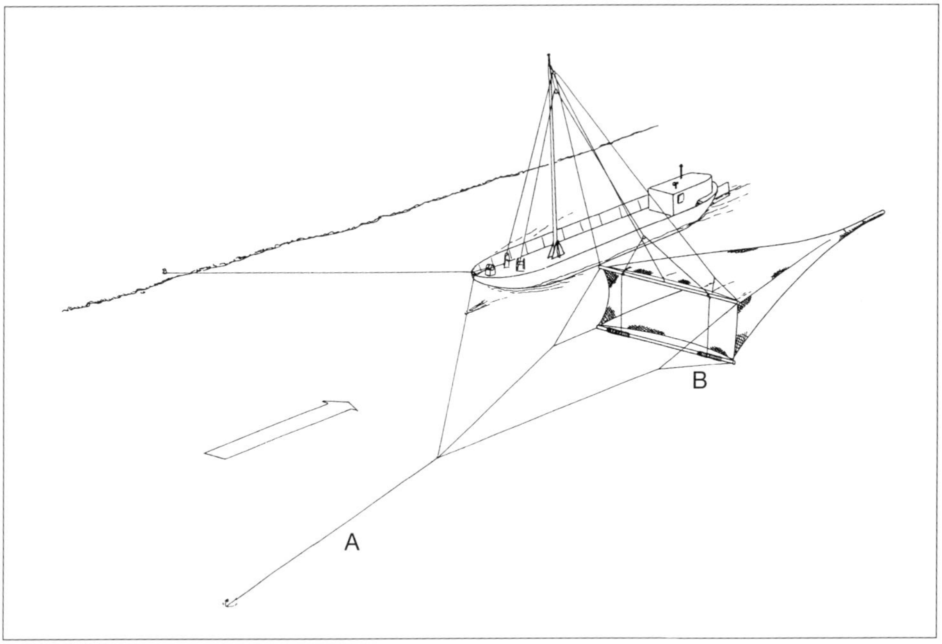

Fig. 5.18 Anchored stow net as on the Rhine with the 'Aalschokker' vessel (after Klust, 1971)

A Anchoring system
B Rigging of the stow net to the vessel

connecting the device to the river bank enables smaller adjustments of location. Cutter locations and fishing seasons are legally regulated. The spreader stow net (German Knüppelhamen) developed by Ems river fisher, Franz Römann, embodies a special anchored stow net that is set either singly or as several adjoining ones anchored with a wire cable loosely stretched across the river (Dahm, 1973). A prerequisite for this is that there be no commercial navigation. The spreader stow net's special design is based on the observation that when an eel on its way downstream meets an obstacle, it gets around it by swimming a short distance against the current and off to the side into deeper water, and then lets itself drift again. Therefore, attached beside the stow net is a conical fyke net to catch detouring eels. An advantage of this is that, when the stow net contains a large amount of drifted debris, the catch is clean and simple to remove.

5.4.3.3 The otterboard stow net

For this most modern form of large stow net, neither frame nor vessel is needed. It originated on the Elbe near Gorleben and is sometimes goes by the name of its inventor: the Koethke stow net. The description of its operation follows (Fig. 5.19).

A net bag about 32 m long with mouth width of at least 3 m and two 20–25 m wings has one of the wings cabled to the shore. At the end of the net bag there is a codend about 3.6 m long into which a funnel projects. The second wing has a large otter board attached, much as a smaller one would be on a trawl. A flat otterboard is about 2.5–3 m high and has a length of about 4 m, including rudder. It is also cabled to the land up ahead. The otter-

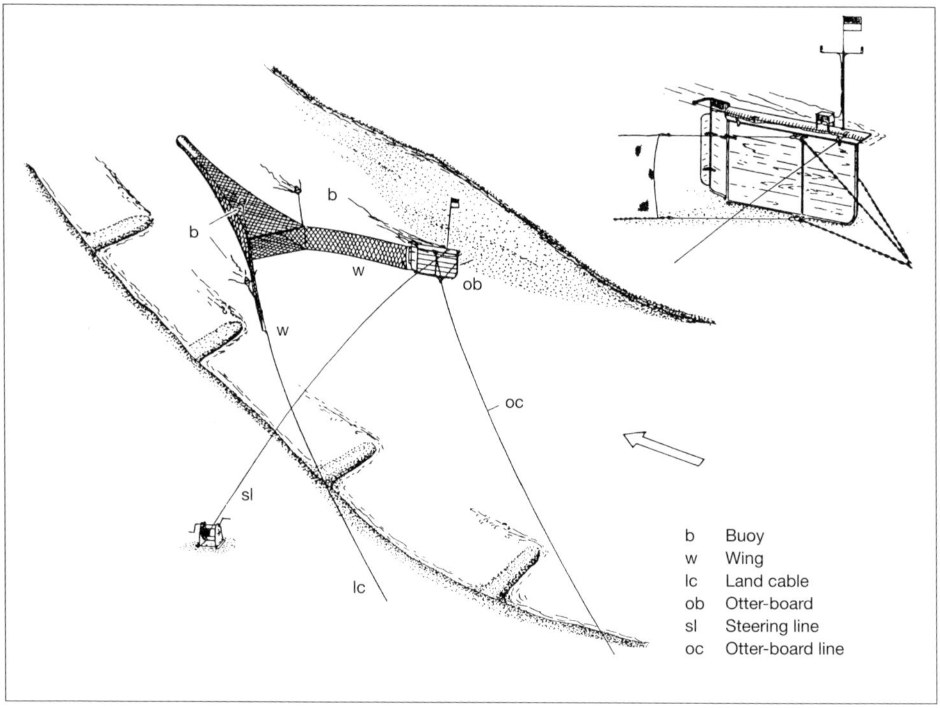

Fig. 5.19 An otter-board stow net installation

board, held vertical by floats, is manoeuvred out into the river by angling it into the current, so as to hold the mouth of the net wide open (blocking about 50 m of the river's width). The otterboard can be brought toward shore quickly, *via* a steering wire connecting the otterboard to land and a steering device, clearing the way for river navigation.

An even larger device like this was built with a curved metal otterboard (Schlieker, 1964a, b).

The otterboard stow net can be kept in the water throughout the April-to-November fishing season. When not in use, it can be held in a waiting position along the river bank. These devices proved themselves to be clearly better eel-catchers than normal stow nets because a greater width of river is fished, and no lines exist in the net mouths. For example, in 1961 near the locality of Graben on the Oder (latitude of Berlin) an otterboard stow net caught 10 t of eels, and the catch of five anchored stow nets nearby totalled 4.6 t. In 1962, the five anchored stow nets were replaced by an otterboard stow net. From then on, they both caught 10 t/year (Schlieker, 1964a). Here, the conversion to otterboard stow nets raised per-person productivity by 700–800%.

It is amazing that it took the otterboard stow net a relatively long time to become generally established as a practice, for masterfisher Hugo Koethke had already made his invention publicly known in 1937. There was also always a question whether conditions of the lower Elbe and Weser rivers would permit the otterboard type of stow net to be introduced in place of the anchored kind (Klust and Schärfe, 1955). There, the anchored stow nets combined with cutters had been justified because the changing tidal current continually required turning the gear around. In addition, the river bank is usually too far away to permit anchoring the device on it. The possibility of being able (or forced) to frequently change the fishing site, posed a further problem.

During 1980–7, use of large stow nets to catch eels was tested on the Danube river by comparing two installations: an anchored 'eel schokker' stow net system, boat included, built according to the latest knowledge, and a Koethke stow net. In this direct comparison, the anchored stow net averaged 3059 kg/year over 7 years, whereas the Koethke gear caught only 1470 kg/year over a 2-year period. The highest daily catch exceeded 1000 kg for the anchored system and was about 200 kg for the Koethke rig. In this comparison, however, costs and technological aspects were also considerations. The outlay for a new anchored stow net system was 160,000 DM, but only 25,000 DM for the Koethke stow net. A further advantage of the latter was that it could be hauled to and from almost most any site with a small truck-trailer. For that, the anchored system needed supplementary tugboat help. In principle, practical suitabilities for the Danube river fishery could be confirmed for both types (Harsanyi and Köthke, 1988).

5.4.4 Stationary eel traps

In streams, as well as in rivers that are not too big and where navigation will not be hindered, it is sometimes possible to install permanent eel trapping facilities. They are applicable where it is certain that stream-flow discharge will not fluctuate too much, where current and water volume are not too great, where pollution will not harm eels, and where drifting debris will not clog a trap too quickly. A stationary eel trap requires complete or partial weiring of the stream or river in order to guide downward-drifting eels into the trapping device. Thus, much the same principles are applied here as in the temporary eel trap, the stow net, where, by means of funnel-shaped wing nets or an actual net funnel, eels are directed into a net bag, which often is also equipped with funnels to prevent escape. Sometimes stow-net-like net bags are also used in stationary eel traps, and their wings consist of massive, permanently installed fences or weirs. Facilities installed in this way are used mainly in rivers. In streams, rigid trap chambers are used almost exclusively, and these consist partially of wood lath (slatted) grating. In the following discussion, eel weirs that stretch across part of the river and have stow-net-like net bags are distinguished from the stiff trap chambers that are used in streams or in association with turbine facilities.

5.4.4.1 Eel trap weirs in rivers

The best-known example of an eel weir is on Northern Ireland's river Bann, which has very favourable conditions. Northern Ireland lies in the path of the glass eels' vanguard arriving from the Atlantic Ocean. The river Bann receives most of its water from the 39,000 ha Lough Neagh, the largest lake in the British Isles. There is less likely to be drifting debris and pollution from a lake's outlet than in rivers receiving water directly from mountains. Stream-flow discharge fluctuates between 9 and 470 m^3/s (Frost, 1950), thus is comparable to the Saar and Aller rivers.

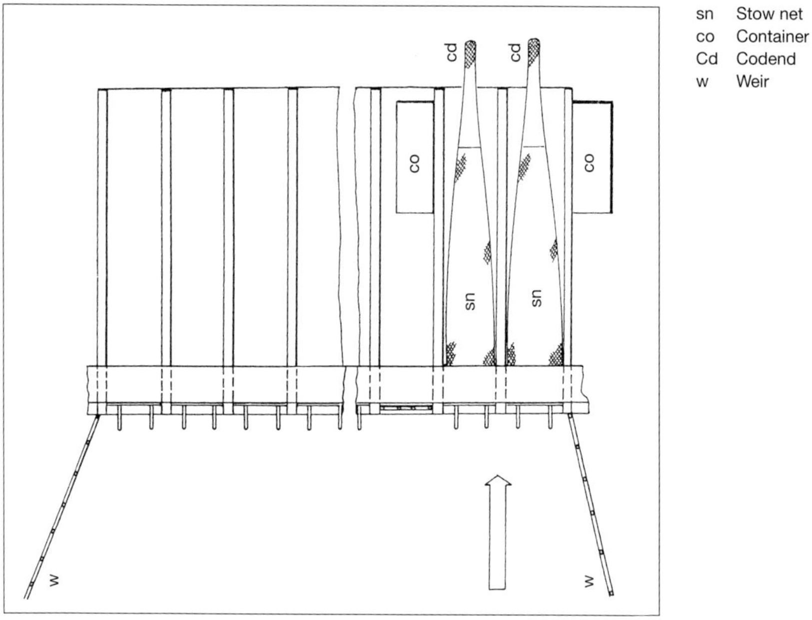

Fig. 5.20 Plan view of the eel weir near Toome on the river Bann, Northern Ireland (after Frost, 1950)

Figure 5.20 shows a diagram of the massive weir below Lough Neagh. It has gates for installing eight stow nets. Gates in which no stow nets are installed can be blocked off with vertical screens so that, for instance, during low flow, the catch stays restricted to just a few nets. Furthermore, flow of water through these gates can also be completely shut off so that during extreme low water, all flow is through the netted gates. The nets are 12 m long, each with an added 6 m of codend containing a funnel. After the codend is lifted, the catch is transferred to floating containers moored alongside the nets. The nets can also be emptied through tubes from the codend directly into the containers.

Halfway between Lough Neagh and the estuary in the Atlantic Ocean there are other stow net weirs of an older type, like the ones common in Canada (Fig. 5.21). These catch only a fraction of what the new installation upstream does. This is not surprising because the new facility presumably catches all the eels coming from the lake before they reach the old weirs.

The eels caught at the older weirs apparently come only from the river Bann and a smaller lake located downstream from the new weir.

The silver eel fishery of the Lough Neagh area, that is, primarily the weir at Toome, caught an average of 240 t/year (over 6 kg/ha/year) during 1960–6 (data, Northern Irish Ministry of Agriculture). Including yellow eel catch in the lake itself, as well, average yield over the same 7 years was 744 t/year. About 32% of this above-average yield was contributed by the eel weir with its extremely low expenditure of labour. Moreover, 96% of the silver eels were caught between August and November (Frost, 1950), so, in fact, it would have sufficed to operate the weirs for just those 4 months. The percentage distribution of annual catch among months:

- June 0.4%;
- July 1.5%;

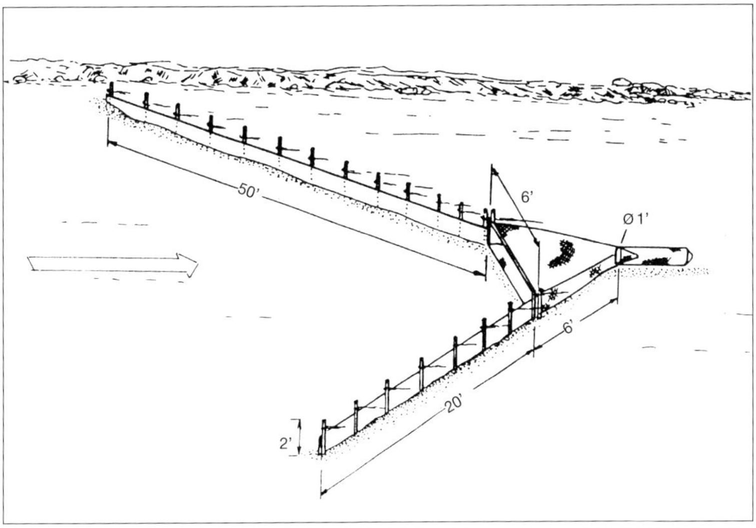

Fig. 5.21 Canadian eel weir, dimensions shown in feet (after Day, 1948)

- August 9.9%;
- September 32.4%;
- October 45.2%;
- November 8.6%;
- December 0.4%;
- January 1.6%.

Thus, peak migration for this area projecting far into the Atlantic Ocean occurs in October, just as in Germany.

The river Bann's silver eel fishery is surely an optimal case. Here, the question arises whether yield would rise or fall if use of fyke nets, long-lines and today also trawls (previously seines; Frost, 1950) were discontinued for yellow eel fishing. Permanently installed eel traps when placed upstream from other nearby fisheries, compete seriously with them. In the Lough Neagh area the relative locations are reversed, but no definite reduction in profitability of the eel traps is yet evident due to the yellow eel fisheries.

Apparently, it is difficult to find suitable places to install eel weirs. In 1964 in the Canadian province of Nova Scotia there were 40–50 eel weirs of the kind shown in Fig. 5.21. However, throughout Canada's maritime provinces, such weirs provided for only 7% of the 1962 eel harvest. At that time, they generally caught less than 2 t/year, though a weir in the province of Quebec is known to have averaged nearly 30 t/year (Eales, 1968). At a cost in those days of about $100,000 for a large installation, it was questionable, of course, whether the catch could repay this large investment when trap effectiveness was so low.

The fishing facilities of the Venice and Commacchio lagoons (e.g. Fig. 5.22), in operation for centuries, must be included among permanent eel trapping weirs. The trap systems of these installations are rebuilt year after year in the form of reed bundle walls fastened on to rows of posts, always at the same sites. It is necessary to dismantle the system so each year's springtime migration of glass eels from the sea can reach the various 'valli' (lagoon sectors separated by dams). Each 'valle' contains one or several trap systems of the sort shown in Fig. 5.22. A single system often is not enough because in one night as much as 100 t of eels can find their way into just one trap chamber. A valle used to cover about 2000 ha on average (Lübbert, 1908). A valle's annual eel catch amounted to 30–40 kg/ha, and the total producing area of valli was 50,000 ha, but it is less today due to land reclamation. About 1000 t of eels are caught annually in the whole lagoon region of northern Adria (D'Ancona, 1961).

Of course, the valli traps are not completely comparable to eel weirs in rivers. Current is much slower in the lagoon canals, so the weir fencing there need not be nearly as strong. Quite modern facilities also exist, such as were built in the northern part of the Venice Lagoon.

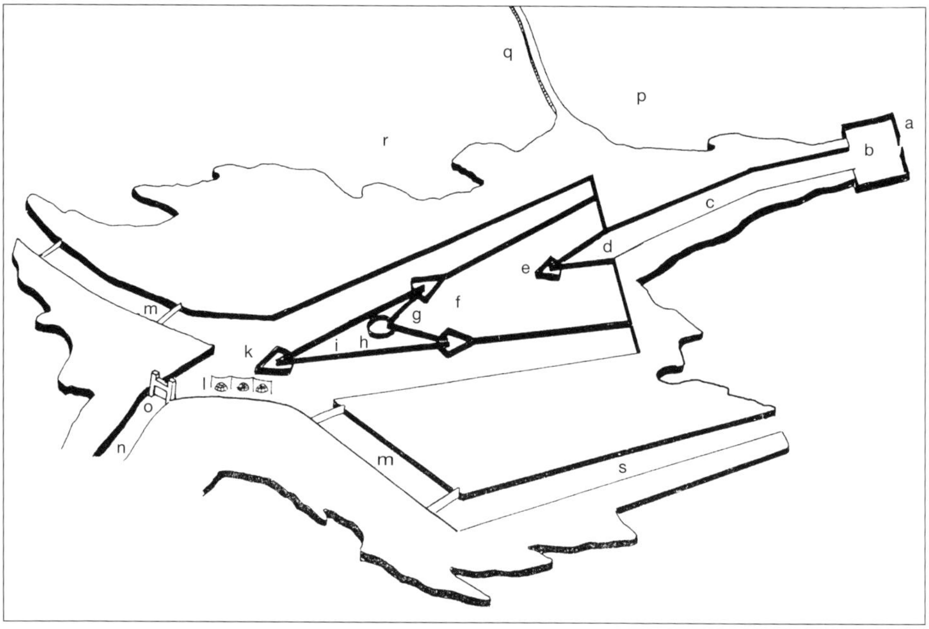

Fig. 5.22 Diagram of a trapping facility at the Lagoon of Comacchio (after Lübbert, 1908)

a Entrance into the trapping system
b Vestibule
c Connecting ditch
d Entrance to the first chamber
e First chamber
f the actual trapping facility
g Entrance to the second chamber
h Trapping chamber for mullet and flatfish (eels can pass through its surrounding barrier)
i Entrance to the eel trapping chamber
k Eel trapping chamber (crib)
l Baskets for holding the catch
m Navigation locks
n Channel to the sea
o Sluice
p, r Two valli (ponds divided by dikes, and where the eels mature)
q Dike
s Channel for lagoon navigation

Eales (1968) also reported on fascine-like weir fencing in the tidal area of the Canadian coast. However, these were not erected in rivers but rather in tidal currents to trap the eels that were moving back and forth. Maximum catches of 15 t/year were quite possible.

5.4.4.2 Eel traps in streams and rivers

Formerly, eel traps were incorporated into existing dams and weirs, especially mill dams. There, a grating was installed where the water plunged from the lip of the dam towards the stream bed, and as water passed through the grate, eels were filtered out (Fig. 5.23). Often such wood or masonry dams were built specifically for eel trapping. Eel trapping without a dam, for example the Swedish eel crate, represents a simpler type that can be set up in small streams. This was common in Germany's Kassel area, for example (Buhse, 1967). A lockable sliding lid is commonly used to prevent theft. Hofmann (1960) described a more modern facility that can also be installed in front of turbines in larger rivers.

Eel traps, especially those in lake outlets, can make substantial catches, 500 kg in one night (Hofmann, 1960) being no rarity. In an eel trap on Germany's Fulda, 2500 kg/year were caught on average, but in a nearby trap on the Werra, only 100 kg. Low values, like those on the Werra were reported from a trap in the Weser area around 1900. Klemke (in Buhse, 1967) estimated in 1935 that the mean annual catch of a stationary eel trap in Hessen (Germany) was 500 kg. In the outlet of a lake in Mecklenburg, average catch for the 15 years before 1900 was 5000 kg. A stationary eel trap built in a tributary of Belorussia's Düna proved clearly superior to the previous stow net facility; annual catch was as much as 6000 kg (Dubovskiy, 1963).

Thus, eel traps could produce 100–6000 kg/year, and in some cases even greater amounts, for example 3000–4000 kg per night in an eel trap at a sluice on Denmark's Ringkoebingfjord (Mann and Klust, 1952) during the silver eel migration. These variable catches were surely attributable to the different sizes of upstream water bodies. Streams flowing from lakes and therefore actually having eel-producing areas of many hundreds of hectares, are clearly superior to simple, flowing streams that are of the same discharge but encompass only a few hectares of surface water. For these reasons, eel traps were known as gold mines, and their catch statistics were closely guarded business secrets. The owners

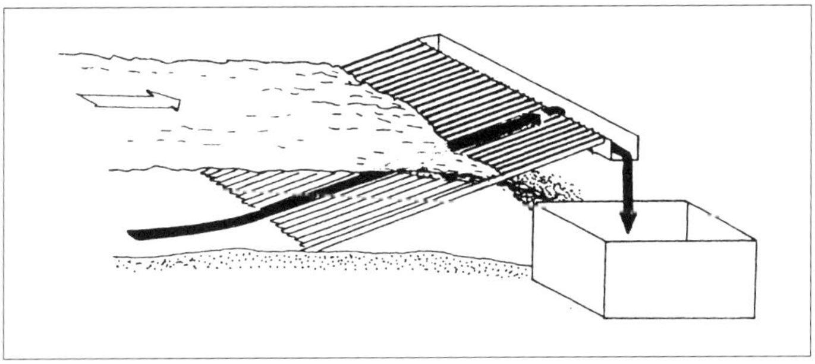

Fig. 5.23 Essential features of a filter trap (after von Brandt, 1984)

- White surface: water current
- Black arrow: the migrating eel's path over the slatted grating into the collection container

of such traps were often envied. Resentment was indeed justified, when the owners' traps caught eels produced in upstream areas that others managed – the management, of course, often including eel stocking. In such situations, the only fair arrangement was cooperative stocking of eels in the system of waters involved, each owner contributing financially to the stocking in proportion to the amount of fishing he did. It was absurd, however, to prohibit construction of eel traps by law (Hofmann, 1960). From an economic standpoint, eel traps and weirs at reasonably suitable places represent the most efficient method of harvest. They often become unfair only on the basis of traditional fishery regulations. However, in many cases, as the aforementioned catch data show, envy became a non-issue, for example, in an area of Lower Saxony (Germany), where many eel traps once existed but eventually had to be abandoned.

Just as for fyke netting, stow netting and weir trapping, fishing seasons vary according to the type of river and the particular area within the river. Whereas, substantial catches were recorded in autumn in the Weser area, good catches occurred throughout the 6 summer months in Mecklenburg and Mazury. This does not seem to have resulted so much, as Walter (1910) guessed, from earlier migration of Mecklenburg eels, as from inclusion of yellow eels in the catch there.

In New Zealand, as well, substantial amounts of both eel species that occur there were caught in eel traps (Fig. 5.24). Primarily the long-finned yellow eel (*A. dieffenbachii*) was recorded throughout the year, but with a peak in spring and another in summer. However, silver eels maintain a pronounced maximum from summer until early autumn.

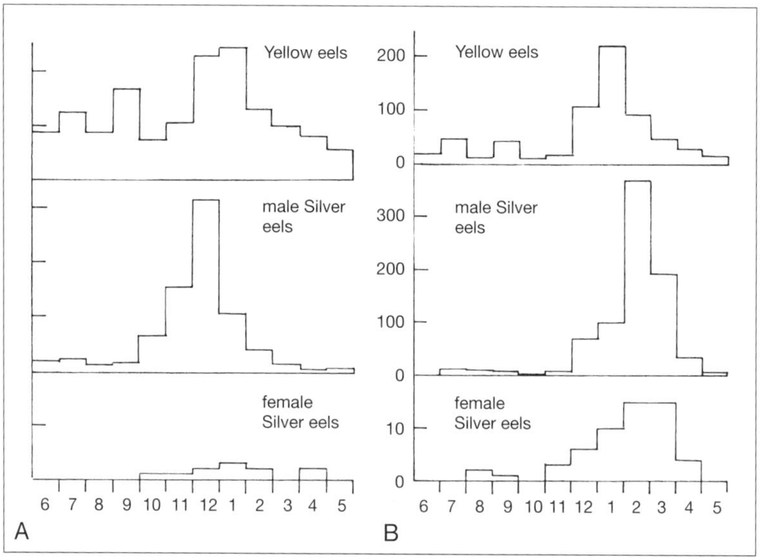

Fig. 5.24 Monthly catches in an eel trap on the South Island of new Zealand (after Burnet, 1969)

A *Anguilla dieffenbachii*

B *Anguilla australis*

5.5 Entangling nets

Trammel nets are not typical eel-fishing gear but are used for this, aided by acoustic or electrical devices to drive (frighten) the animals toward the nets. Figure 5.25 shows a diagram of part of this vertical, three-layered net-wall arrangement and how it works. It consists essentially of a light, fine-meshed net between two heavier, large-mesh nets. When an eel goes through one of the large-meshed outer nets, it pushes the fine-meshed inner net out through the other large-meshed net, forming a pocket that traps the eel.

The so-called pulse fishery, conducted from a boat, uses acoustic devices to drive fish by frightening them. The method stems from the former East Prussia and was later adopted in the western Baltic Sea, as a result of post-World War II developments. In pulse-fishing, the fish are driven toward a wall of netting by noise, that is, by striking the water with a 'pulse stick', commonly in the form of a long-handled rubber plunger, such as is used to unclog plumbing drains.

The outer nets of this three-layered, *c.* 1 m high trammel net wall have mesh bar lengths of 100–200 mm; the inner net has 15 mm mesh bar length. The net wall consists of four to six trammel nets each 30 m in length. The mostly diurnal fishery to water depths of maximum 12 m, is operated in such a way that the net wall consists of four to six trammel net panels, each 30 m long. The fishing, usually done during daytime, proceeds as follows: in water <12 m deep, the trammel net is set in an arc encompassing about 1200 m^2, with the open side facing the shore. Starting at this open end, the boat travels a narrow, zig-zag course or else in a less systematic pattern of moving this way and that, from one end of the

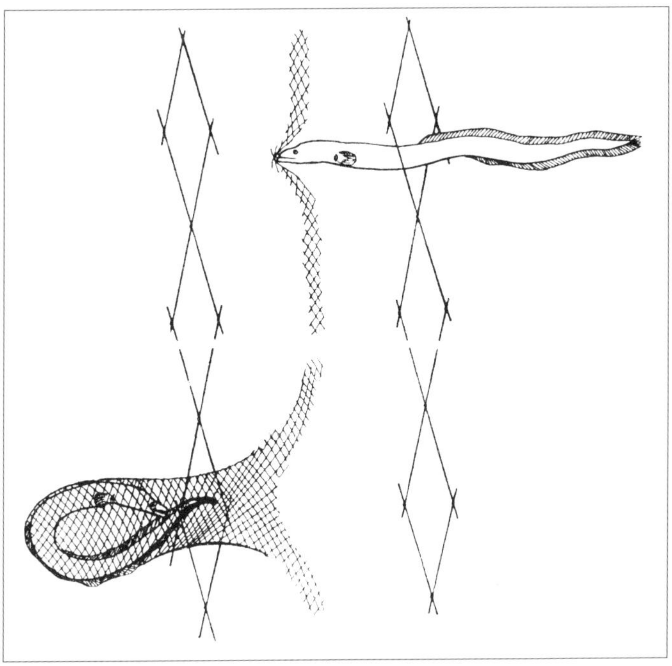

Fig. 5.25 How a triple-walled trammel net functions

arc toward the other, gradually approaching the last net panel, which is at a more or less 90° angle to the shore. While the boat goes back and forth in this confined area, two fishers in the boat take turns doing the pulsing, that is, forcefully thrusting the plunger into the water as close as possible to the side of the boat which serves as a resonator. That is a rather strenuous activity. They stop pulsing after reaching the last panel, and haul the nets starting from that end. When six net panels are used and the catch is 5–10 eels, the process lasts about 25–30 min (Freytag and Mohr, 1973). As soon as the last net panel is hauled aboard, the adjacent area is surrounded with nets and fished in the same way. Such extensive coverage by pulse fishing in near-shore areas not only results in catching almost all the eels present, but may also reduce catch rates of other methods like fyke netting and angling, at least at times.

Another acoustic method consists of pumping up and down with a cross board on a handle to chase eels into a prism-shaped device covered with netting (Section 5.4.3).

Regarding combination of electrofishing gear and trammel nets, which among other applications came to be used for inventorying eel populations in Lake Constance (Berg, 1980), it is noted that larger mesh bar lengths (180 mm) were also used there for the outer nets, and the nets were higher, up to 3 m. Most importantly, this combination of gear resulted in catching larger eels on average than in trammel net alone (Section 5.10).

In addition, it is said that eels also can sometimes become entangled and are caught in simple gillnets set for herring.

5.6 Lift nets and cast nets

Sinking or lift nets and also cast nets are relatively small gear and not typical for eel fishing but occasionally are successfully used in special designs and at particularly favourable sites. Owing to their minor importance, the two types of fishing gear that otherwise differ in mode of effect are described here in the same section for the sake of simplicity.

A sinking or lift net usually consists of a square piece of netting ranging in size from barely one to several square metres. This somewhat sagging net is held open by crossed bow-shaped braces or by a square frame. The net's four corners are suspended from the bowed braces or, if the net is framed, lines lead from each corner to a point above the centre, where the whole apparatus is hung on a line and lowered on to the bed of the water body. If the device is large, it can be deployed by winch from an arm or boom assembly. After some seconds, minutes or even a longer time, the net is pulled up again, and the few eels that may happen to have swum over or settle on it are caught. A place where this method is still used today is, for example, the lagoons of the Italian Adriatic, where a dense eel population makes for a certain degree of success.

A fishing method that also operates on the lift principle is known from the Island of Heligoland (Schnakenbeck, 1953a): the Strieknetz consists of netting stretched between two rods – actually a stake stow net. This was shoved beneath eels that were oriented almost vertically while feeding on insect larvae, and the net was hauled up out of the water almost like a lift net.

Differing from the sinking or lift net, which is, so to speak, shoved under the fish, the cast net is an inverting net. It is clapped down over eels and other fish (in the process being

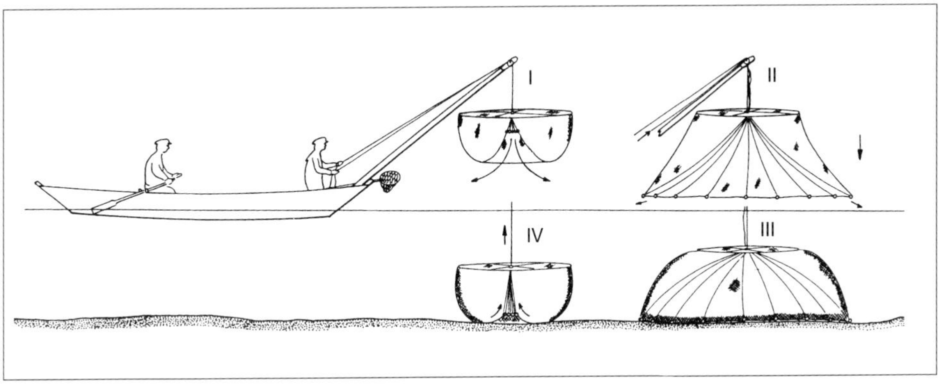

Fig. 5.26 Hamburg cast net. Net ready to release (I), Release (II), Deposited (III), Hauling up (IV).

turned inside out, as it were), and then closed. Among the various cast net designs, a mechanical variety especially designed for eel fishing was used successfully on the lower Elbe. This was the so-called Hamburg cast net (Fig. 5.26). Its framework consists of a 2m-diameter hoop made of 1 cm thick galvanised steel, in which about 10 sturdy wires are stretched radially to stiffen the net's roof. Over this framework, a veil of 9–11-mm mesh netting is drawn. The net's lower edge is weighted with lead balls. Next to each fifth or sixth lead ball, cords are attached to the edge, which converge within the net mantle and join to compose a cord line. Before it is set, the net is first drawn up so far with the cord line that the lead balls lie close together under the hoop that has the net drawn over it. After extending the boom with the cast net out over the intended fishing spot, the cord line is released suddenly, and the lead weights swing down and outwards in a pendular motion, spreading the net mantle as it falls. At that moment, the framework is also let fall, so that the net sinks through the water like a parachute. When the whole net comes to rest like a big disk on the bed, the fisher pulls on the cord line, and all the eels that the net covers are enclosed. As the net is lifted, the frame is turned vertical to decrease the required lifting power. It is returned to horizontal when emptied into a dip net held under it. Cast net fishing begins in springtime when eels become active after their winter resting stage, and it ends in about September.

5.7 Seines

In the past, seines for catching eels, which count as active fishing gear, were frequently operated in the coastal fishery and inland lake region, but their use has declined a great deal because costs for labour and supplies have risen. The towed seine is a special form of encirclement net that envelops a large area. Depending on whether these are deployed from the shore or out in the water from a boat, they are called beach seines or boat seines.

Figure 5.27 shows the main design and features of a seine and the way it is deployed to cover an area when used from a boat. The seine differs from the trawl, which is very similar in design, in that the wings, which are important for envelopment, are longer than the net bag, and in that several-hundred-metre-long hauling lines are used in the

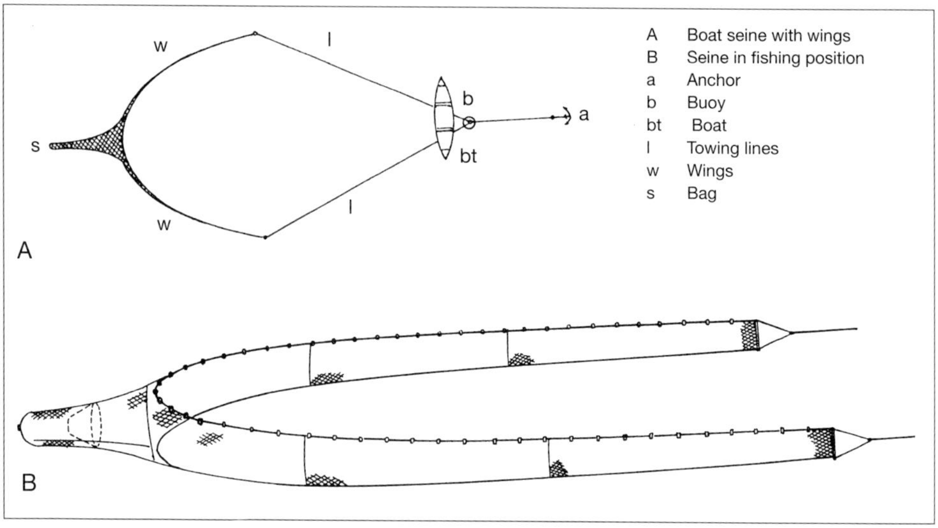

Fig. 5.27 Eel seine

envelopment. In addition, the operation is limited to the enveloped area, and the seine, in its open position, is not dragged a long distance like a trawl is.

Eel seining is done in summer. As the mesh size for eel fishing must be small, the dimensions of the seine are also relatively small. For example, the net bag of the eel seines used in Germany's Flensburger Förde (firth) is only 15–18 m long, when operated by two persons. The appropriate wing length for this is 18–22 m. The depth of the wings is generally calculated to be one-and-a-half times the water depth of the intended fishing site. A funnel or two are usually built into the net bag.

The eel seines used in bays (Bodden) and lagoons (Haffs) around the Isle of Rügen and in the Usedom area of the Oder estuary (Germany/Poland), were usually larger. Their wings were from 150–200 m, and their heights could even reach 16–20 m (Bobzin and Finnern, 1978); for these, two boats and at least four fishers were needed.

In Germany, the commercial seine fishery for eels has continually diminished and become unimportant in favour of the more mechanised and efficient trawl fishery.

5.8 Trawls

In many places, attempts were made to catch eels with commercial fishery's best-known and most useful gear, the trawl. This would employ a fishing device of acknowledged efficiency for an expensive and financially profitable fish. In the Baltic Sea area, the eel trawl became known as the 'Aalzeese'. The German term Zeese came from the sailing ship era and originally indicated a simple wing net, which did not yet have otter boards but was held open by booms extending fore and aft from a boat that drifted broadside with the wind. Fundamentally, an eel trawl is provided with either a funnel or a Flabber to prevent eels from moving towards the net's opening before it is hauled in.

The practical use of small-meshed eel trawls is thus far uncontested only in the open sea beyond the 3-mile-zone, although there, too, ecological misgivings have arisen because the ratio of target species to undesired bycatch is extremely unfavourable (Hahlbeck, 1993; Gabriel, 1997). In coastal waters within the 3-mile-zone, as well as inland waters, there are many legal regulations against trawling. The prohibitions against trawling in coastal and inland waters are based largely on potential damage to other competing methods, for example, fyke netting and long-line fishing for eels. The damage consists of either direct interference with the use of such passive gear or general reduction of catch. Thus, for example, trawl-fishers could fail to see eel long-lines and inadvertently drag them; fyke-net sets could be damaged; or else the annual catch by whatever method could be reduced because the harvest has to be shared by two different types of operation. In addition, overfishing could make exploitation of the eel population unprofitable (Deelder, 1965). A second point to consider in opposition to trawling is direct biological damage. Included in this is destruction of vulnerable juveniles of other species if caught in trawls along with eels, and, among other disruptions of bed materials, the uprooting of aquatic vegetation where fish spawn, seek cover and find food organisms. Protests, as well as threats and prolongation of legal restrictions run throughout the history of the trawl fishery in coastal waters. The biological damage just mentioned is cited as a primary argument. However, protection of passive fishing methods, like fyke netting and angling, stand in the forefront of the matter.

A trawl fishery is undoubtedly unnecessary and undesirable where it adversely affects another profitable kind of fishery, especially if as a result optimal fishing effort is exceeded, that is, overfishing occurs. Such a situation exists nowadays in most large inland waters of Europe, as well as in many coastal waters – and should also be considered from standpoints of labour-saving and energy-use. But in areas far out to sea, trawling today represents the only applicable fishing technique.

How attractive trawling for eels was for individual cutters during the heyday of the North Sea fishery is indicated by the landings (in t) of German North Sea harbours from 1964 to 1978: 10.6, 43.1, 73.1, 165.7, 105.3, 84.4, 72.4, 130.0, 91.7, 75.4, 88.9, 94.7, 72.8, 107.9 and 88.5 (Löwenberg, 1979). In 1972 the limit of optimal fishing intensity was apparently reached in the areas where only eels are fished for between the island of Heligoland and the mainland. According to results of tagging (Aker and Koops, 1970, 1973), the fishery took about 30% of the population annually, so that area's fishing can be considered as having been very intensive in those days. Since then, the results have continually declined. Today, the German trawl fishery for eels in the North Sea has practically no importance.

As Neuhaus (1969) reported, in 1968, even Schleswig-Holstein's shrimp cutters showed increasing interest in eel fishing. Their fishing gear, the shrimp beam trawl resembles a bottom trawl, and on mud flats, there were occasionally eel bycatches of as much as 10 kg. The daily catch of actual eel trawls amounted to 25–75 kg on the near-coast fishing grounds in June 1968, that is, at the beginning of the season (Neuhaus, 1969). At the height of the season 100 kg/day were often caught, sometimes even 150 kg. The annual landings of several cutters were around 5 t, sometimes even about 10 t (Cuxhaven seafish market data). The highest catch rates (kg/fishing hour) occurred in mid-summer (Löwenberg, 1979):

- May 1.9;
- June 5.5;
- July 6.3;
- August 5.1;
- September 4.0;
- October 3.8;
- November 3.3.

It is emphasised that, despite extension of the season far into autumn, silver eels did not compose a significant part of the landings. This was surely because the North Sea warmed up and cooled off later in the course of the year, and because the nets were trawled on the bottom.

In the heyday of this North Sea trawl fishery for eels when high eel populations, and most fishers had converted to trawling, an attempt to catch pelagically migrating silver eels selectively with a mid-water trawl proved successful. In this fishery, the particular gear was a four-panel pair trawl, towed by two cutters of 180–200 hp, which, as a result of a two-panel fore-net could contact the bottom from the 60 mm part onwards. The circumference of the net's forward end was 1200 meshes at 60 mm bar length, which permitted a significantly larger opening than in a normal, two-panel bottom trawl towed by just one cutter (otter-board trawl). The latter, by comparison, has a fore-net circumference of about 500 meshes.

Pelagic trawling's worth is confirmed by its catching about a 55% share of emigrating silver eels, as opposed to only about 18% for bottom trawling. Moreover, yellow eels, too, can occasionally be found in the pelagic zone. This experiment was conducted with two powerful cutters. One of them had a net sonde instrument for measuring the vertical height. In the forward parts of the two big, new experimental nets, mesh bar lengths of up to 100 mm were being used, and net openings of 14 m or more could be measured. Considerable amounts of cod were caught with the eels (Aker, 1974).

From the Baltic Sea's traditional fishing grounds in the Oder Bank area, the trawl cutters landed 250–300 kg per trip in 1969, in one case 1000 kg (Hoffmeister, 1970). This indicated that the daily catches varied within about the same range as those in the southern North Sea at that time. After 1989 the trawl fishery underwent a resurgence there and in the Arkona Basin (Baltic Sea, off the Isle of Rügen's north end). However, that fishery remains controversial due to the aforementioned unwanted bycatch, so there are efforts to replace it as much as possible with other methods like long-lines, baskets and fyke nets (Gabriel and Thiele, 1997). In the western Baltic Sea, experimental trawlings in September–October were rather unproductive (Lamp, 1980). In north German lakes, eel trawls were first used in the fishery around 1960. They caught well at first, but then catches diminished. Only the advent of the pair trawl, corresponding in size to the otterboard trawls of the small-vessel high-seas fishery, increased the catch again. There were 100–150% more eels caught than with the previous employed devices, and, in addition, even larger amounts of other fishes. The pair trawl caught 335 eels in 105 min of towing. Such a catch is seldom even achieved with otterboard trawls in the small-vessel high-seas fishery. Figure 5.28 shows the design of such a pair trawl used in the fisheries of various lakes. The two small motorboats used for towing the trawl each had a 20-hp engine and were just right for operation by one fisher per boat.

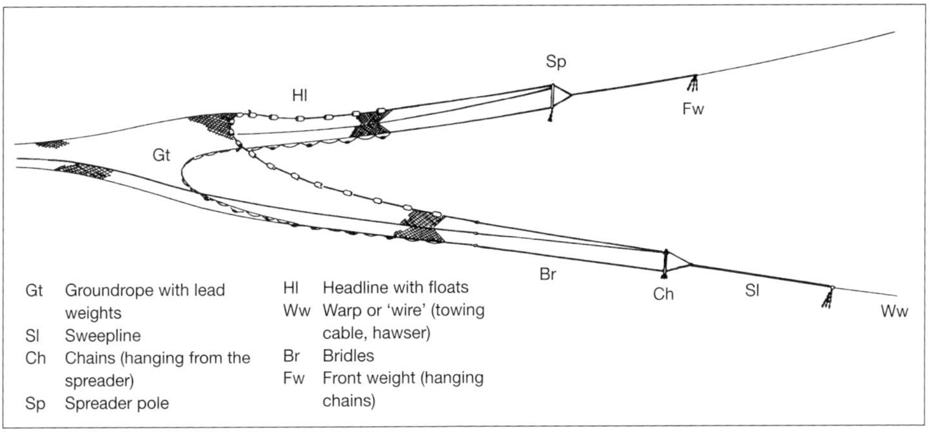

Fig. 5.28 Schematic representation of a pair trawl (bottom trawl) drawn by two boats (after Steinberg, 1964, see also von Brandt, 1984)

By using the trawl with a small-meshed cover over the codend, it can also catch smelt as bait for eel long-lining (Steinberg, 1964). The combination of trawling with electrofishing enables further improvement and still better prospects for catches (Section 5.10).

5.9 Light fishing

The eel's sensitivity to light and its movements in darkness have long been known (Sections 3.3.3 and 3.4.2). In the night-lighting discussed in Section 5.2, light is actually just used to find and flush eels. However, the experience gained from that – having eels flee from light (negative phototaxis) much more slowly over soft bottom than over hard bottom – suggests a further way of putting light to use for eel fishing.

An example is known from the very successful fishing facilities operated for centuries near Comacchio, Italy (Section 5.4.4.1). There, the eels' light sensitivity was put to use in the past by lighting bonfires on the shore to stop their movement temporarily and prevent fyke nets from becoming overloaded to the point of destroying the them (Brandt, 1905). Light causes a fright response in silver eels more than in others, and this can be exploited in devising suitable, mixed-method fishing gear for actually catching the animals, and for doing it more economically.

Around 1900, fishery biologists already knew that young eels migrating from the North Sea into the Baltic Sea move back again, and that the silver eels emigrating from the Baltic Sea do not return there, so the idea arose in Denmark (Petersen, 1906) to have these stocks fished as hard as possible by appropriate methods in the Danish straits, that is, the Belts and the Sound. Centuries of observations at Comacchio amply verified that one could catch out all seaward emigrating eels (male and female) without harming the population. Petersen founded his theory analogously that no impairment of the Baltic Sea eel population need be feared from complete harvest of silver eels in the transition zone to the Kattegat. There, in accordance with the Italian experience, gangs of electric lights were designed as scaring devices to guide silver eels into fyke nets. Preliminary studies mainly with two or three simple stable lamps on a shallow fjord showed that after extinguishing them, 50–100 eels were indeed caught that same night in fyke nets that had been set in their path. Later experiments with an electric cable and light bulbs strung over the water were promising. However, mass protest by fishery groups from outlying regions, who feared for their livelihood put a decisive end to the Danish experiment in 1907 (Henking, 1908).

Further systematic studies on operation of light barriers as guidance weirs to influence eel migration were first done again during World War II in 1941–4 by Lowe, whose results were reported by Frost (1961). These took place in a stream barely 10 m wide in the English Lake District and showed only that when part of a stream was blocked in this way, eels were caught in the other part. The degree of diversion was measured by the number of eels that entered the shaded and lighted sides of the fishing device. The number in the shaded side was higher in all experiments, amounting to as much as 72% of the total catch. Later experiments in Holland are supposed to have shown that lighting the water surface enabled greater catches in eel stow nets. Here also it was assumed that silver eels migrate relatively high in the water column, therefore, surface light forces them into greater depths and into the stow nets set there. Even though a certain effectiveness of the installation is undeniable, it still attained no importance in the fishery (van Drimmelen, 1951 in Bräutigam, 1962).

Experimental light weirs were erected in the East Germany in the 1960s. A major purpose of this experimental fishery, under centrally regimented production requirements, was to balance the interests of different fishery enterprises and cooperatives, as was also being done in Denmark at the time. These experiments showed that good success could be attained with the help of light weirs. The results of an experiment in the 70 m wide Peene near Anklam were quite convincing: if the light was switched off, only 5–10% as many eels could be caught as when the lights were on. The fishing facility was built in the Peene river's estuary area where previously no fish traps could be set because they would have hindered ship traffic. This was the main argument for erecting a light weir instead of a net weir. This relatively small installation caught 7 t of silver eels in 3 years. The light weir was made up of thirteen 200 W lamps and the simplest of materials; no expensive special facilities were needed whatsoever. Here, the following were learned:

1. water velocities of >2.5 m/s are too strong (silver eels break through the light barrier);
2. visible depth should be at least 1 m;
3. the interval between lamps should be chosen so the cones of illumination overlap to form a closed wall of light;
4. the light weir must not be longer than 100 m.

At another site, as well, a light weir that was too long proved detrimental. Presumably eels accommodate to the light intensity after following the light weir too far and then break through. The light bulbs have to be cleaned from time to time because algae grow on them. The catching device used was a three-tunnel fyke net with an outside heart. It was set deep with the codend pointed against the direction of migration because it was assumed that the escape reflex would be toward deeper water and against the current (Hölke, 1964). The main catches were made during the period between the last and first quarters of the moon, that is, at the time of dark nights.

Experiences were similar in a Polish stream only 10 m wide (Swierzowski, 1964). This stream was blocked off by a string of lights in front of the diversion canal for a power plant, and the effectiveness of this barrier was evaluated by trapping in the stream and in the canal. When the lights were not on, the canal's share of the total catch was 15%. With the lights on, 58% of the eels were caught in the canal (42% in the stream), but if the stream's current was increased by opening a dam, the canal's catch fell to 32%. Moreover, the barrier's effectiveness continually declined in the course of the night.

Already at that time, the conditions in coastal waters were significantly more difficult (Bräutigam, 1961a, b, 1962). Favourable sites could apparently yield substantially greater catch if light weirs were used, but when the strings of lights were too long, catch effectiveness was reduced, and only a few sites existed where short light weirs would suffice. However, the large amounts of silver eels that could be caught during their migration on the coast in that era provided great incentive to adopt light weir technology.

To guard against disappointment, though, the basic policy here was that fish trapping weirs should be as much as possible in the form of net leaders, and only those sites too deep for them and that needed to let ship traffic through should be furnished with light weirs. Suitable trapping sites for seaward-migrating silver eels on the Baltic Sea were the outlet of the chain of bays (Boddenkette) behind the Darss peninsula and the connections of Rügen's bays (Bodden) with the sea (e.g. Little and Big Jasmunder Bodden). Three research facilities, one at each of the above sites, have more or less clearly shown that light weirs can improve silver eel catch very effectively.

The experimental light weir installation at the outlet of the chain of bays behind Darss is diagrammed in Fig. 5.29. Its string of lights was 80 m long. The trap was a large fyke net like the one described in Section 5.4.2.2. Further fyke nets with conventional leaders existed on the inland and seaward sides of the experimental, mixed-method installation. Catches of the latter were compared with those from one fyke-net facility located in front of it on the inland side, and one with those from a seaward fyke net behind it. The experimental light weired fyke net continually caught more than the double fyke net in front of it, which should have caught more because it actually consisted of two fyke nets. The fyke net on the seaward side behind it caught less yet. On the best day of eel trapping (12 September 1960) the trap with the light weir caught 1000 kg, the double fyke net caught 250 kg, and the seaward fyke net 100 kg. Thus, the light weir proved its effectiveness here. The facility at the outlet of Rügen's chain of bays likewise proved effective. It existed, how-

a Light weir
b Power line
c pound net
d Net weir

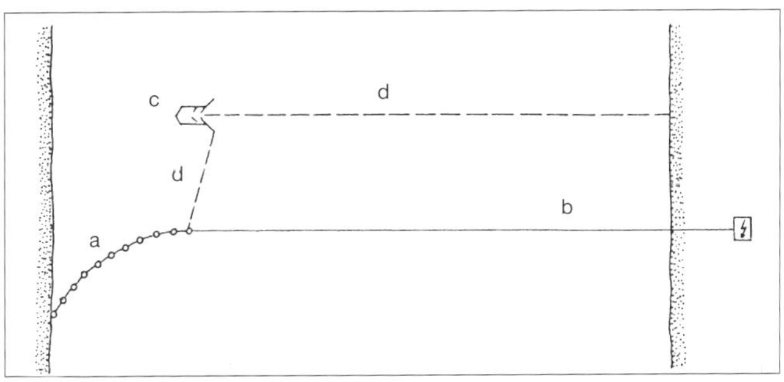

Fig. 5.29 Plan view of a light weir at the outlet of the Darss chain of lagoons (Bodden)) near Barhöft, Germany (after Bräutigam, 1961b)

ever, at an unfavourable site and had a 250 m light weir, which was too long, so the results here were less definite (Bräutigam, 1961a, b).

A further case, likewise at the outlet of one of Rügen's bays, showed very high effectiveness (Bräutigam, 1962). Here, a wedge-shaped light weir was used, its apex pointing against the direction of the eels' migration. The light weir was located at the deepest place in the bay's outlet, and each side had a net weir connecting it to the shore and to a fyke net. Thus, in contrast to the previously described installations, the seaward-migrating eels here were forced away from the middle of the stream towards shore and guided into fyke nets, which stood close to shore and had their codends pointed towards midstream. These devices were, therefore, set like those described for entrances to rivers (e.g. the Peene), but here both shores were put to use. Those eels that passed through the mixed-method installation were caught in a trapping facility that stood behind it and blocked the bay outlet's entire width with a net weir. While the lights were turned on, 12% of the silver eels passed through the mixed-method installation and were caught in the control facility in the rear, but while the lights were off, the figure was 65%. Thus, a fifth of the silver eels at most succeeded in passing through the light weir. The result for yellow eels that were moving by was quite different. Over 70% of them passed through the light weir, whether the lights were on or off. Flashing light increased the light weir's frightening effect on silver and yellow eels somewhat. However, no definite improvement of blocking effect on yellow eels was detected, so using light weirs to catch yellow eels seems pointless. Much the same can be inferred from results described for use of a light weir in the Peene near Anklam (Hölke, 1964). At that time, such installations enabled a catch increase of 150%, according to results of the experiments just described (Bräutigam, 1962). At the outlet of the chain of bays behind the Darss/Zingst Peninsula about 15 t of eels were caught per year at that time. Theoretically, a light weir could have raised this to 23 t. Just 1 t of eels would repay the greater financial outlay for a light weir with electrical supply and connecting weir. There was no concern about negative effect on catch of other fishes. Positively phototactic fishes even concentrated behind the light weir, so it became possible to catch more of them in the devices there. These species (Bräutigam, 1962) included roach (*Rutilus rutilus*), perch (*Perca fluviatilis*), pike (*Esox lucius*), pike perch (*Stizostedion lucioperca*) and herring (*Clupea harengus*).

5.10 Electrofishing

Long before human beings discovered electric current and put it to use, a kind of electric fishing already existed in the electric eel (*Electrophorus electricus*), which generated current and used it for orientation and for repelling enemies, as well as for stunning and capturing prey. Intentional use of electric current by humans did not happen in fisheries until after World War II. The eel is especially well suited to the technique, so electrofishing became the most important form of mixed-method eel fishing. Now this method is used mainly for research purposes, such as making population estimates, and is uncommon in commercial fishing.

The term electrofishing encompasses capturing fish with electric equipment devised for that purpose, as well as using electric current to repel, attract, stun or kill fish in order to increase the effectiveness of any of the previously described common methods like fyke netting, stow netting and trawling. In electrofishing, the direct influence of electric current in the water on the fish is exploited. The simplest electrofishing situation involves use of direct current (DC) and proceeds as follows. If an anode and a cathode are held some distance apart in water at sufficient voltage (150 V or much higher, depending on conductivity of the water), fish that are close to the anode are impelled towards it. Those near the cathode are repelled from that pole. Farther from the anode, fish feel shock but can get away. If a fish enters the strong zone of the electric field near the anode, it is forced to swim further toward the anode, a seeming attraction called galvanotaxis. As the fish comes closer yet to the anode, hence into even stronger parts of the field (or stays at the same distance a short while longer), it becomes stiff and unconscious, a condition called galvanonarcosis, and tends to roll over in the water. During the progressive galvanotaxis-to-narcosis process, and particularly at the latter stage, the fish can be caught with a dip net or other conventional gear. Risk of injuring the fish is less if it is netted out of the water quickly before narcosis. Galvanotaxis brings eels to the surface in shallow water (around 1.5 m or less), so a dip net suffices there. In streams, in small rivers and in the shallow areas between the banks and drop-off zone of lakes, as well as in other shallow, standing waters, hardly any other fishing technique is as good as the combination of electricity and dip net. (Often, the anode itself is designed to be the net, but this holds fish at the anode too long, promoting injury and death.) This classical form of electrofishing was then further developed stepwise within the fishery field and now varies much worldwide. Early steps involved increasing efficiency by using several anodes and anodes of different sizes and shapes, depending on size and other characteristics of the water body. In small streams, where the fishers wade, they can use back-pack shockers or, for greater effectiveness, tow a larger generator in a boat, or place it on the bank (see below). Wading fishers use anodes on wooden or fibreglass handles. The cathode can be a metal sheet on the boat bottom, or a metal object dragged behind. Where electrofishing is done from within a boat, similar anodes can be used, or they can be hung in various forms from booms. Two or more anodes can be fished from one boat. Cathode shape has also been improved because the early, simple ones let too many fish escape. An advanced cathode design, consisting of a cable stretched between boat-mounted booms and with wires hanging from it into the water, is an effective barrier against escape. If the fishing is not done from a boat, one can use a bamboo rod on which steel or copper cables are suspended at 5-cm intervals. In various circumstances, for example, in streams too small for a boat, or where the river, lake or pond bed is densely vegetated, the so-called long-lining method of electrofishing may be best. For this, the earth is used as the cathode by driving a steel stake into it (or laying a large coil of wire on the water body's bed), and connecting this to the negative pole of the generator, which is positioned on the bank. Then by boat (Halsband, 1980) or by wading, one can carry the anodes on long electric lines far out into the stream or lake. Distances of at least 500 m are possible.

The traditional power source is a portable, petrol-driven DC generator. Batteries can also be used, as can DC power produced by rectifying alternating current (AC) from a standard electricity distribution system (power lines or mains) or from a portable, petrol-powered alternator. Regardless of source, straight (unpulsed) DC, is generally most effective for catching fish (Lamarque, 1967) and is the type of current least likely to injure them. Pulsed DC is a feature that conserves energy, allowing use of a lighter generator or batteries; this is especially practical for fishing steep or poorly accessible streams. However, pulsed DC damages internal tissues of fish more than straight DC does. Muscle haemorrhaging and spinal dislocation are the major consequences. For this and other reasons, operating pulsed DC gear requires special technical knowledge. Modern back-pack electrofishing units with pulsed DC have overheating and short-circuit protection, as well as a dead-man switch and weigh only 10–20 kg. From observations in Germany's Eder Lake it was concluded that eels quickly learn about this very sophisticated fishing method and can instinctively adapt with evasion tactics (*Fisch und Fang* **5**, 1976).

As in Germany (Schiemenz, 1962), a special permit from the appropriate authority is likely to be required for electrofishing. It is also essential that those who engage in it be qualified by training, especially with regard to safety measures (Meyer-Waarden and Halsband, 1975). One can usually get information about types and sources of electrofishing equipment from the responsible government agency or, better yet, from the fishery department of a college. It is absolutely inadvisable to build electrofishing gear yourself; commercially available equipment generally has the necessary safety devices already, so this can avoid difficulties with regulatory agencies or even endangerment of life and limb.

A team of skilled fishers, using a single electrofishing unit of proper design in a stream reach that is as much as 10 m wide, is not too deep or steep, and has reasonable conductivity, should be able to easily capture the majority of the population of eels >30 cm in a single pass. The larger an eel is, the more likely it will be caught, as electrofishing is very size selective. In lakes, of course, eels are harder to catch. Nevertheless, Table 5.2 indicates how electrofishing appears to diminish a lake's population. Within a year after this method was begun in Germany's Sakrower Lake, it was catching several times more than all other methods combined, but then the success declined. Presumably, the part of the eel population susceptible to that gear was reduced more than the part living in the middle of the lake where harvest was by long-lining and jugging. That the eels caught by electrofishing differed, at least in size, from those caught by other methods is evident in a between-methods comparison of mean weights for 1954 and 1955, combined (Rahn, 1957b) from:

- seining: 793 g;
- fyke netting: 405 g;
- eel long-lining: 380 g;
- jugging: 478 g;
- electrofishing: 245 g.

Table 5.2 Results of eel fishing (kg) with various gears in the 100 ha Sakrower Lake (Potsdam) when electrofishing was introduced in 1950 (after Rahn, 1957b).

	1949	1950	1951	1952	1953	1954	1955
Electrofishing gear	–	45.0	341.5	193.5	161.5	112.0	19.0
Seine	8.3	1.0	–	8.0	11.0	4.5	5.5
Fyke nets	20.0	22.5	8.0	9.0	13.5	4.0	43.5
Eel lines and jugs	28.8	22.5	16.0	12.0	13.0	63.5	66.5
Total	57.1	91.0	365.5	222.5	199.0	184.0	134.5

Thus, the eels from the rest of the fishing methods averaged 1.5–3.2 times heavier than the electrofished ones.

The eels caught by electrofishing were smaller and younger because it was done in the littoral zone, thus in beds of emergent plants such as reeds, and in the shallow, offshore bank areas. The smaller eels generally inhabit shallow areas. Therefore, apparently it was not so very much a previously unharvested part of the eel population that was fished out here, but rather those eels which, at least to some degree, would later be caught with other gear. The Lake Sakrow example thus shows that by augmenting other harvest methods with electrofishing, an entire eel population might be completely caught out. In this case, electrofishing could have been done less frequently in order to reduce fishing effort per surface hectare. How much this would reduce labour depends largely on the extent to which other fishing methods can be restricted.

In deeper water a dip net does not suffice for catching the narcotised eels. It was not until 20 years after World War II that trawling and electric current were combined to exploit such areas. This development was so late because use of trawls in fresh water was frowned upon or even prohibited due to their potential adverse effects on aquatic flora and fauna. Under pressure to make operations of the freshwater fishery more profitable and get high yields with the least possible expenditure of labour, studies were conducted on practical application of electro-trawls (Hattop and Predel, 1969; Freytag and Horn, 1970). In the 1970s, and especially in the former East Germany, these investigations showed that the method could be very successful.

Figure 5.30 shows the arrangement of electrodes on the headline and groundrope of a trawl. It is important that the effective electric field at the groundrope extend a short distance in front of the trawl opening and down into the lake bed in order to flush eels out of the bed in front of the approaching trawl. Normal AC, sychronous AC and pulsed DC can each be used.

In electro-trawling, the fish need not be drawn to a particular electrode. In AC, which is physiologically more effective than DC, for example, the eel aligns itself at right angles to the lines of force during galvanotaxis, but is almost immediately galvanonarcotised, so ends up in the trawl (Fig. 5.30A). Electrified otterboard trawls are also possible but were less often used. For effective electrotaxis electro-trawls must be towed only at low speed, that is, 0.25 m/s, about 20% of the usual speed; for this, boats with 6-hp outboard motors will do. This could provide an apparent alternative to the normal trawl's higher required towing speed and the consequent injury to young fish.

Production of 50 Hz AC can be accomplished on the boat by a petrol-driven alternator. The current was at first supplied to the trawl *via* a cable attached to one of the tow lines; later, this was often replaced by a durable coaxial cable that served as a tow line at the same time (Freytag and Horn, 1975).

Electro-trawling is usually done during daytime but in principle can also be carried out at night. It is done above all outside the actual fishing season. Next to greater yield, this aspect was the most important motivating force behind the rapid innovations in the field. For example, Freytag and Horn (1972) reported on a successful eel fishery even when the water was around +2°C, a temperature at which any other fishing method would have been pointless.

The degree of success for combined operation of trawl and electricity in lakes is apparent in that after this method's introduction in the larger lakes of northern Germany's Brandenburg, like the Ucker Lakes, it took 40% of the total eel catch (Hattop and Predel, 1969). The analysis of the device from 12 lakes showed, in addition, that on average about 30 kg of eels (besides 20 kg of other fishes) per hour of operation could be expected. In Lake Schwerin, an operation involving two boats (10 and 6 hp, respectively) pulling an 8 m long electro-trawl caught 4.8 t of eels in 1967 and 3.0 t in 1968, and using a 12 m long electro-trawl in 1973 resulted in 29 t (Woest, 1977). In the course of an approximately 7 year electro-trawl fishery, three types (8, 12 and 16 m trawls) were developed in East Germany. Using these, many fishing operations accounted for as much as 60% of the eel catch (Predel and Gollub, 1973).

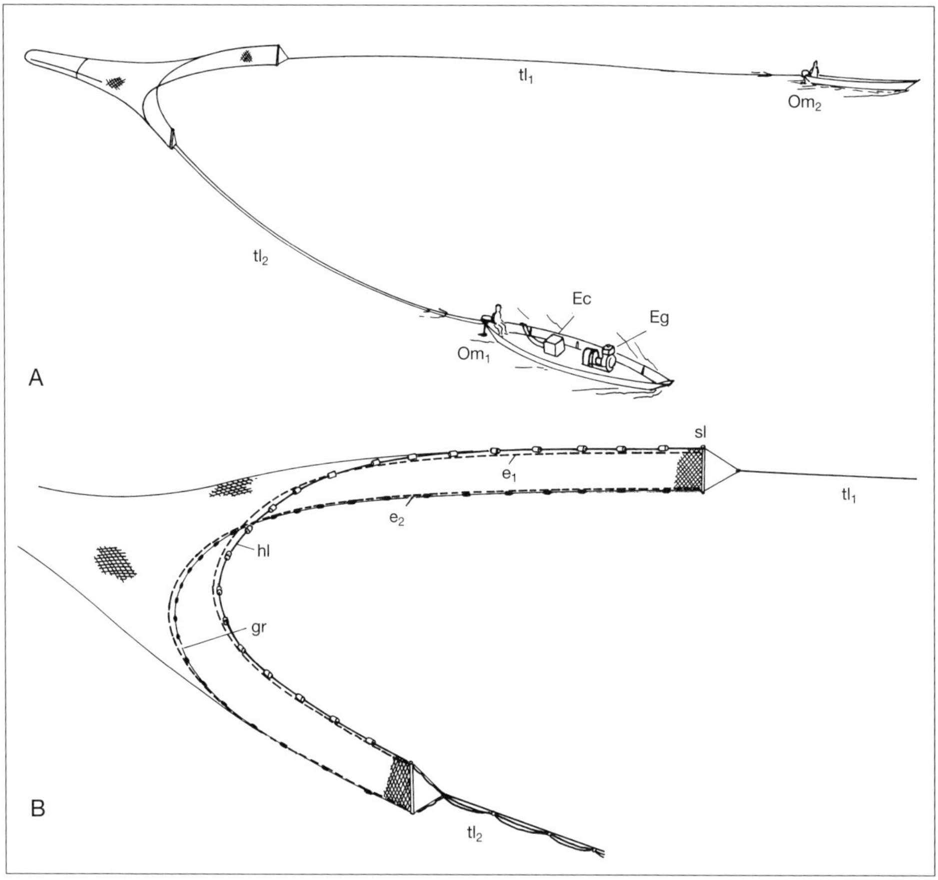

Fig. 5.30 Electro-trawling with two boats (after Hattop and Predel, 1969)

A Overall view
Eg Electric power source (generator or alternator)
Om_1, Om_2 Outboard motors, 6 hp
Ec Electricity control unit
tl_1 Warp (tow line)
tl_2 Warp with two insulated electricity lines attached (can be omitted if a strong enough coaxial cable is used at the tow line)

B Arrangement of the electrodes on the trawl's headline and groundrope
sl Spreader pole
e_1, e_2 Flexible, braidlike copper electrodes
hl Headline with floats
tl_1, tl_2 see A
gr Groundrope with lead weights

The idea of electrified trawling was taken up in Poland at almost the same time as in East Germany. The Polish nets had short wings and a horizontal opening of about 8 m. The codend was situated about 15–20 m behind the opening. Towing speeds were *c.* 0.4 m/s, so in 1 h about 1 ha could be fished. The best eel catch on Gardno Lake was 38 kg/h (Dembinski and Chmielewski, 1971). Swierzowska and Dembinski (1978) also reported on the amounts and profitability of electro-trawl catches in Poland. An electrified shrimp beam trawl was also used in Holland's IJsselmeer to catch eels for scientific purposes (Boonstra and Deelder, 1975). Here, they operated with pulsed DC, which finally was also used in the electro-trawling. The anode (6 mm^2 copper braid) was attached behind the headline to pull eels into the trawl; fastening it directly to the headline might frighten eels away before the trawl could envelop them.

Of course, this kind of fishery is worthwhile only if a large-enough surface area of water is available. Entering into this also is the extent to which other fishing methods remain in profitable operation or must be restricted. After electrofishing experiences in small streams, most other fishing methods immediately ceased to be competitive there. Electrofishing had a virtually revolutionary effect on scientific stock assessment.

For eel fishing, electric current can be combined not only with dip nets and towed nets, but also with other gear. Another example of electrofishing's application is to supplement trammel netting in lakes, as mentioned in Section 5.5. For this, a 2.5 kW DC unit was used. The purpose of this combined operation of electrofishing gear and trammel net was to get the most representative and complete samples possible of eel populations of shallow lakes or near-shore areas. The results were that between 34% and 88% by weight of the eels were caught with trammel nets rather than with anode-mounted dip nets. Anodal attractive and cathodal repulsive effects were both used in doing this (Berg, 1980).

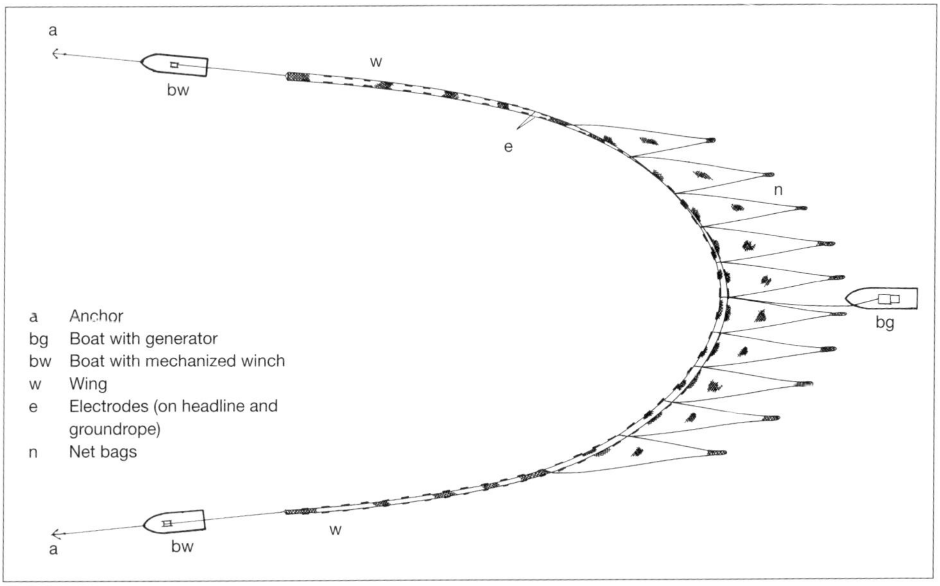

Fig. 5.31 Electrified seine (after Dembinski and Chmielewski, 1971)

A seine was combined with electricity in a Polish eel fishery (Dembinski and Chmielewski, 1971). Initially, simply a connecting piece about 30 m long between two wings of the seine was electrified. Although this did a good job of capturing eels, it was also noted that they were subjected to electric current too long because it took a long time to haul in the 400 m wings.

Therefore, a special electro-seine was developed, consisting of 50–75 m wings and a series of smaller net bags 25–30 m long (Fig. 5.31), so eels would not have to stay in electric current until the very last phase of hauling the gear to concentrate and catch them. Operating such electro-seines at speeds of 0.27–0.55m/s could cover 10–20 ha. The labour for this stayed within the limits of four to five persons.

Electro-seining operations in East Germany involved so-called ring-electro-seines. These had rings at 1 m intervals on the upper line, and through these ran a double braided rope for pursing the net. The seines had as much as 500 m total wing length and were successful to an operational depth of as much as 8 m (Hattop, 1979).

Another very effective combination, described by Halsband (1971a), is an electrified stow net deployed from a river vessel. (Coincidentally, such a stow net system [Fig. 5.18], *whether electrified or not* is called an 'Aalschokker' in German fisher-parlance.) These stow net systems (Section 5.4.3), usually set in the outer curves of rivers because water is deeper there, apparently create slow current in front of the net mouth, allowing a large share of the fish to swim away. Electrified leaders can block the way around the net mouth and guide the fish into it. Such electrification enlarged the stow net's collecting area from 72 to 250 m^2, resulting in 80– 100% increase in eel catch. The 12.5 kW pulse unit and generator were placed on the bank (Fig. 5.32). The two leads serve as cathodes, the iron chain as anode.

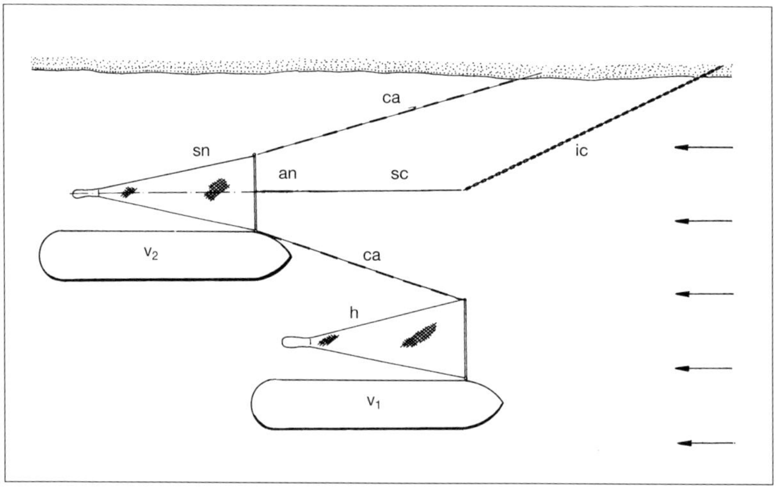

Fig. 5.32 Anchored, electified stow nets (after Halsband, 1971)

sn Stow net
an Anode
sc Synthetic-fibre cable
ic Iron chain
v_1, v_2 Vessels
ca Cathode

Such electrified leaders have been used in various ways in combination with fyke nets (e.g. McGrath *et al.*, 1976).

The so-called Hager cathode net was also used for eel fishing in a north German lake (Halsband, 1971a). This involves a gill net, the head- and groundlines of which are each equipped with an electric cable. The cables are connected to the negative pole of an electrofishing unit and, therefore, function as electrodes. Moving the net across the water with two boats flushes eels and can drive them into shallows where they are caught with normal electrofishing gear.

All the combined gear described above pertain essentially to fresh water. Successful use of such devices in brackish or salt water would require pulsed DC of greater strength because such water is more conductive. Such use is as yet unknown for eel fishing.

6
Eel culture

Revised by A. Kamstra

6.1 Current status of eel culture

Global demand for eels is met largely by aquacultural production of two species, the Japanese eel (*Anguilla japonica*) and the European eel (*A. anguilla*). Production of the American eel (*A. rostrata*), the short-finned eel (*A. australis*), and the long-finned eel (*A. dieffenbachii*) is of minor importance. Total production of eels (including fisheries) was estimated to be around 205,000 t in 1995 (FAO, 1997). Of this, cultured eels, at 188,401 t, composed about 92%. The market price of eels is high, so the economic value of cultured eels is considerable. This value was estimated at US$3.1 billion in 1995 and made up about 12% of the total value of all fish cultured. The fact that eel farming is such a big industry is even more remarkable in view of the fact that stocking material (glass eels) must still be collected in nature by fishing.

Eel culture in the mid-1990s was by no means static (Table 6.1). The increase in Chinese production, though difficult to quantify, has been spectacular, while production in Taiwan and Japan has been declining recently. Eel culture in Europe is also slowly on the rise due to recirculation technology. In Asia, production is dominated by *A. japonica*. China has imported *A. anguilla* increasingly since the early 1990s. In 1996 alone, an estimated 250 t of *A. anguilla* glass eels were exported from Europe to China! Such an amount can result in production of at least 50,000 t of marketable eels. Therefore, the importance of the European eel as a cultured species has risen considerably in recent years. On the other continents, eel culture utilises the locally available species, for example, *A. anguilla* in Europe, *A. rostrata* in North America, and *A. australis* in Australia and New Zealand.

6.2 Historical development

6.2.1 Introduction

Culture of fish implies control over at least part of the life cycle and typically involves breeding, feeding, disease control, and protection from predators. Artificial breeding and rearing of eel larvae is not yet possible; this problem can be considered the 'Holy Grail' of

Table 6.1 The global status of eel culture in 1995/6 (t).

Asia	
Japan	29,131
Taiwan	25,546
China	120,000
South Korea	2345
Malaysia	2969
Indonesia	610
Europe	
Belgium	140
Denmark	1200
Germany	100
Greece	550
Italy	3000
The Netherlands	1800
Norway	200
Portugal	110
Spain	250
Sweden	200
Others	50
North America	100
Australia/New Zealand	100
Total	188,401

Sources: FAO (1997a), FAO (1997b), Anonymous (1997a).

eel research. Although some progress has been made recently in artificial propagation of eels (Anonymous, 1997b; Lokman and Young, 1995), the remaining problems in this field are formidable and would not be solved easily. However, stocking material, for example, glass eels or elvers, is readily available in many coastal regions of Southern and Western Europe, and some locations already have a long history of stocking and extensive rearing of eels in ponds. The lagoons of Comacchio in Italy and Arcachon in France are good examples, with roots going back to the 18th century. The productivity in these kinds of systems is low and depends completely on natural production of the water body involved.

This brings us to the second aspect of culture, feeding. Development of feed, and especially that of formulated dry feed, has been a major breakthrough in the development of eel culture, as we will later see in the example of Japan. It was recognised at the end of the 19th century that productivity of ponds could be improved by feeding eels, for example, with trash fish (Matsui, 1986). Feed intake of fish depends largely on water temperature, and this brings us to another important aspect of eel culture.

Eels are essentially a warm-water species showing optimal growth at 23–26°C and ceasing to grow at around 13–15°C. Therefore, eel growth, even when food is abundant, is very limited when water temperature is optimal only part of the year. This explains why eel culture in northwestern Europe, despite early interest, had a difficult start and is technically feasible only by using waste heat (e.g. from power stations) or advanced recirculation technology. Control over water temperature is essential for high productivity and economy of the operation. Many technical developments in eel culture in the last 30 years have been focused on control of water temperature, as we will see in the sections on developments in Japan and Europe.

The other aspects of culture are control of diseases and protection from predators. It is usually easy to protect against predators in a rearing system. However, infectious diseases and problems with water quality are one of the major factors affecting profitably and deserve constant attention. As we shall see, disease problems have seriously hampered development of eel culture at some stages in the past. Eel diseases are treated in Chapter 7.

6.2.2 Japan

Eels have been highly regarded in Japan since ancient times and occupy a special place in this country's tradition. Eel consumption peaks during the Ushinoshi feast in summer and reputedly gives strength in the hot summer months. It is, therefore, no coincidence that Japan can be considered the cradle of modern eel culture, being the leading producer of eels until the mid-1980s and the major innovator of culture techniques. A close look at historical development of eel culture in Japan can serve as a good example of the problems that have to be solved when culturing eels (Fig. 6.1).

Farming of eels on a commercial scale started in the period 1880–1900 in the prefectures Shizuoka, Mie and Aichi. In this period, large ponds covering several hectares had eels stocked in polyculture with carp (Matsui, 1986). Regional fisheries research stations already started research on eel culture in this phase. During the agricultural recession of 1912–26, the price of rice fell sharply, and farmers were motivated to convert their land into eel ponds.

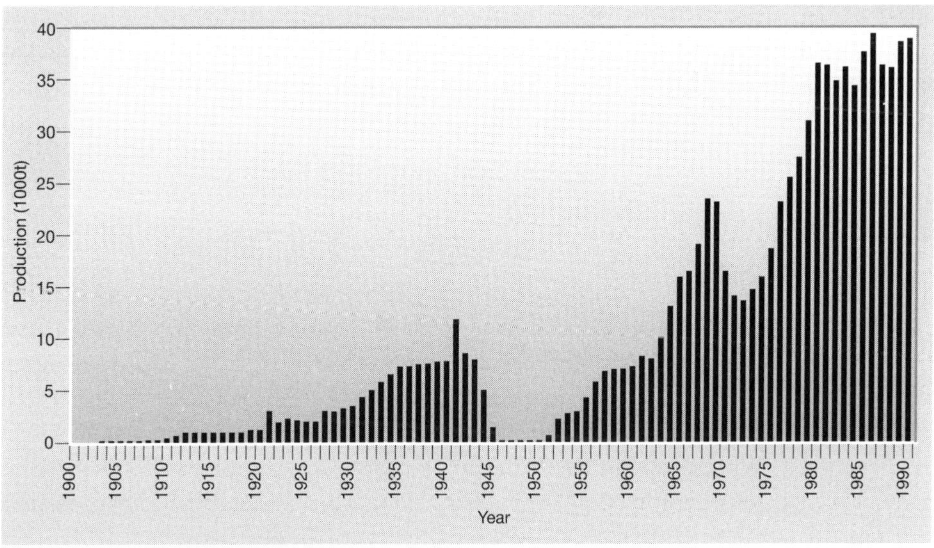

Fig. 6.1 Aquacultural production of eels in Japan, 1910–90

In the early 1920s, two developments forced the industry to take new directions. Silkworm pupae, until that time the sole feed used, became expensive and were soon replaced by sardines, thereby reducing cost. Shortages and excessive prices of stocking material forced the industry to develop methods for rearing glass eels. Production rose steadily to 10,000 t in 1940 but came to a complete standstill during World War II. In this period, many ponds had to be converted into rice paddies for domestic production. Restrictions on use of raw fish hampered development immediately after the war.

In the early 1950s, production started to rise again and a period of consistent growth lasted until 1969 when production reached *c*. 24,000 t. Major innovations in this period, enabling higher productivity and lower costs, were the introduction of the paddle wheel aerator in the mid-1950s and artificial feed in the 1960s. From 1969 to 1972, production decreased sharply to a mere 14,000 t. A number of factors were responsible for this collapse:

- Infection with the fungus, *Saprolegnia* sp. (Usui, 1974), gill rot disease and a condition called branchio-nephritis caused losses of over 40% in culture. In the latter disease the EVE-virus is implicated (Sano *et al*., 1981) which might have been introduced with imports of *A. anguilla* from Europe.
- Catches of glass eels declined sharply at the end of the 1960s. Moreover, glass eel imports from Taiwan were progressively restricted starting in 1965, culminating in a complete ban on exports from this country in 1972.

Through an intensive research effort, solutions to most of the disease problems were developed in a few years. Low water temperature in winter was implicated as a major negative factor for eel survival. This conclusion greatly stimulated development of the greenhouse technique in which water was heated and energy conserved. In the late 1970s, this innovation was commonly applied in Japan, especially for rearing the first stages. Large imports of glass eels of European eel (1972: 220 t) in the late 1960s and early 1970s to alleviate shortage of stocking material were not successful and continued at a level of only 10–15 t/year.

Production increased again rapidly from 14,000 t in 1972 to 37,000 t in 1979 and has been more or less stable since then. Growth of eel culture in Japan is currently restricted (Gousset, 1992) by:

- low supply of *A. japonica* glass eels and correspondingly high prices for stocking material;
- intense competition with eel producers in China and Taiwan resulting in relatively low prices for the end product.

In summary, the development of eel farming in Japan is a perfect example of the complex forces shaping the industry. The large domestic market has given strong impetus to developing eel culture in this country. The specific character of eel farming as an industry relying completely on wild stocking material is perfectly illustrated by the Japanese problems encountered in this respect. Moreover, the development of eel farming in Japan serves as a good example of the way in which technical innovation and research can change existing practices.

6.2.3 Europe

In Europe, eel farming was of minor importance in the past, despite the fact that eels are much appreciated in many countries. In the past, only Italy had some farming, the extensive valliculture in the Comacchio area near Venice. This production was based on stocking glass eels or larger pigmented eels in polyculture with mullet and sea bass. There was supplementary feeding in some cases but always with low productivity. In the early 1970s, *c.* 3400 t of eel were produced this way. In the 1970s, parasitic diseases, mainly *Argulus giordani* and white spot disease (*Ichthyophthirius multifiliis*), caused severe problems that were difficult to control due to the large areas of water involved (Saroglia *et al.*, 1985). In 1980, valliculture production was reported at less than 1500 t. In the meantime, attention shifted to intensive, Japanese-style pond culture, which developed into a business producing roughly 3000 t/year. Apart from earthen ponds, concrete tanks and systems using heated effluents are employed. Currently, the production level seems to be stable and restricted by availability of wild fingerlings from fisheries in other parts of Europe.

Since the early 1980s, eel farming also has started in northwestern Europe. In this area, lacking a suitable climate for pond culture, eel farming is made economically viable by recirculation technology, which enables heating of water and conservation of energy (Kamstra and Davidse, 1991). This technique requires skilled personnel and high investments, and has developed particularly in The Netherlands and Denmark. In some of the early European attempts to farm eels, waste heat from power stations was used to warm the water (e.g. Koops and Kuhlmann, 1980a). However, sites having waste heat of good quality are rare, and using power station effluents involves some risks. The development of dependable recirculation systems has largely replaced use of waste heat.

6.3 Rearing systems

An eel rearing system should provide an adequate environment for optimal growth, require little investment, and be easy to operate. A rearing system also provides living space for the fish. Given the fact that eels can easily be kept at high density (200 kg/m^2) when water quality (oxygen) is optimal, rearing space alone is hardly ever a limiting factor for production. Under some circumstances, low densities can even result in aggressive behaviour among individuals and mortality (Peters *et al.*, 1980). Rearing systems can be classified according to the amount of control exerted over water quality, which is closely related to the productivity of the system (Table 6.2).

Water temperature is first of all a determinant for growth. In pond culture, water temperature depends on climate and latitude. Thus, these factors determine the length of the growth cycle (Gousset, 1990). This explains the rise in production outside Japan in countries like Taiwan and China, which have a more favourable climate. In greenhouses and recirculation systems, it is economically feasible to heat water, making climate less influential. Feeding eels in summer reduces the amount of dissolved oxygen in the water, and this limits production. Therefore, paddle wheels are often used to aerate ponds, and bottled oxygen is commonly injected in recirculation systems.

Table 6.2 A comparison of characteristics of rearing systems used for eel farming.

Location	Japan	Japan	NW Europe
Species	*A. japonica*	*A. japonica*	*A. anguilla*
Type of system	Greenhouse	Pond	Recirculation
Feeding	Paste	Paste	Pellet
Water supply (m^3/kg per feed)	4–5	6	0.5
Oxygen	Paddle wheel aeration	Paddle wheel aeration	Technical oxygen
Ammonium removal	Algae/biofilter	Algae	Biofilter
Productivity (kg/m^2 per year)	8–10	1	120–170
Reference	Gousset, 1992	Gousset, 1992	Heinsbroek and Kamstra, 1990

When dissolved oxygen requirements are satisfied, ammonia produced by the fish becomes the limiting factor and must be removed. This can be done by:

- flushing with large amounts of water (flow-through systems using waste heat);
- removal *via* the pond's primary production (algae); or
- bacterial conversion (nitrification) in a biological filter as applied in recirculation systems.

To remove minerals (nitrate) and other substances that accumulate in the rearing environment, some degree of water supplementation is always necessary. Water quality is almost completely controllable in rearing systems, but doing so has a price. Experience in northwestern Europe has shown that advanced water treatment systems require relatively high investment and skilled personnel, and can be risky.

6.3.1 Water quality requirements

Eel farming requires detailed knowledge of the necessary water quality and means to control it. Table 6.3 shows the most important water quality parameters. The preferred water temperature for culturing Japanese eels is somewhat higher than that for European eels. European eels reputedly have lower tolerance for high summer temperatures (Gousset, 1992). The eel's oxygen consumption is closely related to feed intake, which, in turn, depends on body size and water temperature. In pond culture, algal oxygen consumption is also an important factor to take into account. Ammonia is the main waste product excreted by fish as a result of protein catabolism. Compared with most fresh water fish species, eels are very tolerant of environmental nitrite.

6.3.2 Design features of rearing systems

The physical design of eel farms shows enormous diversity in location and tradition. Usui (1974) gave a detailed description of the construction of classic outdoor ponds for culture of eel, now mainly in use in China and Taiwan. These outdoor ponds range in size from 0.5 to 2.5 ha, are equipped with paddle wheels for aeration (Fig. 6.2), and have special places for feeding the eels (Fig. 6.3). Fish are harvested for sale or grading by seining or draining the pond.

Table 6.3 An overview of the most important water quality requirements for eel culture.

Parameter	Level	Remarks	References
Temperature	24–28: *A. japonica* 23–26: *A. anguilla*	Glass eels prefer temperatures on the right side of the scale; large eels those on the left.	Gousset (1992); Kuhlmann (1974)
Oxygen	>5 mg/l	For optimal feeding behaviour	Rowchai *et al.* (1986)
Ammonia	<0.05–0.1 mg/l	Ammonia (NH_3) is toxic form; equilibrium NH_4/NH_3 strongly pH-dependent	Yamagata and Niwa (1982); Sadler (1981)
Nitrite	<30 mg/l	Toxicity dependent on pH and chloride-concentration	Yamagata and Niwa (1979); Kamstra *et al.* (1996)

In Japan, most eels are produced in greenhouse culture nowadays. Pond area in greenhouses ranges from 50 m^2 for glass eels to 500 m^2 for larger eels, and ponds are usually built of concrete. In Japan, farms in areas having high costs for water and energy make much use of water recirculation, which necessitates extensive water treatment (Gousset, 1992). Sedimentation tanks are often used to remove faeces and other suspended matter, and in some cases, a biological filter is employed to remove ammonia. About 5–30% of the water volume is replaced daily in the system (4–6 m^3/kg feed).

In Europe, farms employing recirculation technology are situated in well-insulated buildings and use advanced water treatment. Rearing tanks vary in size from 4 m^2 (Fig. 6.3) for glass eels to 50 m^2 for larger eels. Suspended solids are continuously removed from the recirculating flow by mechanical filters with fine mesh screens (60–150 μm). Much of the recirculating flow is pumped over a biological filter, which can be a trickling filter or a submerged up- or downflow filter filled with a medium that has a high specific surface area. The biofilter converts the toxic ammonia into relatively harmless nitrate. In this type of farm, pure oxygen is usually added to the main flow in order to reduce pumping costs (Heinsbroek and Kamstra, 1990). Some farms use denitrification to remove nitrate; in those cases water consumption can be as low as 100 l/kg of feed instead of the 500 l/kg normally applied in farms using recirculation systems. In some cases, UV-light or ozone is used to disinfect the culture water.

Fig. 6.2 Eel culture pond in a Japanese greenhouse; paddlewheels installed for aeration (photo: Tesch)

Fig. 6.3 Left: feeding with a paste formulation on a screen as in Japanese and Chinese pond and greenhouse culture. Right: automatic feeding machine distributing pelleted feed in a Dutch circulating tank (photos: Kamstra).

6.4 Rearing technology

6.4.1 Stocking and grading practices

The eel rearing cycle always starts by stocking glass eels obtained from specialised commercial fisheries in December–May (Sections 3.1.3 and 3.2.2). Stocking with yellow eels caught in the fishery is seldom done but is technically quite possible. *A. japonica* glass eels (*c*. 5000/kg) are much smaller than *A. anguilla* glass eels (*c*. 3000/kg) (Table 2.4) and command high prices (Section 8.3). There is a lively global trade in glass eels, which involves, for example, major shipment of *A. anguilla* from Europe to Asia. Glass eels are often transported in styrofoam boxes containing *c*. 5 kg of fish, kept cool and moist with ice. Trucking in water is common in Europe.

Individual variation in eel growth (Section 3.3.2.5) is considerable under farming conditions, and frequent grading is needed to keep the sizes separated. Practices of stocking, grading and harvesting differ considerably according to the type of system and market demands. Table 6.4 gives examples for Japanese and European eels. It shows that culture of *A. japonica*, even in a greenhouse is relatively extensive with regard to densities and productivity. In culture of *A. japonica*, the expensive glass eels are stocked at low density and graded regularly, resulting in high survival of 70–90%. In Europe, survival of *A. anguilla* usually ranges from 50% to 70%. In recirculation systems, high productivity is an economic necessity, given the high investments in a rearing system. The growth rate of *A. anguilla* is somewhat lower than that of *A. japonica* but most characteristics of these species are similar. The large-scale introduction of *A. anguilla* in Japan in the early 1970s showed that this species is sensitive to a number of the same diseases that normally afflict *A. japonica*. Especially the gill worm (*Pseudodactylogyrus* sp.) and swimbladder worm (*Anguillicola crassus*) have proven harmful for European eel (Egusa, 1979). In Japan, some farmers specialising in *A. anguilla* have developed a rearing method suitable for it (Heinsbroek, 1991).

A peculiar aspect of eel farming is that, for reasons unknown, most individuals produced are males (Matsui, 1952; Egusa and Hirose, 1973). This strongly affects size and

Table 6.4 An outline of a typical stocking, grading and harvesting practice for *A. japonica* in greenhouse culture (after Chiba, 1983) and *A. anguilla* in recirculation systems (Gousset, 1990).

Activity/species	*A. japonica*	*A. anguilla*
Stocking density	0.2–0.5 kg/m^2	5–15 kg/m^2
First grading	20–25 days	1–3 months
Density of fingerlings (1–20 g)	5–20 kg/m^2	20–80 kg/m^2
Frequency of grading fingerlings	Every 15–25 days	Every 1–3 months
Density of on-growing	5–40 kg/m^2	60–200 kg/m^2
Total rearing period	6–18 months	12–24 months
Productivity	8–10 kg/m^2 per year	120–170 kg/m^2 per year

quality of farmed eels; most individuals sold are in the silver stage and range from 120 to 200 g. Technically, it is possible to change a cultured eel population's sex ratio to almost 100% females by adding oestrogens to the diet of elvers (Degani and Kushnirov, 1992; Andersen *et al.*, 1996).

6.4.2 Feeds and feeding

Eels have a peculiar feeding behaviour, and feed intake depends largely on their formidable abilities to smell and taste. A number of feeding stimulants have been discovered for eels, among which some L-amino acids are important. Inducing glass eels to feed is especially difficult and still depends mostly on using *Tubifex* (Asia) or roe from cod or plaice (Europe) for the first few days. Nowadays, however, good artificial feeds for start-feeding of glass eels are becoming available, and use of 'natural' feeds is decreasing. Table 6.5 gives the composition of feeds used in commercial eel farming. In Asia and Italy most of the feed is fed as a paste (Fig. 6.3) containing 20–45% dry matter, which is made from dry meal and mixed with water and oil on the farm. In Europe, pelleted and extruded feeds are applied; these have major advantages in terms of distribution, storage and handling.

Table 6.5 An overview of some important characteristics of commercial feeds used in eel culture (after Heinsbroek and Kamstra, 1995).

Feed type	Starter *A. japonica*	Grow-out *A. japonica*	Starter *A. anguilla*	Grow-out *A. anguilla*
Feed size (mm)	Paste	Paste	0.3–0.5	2–4.5
Fish size (g)	0.15–0.3	10–200	0.2–0.5	20–500
Feed composition				
Dry matter (dm) (% as fed)	19–31	36–45	90	92
Crude protein (% of dm)	63–76	50–75	59	42–48
Crude fat (% of dm)	4–19	5–8	8	15–25
NFE (% of dm)	5–11	18–24	14	15–23
Ash (% of dm)	5–13	18–20	8	8–12
Crude fibre (% of dm)	0.5–2	1.1–1.7	1	0.5–2
Phosphorus (% of dm)	No data	1.1–1.9	1.6	1.2–1.4
Feeding level (% dm of BW/day)	3–10	2–3	2–5	0.5–1.5
Feed conversion (g dm/g fish)	1.1	1.2–1.8	0.9–2.5	1.2–2
Feed price US$/kg	35–45	1.5–2.5	5	0.8–1

A notable difference in the composition of feeds in Asia and Europe, is the amount of fat (fish oil). European feeds for *A. anguilla* contain up to 25% fat, while those for *A. japonica* in Asia only contain up to 8% fat. Feed conversion depends largely on the growth rate of the fish. For small fish, conversions below 1 (conversion of dry weight of feed to wet weight of eel) are possible, whereas, feed conversion of slow growing fish can easily increase to 2. The eel's requirements for essential amino acids, fatty acids, vitamins and minerals are relatively well studied and a large body of literature is available (e.g. Degani and Gallagher, 1995; Arai, 1991).

6.5 The economy and future perspectives of eel culture

Developments in eel farming are continually driven by the farmer's need to maintain a satisfactory margin between costs and benefits. Insight into the economy of eel farming is necessary to understand and predict these developments. Figure 6.4 serves as an example of eel production costs and market prices in different countries. However, it is extremely difficult to compare production costs between countries because statistics on changes in currency exchange rates are inadequate, and large variation between farms exists. Figure 6.2 shows that eel production costs in Japan are relatively high mainly due to costs for glass eels and feed. The costs of glass eels amount to 25–40% of total production costs (Gousset, 1992) and threaten profitability. Glass eel prices vary considerably year to year as their catch fluctuates. Currently, catches of *A. japonica* glass eels are too low to support the industry in Asia. This has prompted large imports of *A. anguilla* from Europe and a strong research effort to solve the problem of artificial reproduction. Feed costs are relatively low in north-western Europe, and water recirculation now keeps energy costs from being excessive. Major investment in water treatment makes capital costs relatively high on Dutch farms. Figure 6.4 shows only a small margin between total cost and market price in all cases.

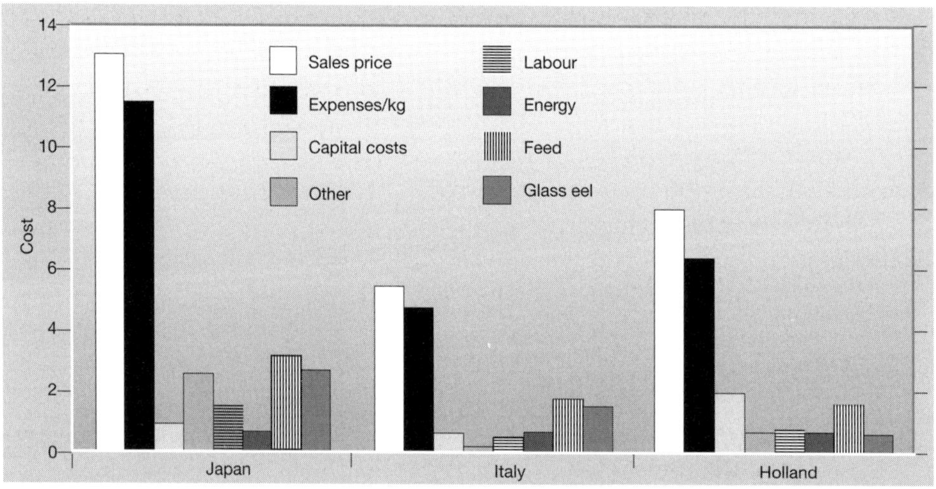

Fig. 6.4 Comparison of eel culture costs (US$/kg) in three situations: Japan (Gousset, 1992), Italy (Saroglia *et al.*, 1991) and Holland (van Rijsingen, 1995). Conversion to US$ according to the December 1997 exchange rate.

Market prices are strongly influenced by supply and have begun to decline mainly due to increased Chinese production. In Europe, farmed *A. anguilla* has filled the gap left by declining production from fisheries to some extent, and prices have been relatively stable.

Regarding future prospects for eel farming, it is obvious that instability of the *A. japonica* glass eel supply will severely restrain the industry's development in Asia. In this respect, prospects for culturing *A. anguilla* are promising because a large surplus of glass eels exists in Europe even in years of relatively low catch. Although Japan was traditionally the world's eel farming centre, this changed dramatically during the last decade. Japanese eel farms, producing at relatively low cost, will continue to compete strongly with farms in Taiwan and China. In Europe, the market for farmed eels will largely determine the industry's growth. Interesting European developments involve recent investments in processing and export of *A. anguilla* to Japan. Moreover, environmental constraints on the use of water and discharge of wastes by eel farms are increasingly felt in northwest Europe. These matters must be addressed if eel farming is to expand.

6.6 Live storage

It is necessary when catching and selling eels to keep them alive for some time, as for instance, when delaying the sale may mean higher prices later, or, some eels have an unpleasant flavour when first caught; it may be necessary to keep them until this has disappeared. A well-known example is the so-called phenol flavour. This taste does not originate from the phenols themselves, but from accompanying substances; often, it is produced by detergents. In a live storage experiment eels were kept for three weeks in water at 16–20°C; this led to a marked, but still unsatisfactory decrease in the unpleasant flavour (Mann, 1960). After 3 months the tail region, which is particularly susceptible, still tasted slightly unpleasant in a few of the fishes. It is thought that feeding the eels leads to a more rapid improvement in flavour.

Feeding has other advantages too, as it counteracts the otherwise unavoidable weight loss which occurs while the eel is in captivity. In connection with this, experiments have been carried out in a small pond (area 100 m^2) in the lower Elbe area (Koops, 1965), where 800 kg of eels, with individuals ranging in weight from 0.4 to 1.5 kg were kept at a temperature of 16°C. Previous experiences had shown that a weight loss of 10–20% could be expected within a period of 3–6 weeks. In this experiment, the eels were given 16 kg of food daily; the weight loss after 5–6 weeks was only 0.5 kg. Since the daily food ration amounted to about 2% of the body weight, this percentage can be regarded as the minimum quantity of food necessary to maintain body weight. Japanese experience showed that the loss in body weight is strongest during the first 10 days after capture (0.8%/day). Later it decreases to 0.1%/day in winter and 0.2%/day in summer (Matsui, 1980).

In addition to losses in weight, losses in number of the stored individuals have to be considered. Eels kept in holding tanks are aggressive and this aggression becomes stronger with decreasing density. Biting attack showed in some cases that the eels died after showing various diseases: reduced spleen and blood quantity, but increased leucocrits and plasma cortisol concentration; decreased liver glycogen but rising blood glucose and lactate; and detection of haemoglobin in the skin mucus (Peters *et al.*, 1980).

7
Diseases, parasites and bodily damage

Newly reworked up to 1999: L.W. Reimer

7.1 Introduction

For economically important fishes, diseases are well known today in large part because intensive studies were required to avoid major damage to pond culture and aquaculture. Eels often occurred in dense populations that were frequently created by stocking or intensive pond management. These situations involved various serious diseases. Today the problems have shifted somewhat. Population density of wild eels has declined, hence also the danger of infectious diseases and directly transmissible parasites. On the other hand, aquaculture has developed further in fresh and salt water in recent decades, and this has involved new experiences. Problems increased through worldwide, human-generated dissemination of disease agents and parasites.

Therefore, the general literature on fish diseases and fish parasites deals significantly with diseases and parasites of eels, for example, Schäperclaus (1979), Amlacher (1986), Reichenbach-Klinke (1969, 1980), Kinne (1980–90) and Möller and Anders (1986).

In this chapter, only economically important diseases are discussed. Further examples are dealt with in the form of bibliographic tables. For diagnosis and treatment, the reader is referred to the publications mentioned. A guide to the special diseases and parasites of the eels in New Zealand was published by Hine and Boustead (1974). For the parasites of *Anguilla rostrata* in Canada, a *Guide to the Parasites of Fishes of Canada* has been issued by Margolis and Kabata since 1979.

Damage means injury only to individual fish. Eel damage on a population basis was described in Chapters 3 and 4 on ecological and fishery matters (e.g. Sections 3.13, 4.24). Situations in which the eel itself is the damaging agent, for example, in eel competition with and predation on salmonid and coregonid populations (Sections 3.3.1.6; 4.4.2.2; 4.4.2.3) are omitted here also. Counteracting and preventing such damage (e.g. Smith, 1941, 1955, 1966) should be understood primarily as human interventions into natural biotic communities.

7.2 Viral and bacterial diseases

7.2.1. Viral diseases

Epidemic diseases that noticeably plague eel populations have been caused for the most part by viral and bacterial attack.

The three viral diseases at issue are known by their abbreviations as

- EV2 (cauliflower disease);
- EVA (eel virus America);
- EVE (viral kidney disease of eels).

7.2.1.1 Cauliflower disease (EV2)

The cause of cauliflower-like growths are viruses of the Orthomyoxovirus group, which is specific to *A. anguilla* and abundant in Germany. A first occurrence was reported in 1907 (Nagel, 1907). The first scientific studies took place in 1950 (Christiansen and Jensen, 1950; Schäperclaus, 1953). Besides the eel, cod and bleak can be infected also (Amlacher, 1986). The pathogen of the stomatopapilloma of eels is 80–140 nm in diameter and has RNA as genome. In addition to EV2, the blood of eels with this cauliflower growth contains a 55-nm polyhedral particle that is called papovavirus (EV1). The causative connection with stomatopapillomata could not yet be proved (Ahne and Thomsen, 1985).

Insufficient knowledge is available about transmission in fresh water or the sea and about incubation. Even though this eel disease was first noticed in Germany, similar isolates could actually be obtained in North America.

Diseased individuals form skin tumours on one or more places anywhere on the body surface, but in most cases on the upper and lower jaws (Fig. 7.1). Their appearance resem-

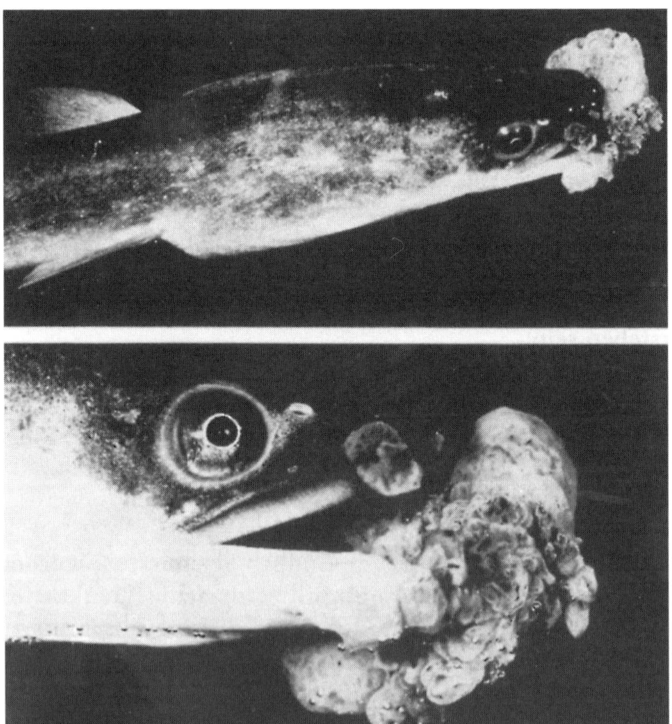

Fig. 7.1 Eel with cauliflower tumour (photo: Marschall)

bles cauliflower, thus the disease's name (Schäperclaus, 1953). At first the growths are whitish, later light brown and with progressive pigmentation also darker coloured. The swollen mass can be so large that its protrusion from the tip of the head can exceed the head's length. On larger eels it can be the size of a chicken egg. Infected eels are often severely emaciated. In severely diseased 19–29 cm eels, fat content declines from a level of 1.6–5.0% to 1% and less, and mean weight decreases by about 35% (Lühmann and Mann, 1956; Koops and Mann, 1966). Although major mortality of these animals was not observed in captivity, it is said that affected eels in traps are more likely to die than healthy ones there.

Histologically, the following is seen: the highly differentiated epidermis is replaced by largely undifferentiated tissue. The change or transformation starts in the stratum basale and progressively supplants the normal structure of the epidermis (Schmid, 1969). 'The abnormal growths are referred to as fibro-epithelial tumours or as papillomata (i.e. villous formations are involved) in the connective tissue of which, protruding papillae have small, lateral, secondary, tertiary, etc. papillae covered by epithelium-like tissue' (Schäperclaus, 1953).

In cell cultures, it was possible to show the virus infection in several successive cell divisions (Pfitzner, 1969; Schwanz-Pfitzner, 1973). Blood transfusions from infected to healthy eels had a positive result. Apparently the virus is virulent only a short while in water. Sites of skin injury are particularly at risk of infection, an important problem especially in intensive eel culture (Pfitzner, 1969).

Another possibility for pathogenesis of cauliflower disease is the effect of wastewater substances on eel skin. Fluoride ions and mono-iodine acetate produce tumour-like cell multiplication in eel epidermis within 24 h (Peters and Peters, 1970). Apparently, *via* chemical effect, respiration is reduced and glycolytic metabolism increased here. The effect of these inhibitors is negated by inorganic diphosphate and presumably also by a resultant better energy supply to the cells. It was possible to destroy and resolve the cauliflower growths in the same way chemically, and from this a therapeutic treatment of the disease resulted.

The pock-like growths found by Bremer (1974) on the skin of glass eels presumably have other causes and are probably attributable to osmotic insufficiency.

In ponds, as much as 24% of juvenile eels could be sick with this disease (Koops and Mann, 1969).

Since its first appearance in the Baltic Sea coast area at the beginning of the 1900s (Schäperclaus, 1953), the disease gradually spread over most areas of the Baltic and North Seas. Now cauliflower disease occurs on the Scandinavian west coast also (Gullestad, 1972). Thulin *et al.* (1989) spoke of $<1\%$ infection rate on the whole Baltic Sea coast but of a high rate (39%) in The Sound (Denmark). Radulescu and Angelescu (1972) reported only sporadic infection in the Black Sea. Cauliflower disease is unknown in Belgium and the British Isles. In the lower Elbe river, 5.6% of the eels were infected in 1957/9, 12% in 1967 (Koops and Mann, 1969) and 91% in 1981/2 (Anders and Möller, 1991). The last mentioned authors determined 6% on the mud flats in 1988/9. Observations of Tesch (1983) indicated that the disease increases from spring to autumn. According to Peters (1977), the tumours are most prevalent in August and September.

7.2.1.2 Viral kidney disease (EVE)

Viral gill disease of eels is caused by an IPN-like virus (Birnaviridae). As it was first identified in European eels, it is also referred to as EVE (Eel Virus European).

The disease occurs in *A. anguilla* and *A. japonica*, though until now only in Japan and specifically in the cool season. Symptoms are swollen gills and hypertrophied kidneys.

Mortality of juvenile eels can be very high (Wolf, 1984).

7.2.1.3 Eel virus American (EVA)

The abbreviation EVA stands for Eel Virus America, but it was isolated in Japan (Sano, 1976; Hill *et al.*, 1980) and in fact from *A. anguilla* and *A. rostrata*. Juvenile eels show signs of haemorrhages and necroses in the musculature.

7.2.2 Bacterial diseases

7.2.2.1 Freshwater eel disease and similar sicknesses

This epidemic disease is caused by the bacterium *Aeromonas punctata*, and is known as red disease or red fin disease. The pathogens belong to the Spirilaceae and are $0.5 \times 0.8–2.0\,\mu m$ in size, coccus- to stick-shaped with a flagellum.

The disease's symptoms according to Schäperclaus (1979) were:

- Eels without external evidence of disease swimming around listlessly; shortly before death lying contorted at the water surface. Dying rapidly in water bodies or just after being caught.
- More or less strong, general, stripe-, blotch- or dot-like red colouration of the body (especially the belly), the fins, and the anus.
- White, often bluish patches on the epidermis and in parts of the dermis, caused by loss of mucus.
- Raw, pale sites as in spot epidemics. The musculature lies exposed and is surrounded by reddened or white, dead dermal tissue. Such sites are often massive on the tail, which can look as if it has been eaten.
- Boil-like, pea- to walnut-sized sores, occurring especially on the head and in springtime. They can burst, become washed out, and, upon formation of an initially white scar, heal again or, in the event of sizeable expansion, lead to general infection and to the eel's death.
- Blindness.
- Internal haemorrhages, bleeding in the liver and inflammation of the large intestine.

Diseases having similar symptoms can also be caused by other pathogens, in Europe above all by *Pseudomonas fluorescens*, a causative agent that also appears in compound infections (Reichenbach-Klinke, 1980).

Pseudomonas anguilliseptica causes a syndrome resembling freshwater eel disease (*A. japonica*) (after Miyazaki and Egusa, 1977; Kuo and Kuo, 1978). For Japan, *Aeromonas hydrophila* (Kanai *et al.*, 1977) and *Aeromonas liquifaciens* (Usui, 1974) are mentioned, as well. In freshwater culture of *A.japonica*, *Edwardsiella tarda* has become especially well known due to great losses.

7.2.2.2 Saltwater eel disease

This eel disease occurs in the warmer seasons in brackish and sea water. It has been known in Italy since 1718 and is now present on the Dutch, Danish, Swedish and German coasts of the North and Baltic Seas. In 1931–2, there was a major die-off of eels in which 1000 dead eels were picked up along 1 km of the Isle of Vilm's coastline. In 1949 and 1959, the death rates were not quite as bad.

In the North Sea, the eels landed at Büsum in 1969, 50% were sick in August and 62% in September. There were also infections of 30.4% in 1973, 10.3% in 1974 and 26.4% in 1975 (Aker, 1975, 1976). As said before, the conducive condition is water temperature of at least 19–22°C. The optimal salinity range for infection is 15–35‰, but it can also occur at salinities as low as 2.5‰.

Symptoms:

- Sudden death without external signs of disease. Just before death the eels come to the surface and convulse spasmodically from time-to-time.
- Blotchy reddening of the skin, fin margins and anus.
- Swelling and reddening of musculature in the heart region.
- Tetanic spasms that produce contraction of the anus.
- Red, often boil-like sores, which can be covered by a glassy skin growth.
- According to Schäperclaus (1927) and Nybelin (1935), the liver often shows 1 mm haemorrhages (i.e. a slight reddening).

Five biotypes can be distinguished according to biochemical reactions (Table 7.1). According to Miyazaki *et al.* (1974), the disease is histologically represented for *A. japonica* as follows:

- Haemorrhagic patches on the body surface or the caudal fin, later lesions and necroses.
- Liver with signs of digestion, spleen swollen and deep red; kidneys necrotic; visceral blood vessels expanded, intestine red with detachment of the epithelium.

Vibrio anguillarum can also occur in pike, rainbow trout and other fishes. In captivity, if no resistance of the pathogen exists, therapy is possible with antibiotics, for example,

Table 7.1 Biotypes of *Vibrio anguillarum* (after Nybelin, 1935; Anderson and Conroy, 1970; Conroy, 1984).

Biotype	Acid from Mannitol	Acid from Saccharose	Formation of Indol	Type name
A	+	+	+	forma *typica*
B	–	–	–	forma *anguillicida*
C	+	+	–	forma *ophthalmica*
D	–	+	+	–
E	–	–	+	–

Table 7.2 Economically damaging eel epidemics caused by fungi, Sporozoa, Ciliata, Helminths and Crustacea.

Parasite	Host animal	Water body	Proportion of animals infested	Sizes and proportion of dead animals	Behaviour	Kind of infestation and damage	Time of infestation, temperature	Treatment and cure	Remarks	Sources
Saprolegnia parasitica	*A. japonica*	eel ponds, Japan	19.4% of 1442 kg		lethargic	heavy fungal growth mainly on head and tail area	III–V 14–21°C Ø 16°C	baths: 20–90 s malachite green 1:20,000 to 50,000; 40 min potassium permanganate 1:10,000; 4–40 min formalin 1:1000 to 1:3000	partial secondary infection with *Pseudomonas*	Hoshina and Ookubo,1956
Myxidium matsui	*A. japonica*	eel ponds, Japan	100 kg per pond	small, 33–38 cm long	almost normal	numerous white cysts on whole body surface (not on head and gills)	VI, 23 °C		duration 20 days, fish very unsightly	Hoshina, 1952
Ichthyophthirius multifiliis	*A. anguilla*	Klingenberg Reservoir, Saxony, very low water level		several hundred specimens, small eels, 250 kg		entire outer skin destroyed and shed	VI	Therapy: Trypaflavin, Chloramin, Ichthyrapid, Salt content elevation		Timmermann, 1939; Schäperclaus, 1954; Reichenbach-Klinke, 1966 Tesch, 1968a
Ichthyophthirius multifiliis	*A. anguilla*	small ponds in Gorleben and Wielenbach		juvenile eels		entire outer skin is destroyed and shed		0.15–0.18 g/m^3 malachite green		Koops, 1965; Bohl, 1968

Pseudo-dactylogyrus sp.	*A.japonica* *A.reinhardtii* *A.anguilla*.	aquaculture		juvenile eels	gill epithelium destroyed, haemorrhages	20–25 °C	Mebendaazol 1 mg/l, Praziquantel 10 mg/l		Buchmann & Bjerregaard, 1990
Goezia sp. larvae	*A. anguilla*	holding facilities, lower Elbe river		20–30 cm	mainly stomach, also intestine, corium, musculature, tubercles	V			Denecke, 1935
Argulus sp.	*A. anguilla*	lagoon ponds, Venice 9–29‰ salinity	9–29% death at >25°C water temp.	Considerable economic consequences	body surface, as many as 1000 specimens per eel, particles in gill cavities	infestation occurred at 10–15 °C	no treatment in ponds; otherwise lysol, potassium-permangenate, vermicide or Neguvon, Trichlorphon	Parasite tolerates as much as 40‰ salt concentration	D'Ancona, 1960

chloramphenicol (which is now prohibited because of harmfulness for man), oxytetracyclene, terramycin and sulfamethazine (Schäperclaus, 1979; Reichenbach-Klinke, 1980). Immunisation is also possible (Anders, 1978).

Chondrococcus columnaris causes injury of the gills in *A. japonica* (Usui, 1974).

7.3 Infection with fungi, protozoa, and metazoa

7.3.1 Examples

Among eel diseases, viral and bacterial infections have the foremost economic and ecological effects. Less important, though significant in individual situations, are fungal, protozoan and metazoan diseases (Table 7.2)

The fungus, *Ichthyophonus hoferi*, is widespread, but no serious symptoms are known in eels.

There is an external fungus that can be called an eel disease, *Saprolegnia parasitica* (Table 7.2), which usually occurs as a result of injury and as a secondary disease. In 1965 and 1968, it was said to have destroyed several hundred tons of eels in Japanese pond culture (Usui, 1974).

In pond culture and aquaculture, the ciliate, *Ichthyophthirius multifiliis* can cause a lot of trouble. This skin parasite of numerous fish species grows to maturity in the skin and reproduces in swarms that reinfect the fish from the beds of water bodies. For this reason, any time fish are held in shallow waters that lack strong current it can be risky. As eels live on the bottom and in cavities and hollows, they are particularly susceptible to this parasite.

A related species, *Cryptocaryon irritans*, occurs in sea water at an optimum of 25–30°C (Cheung *et al.*, 1979), but there is no evidence of eel infestation. Besides that, *Ichthyophthirioides browni* was described, a parasite similar in behaviour and having a rod-shaped, not horseshoe-shaped nucleus.

Besides the treatments shown in Table 7.2, a combination of 0.1 g/m^3 malachite green with 15 g/m^3 formalin is recommended as especially effective (Loyacano and Crane, 1977).

An overview of the rest of the more or less damaging parasites is found in Table 7.3. The number of relevant parasites has increased since the mid-1980s because parasites were brought into Europe, especially from East Asia, by actions that violated safety measures. Included are primarily monogenea, especially from the genus *Pseudodactylogyrus*. These monogenans are particularly significant in the culture of juvenile eels (Table 7.2). Environmental influences also play a role in parasite infestation. Thus, Reimer (1995) attributed the greatly increased infection with *Deropristis inflata* (Digenea, Table 7.3) in the course of the last 40–50 years in the southwestern Baltic Sea to eutrophication. The infection climbed from 3.4% in the 1950s to about 60% at the beginning of the 1990s. Since then, the infestation has declined slightly.

Also introduced into Europe and North America (Fries *et al.*, 1996) since then were swim bladder nematodes of the genus *Anguillicola* (Fig. 7.2), which feed on blood. They damage the swim bladder wall, an organ that is of major importance for equalising pressure, especially in spawning migration. Various authors evaluate the infestation differently. Sprengel and Lüchtenberg (1991) observed an 18.5% reduction in swimming

Table 7.3 Parasite infestations that usually do little damage in eels. In all cases where the host is not indicated, *A. anguilla* was involved. Examples without statement of source were taken from Reichenbach-Klinke (1980).

Kind of parasite	Host, infested body part, geographic location, reference
Fungi	
Saprolegnia parasitica	very frequent (e.g. Table 7.2)
Ichthyophonus hoferi	
Flagellata	
Trypanosoma granulosum	reproduces by fragmenting, Hirudinea are transmitters
Trypanosoma murmanensi	*A.rostrata*,experimental (Margolis and Arthur, 1979)
Spironucleus anguillae	liver, kidney, spleen, muscles, blood, in Poland (Einszporn-Orecka, 1979)
Cryptobia markewitschi	'somnolency', transmitted by Hirudinea (Nolte, 1975)
Sporozoa	
Epieimeria anguilla	*A. anguilla* and *A. rostrata*, intestine, estuaries (Hanek and Molnar, 1974)
Myxosporea	
Myxidium giardi	forms white cysts in the kidney (Cépède, 1906) and for *A. anguilla* and *A. rostrata* also cysts on gills and skin (Ghittino *et al.*, 1974) (Tesch, unpublished), or infestation of liver and intestine (Hanek and Molnar, 1974)
Myxidium zealandicum	*A. rostrata*, Canada (Margolis and Arthur, 1979)
Myxidium enchelypterygii	*A. japonica*, fins, 5–11 white cysts on dorsal or anal fins, Japan (Hoshina, 1952)
Myxidium matsui	(e.g. Table 7.2)
Myxidium truttae	gills, gallbladder and intestine; ponds near Munich (Bohl, 1968)
Myxobolus sp.	*A. rostrata*, gills and skin, Canada (Margolis and Arthur, 1979)
Hoferellus gilsoni	urinary bladder, Belgium and Hungary (Lom *et al.*, 1986)
Myxosoma (Lentospora) dermatobia	*A. japonica*, cysts on entire body surface, Japan, marine (Hoshina, 1952), Russia (Shulman, 1966)
Sphaerospora sphaerocapsulara	urinary bladder, Poland (Wierzbicka, 1986)
Microsporidia	
Plistophora anguillarum	*A. japonica*, muscle tissue, Japan (Usui, 1974)
Haplosporidia	
Dermocystis anguillae	rearing in Japanese ponds, gills (Spangenberg, 1975; Hatai *et al.*, 1979)
Ciliata	
Trichodina anguillae	gills, southern Bohemia (Lom, 1970; Nolte, 1975)
Trichodina epizootica	
Trichodina fultoni	mass infestation in Polish aquarium (Markiewicz and Migala, 1980)
Trichodina jadranica	Russia (Shulman, 1984)
Ichthyophthirius multifiliis	skin, gills (e.g. Table 7.2, see also Mann, 1962)
Monogenea	
Gyrodactylus anguillae	gills, Lausitz, Albania (Nolte, 1975)
Gyrodactylus nipponensis	*A. japonica* gills, Japan (Ogawa and Egusa, 1978)
Pseudodactylogyrus anguillae	*A. japonica* China, Japan, Taiwan, *A. reinhardtii* Australia, since 1983 known in Europe (Molnar, 1984)
Pseudodactylogyrus bini	*A. japonica* Japan, China, Taiwan, *A. reinhardtii* Australia; since 1977 known in northwestern Russia; now south and central Europe to Denmark (Golowin, 1977; Koie, 1988)
Cestoda	
Proteocephalus macrocephalus	*A. anguilla and A. rostrata*, intestine, rivers and coastal waters of Spain, Lake in Hungary, Canadian estuary, Greifswald Bay (Germany) 1958 8.2%; 1996, 15% (Gandolfi-Hornyold, 1929; Baer, 1948; Doby and Jarecka, 1966)

Table 7.3 (*Cont'd*)

Kind of parasite	Host, infested body part, geographic location, reference
Proteocephalus percae	Valencia/Spain (Gandolfi-Hornyold, 1929)
Triaenophorus nodulosus	intestine, Greifswald Bay 5% (Engelbrecht, 1958)
Bothriocephalus claviceps	*A. anguilla* and *A. rostrata*, intestine, Greifswalder Bay 1956 1.5%, 1996 25% (Engelbrecht, 1958 and data of the author); rivers in Spain and Canada, Lake in Hungary (Gandolfi-Hornyold, 1929; Bangham and Venard, 1946; Murai, 1971; Hanek and Molnar, 1974)
Bothriocephalus scorpii	Baltic Sea, Lagoon, 1993–6 (Reimer, unpublished) *A. rostrata* Canada (Margolis and Arthur, 1979)
Bothriocephalus japonicus	*A. japonica*, Japan (Yamaguti, 1959)
Bothriocephalus travassosi	*A. marmorata* Philippines (Yamaguti, 1959)
Diphyllobothrium latum larv.	musculature, Neva Bay Baltic (final host: human) (Dogiel, 1936)
Scolex pleuronectis	Denmark, (Koie, 1988)
Tentacularia spec., larv.	Black Sea (Gaevskaya *et al.*, 1975)
Nybelinia anguillicola	*A. japonica*, encysted in submucosa of the intestine, Japan (Yamaguti, 1952)
Rhynchobothrium heterospine l.	*A. rostrata* (Linton, 1897)
Grillotia erinaceus larva	Rhine (Heitz, 1918)
Digenea	
Diplostomum spathaceum Metac.	Wurmstar, Rügen (Reimer, 1966, 1970), *A. rostrata* Canada (Gibson, 1996)
Diplostomum flexicaudum Metac.	*A. rostrata*, Canada (Gibson, 1996)
Bucephalus polymorphus	fresh water (Manter, 1955)
Azygia longa	*A. rostrata*, Canada, fresh water (Gibson, 1996)
Azygia anguillae	*A. japonica*, Japan (Manter, 1955)
Azygia lucii	Europe (Yamaguti, 1971)
Azygia acuminata	*A. japonica*, Japan (Yamaguti, 1971)
Hemiurus communis	Baltic Sea, Rügen (Reimer, 1966, 1970)
Hemiurus levinseni	*A. rostrata*, Canada (Gibson, 1996)
Brachyphallus crenatus	Baltic Sea, Rügen (Reimer, 1966, 1970) *A. rostrata* (Manter, 1955)
Tubulovesicula pinguis	*A. rostrata* USA, sea water (Manter, 1955)
Tubulovesicula anguillae	*A. japonica*, Japan, sea water (Manter, 1955)
Tubulovesicula angusticauda	*A. reinhardtii*, Australia (Bray *et al.*, 1993)
Lecithochirium rufoviride	sea water (Manter, 1955)
Lecithochirium gravidum	sea water (Manter, 1955)
Lecithochirium microstomum	*A. rostrata* USA (Yamaguti, 1971)
Lecithochirium (=Sterrhurus) grandiporum	sea water (Manter, 1955)
Sterrhurus musculus	intestine, Heligoland (Tesch, 1983) *A rostrata*, Florida (Yamaguti, 1971) *A. japonica*, Japan (Yamaguti, 1971)
Tricotyledonia (=Grassitrema) prudhoei	*A. dieffenbachii* or *A. australis*, New Zealand (Yamaguti, 1971)
Derogenes varicus	*A. rostrata*, NW Atlantic (Stafford, 1904, 1907)
Paracordicoloides yamagutii	*A. reinhardtii*, blood vessels, Australian rivers (Martin, 1974)
Phyllodistomum anguillae	*A. japonica*, China (Yamaguti, 1971)
Phyllodistomum lesteri	*A. japonica*, China (Yamaguti, 1971)
Bunodera luciopercae	*A. rostrata*, Canada (Gibson, 1996)
Crepidistomum cornutum	*A. rostrata*, Canada (Gibson, 1996)
Crepidistomum brevitellum	*A. rostrata*, Canada (Gibson, 1996)
Opegaster anguilli	*A. bengalensis*, India (Manter, 1955)
Opegaster jamnica	*A. bengalensis*, India (Yamaguti, 1971)
Plagioporus angulatus	Baltic Sea, Rügen (Reimer, 1966, 1970)

Table 7.3 *(Cont'd)*

Kind of parasite	Host, infested body part, geographic location, reference
Plagioporus angusticollis	Ukraine (Yamaguti, 1971)
Helicometra fasciata	marine water (Manter, 1955)
Podocotyle atomon	intestine, Baltic Sea, 11.14% (Reimer, 1966, 1970 and unpublished)
Podocotyle ollsoni	*A. rostrata*, sea water (Manter, 1955)
Nicolla allahababensis	*A. bengalensis*, India (Yamaguti, 1971)
Nicolla gallica	Egypt (Yamaguti, 1971)
Nicolla indica	*A. bengalensis*, India (Yamaguti, 1971)
Sphaerostoma bramae	intestine, Baltic Sea estuary, Kleines Haff (Reimer, unpublished)
Crowcocaecum skrjabinii	intestine, Hungarian lake (Murai, 1971)
Proctotrematoides pisodontophicis	*A. japonica*, (experimental) (Yamaguti, 1971)
Deropristis inflata	Baltic Sea, (Reimer, 1966, 1970, 1995) *A. rostrata* North-West Atlantic (Gibson, 1996)
Stegodexamene anguillae	*A. dieffenbachii* and *A. australis* New Zealand, fresh water (Manter, 1955) also brackish water of New Zealand (Hine, 1980)
Stegodexamene callista	*A. reinhardti*, intestine, Australia (Watson, 1984)
Tetracerasta blepta	*A. reinhardti*, intestine, Australia, (Watson, 1984)
Telogaster opisthorchis	*A. dieffenbachii, A. australis*, fresh water of New Zealand (Manter 1955, Hine, 1980a)
Centrovarium lobotes	*A. rostrata*, Canada (Manter, 1955, Gibson, 1996)
Tubanguia anguillarum	*A. marmorata*, Philippines (Yamaguti, 1971)
Microphallus opacus	*A. rostrata*, Canada (Gibson, 1996)
Stephanostomum tenue	*A. rostrata*, intestine, Canada (Gibson, 1996)
Haplorchis anguillarum	*A. marmorata*, Philippines (Manter, 1955)
Nematoda	
Eustrongylides mergorum larv.	lake in Denmark (Koie, 1988)
Contracaecum microcephalum larv., (syn. *C. squali*) and *C. rudolphi*	Germany and Denmark (Schäperclaus, 1979; Koie, 1988)
Hysterothylacium aduncum larv.	Baltic Sea, Greifswald Bay (Engelbrecht, 1958), North Sea, Heligoland (Tesch, 1984), brackish water of Denmark, 6.7% and 11.3% (Koie, 1988), Elbe estuary 5% (Lick, 1991)
Pseudoterranova decipiens larv.	Elbe estuary, 9–11.8% (Lick, 1991; Kerstan, 1992)
Raphidascaris acus	intestine, Baltic Sea: Greifswald Bay 5% (Engelbrecht, 1958) Oder Estuary 13% (Reimer, unpublished), Finnish Gulf 38–75% (Schneider, 1904; Levander, 1909), Salzhaff (Germany) 8.7% (Reimer, 1987) fresh and brackish water of Denmark (Koie, 1988)
Goezia sp.	(see Table 7.2), Elbe estuary (Lick, 1991)
Paraquimperia tenerrima	intestine, Great Britain, Germany, Denmark, Spain, Czech area (Moravec, 1966; Koie, 1988)
Paraquimperia aditum (syn. Haplonema aditum)	*A. rostrata*, Canada (Margolis and Arthur, 1979)
Paraquimperia anguillae	*A. bengalensis*, intestine, India, Poona and Nagpur (Moravec, 1966)
Haplonema hamulatum (syn. *Ichthyobronema gredini*)	intestine
Cucullanus anguillae	*A. japonica*, China (Puqin and Xiumin, 1976)
Johnstonmawsonia sp.	*A. rostrata*, freshwater Guadeloupe (Fetter *et al.*, 1977)
Ortleppina longissima	*A. mossambica*, South Africa, Buffalo river (Jubb, 1961)
Campanarougetia campanarougetiae	*A. marmorata*, intestine, South Viet Nam (Hoa and Khue, 1968)
Procamallanus armatus	*A. bicolor*, Madagascar (Campana-Rouget and Therezien, 1965)
Camallanus lacustris	intestine

Table 7.3 *(Cont'd)*

Kind of parasite	Host, infested body part, geographic location, reference
Camallanus truncatus	lake, Hungary (Murai, 1971)
Philometra abdominalis	body cavity, Elbe, frequent (Mann, 1962)
Philometra ovata	fresh water of Denmark, 10% (Koie, 1988)
Philometroides anguillae	*A. japonica*, eye (Ishikawa, 1916)
(syn. *Filaria anguillae*)	*A. reinhardtii*, Queensland (Ivashkin *et al.*, 1977)
Ascarophis arctica	Denmark, Isefjord (Koie, 1988)
Cystidicoloides tenuissima	*A. rostrata*, Canada (Margolis and Arthur, 1979)
Metabronema salvelini	*A. rostrata*, stomach, gut Canada (Margolis and Arthur, 1979)
Spinitectus inermis	Slovakia (Moravec, 1977)
Hedruris spinigera	*A. dieffenbachii, A. australis*, intestine, New Zealand, river and brackish water, temporary transmission by prey fish (Hine, 1980a)
Rhabdochona aegypticus	Nile river, Egypt (El-Naffar and Saoud, 1974)
Rhabdochona anguillae	Bulgaria (Kakaceva-Avramova, 1983)
Paracuaria adunca larv.	Larvae of gull nematodes, Denmark (Koie, 1988)
Cosmocephalus obvelatus larv	adult in gulls, Denmark (Koie, 1988)
Daniconema anguillae	swimbladder, Denmark, especially in Lake Esrum (Moravec and Koie, 1987)
Anguillicola crassus	*A. japonica*, swimbladder, Japan, China, Korea, since 1982 in Europe: Italy, Hungary, Austria, Germany, France, Belgium, Netherlands, Great Britain, Denmark, Sweden, Czech Republic, Poland, Russia, Portugal, Spain and Greece. Infestation up to 90% (Moravec and Taraschewski, 1988; Koie, 1988)
Anguillicola globiceps	*A. japonica*, swimbladder, Japan and China (Nagasawa *et al.*, 1994; Kim and Hirose, 1994)
Anguillicola australensis	*A. reinhardtii*, Australia (Moravec and Taraschewski, 1988)
Anguillicola novaezelandica	*A. dieffenbachii,* New Zealand (Moravec and Taraschewski, 1988)
Anguillicola papernai	*A.mossambica*, SE Africa (Moravec and Taraschewski, 1988)
Pseudocapillaria tomentosa	intestine, Denmark (Koie, 1988), Baltic Sea (Möller, 1975a)
(*Capillaria* spec.)	Weser (Reimer, unpublished)
Acanthocephala	
Neoechinorhynchus rutili	intestine, Oder estuary 6% (Engelbrecht, 1958) Finnish Gulf 8–25% (Schneider, 1904; Levander, 1909)
Pseudoechinorhynchus clavula	intestine, Germany, France, Sweden, intermediate hosts *Gammarus pulex*, *Pontoporeia affinis* (Golvan, 1969)
Metechinorhynchus salmonis	intestine, Eurasia (Golvan, 1969)
Metechinorhynchus coregoni	*A. rostrata*, intestine, North America, Great Lakes (Yamaguti, 1963a)
Acanthocephalus clavula (syn. *Metechinorhynchus clavula*)	rivers of Spain (Gandolfi-Hornyold, 1929)
Acanthocephalus anguillae	Europe, Greifswald Bay 0.7%, Oder estuary 60% (Engelbrecht, 1958) Oder estuary 1982: 1% (Reimer, 1987)
Acanthocephalus lucii	intestine, Finnish Gulf 25–54% (Schneider, 1904; Levander,1909), England 75% (Mishra and Chubb, 1969), Baltic Sea, Oder estuary 19% (Engelbrecht, 1958) 1% (Reimer, 1987)
Acanthocephalus jacksoni	*A. rostrata*, intestine, northeastern USA, intermediate host *Asellus* spec. (Yamaguti, 1959)
Acanthocephalus gotoi	*A. japonica*, intestine, Japan (Yamaguti, 1959)
Acanthocephaloides propinquus	intestine, Mediterranean Sea, French Atlantic coast (Golvan, 1969)
Acantocephaloides incrassatus	intestine, Mediterranean Sea, Black Sea (Golvan, 1969)
Leptorhynchoides thecatus	*A. rostrata*, Canada and northern USA, intermediate host *Hyalella azteca* (Golvan, 1969)
Echinorhynchus lateralis	*A. rostrata*, intestine, Canada (Arai, 1989)

Table 7.3 *(Cont'd)*

Kind of parasite	Host, infested body part, geographic location, reference
Echinorhynchus gadi	intestine, Kiel Bay (Germany) (Möller, 1975a), Wismar Bay 3%, Greifswald Bay 5%, western Oder estuary 2.4% (Reimer, 1987) 1996 11.8% (Reimer, unpublished)
Pomphorhynchus laevis	intestine, Greifswald Bay 6% (Engelbrecht, 1958), 18–22% (1954–63 Reimer, unpublished), western Oder estuary 1.5% (Reimer, 1987), Baltic Sea 61% (Pohl, 1956), Kiel Firth 2.7% (Möller, 1975a), North Sea coast 5% (Saint Paul, 1977)
Longicollum alemniscus	*A. japonicus*, intestine, Taiwan (Golvan, 1969)
Paratenuisentis ambiguus (syn. *Tanaorhamphus*)	*A. rostrata*, intestine, east coast of USA (Manter, 1955), now also *A. anguilla*, Weser (Taraschewski *et al.*, 1987), brought in with intermediate host *Gammarus tigrinus*
Telosentis australensis	*A. reinhardtii*, Australia (Golvan, 1969)
Corynosoma semerme larv.	Baltic Sea, Isle of Hiddensee (Reimer, unpublished)
Hirudinea	
Cystobranchus respirans	Poland (Sitowski, 1937)
Piscicola geometra	Denmark (Koie, 1988)
Mollusca	
Anodonta spec., *Glochidium*	Denmark (Koie, 1988)
Crustacea	
Argulus foliaceus	(example: Table 7.2) Denmark, Kesrum Lake 8.3% (Koie, 1988)
Ergasilus gibbus	gills, Eurasia, Denmark 8.3–56.7% (Koie, 1988) Baltic Sea, Salzhaff 45–75% (Reimer and Walter, unpublished)
Ergasilus fryeri	gills, Bafa Lake, Turkey, 60%
Ergasilus sieboldi	Denmark, Sjaelso 25% (Koie, 1988)
Ergasilus caeruleus	*A. rostrata*, USA and Canada (Margolis and Arthur, 1979)
Ergasilus celestis	*A. rostrata*, Canada (Margolis and Arthur, 1979)
Lernaea cyprinacea	*A. japonica* and *A. anguilla* oral cavity and skin, Japan, China (Yamaguti, 1963b; Usui, 1974)
Lernaeocera branchialis Copepodite stage	Denmark, Kattegat 13.9% (Koie, 1988)

movement compared with uninfested specimens. In contrast, Möller *et al.* (1991) detected no changes in condition factor for eels having up to 20 parasites each. Nagasawa *et al.* (1994) reported hardly any inflammations, no pathological changes and no clear differences in condition factor in *A. japonica*. According to Egusa (1978, 1979), much injury occurred in European eels reared in Japan. The diseased eels' abdomens were swollen by enlargement of the heavily infested swim bladder. They lost appetite and liveliness and became emaciated. When the swim bladders burst, they died.

Heavily infested *A. anguilla* specimens had heightened susceptibility to bacterial infections and greater mortality (Hartmann, 1987). The swim bladder wall can become inflamed and thickened. It can also burst when heavily infested (Mellergaard, 1988), as Egusa (1978) had already observed. Koie (1991) and other authors (e.g. Hartmann and Nellen, 1998), doubt that eels heavily infested with *Anguillicola* are successful in their spawning migration to the Sargasso Sea. For infested eels, Fontaine *et al.* (1990) studied the functional ability of swim bladders in experiments with such extreme pressure as the

Fig. 7.2 The nematode *Anguillicola crassus* in an eel swimbladder. The largest specimen is female; middle-sized specimen male; smallest specimen pre-adult (photo: Hartmann).

eel would face in diving to a depth of 800 m. Damage from *Anguillicola* was judged to be insignificant because performance of the experimental animals did not differ from that of uninfested eels. In The Netherlands, juvenile eels 6–15 cm long had 80–100% infestation, and 60% of these showed inflammation of the swim bladder (Banning and Haenen, 1990).

For farmed eels, loss of appetite, floating at the water surface, open skin sores and a red, swollen anus were observed. After the swim bladder bursts and after *Aguillicola crassus* specimens occupy the body cavity, there are secondary bacterial infections by *Aeromonas hydrophila*, and in marine aquaculture, by *Pseudomonas* sp. Thus, direct or secondary consequences can raise mortality by about 10–20% (Liewes and Schamine-Mann, 1987). Knopf *et al.* (1998, 1999) observed that low water temperatures slowed the development of *Anguillicola crassus* larvae and adults. After 4 months at 4°C, third-stage larvae were unable to invade the swim bladder, and adults were damaged. From this, the authors inferred that temperature determined a natural distribution boundary for Europe.

Reimer *et al.* (1994) found an infestation of small fishes with *Anguillicola crassus* larvae in the Baltic Sea. Whereas copepods serve as the first intermediate hosts, numerous fishes have been recorded as second intermediate hosts, such as *Abramis brama*, *Osmerus eperlanus*, *Gymnocephalus cernuus*, *Perca fluviatilis*, *Stizostedion lucioperca and Gasterosteus aculeatus* (Haenen and van Banning, 1990). It was added that infestation was 9.6% for *Syngnathus typhle* and 0.7% for *Gobius niger*.

No major damage is yet known from the likewise introduced acanthocephalan, *Paratenuisentis ambiguus*, which came from North America (Table 7.3).

Hine and Kennedy (1974) observed for *A. anguilla* and some other fishes in England's river Avon, in contrast to *Leuciscus cephalus* and *Barbus barbus*, that the specimens of *Pomphorhynchus laevis* do not mature. There is no confirmation of this for the Baltic Sea (Engelbrecht, 1958; Möller, 1975b; Koie, 1988a; Reimer, unpublished). It must be tested whether Hine and Kennedy's (1974) data can be generalised.

7.3.2 Parasite infestation and eel origin

Some conclusions should be discussed, based on data in Table 7.3. Manter (1955) tried to use parasites to make statements about the origin of species of the genus *Anguilla* in the Atlantic and Pacific Oceans. He found 20 species of trematodes in Atlantic eels, 14 of which occurred in other marine fishes, five in freshwater fishes, and only one species in eels alone. In the Pacific Ocean, on the other hand, three genera of trematodes (eight species) occur only in eels. The large number of shared parasite species for Atlantic fishes is attributable to longer residence in the ocean. These statements cannot be refuted in principle but should be modified.

First, regarding origin of the genus *Anguilla*. The geographic centre, where this process could have gone on was presumably the Tethys, the so-called tertiary Mediterranean Sea. Based on the former distribution, a spreading of the eel species to their present range is conceivable (Fig. 7.3). Manter (1955) identified the eel-parasitising trematodes in particular. If we add in cestodes and nematodes from the helminths, then, aside from exceptions, parasite species are involved that are included with such secondary intermediate hosts as polychaetes, bivalves, crustaceans (especially amphipods and isopods) and with small benthic fishes. The higher number of marine species is indeed attributable to the fact that in such European waters as the North and Baltic Seas, eels can remain longer during their feeding stages under conditions of higher salinity. Thus, this has nothing to do with the longer spawning migration, during which they no longer consume food. In the southern part of the Baltic Sea, it can even be seen (Reimer, 1970, 1987) that the eels have a

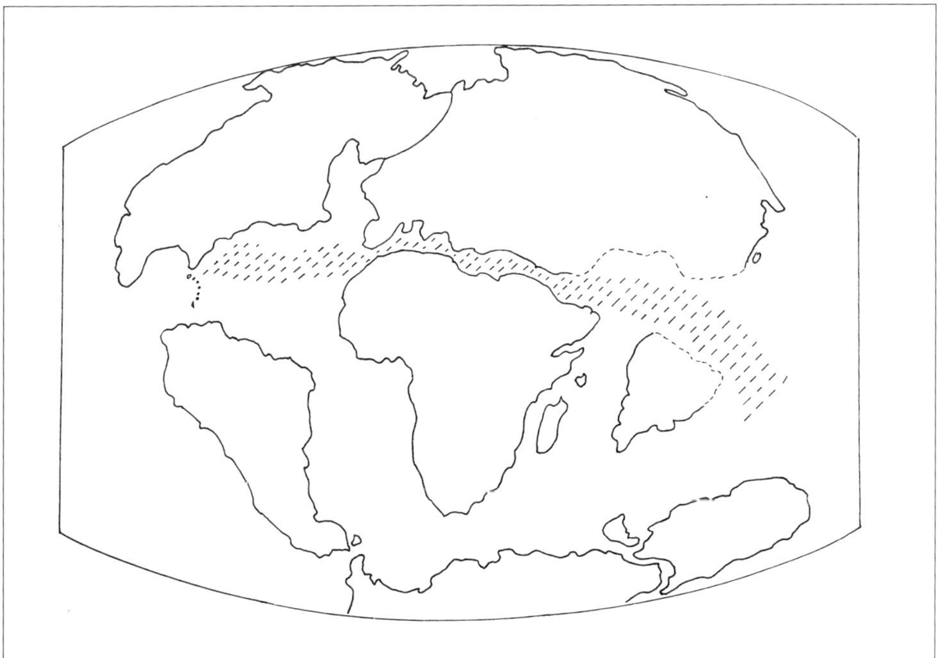

Fig. 7.3 Tethys 65 million years ago (cretaceous period/tertiary) and presumed area where the genus *Anguilla* developed (after Reimer, 1987; map based on Dietz and Holden, 1970)

certain eastward drift. In various areas they contain parasite species, whose intermediate hosts occur only further to the west (*Deropristis inflata*). On the other hand, the immigration of eels from fresh water is made clear by abundance of the cestode species, *Proteocephalus macrocephalus* and *Bothriocephalus claviceps*, and by infestation with the copepod, *Ergasilus gibbus*.

What do the two Atlantic eel species have in common, and what differentiates them? In considering the parasite species that occur in *A. anguilla* or *A. rostrata*, this narrows the list in Table 7.3. Among protozoans these are *Epieimeria anguillae* and *Myxidium giardi*; among cestodes; *Proteocephalus macrocephalus* and partially *Bothriocephalus claviceps*. For *Anguilla rostrata* in North America, in contrast to *A. anguilla*, there is still a number of other species known to host the cestodes just mentioned. Among trematodes, only *Deropristis inflata* can be mentioned as a shared species. However, this shared characteristic of freshwater cestodes, the marine trematodes of the genus *Deropristis*, and the above protozoans supports the theory that the two Atlantic eel species had a common origin and developed in parallel with the splitting of the northern continents from Laurasia.

Can parasites also be brought to bear in clarifying relationships of the rest of the species in the genus *Anguilla*? As Manter (1955) already determined, most eel parasites are not species specific, that is, they occur simultaneously in quite different species of fish.

Among cestodes, the genus *Bothriocephalus* was just mentioned. Whereas *B. claviceps* occurs in the Atlantic Ocean, we find *B. travassosi* in *A. marmorata* in the Philippine Islands, and *B. japonicus* in Japan's *A. japonica*. The last two *Bothriocephalus* species mentioned show a high degree of species specificity.

In the trematode genus *Azygia*, several species occur in *Anguilla*:

- *Azygia lucii* in *A. anguilla*;
- *Azygia longa* in *A. rostrata*;
- *Azygia acuminata* and *Azygia anguillae* in *A. japonica*.

However, except for *Azygia anguillae*, these trematodes show no pronounced species specificity.

For nematodes there are two genera to be considered in this respect. First, the genus *Anguillicola*. If we ignore the part of its present distribution that resulted from human-generated transfer, Asiatic commercial eels have the following parasitic species:

- *Anguillicola crassus* and *A. globiceps* are found in *Anguilla japonica*;
- *Anguillicola australiensis* is in *A. reinhardti*;
- *Anguillicola novaezelandica* is in *Anguilla dieffenbachii*.

From the standpoint of host choice, the nematode genus *Paraquimperia* exhibits a more occidental range:

- *Paraquimperia anguilla* has *Anguilla bengalensis* as a host;
- *P. tenerrima* has *A. anguilla*;
- *P. aditum* has *A. rostrata*.

However, the data of Cone *et al.* (1993) that *Paraquimperia tenerrima* are also evident in *Anguilla rostrata* in Canada's Nova Scotia, fall outside this scheme. A misidentification may be involved here.

Unfortunately, for some *Anguilla* species the parasites are known insufficiently or not at all, so a comprehensive analysis is not possible. But based on present knowledge it can be inferred that, among the Indopacific eels, *A. bengalensis* is parasitologically the most consistent with the two Atlantic eel species, and *A. mossambica* is much less so. This favours the Tethys theory, not that of a movement around South Africa. There is parasitological consistency between *A. japonica* (Japan, Korea and China) and *A. reinhardti* of the Australian region. Within this region, of the two Australian species, *A. australiensis* has greatest parasitological consistency with the New Zealand species, *A. dieffenbachii*; large populations of *A. australiensis* likewise occur in New Zealand. *A. marmorata* shows a certain parasitologic independence.

7.4 Deformities, anomalies and other damages

As in other fishes, morphological changes of various body organs also occur in eels, and the causes cannot always be explained. Frequently, they are results of diseases from which eels recovered, but often are also effects of injuries that other species of fish would not be expected to survive. Deformed organs and body parts are, therefore, relatively frequent in eels. In other cases that do not involve such injuries, developmental disruptions or serious anomalies can exist instead.

Regarding changes of inner and sensory organs that may have been caused by diseased conditions, the following can be mentioned:

- kidney tumour (Plehn, 1924; Scheer, 1934);
- ovarian cyst (Eichler, 1935);
- cyst-shaped fibromas in the body cavity (Schäperclaus, 1954; Mattheis, 1964b);
- mesenterial-fibrosarcomas (Wolff, 1912; Plehn, 1924; Mattheis, 1964b);
- swimbladder inflamation (Flemming, 1954);
- formation of a stone in the body cavity in two different cases (Reichenbach-Klinke, 1956; Tesch, 1958);
- macrophthalmia or exophthalmia (Mercier and Poisson, 1927; Vladykov, 1973).

Unfortunately, in most of these cases only the pathological–morphological situation has been described, and no cause for the disease whatsoever.

The virally caused cauliflower disease has already been discussed. Formation of adenomatose polyps of the stomach's mucus membrane is known in aquacultural *A. japonica* in Japan. They develop as bushy outgrowths on ridges of the stomach folds, and are the size of a drawing pin head (Kubota *et al.*, 1974; Kimura *et al.*, 1976). Dry pellet feeding and aflatoxin among other things are being discussed as causes.

Kidney tumours in eels are supposed to occur more frequently, as can be seen from the cited literature. In the previous case, growths of the kidney's connective tissue were involved, these being so severe that the actual kidney tissue was completely replaced.

Ovarian cysts appear to be of similar character, and the same goes for swimbladder inflammation, in which the swim bladder was enormously enlarged and was filled with spongy tissue and fluid instead of gas.

Concerning cyst-shaped sarcomas, the majority seem to have involved benign tumours, as their causes are often regarded as local processes, such as wounds and excessive regeneration. Remarkably, the regenerations accounted for 34–49% of body weight, and the nutritional condition of the affected eels was normal for the most part (Mattheis, 1964a).

It is difficult to explain the two cases of a stone (calcium phosphate) found in the body cavity. One stone had a maximum diameter of 13.5 mm, the other 20 mm. It is to be considered whether the stones had developed from ovarian cysts. Reichenbach-Klinke (1956) spoke of a 'stone formation on the inner wall of the body cavity'.

Exophthalmia can result from various diseases.

7.4.1 Gross spinal abnormalities

Spinal contortion (plecospondyly), or wavy curvature, is among the best known deformities of eels. Schräder (1930) alone studied 11 cases. Five other such abnormal eels were described by other authors (Scheuring, 1927; Schäperclaus, 1954; Stolck, 1956; Kokhnenko, 1958; Wunder, 1969). The unusually high number of cases known for eels is presumably because a wave-like vertebral column is more obvious in the elongated eel than in other fishes.

The spinal column's undulating form is expressed in the body as well (Fig. 7.4A). The individual vertebrae are deformed. Wunder (1969) explained this kind of deformity as stemming from inhibited growth of the body's main tube due to injury undergone by the juvenile eel. The normal number of vertebrae still forms, and the growing vertebral column must bend into many curves so as to fit the insufficient available body lumen. It might be mentioned here that electrofishing with pulsed direct current, often causes spinal deformation in fishes, so this can be another reason for crooked eels (White, personal communication). Concerning frequency of this affliction, there are data from the North Sea mud flats that 0.2% of 598 eels examined had crooked spinal columns (Anders and Möller, 1991).

Spinal shortening, which is much rarer, has other characteristics (Wunder, 1968). In certain regions of the body the vertebrae are compressed, shortened and heightened (Fig. 7.4C); the total vertebral count for this specimen was 110, as compared with 114 for normal eels, and the heightened appearance of the eel's body is reminiscent of its larval body form (Figs 2.2 and 2.3).

This sort of eel deformity also certainly resembles the compressed vertebrae and associated body shape of carp, caused to some extent by artificial breeding. For the eel it is, therefore, attributed to a disruption of vertebral column differentiation by presumably substantial causes during a very early developmental stage.

7.4.2 Colour anomaly

Reports of colour anomalies are not infrequent (e.g. Pavesi, 1894; Neubaur, 1924; Rumphorst, 1929; Thumann, 1953; Jones and Pantulu, 1954; *Fisch und Fang* **8**, 96, 1967;

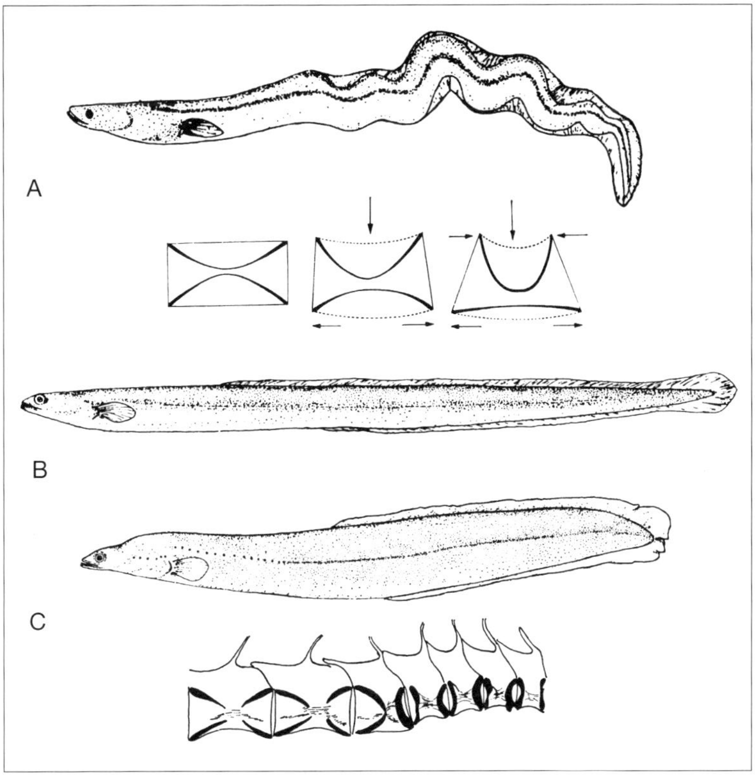

Fig. 7.4 Eels with deformed spinal columns (after Wunder, 1968, 1969)

A Wave-like contortion and vertebrae in three conditions
Left: normal vertebra
Middle: vertebra extended by pressure on one side and by tension on the other side
Right: somewhat more strongly pronounced distortion of a vertebra
B normal eel
C eel with shortened vertebrae, and a section of the spinal column

Schulze, 1980). The author also knows of further cases. The anomalies range from nearly white animals (albinism: Walter, 1910) through yellow (xanthochromatism) and golden-yellow specimens, to eels that have black spots or marbling.

As in abnormally light-coloured animals there is insufficiency of melanophores or chromatorphores, or of their pigments, it can be assumed that a heritable anomaly exists, and that this is possibly brought about by mutation. However, Thumann (1953) reported on four xanthochromatic eels that were held for a long time in a brightly lit aquarium. After 8–10 weeks, these animals took on the colouration of normal eels, and did not regain their previous colouration when held in darkness. Thus, it appears that the xan-throchromism was only temporary (see also Ehrenbaum, 1930), and it is doubtful that such colour anomalies can be considered generally as heritable.

Note that the term 'yellow eel' is used for a normal developmental stage in eels; thus, the abnormally yellow-coloured eel is quite another matter. The same applies to the term 'golden eel', which is sometimes used in English instead of 'yellow eel'.

7.4.3 Deformities caused by injury

There is a great diversity of deformities caused by injuries during relatively late stages of development. They range from alterations for which the cause is uncertain, to clearly identifiable wounds.

Among the deformities very probably attributable only to a wound is constriction of the stomach and its transformation into a fluid-filled balloon (Wunder, 1967). Injury by ingested fish-hooks is regarded as the cause. Presumably, the stomach fundus is torn and closes off into a functionless sac or stomach cyst. The cardiac and pyloric section fuse into a shortened but functional stomach.

Even a case of haemoglobin deficiency was caused presumably by a fish-hook (Steen and Berg, 1966). The corpuscles of Stannius enlarged due to artificial injury (Lopez and Fontaine, 1967).

Unfortunately, eels are often caught that have swallowed fish-hooks. It cannot be emphasised enough that anglers and long liners should reduce such cases to a minimum by using proper leader material and secure knots. During electrofishing surveys in the middle reaches of the Weser river, the author often caught such damaged specimens (Fig. 7.5). Close examination of the photo will reveal that the synthetic leader still protrudes from the eel's mouth and has caused a wound in the corner of the mouth, which has already healed with a highly pigmented scar.

7.4.4 Damage by turbines

Updated by R. Berg

Reports and complaints about damage to migrating eels by hydropower turbines are numerous and date back as far as the power plants themselves (e.g. Lundbeck, 1927; Kroezus, 1954; von Raben, 1955, 1958; Butschek and Hofbauer, 1956; Schiemenz, 1958, 1960a; Müller, 1962). Eels are often injured in turbines not only because of their elongated shape, but especially because as obligatory migrants they follow the main current and inevitably pass through hydropower facilities. Screens installed in front of these facilities are limited to turning away fish of the usual body shapes and usually cannot hold back

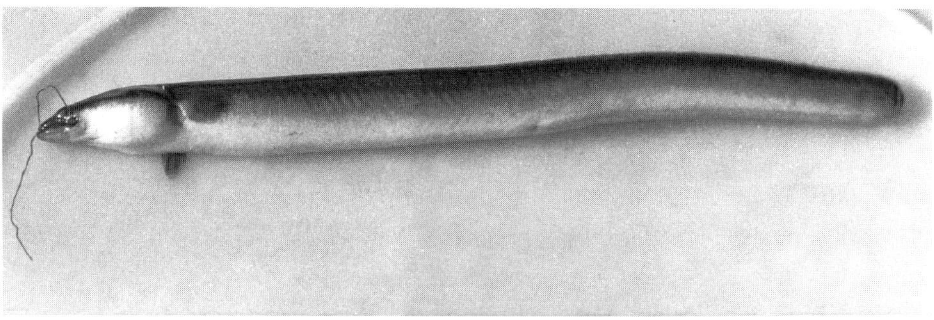

Fig. 7.5 An eel from the middle section of the Weser with part of its tail cut off and containing an angling hook and broken leader (photo: Tesch)

eels because they are so thin. Only if the screen has bar spacing <2 cm will it be effective in blocking emigrating eels. In large rivers, as a rule, power plant screens with such bar spacings are not used and cannot all by themselves suffice as the suitable device for reducing injury. Recent studies showed that great numbers of migrating silver eels can perish on power plant screens that have such narrowly spaced bars if no alternative possibilities for emigration exist, such as bypass channels (Rathcke, 1994; Berg, 1996a).

Knowledge about turbine injuries to eels is based mainly on studies of Kaplan turbines, which predominate in run-of-river power plants. The percentage of lethally injured eels at these power plants lies between 5% and about 75%, depending on construction design and the timing of the study (Rathcke, 1995). Studies in southern Sweden even yielded injury rates of 40–91% (Svensson, according to Müller, 1962). The degree of damage is highly dependent on streamflow discharge and the consequent adjustment of the turbine vanes. Besides that, the circumference speed of the turbine rotor is crucial. The importance of other technical characteristics like reservoir head, intake conditions or positions of the turbine are largely unclear. Having numerous power plants in close sequence substantially reduces silver eel emigration in many river regions (Berg, 1987a). In studies of run-of-river power plants in southwest Germany, more favourable possibilities for emigration were evident during high water events. Under those conditions turbine vanes are opened wider, the facility's operation is often restricted, with the high water flows bypassing it, thus opening up otherwise non-existent possibilities for emigration. Lethal injuries in turbine facilities that were studied on the Neckar river declined a great deal at least temporarily during high water (Berg, 1996b), the estimates ranging upward from <5%. The highest number of emigrant eels occurred in autumn, especially during such flow conditions. Due to methodological difficulties, exact determinations of damage to eels have not yet been undertaken during severe high water flows. Therefore, knowledge about the actual extent of damage is very incomplete and remains unsatisfactory for estimating damage adequately.

Several authors tried to characterise and estimate probability of injury rates with formulae based on turbine-specific rating values and prevailing flow conditions (von Raben, 1958; Monten, 1985; Larinier and Dartiguelongue, 1989). To some extent, satisfactory percentage values for the injuries that occurred were determined by means of the calculation procedures used. Sometimes the calculations estimated injury rates adequately. However, because the estimation of total losses depends not only on technical characteristics, but also on the number of eels migrating at the particular time, the extent of injury cannot be quantified by means of theoretical formulae based on technical ratings.

In some studies, rather than trying to further refine estimates of injury rate, the investigators emphasised testing of methods to reduce it. This involved trying to guide eels away from power plant intakes by means of frightening devices installed to augment the screens. Submerged light barriers, electro-barriers and producing noisy gas explosions in front of the power plant intake were unsuitable for this (Halsband, 1989; Hadderingh, 1993; Berg, 1996b) to the extent that all the discharged water was directed through the turbines during night-time, as is usually done. Therefore, in, the meantime, attempts to minimise injury involve bypasses for emigrating eels or temporarily reducing operations during the main migration period.

7.4.5 *Effects of pollutants on eels*

K. Söffker

Studies of the effect of cadmium on juvenile stages of the European eel revealed structural and functional alterations of the gills, hence an influence on osmoregulation (Lemaire-Gony, 1993). These injuries proved to be reversible, but the liver lesions that appeared were, in contrast, irreversible. According to experiments by Lemaire-Gony and Lemaire (1992), cadmium that was administered with the water produced perivascular fibroses, whereas, one-time injection of benzo(a)pyren led to decrease in glycogen content and to accumulation of lipid globules. The livers of eels exposed to both cadmium and benzo(a)pyren showed complete disorganisation of the parenchyma and nuclear degeneration.

Hwang *et al.* (1996) dealt with the influence of copper and zinc on isolated hepatocytes. They determined that the amount of accumulated metals and the mortality of the hepatocytes increased with rising metal concentration if only one of the two metals was applied in the experiment. In contrast, if copper and zinc influenced the liver cells simultaneously, an antagonistic effect could be observed; the amount of the two metals accumulated then appeared to be less significant.

Exposure to chemical waste waters of the Rhine damaged eel spleens, reducing this organ's main functions. Changes in the mitochondria, lysosomes, endoplasmic reticulum, cytoplasm and blood cells were detected (Spazier *et al.*, 1992). PCBs may remain as toxic contaminants in the lipids of fasting and migrating eels, which may have noxious effects during migration and maturation. However, no noxious effects were found during the experiment (Duursma *et al.*, 1991). In contrast, administration of Aroclor 1242, a technical mixture of polychlorinated byphenyls, affected the heart rate of the experimental eels (Mourad *et al.*, 1992). Lindane mainly caused bradycardia and hyperventilation, subsequently a diminution of ventilation rate, but this substance also influenced the cardiovascular and respiratory systems, leading to considerable damage to heart muscle (Mourad, 1990, 1991). Waste water from the cellulose processing industry triggered kidney damage and skin alterations in yellow eels (Santos *et al.*, 1990a). Likewise, skin lesions were noted after application of mercury (Santos and Hall, 1990b).

In juvenile Japanese eels, various heavy metals disrupted growth, cadmium being a greater hindrance than copper and zinc (Yang and Chen, 1996). The organophosphate insecticide, Diazinon, also showed toxic properties *vis-à-vis* fish; for eels, even relatively low concentrations caused vertebral malformations (Sancho *et al.*, 1993).

Carbohydrate metabolism can be significantly disrupted by harmful substances, such as the pesticides Fenitrothion (Sancho *et al.*, 1996), Lindane (Ferrando and Andreu-Moliner, 1991a) and Endosulfan (Gimeno *et al.*, 1994, 1995), likewise heavy metals (Santos and Hall, 1990a, b). Among other problems in exposed eels, as opposed to unexposed controls, the following arose: reduced glycogen content of the liver or musculature, elevated glucose content in blood and muscle tissue, and decreased lactate in gills, liver, musculature and blood. Beyond that, the composition of the blood changed as follows: after exposure to waste water from, for example, the cellulose processing industry, the red blood cell count increased (Santos *et al.*, 1990), whereas, heavy metals, above all cadmium, and to some extent also lead and mercury, caused, on the one hand decreases in erythrocyte count, haemoglobin, and haematocrit and on the other increases in lymphocytes, leucocytes and leucocrit (Santos and Hall, 1990a, b; Gill and Epple, 1993; see also Halsband and Halsband, 1971). Cholesterol values fell as a result of influence by mercury or Lindane (Santos and Hall, 1990b; Ferrando *et al.*, 1991).

Environmental contaminants can also influence ATPase activity. *In vitro* tests of cadmium on gill membrane microsomes revealed an inhibition of Mg^{2+}-ATPase; the decrease amounted to as much as 40% (Lemaire-Gony and Mayer-Gostan, 1994). Fent and Bucheli's (1994) studies of fishes, among them eels, showed that tributyl tin and triphenyl tin hampered EROD-activity in proportion to the concentrations of these two organo-tin compounds. In contrast, van der Oost *et al.* (1991) could show in eels from Dutch waters that the harmful substances occurring there induced malfunctional oxygenases of the 3-methylcholanthren and phenobarbital types.

Within the organism, it is generally known that various environmental contaminants are transformed in part into reactive compounds that can react with various macromolecules of the cells, also with DANN, and have mutagenic, even carcinogenic effects. In livers of eels from Dutch waters, for example, aromatic DNA-adducts were detected, and, in fact, significant correlations existed between the concentration of aromatic DNA-adducts and loading of sediments with polycyclic aromatics (van Schooten *et al.*, 1995). Residues of DDT, including metabolites, hexachlorbenzol, polychlorinated biphenyls, polychlorinated dibenzofuranes and dibenzo-p-dioxins, were likewise taken up by eels with the sediment burden (van der Oost *et al.*, 1996; see also Janicki and Kinter, 1971). Further results on contamination of eels are found in the following, among other sources: Reynier *et al.* (1970); Holmberg *et al.* (1972), Muzykiewicz (1978), Ogata and Miyake (1979), Wegrzynowicz *et al.* (1979), Narva and Engelhardt (1980), Barak and Mason (1990), Ferrando *et al.* (1992), de Boer *et al.* (1993), Hendriks and Pieters (1993), Atuma *et al.* (1996), Batty *et al.* (1996), Linde *et al.* (1996) and Tulonen and Vuorinen (1996).

8 World trade and processing

Eel prices are high and, therefore, their worldwide economic value is substantial. Total world eel harvest from fishing and aquaculture amounted to about 205,000 t in 1995 (FAO, 1997b). Of this, 188,401 t were aquacultural production, at an estimated wholesale market value of US$3.1 billion, which made up about 12% of the entire product value of the world's aquaculture that year. In Europe, where, in the mid-1990s, the amount of eels taken by fishing significantly exceeded aquaculture production, a committee of the European Inland Fishery Commission (FAO and ICES) estimated that continent's eel harvest at 20,000–30,000 t, valued at DM 351 million or *c.* US$225 million (January 1995 conversion) – instead of the official 14,000 t reported harvest (Moriarty, 1996). The 800–1200 t of glass eels harvested were valued at DM 116 million or US$74 million (Kuhlmann, 1998).

Eels are traded in various forms, the weights of which must be multiplied by the following factors to convert to live weight (after Jellyman, 1993):

- frozen: 1.0;
- dressed: 1.1;
- dressed without gills: 1.2;
- dressed without head: 1.5;
- fillet: 2.5;
- smoked: 1.5;
- smoked half-eel: 2.3;
- smoked fillet: 3.0;
- skinned: 1.5.

8.1 International trade

Next to salmon and trout, there is probably hardly any other kind of fish that undergoes so much worldwide marketing, live or deep-frozen, as the eel. With respect to live shipping (Section 3.1.3), it probably ranks as the absolute leader. The first reports about intercontinental shipping stemmed from 1923 (*Fischerbote* **17**, 64–65, 1925 and **21**, 357, 1929), when eels were sent live from Canada to Holland and Germany. However, only about 100 kg were involved, whereas, Canadian exports into the USA then amounted to 7000 kg. After 1949, the main importers of Canadian eels, besides the USA, were West Germany and the Benelux countries.

The first shipments from Canada to Europe in the 1920s were received with much uneasiness. The German market was not so import oriented in those days, thus it feared competition. After World War II, people were much more relaxed about such external influences. In West Germany, only about 10% of the eels consumed came from within the country.

Fig. 8.1 Production of Japanese-style fried eels on a conveyor belt at the earlier Kingman firm, not far from Canton, China. In 1996, with two facilities, this fishery business was capable of processing 8000 t of eels, which had been produced in its own ponds (photo: Fa. Kingman).

In 1955, West German eel imports amounted to 2272 t, then climbed to 5000 t in 1978–79, at which time Germans were eating the most eels in Europe. Table 8.1 shows the countries that recently have been especially active in importing and exporting live eels, which are, of course, only a share of all eels marketed. From this it can be seen that Japan imports the most eels of any country, and it is known that these, combined with Japan's inland production, account for the world's highest consumption, too. The exporters to Japan are Taiwan and China (Table 8.1). Japan's actual consumption is even greater, if imports of *processed* eels from China and Taiwan are taken into account as well. Figure 8.2 shows an eel-processing plant near Canton, China that was equipped according to Japanese technology. Supplied from Chinese aquaculture, it produces Japanese-style fried eel (Kabayaki) for export to Japan. Also in Germany and other lands, live eel imports account for only part of the amount traded. In the 1970s, a third consisted of deep-frozen eels. Smoked eels are internationally marketed, as well.

The foremost eel exporter in Europe is Denmark (Table 8.1); at the same time, however, it also imports considerable amounts. Germany is Denmark's primary customer, even though it is often uncertain whether the eels really come from Denmark. This is important because, in the past, the Danish exports were mainly Baltic Sea eels, which are the kind most preferred in Germany. Today Danish exports include aquacultural eels

Table. 8.1 Prices (US$) and amounts (t) of live eels of the most important importing and exporting countries (after Yearbook Fish. Statistics FAO).

	Imports					
	1994		1995		1996	
Country	t	$/kg	t	$/kg	t	$/kg
Japan	15636	18.0	11969	20.1	11442	16.4
Hongkong					2693	13.6
Holland	3101	8.8	2615	8.8	3044	10.4
Italy	2218	8.0	1032	10.2	1032	11.7
Germany	1956	10.5	1529	12.0	1740	12.4
Denmark	1795	8.4	1635	10.1	1255	11.6
Belgium	1580	8.1	949	10.1	862	10.1
	Exports					
Taiwan	9495	15.5	7444	18.4	10288	15.0
China	6843	9.0	5605	12.8	3917	11.3
Denmark	2082	10.2	1996	11.7	1639	13.4
Sweden	1771	5.3	971	10.0	955	11.0
Holland	1396	20.5	927	13.1	1724	11.7
England (GB)	737	12.3	894	16.9	801	22.1
France	636	38.1	523	36.8	534	52.0

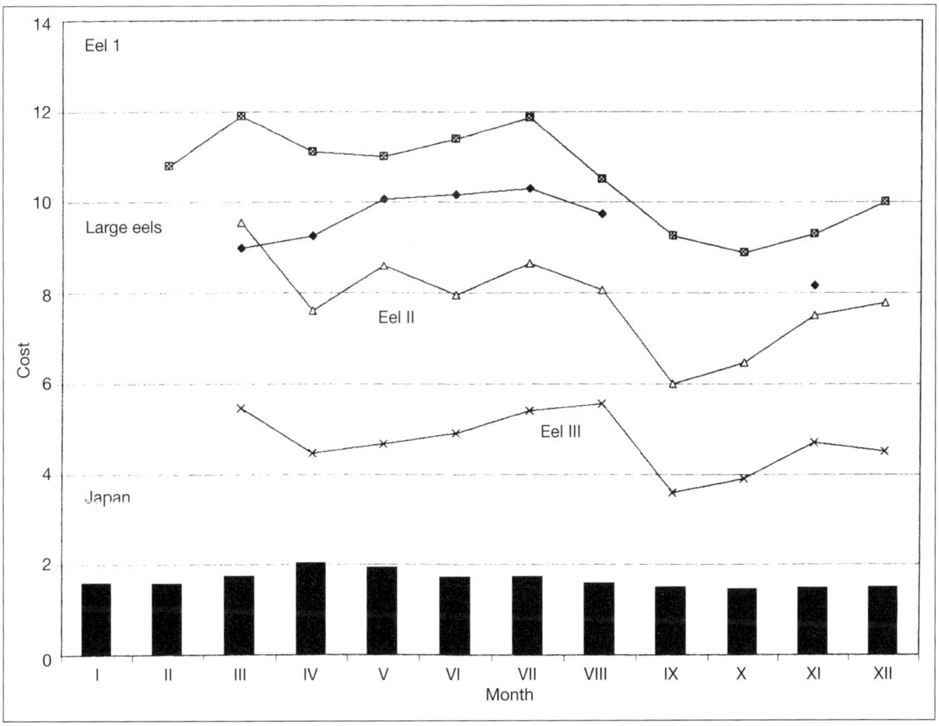

Fig. 8.2 Mean monthly price of the various grades of eels on the Kiel, Germany, seafish market in 1967 in DM/kg (after Herrmann, 1967) and on the wholesale market in Tokyo in US$/kg (after Brown, 1969 from Fritzsche, 1970)

(Table 6.1), which is not necessarily a sign of inferior quality (Section. 8.4). The greatest share of European imports is composed of European eels, even though American, New Zealand and other eels were involved temporarily. There are individual reports that, for example, 4.5 t of *A. australis* were exported from Australia to Europe in 1965. New Zealand shipped 100 t (*Dtsch. Fisch. Ztg.* **13**, 280, 1967). Further data on New Zealand exports can be found in Teklenberg (1972). From the three Egyptian Lakes Edku, Menzaleh and Birollos near the Mediterranean coast, a Dutch firm caught 'African' eels (*A. anguilla*, of course) – 25 t in 1962/3, 58 t in 1963/4 and 90 t in 1965/6 – and live transported them by sea to Europe (*Inform. Fischw. Ausland.* **16**(6), 28, 1966). Also the most frequently imported American eels were shipped by air after the war.

8.2 Trade within Germany

The paths for eels from fisher to consumer vary greatly. In Germany, a very large amount is routed through processing facilities, especially the smokehouse (see below). A significant amount goes neither through the market nor any middlemen, but instead directly from fisher to consumer, especially in areas where only small amounts are caught. This route is more typical for eels than for other kinds of fish. The higher price of eels makes it worthwhile to deal with customers even on a small scale. Therefore, this will taint any official statistics for a region (Section 4.1) with mistakes because the figures are based mainly on data from fishers having the larger catches.

On the German Baltic Sea coast, cooperatives collected the small individual catches in the past, and, to some extent, still do so today. Smaller catches in the northwest German region are often brought to consumers by dealers. However, a significant portion was traded *via* sea-fish markets, and this was especially so at the period for the abundant catches in the North Sea. In Germany, it is only through the sea-fish markets that we are informed about amounts caught and prices paid, as these data are published in the fishery trade sheets.

8.3 Prices

8.3.1 Price fluctuations

The price of eels, of course, follows the general upward trend of prices.
However, average prices fluctuate from year to year depending on total amounts landed, and they clearly reflect supply and demand. Statistics compiled in Canada, using that country's currency, reflect the long-term price rise, especially if the World War II years are disregarded (Eales, 1968):

- 1920: 8.9;
- 1925: 8.9;
- 1930: 8.7;
- 1935: 6.4;

- 1940: 4.9;
- 1945: 9.6;
- 1950: 8.4;
- 1955: 14.8;
- 1960: 18.7;
- 1965: 31.7.

Statistics compiled in US$ from the *Aquaculture Prod. Stat.* for recent years are as follows:

- 1984: 7.4;
- 1985: 6.8;
- 1986: 8.8;
- 1987: 9.5;
- 1988: 10.6;
- 1989: 6.9;
- 1990: 6.0;
- 1991: 5.6;
- 1992: 6.2;
- 1993: 6.5;
- 1994: 7.4;
- 1995: 7.3;
- 1996: 5.8.

From this display, one perceives mainly fluctuations, and in regression calculations at best a slight upward trend.

Eels of different sizes or grades command different prices, except that quality is also differentiated, as well as between narrow- and broad-headed, yellow and silver eels. Narrow-headed and silver eels are more sought after in Germany, so get higher prices there. Size-rating systems having grades I to III or IV are used in various countries, but the same grade can mean a different size range in each country. For example, grade I means eels of 200–500 g in Germany, but eels over 750 g in Poland. In France and Italy, eel size is generally expressed as number per kilogram, but some other places trading is still done in pounds per score (20 eels). The latter measure came from Denmark and was common especially among dealers who imported eels from there. 'Large eels,' for example, may be those for which a score weighs 30 pounds (750 g each); a score of 'small eels' may weigh 4 pounds (100 g each). Smoked eels are graded in Germany by a different system than are live or fresh eels (Struck, 1965).

- grade I (giant): >500 g each;
- grade II: 375–499 g each;
- grade III: 250–374 g each;
- grade IV: 175–249 g each;
- grade V: 125–174 g each.
- 'bunch eel': several eels, each weighing 100–125 g, bound in a bundle with raffia fibres.

Japanese eels are sold at different prices according to whether they originated from rivers, lakes or pond culture. The price per kilogram was around DM 3.00 (US$0.75) from 1955 to 1958, but pond eels were about DM 0.40 (US$0.10) more expensive than were eels from natural waters (Koops, 1966b): 'The higher price in the statistics for pond eels is probably explained by the fact that the prices for eels from river and lake fisheries are subject to the seasonally dependent price fluctuations.' Pond eels are richer in fat than are eels from lakes and rivers (see below), which, however, in Japan is not necessarily considered a sign of better quality. In Germany, origin likewise has a certain influence on price; there, Baltic Sea eels are regarded as qualitatively the best. The osmotic conditions of the Baltic Sea's salinity might play a role in this (Section 1.7). Similar relationships are known for the flat fishes (Tesch, 1956).

Reasons other than qualitative differences underlie the price fluctuations for table eels during the year (Fig. 8.2). Mainly, it is seen that regardless of the differing relative prices of the various grades of European eels, prices of all grades undergo rather similar patterns of fluctuation in the course of a year. Eel prices tend to be relatively low in April/May but lowest of all in September/October. These are the periods when most are caught (Section 4.3), when the eels move between winter and summer habitats (Section 3.3.3.4) and when the spawning migration begins (Section 3.4.1). Here, one can really speak of an 'ecology of price,' in contrast to economics, which is determined by non-biological, purely financial conditions, for example, currency value, competition from other foodstuffs, imports, weather and so forth. The price fluctuations produced by ecological relationships surely affect eels less than other fishes. Eels, though, can have their sales season extended for a relatively long time by the practice of keeping them alive in holding tanks (Section 6.6). In addition, their high market value makes preservation by deep-freezing acceptable.

In Japan, there is a peak price relative to the seasonal price level. It occurs during the months of March through July, with its highest value in April/May. This is because supply is too low in springtime. In the subsequent summer months eel sales reach their annual maximum, so prices then fall correspondingly (Fig. 8.2; Gousset, 1992). In Japan, the peak of eel production is determined by pond culture (Section 4.2), which can deliver eels continually during summer. In Japan, the lake and river fisheries, with their peak catches in springtime, lag far behind.

8.3.2 Price levels in different countries including glass eels

Compared with German eel prices, those in Japan formerly were a little lower, and have been significantly higher in recent years (Table 8.1). If it is considered that the eels produced in Japan are primarily what would be grade III in Germany, and if the German prices for grade-III eels are compared with the Japanese mean prices, it is evident that eels were more expensive in Japan than Germany in the past, too. From 1955 to 1958, eel prices in Japan were still half as high as in West Germany (Koops, 1966b). In the mid-1990s, they were twice as high as in Europe (Table 8.1).

In countries that export eels to Japan, the prices, likewise, are significantly higher than in Europe. In Canada, in 1965, the eel prices were only a third as high as in Germany, and in earlier years even lower. Recent average prices in the USA (US$/kg) were as follows (from *Aquaculture Prod. Stat.*):

- 1990: 2.6;
- 1991: 3.1;
- 1992: 3.2;
- 1993: 3.3;
- 1994: 3.6;
- 1995: 3.5;
- 1996: 3.6.

Thus, judging by Table 8.1, they were still only a third as high as in Europe, despite a rising trend. The reason for this, according to information from Japan (pers. comm., Tabeta), might be that, compared with the European and Japanese species most other eels including *A. rostrata* are of less desirable culinary quality. This could have contributed to the lower price of *A. rostrata* in Germany, too. Chemical studies might clarify this. As is shown in section 2.5, the European and American eels are genetically so similar that biochemical differences between them may be not significant.

The eel is considered a luxury food primarily in central Europe, especially Germany, and also in Japan. There is also above-average consumption in Denmark, Holland, England and Italy, perhaps likewise in France and Spain, if one takes the Spaniard's predilection for glass eels into account. In England, consumption in 1918 is supposed to have been about 7000 t, of which only 1000 t came from that country's own catch (Koops, 1967a). This is amazing because the eel is regarded in England as a rather inferior fish, a pest in salmonid waters and, therefore, by no means a luxury item.

Glass-eel prices are significantly higher – tremendously so in East Asia. The following comparison of glass-eel prices elucidates the development. In Germany, the market prices for *A. anguilla* glass eels amounted to (*Die Aalpost* 1, 1997; *Fisch und Fang* (11), 20, 1997):

- 1970s: 18–22 DM/kg (1972: 30 DM/kg on average);
- 1996: 600 DM/kg on average (end of 1997: as much 800 DM/kg).

In Japan, consumers paid the following for glass eels of *A. japonica*, which are very valuable there:

- 1973: 12–18 £ sterling/kg (*c.* 60 DM/kg – Forrest, 1974);
- 1996: 3000 DM/kg on average.

In March 1998, according to *U.K. Glass Eels* (Glass Eels Ltd., Gloucester, England), there was even talk of US$6000/kg. In 1996/7, dealers came from the Far East to buy in Europe and, therefore, drove the prices for European glass eels so tremendously high. Similarily, during about 1995/6, Japanese dealers sought glass eels in North America, offering unheard-of high prices. This stimulated chaotic fishing in eastern US streams, resulting in very restrictive official regulations on glass-eel harvest (White, personal communication). Nevertheless, at the end of 1997, the glass-eel price rise seemed to have ended. Japanese annual demand for table eels was about 120,000 t. East Asian production trended higher, however, so prices fell (*Die Aalpost* 1, 1998). However, further development is a matter of the future catch of the most sought-after eel species.

Table 8.2 Composition of bodies and body parts of eels differing in origin, size and developmental stage: various body parts or sizes (origin of the animals: Oder estuary) and smoked eels.

Body part	Eel weight (g)	Fat (%)	Water (%)	Protein (%)	Ash (%)	Carbo-hydrate (%)	Sources
Head area	722	18.3	61.1	16.7	2.5	–	Meyer, 1943
Anal area	722	25.9	56.1	8.8	1.3	–	
Tail area	722	37.2	51.7	14.1	0.9	–	
Innards	722	19.3	63.7	17.5	1.1	–	
Head area	30	12.4	67.4	10.4	2.1	–	
Anal area	30	13.4	67.6	16.1	1.7	–	
Tail area	30	14.6	65.7	13.1	2.0	–	
Innards	30	9.6	73.3	–	2.0	–	
Adjacent. zone	722	30.5	62.9	–	–	–	
Under the skin	722	40.6	45.6	–	–	–	
Skin	722	9.7	56.4		–	–	
Smoked eel	–	26.4	52.7	18.7	1.4	0.8	Kraut from Meyer-Waarden, 1965

Different developmental stages and forms

Developmental stage or form	Weight	Fat (%)	Dry substance (%)	Sources
Eel fry	168 mg	3.2	22.2	Reuss and Weiland, 1913 from Wiehr, 1932
Eel fry	–	4.1	21.1	König and Splittgerber from Wiehr, 1932
Glass eel (montée)	–	2.6	20.1	König and Splittgerber from Wiehr, 1932
Juvenile eel	14 g	5.4	19.6	Wiehr, 1932
Narrow-headed eel	–	27.3	40.9	Wiehr, 1932
Broad-headed eel	–	11.9	54.7	Wiehr, 1932
Origin: river	–	28.4	–	König and Splittgerber from Wiehr, 1932

Different origins in Germany

Origin	Length (cm)	Weight (g)	Fat (%)	Remarks	Sources
Wismar Bay (Baltic)	25	25	3.5	narrow-headed males	Meyer-Waarden, 1965
	28	30	4.1	narrow-headed males	
	29	–	–	narrow-headed males	
	31	45	11.1	narrow-headed males	
	33	70	28.1	narrow-headed males	
	35	75	29.0	narrow-headed males	
	37	90	27.8	narrow-headed males	
	39	110	28.8	narrow-headed males	
Dassow lake	25	26	6.8	narrow-headed males	Meyer-Waarden, 1965
	28	35	4.3	narrow-headed males	
	29	–	–	narrow-headed males	
	31	50	9.8	narrow-headed males	
	33	60	26.8	narrow-headed males	
	35	80	10.7	narrow-headed males	
	37	107	23.1	narrow-headed males	
	39	113	14.1	narrow-headed males	

Table 8.2 (*Cont'd*)

Origin	Length (cm)	Weight (g)	Fat (%)	Remarks	Sources
Greifswald Bay (Baltic Sea)	25	19	4.2	narrow-headed males	Meyer-Waarden, 1965
	28	29	7.3	narrow-headed males	
	29	30	8.3	narrow-headed males	
	31	35	10.7	narrow-headed males	
	33	37	1.8	narrow-headed males	
	35	34	2.7	narrow-headed males	
Baltic Sea	–	–	31.0	fresh weight of silver eel fillets	Meyer-Waarden and Koops, 1968
River Mosel	–	–	21.9	fresh weight of silver eel fillets	
Pond eels, river Elbe	–	–	27.0	fresh weight of silver eel fillets	
Pond eels, river Mosel	–	–	24.1 + 26.6	fresh weight of silver eel fillets	
Pond eels, Japan	–	–	23.8 + 24.8	fresh weight of silver eel fillets	

8.4 Quality of the commercial product

Aside from commercial values reflected in prices, it is also possible, of course, to evaluate eels based on food chemistry and taste-testing. The crucial criterion for most fish species is fat content, and this applies to eels as well. As more than 80% of the marketed eels in Germany are smoked (Struck, 1965), their high fat content, which is advantageous for that process, enters into determination of value.

Table 8.2 presents the fat content according to various studies. Eels <30 cm long are relatively low in fat; usually little more than 5%. However, their water content is relatively high. Narrow-headed eels have a greater share of fat than do broad-headed eels (Thurow, 1958; Piatek, 1970; Dahl, 1973). Therefore, narrow-headed eels are favoured for smoking if silver eels are not available; silver eels are fatter than narrow-headed eels. Of course, silver eels that supposedly have a minimal fat content of 20% (Boetius and Boetius, 1980) are fatter than yellow eels. Eels from the Baltic Sea are especially fat. Verifying this was a test undertaken with river and pond eels (Table 8.2; Meyer-Waarden and Koops, 1968), as well as a comparison of eels from Wismar Bay (Baltic Sea) and the inland Dassow Lake. Eels from the inland lake were rather heavier than those from the Baltic Sea: greater water content of the freshwater eels is inferred from this. According to Table 3.12, as well, coastal eels usually weigh less than inland eels of equal length. On the other hand, pond eels are heavier than any other eels (Meyer-Waarden and Koops, 1968; Koops, 1971). However, their fat content is not exceptionally high. This applies to both Japanese and European eels.

The eel's caudal area and the areas immediately under the skin have the highest fat content of any part of the body (Table 8.2). The most deficient in fat are the innards and the central parts of the vertebral column. Smoked eels, which of course lack innards, are understandably low in water and rich in protein, but like the present example, probably not exceptionally fatty. The eel exceeds most other smoked fishes in fat content, the main reason for its high culinary and commercial value in Europe.

With respect to fatty acids, eel fat contains:

- 86.5% total fatty acid;
- 66.2% unsaturated fatty acid;
- 20.3% saturated fatty acid;
- 11.4% glycerine;
- 0.7% unsaponified substances (cholesterol derivatives);
- 2.8% lecithin.

The iodine count of the fat is 97.4 for narrow- and 105.8 for broad-headed eels. For other fishes, it is usually >100, as much as 161 for salmon. From this it can be inferred that eel fat is not composed to such great extent of the glycerides of unsaturated fatty acids, as is that of other fishes. In contrast, the differences between narrow- and broad-headed eels are not particularly large, and obviously can be attributed to the prey (Section 3.3.1.4) eaten by each of these two forms (Wiehr, 1932).

Data are available from Japanese eels on fatty acid composition in relation to various feeds (Ando, 1968). Dabrowski *et al.* (1970) studied the cholesterol content of European eel muscle tissue and fat in comparison with 10 other freshwater fish species. With respect to composition of fatty acids in muscle lipids, Reichwald and Meizies (1973) and Meizies and Reichwald (1973) compared the eel with freshwater fishes, and fresh and smoked eel with sturgeon, herring and spiny dogfish (*Squalus acanthias*). These authors found the eel to have above-average olein content and much higher content of eicosenoic acid than any of the seven species studied.

Smoking the eel improves the meat's quality and/or flavour, and at the same time preserves it (Gehring and Wünsche, 1965). The smoking process has often been described (e.g., Breitenstein, 1954; Visurgis, 1954; Struck, 1965; Meike, 1970; Rehbronn and Rutkowski, 1997). In addition, smoked eel is frequently canned to make it exportable (Lantz, 1966). Besides that, there are many common ways of preparing eels to eat, which likewise often involve methods for making the meat keep longer (Sparrenberg, 1934; Struck, 1965). Teklenberg (1972) described the methods used in New Zealand.

References

Ahne, W., Thomsen, I. (1985): The existence of three different viral agents in a tumor bearing European eel (*Anguilla anguilla*). Zentralbl. VetMed. B, **32,** 228–235.

Aker, E. (1974): Pelagische Aalfischerei in der Deutschen Bucht. Fischerblatt **22,** 285–292.

Aker, E. (1975): Infektion mit *Vibrio anguillarum* bei Aalen aus der Deutschen Bucht 1975. Infn. Fischw. Auslds. **22**(1), 13–14.

Aker, E. (1976): Infektion mit *Vibrio anguillarum* bei Aalen aus der Deutschen Bucht nimmt 1975 wieder zu. Inform. Fischwirtsch. **23**(1), 8–9.

Aker, E., Koops, H. (1970): Das Wanderverhalten der Aale in der Nordsee und die Befischungsintensität. Infn. Fischw. Auslds. **17**(2), 46.

Aker, E., Koops, H. (1973): Untersuchungen über Aalbestände in der Deutschen Bucht. Arch. FischWiss., **24,** 19–39.

Aker, E., Koops, H. (1976): The status of the German trawl fishery on eels in the German Bight. ICES/EIFAC Symp. On Eel Res. and Management, No. **11,** Helsinki, Publ.: 1979: Rapp. p.-v. Réun. Cons. int. Explor. Mer. **175,** 7–9.

Albrecht, G. (1975): Aalerträge unserer Seen in der Großbrigade Seddin. Z. Binnenfischerei **22,** 180–183.

Albrecht, M. L. (1964): Die Oder als Fischgewässer, Z. Fisch., **12,** 479–506.

Alexander, R. Mc N. (1969): The orientation of muscle fibres in the myomeres of fishes. J. mar. biol. Ass. UK **49,** 263–290.

Alexander, R. Mc N. (1970): Mechanics of the feeding action of various teleost fishes. J. Zool., London, **162,** 145–156.

Ali,, M.A., Dutil, J.D., Fortier, L. (1987): Retinal response in the postmetamorphic American eel (*Anguilla rostrata*) (Pisces, Teleostei). Helgoländer Meeresunters. **41**, 437–441.

Al-Kharrat, H., Weiss, U., Tran, Q., Nibbio, B., Scholz, S., Epple, A. (1997): Cholinergic control of catecholamine release in the eel. Gen. Comp. Endocrinol. **108**, 102–108.

Allen, K.R. (1951): The Horokiwi stream: a study of a trout population. Fish. Bull. Wellington, N.Z. **10,** 1–231.

Allen, K. R. (1957): Freshwater fish. In Science in New Zealand, Handbook for the 1957 A.N.Z. A.A.S Conference. A. H. and A. W. Reed, Wellington, New Zealand.

Alnaes, E. (1973a): Two types of lateral line afferent in the eel *Anguilla anguilla*. Acta physiol. scand. **87,** 535–548.

Alnaes, E. (1973b): Peripheral inhibition of eel lateral line receptors as caused by antidromic sensory terminal invasion. Acta physiol. scand. **88,** 35–48.

Alnaes, E. (1973c): Lateral line input to the crista cerebellaris in the eel. Field potentials and histology. Acta physiol. scand. **88,** 49–61.

Amlacher, E.·(1986): Taschenbuch der Fischkrankheiten. 5. Aufl., VEB Gustav Fischer, Stuttgart. 478pp.

Anders, E. (1978): Versuche zur oralen Immunisierung von Regenbogenforellen und Aalen gegen *Vibrio anguillarum.* Fisch. Forsch., Wiss. Schriftenr. **16**(3), 59–60.

Anders, K., Möller., H. (1991): Epidemiologische Untersuchungen von Fischkrankheiten im Wattenmeer. Ber. Inst. Meereskunde Univ. Kiel, Nr. 207.

Andersen, D., Boètius, I., Olesen Larsen, L., Hjulmann Seidler, P. (1996): Effects of oestradiol-enriched diet and of feeding with porcine testicular tissue on macroscopic gonadal sex in European eels. J. Fish Biol. **48,** 484–494.

Anderson, J. I. W., Conroy, D. A. (1970): *Vibrio* disease in marine fishes. In: Snieszko, S. F.: A Symposium on Diseases of Fishes and Shellfishes. Amer. Fish. Soc., Spec. Publ. **5,** 266–272.

Ando, K. (1968): Biochemical studies on the lipids of cultured fishes. J. Tokyo Univ. Fish. **54,** 61–98.

Ando, K. (1975): Intestinal water transport and chloride pump in relation to sea-water adaptation of the eel *Anguilla japonica*. Comp. Biochem. Physiol. **52A,** 229–233.

Ando, M. (1974): Effects of cortisol on water transport across the eel intestine. Endocrinol. Jap. **21,** 539–546.

Anonymous (1979): Report of the EIFAC/ICES Working Group on Eels 1979. ICES CM 1979/M1.

Anonymous (1997a): Eel research needs kept up – NZ project latest in breeding studies. Fish Farming International, October 1997, 36–37.

Anonymous (1997b): Report of the EIFAC/ICES Working Group on Eels 1997. ICES CM 1997/M1.

Antunes, C., Tesch, F.-W. (1997a): Eel larvae (*Anguilla anguilla* L.) caught by R. V. Heincke at the European continent slope in autumn 1991. Ecology of Freshwater Fish **6,** 50–52.

Antunes, C., Tesch, F.-W. (1997b): A critical consideration of the metamorphosis zone when identifying daily rings in otoliths of European eel, *Anguilla anguilla* (L). Ecology of Freshwater Fish **6,** 102–107.

Anwand, K., Valentin, M. (1981a): Die Zusammensetzung der Aalfänge aus Seen und Fliessgewäsern nach Kopfformtypen, Geschlechtsreizezuständen sowie Geschlechtern. Z. Binnenfischerei **28,** 55–60.

Anwand, K., Valentin, M. (1981b): Aalbesatzmassnahmen als Voraussetzung für eine intensive Aalwirtschaft. Z. Binnenfischerei **28,** 237–240.

Aoyama, J., Hissmann, K., Yoshinaga, T., Sasai, S., Uto, T., Ueda, H. (1999): Swimming depth of migrating silver eels *Anguilla japonica* released at seamounts of the West Mariana Ridge, their estimated spawning sites. Mar. Ecol. Prog. Ser. **186,** 265–269.

Aoyama, J., Kobayashi, T., Tsukamoto, K. (1996): Phylogeny of eels suggested by mitochondrial DNA sequences. Nippon Suisan Gakkaishi **62,** 370–375.

Aoyama, J., Tsukamoto, K. (1997): Evolution of the freshwater eels. Naturwissenschaften **84,** 17–21.

Aoye, H. (1977): Formula feed for eels. In: Noda, U. (Ed.), Fisheries in Japan. Publ.: Japan Mar. Products Photo Material Assoc. 601–5–17–1, Hyakunincho, Shinjuku-ku, Tokyo 160 Japan, 217–721.

Apollova, T. A. (1969): Ob ugre Kurskogo zaliva (On the eels of the Courland Lagoon). Trudy. atlant. nauchno issled. Inst. Kyb. Khoz, Okeanogr. **21,** 185–191.

Aprahamian, M.W. (1988): Age structure of eel *Anguilla anguilla* (L.) populations in the River Severn, England, and the River Dee, Wales. Aquacult. Fisheries Mgmt. **19,** 365–375.

Arai, H. P. (1989): Acanthocephala. In: Margolis, L., Kabata, Z. (Eds): Guide to the parasites of fishes of Canada III. Canad. Spec. Publ. of Fish. and Aquat. Sci. **107**.

Arai, S. (1991): Eel. In Wilson, R.P. (ed.): Handbook of Nutrient Requirements of Finfish. CRC Press, Boca Raton, USA, 69–75.

Arai, S., Nose, T., Hashimoto, Y. (1971): A purified test diet for the eel, *Anguilla japonica*. Bull. Freshwat. Fish. Res. Lab. Tokyo **21,** 161–178.

Arai, T, Otake, T., Tsukamoto, K. (1997): Drastic changes in otolith microstructure and microchemistry accompanying the onset of metamorphosis in the Japanese eel *Anguilla japonica.* Marine Ecology Progress Series **161,** 17–22.

Arakawa, E., Kaneko, T., Tsukamoto, K., Hirano, T. (1992): Immunocytochemical detection of prolactin and growth hormone cells in the pituitary during early development of the Japanese eel, *Anguilla japonica*. Zool. Sci. **9,** 1061–1066.

Arbeiten Deutsch. Fisch. Verb. **69,** 47–61.

Ardizzone, G. D., Corsi, F. (1985): Eel population structure, dynamic and fishing yield in a Mediterranean coastal lagoon. Oebalia **11,** 547–560.

Arnold, G. P., Cook, P. K. (1984): Fish migration by selective tidal stream tansport: first results with a computer simulation model for the European Continental shelf. In McCleave J.D. et al., (eds.): Mechanisms of Migration in Fishes. Plenum, New York, 227–261.

Arratia, G. (1997): Basal teleosts and teleostean phylogeny. Palaeoichthyologica **7,** 1–168.

Arratia, G., Schultze, H. P. (1990). The urohyal: development and homology within osteichthyans. J. Morph. **203,** 247–282.

Arvy, L., Fontaine, M., Gabe, M. (1959): La voie neurosecretoire hypothalamohypophysaire des téléostéens. J. Physiol. Paris **51,** 1031–1085.

Asano, H. (1962): Studies on the congrid eels of Japan. Bull. Misaki Mar. Biol. Inst. **1,** 1–143.

Asawaka, M. (1974): Sialic acid-containing glycoprotein in the external mucus of eel, *Anguilla japonica* Temminck et Schlegel. Carbohydrate and amino-acid composition. Bull. Jap. Soc. Sci. Fish. **40,** 303–308.

Ask, L., Berntsson, K.-E., Öhlund, S. 0. (1971): Undersökningar orn gulålens alder, kon och tillvaxt. Meddn Havfiskelab, **108,** 1–7.

Ask, L., Erichsen, L. (1976): Blankålmärkningar vid Svenska Östersjökusten 1941–1968. Medd. Havsfiskelabor. Lysekil 199.

Atuma, S. S., Linder, C. E., Andersson, O., Bergh, A., Hansson, L., Wicklund-Glynn, A. (1996): CB 153 as indicator for congener specific determination of PCBs in diverse fish species from Swedish waters. Chemosphere **33,** 1459–1464.

Aust, S. (1936): Die Epidermis von *Anguilla vulgaris* Turt. während seiner Entwicklung. Zool. Jb. **62,** 1–44.

Avise, J.C., Helfman, G.S., Saunders, N.C., Hales, L.S. (1986): Mitochondrial DNA differentiation in North Atlantic eels: population genetic consequences of an unusual life history pattern. Proc. Natl. Acad. Sci. USA **83,** 4350–4354.

Avise, J. C., Nelson, W. S., Arnold, J., Koehn, R. K., Williams, G. C., Thorsteinsson, V. (1990): The evolutionary genetic status of Iclandic eels. Evolution **44,** 1254–1262.

Baer, J. G. (1948): Les helminthes, parasites des vertèbres. Relations phylogénétiques entre leur évolution et celle de leurs hôtes. Conséquences biologiques et médicales. Ann. Franche-Compté Univ. Besançon. 15pp.

Bagenal, T. B., Tesch. F.-W. (1978): Age and growth. In: Bagenal, T. (ed.), IBP Handbook No. **3,** Third Edition, Methods for Assessment of Fish Production in Fresh Waters, Blackwell Scientific Publications, Oxford, 101–136.

Baker, B. I. (1969): The response of teleost pituitary thyrotrophs to thyreosin in vitro: a histological study. Gen. Comp. Endocr. **12,** 427–437.

Baker, B. I. (1972): The cellular source of melanocyte-stimulating hormone in *Anguilla* pituitary. Gen. Comp. Endocrinol. **19,** 515–521.

Baker, B. I., Rance, T. A. (1983): Further observations on the distribution and properties of teleost melanin concentrating hormone. Gen. Comp. Endocrinol. **50,** 423–431.

Ball., J. N. (1969): Prolactin (Fish prolactin or paralactin) and growth hormone. In: Hoar, W. S., Randall, D. J. (eds) Physiology of Fishes **2,** 207–240. Academic Press, New York and London.

Ball., J. N., Baker, B. I. (1969): The pituitary gland: anatomy and histophysiology. In: Hoar, W. S., Randall, D. J. (eds) Fish Physiology **2,** 1–111. Academic Press, New York and London.

Ball, J. N., Baker, B. I., Olivereau, M., Peter, R. E. (1972): Investigations on hypothalamic control of adenohypophysial functions in teleost fishes. Gen. Comp. Endocr., Suppl. **3,** 11–21.

Ball, J. N., Chester Jones, I., Forster, M. E., Hargreaves, G., Hawkins, E. F., Milne, K. P. (1971): Measurement of plasma cortisol levels in the eel, *Anguilla anguilla* L. in relation to osmotic adjustments. J. Endocr. **50,** 75–96.

Ball, J. N., Olivereau, M. (1964): Rôle de la prolactine dans la survie en eau douce de *Poecilia latipinna* hypophysectomisé et arguments en faveur de sa synthèse par les cellules érythrosinophiles de l'hypophyse des téléostéens. C. r. hebd. Séanc. Acad. Sci., Paris **259,** 1443–1446.

Balligand, P., Tronel-Peyroz, J., Descamps, B., Grauby, A., Foulquier, L., Dumas, M. (1980): Réalisation d'un pilote industriel utilisant des eaux de réfrigération d'une usine pour le grossissement des anguilles. Symposium CEGPI/CIEM sur les recents dévelopements de l'utilisation des eaux réchauffées et des eaux recyclées en aquaculture intensive. **22,** 1–13.

Balon, E. K. (1971): The eels of Siengwazi Falls (Kalomo River, Zambia) and their significance. Zambia Museums J. **2,** 65–82.

Balon, E. K. (1975): The eels of Lake Kariba: distribution, taxonomic status, age, growth and density. J. Fish. Biol. **7,** 797–815.

Bandayan, M., Sanubokn, E., Rasio, E. (1974): The capillary endothelium in the rete mirabile of swimbladder of the eel (*Anguilla anguilla*). Functional and ultrastructural aspects. Can. J. Physiol. Pharmacol. **52,** 613–623.

Bandayan, M., Sanubokn, E., Rasio, E. (1975): Studies of the capillaries basal lamina: 1. Ultrastructure of the red body of the eel swimbladder. Lab. Invest. **32,** 757–767.

Bandt, H. J., Nolte, W. (1934): Über die Tötung der Aale vor dem Räuchern. FischZtg. Neudamm, **37,** 505.

Bangham, R. V., Venard, C. E. (1946): Parasites of fish of Algonquin Park lakes. Publs. Ont. Fish. Res. Lab. **65,** 31–46.

Banning, G. P. van, Haenen, O. L. M. (1990): Effects of the swimbladder nematode *Anguillicola crassus* in wild and farmed eel, *Anguilla anguilla.* In: Perkins, F. O., Cheng, T. C. (Eds.): Pathology in Marine Science. Academic Press, New York, 317–330.

Barak, N. A. E., Mason, C. F. (1990): A survey of heavy metal levels in eels (*Anguilla anguilla*) from some rivers in East Anglia, England: the use of eels as pollution indicators. Int. Rev. Gesamt. Hydrobiol. **75,** 827–833.

Barbin, G.P., McCleave, J.D. (1997): Fecundity of the American eel *Anguilla rostrata* at 45ºN in Maine, USA. J. Fish Biol., **51,** 840–847.

Barker, D.E., Cone, D.K. (2000): Occurrence of *Ergasils celestis* (Copepoda) and *Pseudodactylogyrus anguillae* (Monogenea) among wild eels (*Anguilla rostrata*) in relation to stream flow, pH and temperature and recommendations for controlling their transmission among captive eels. Aquaculture **187,** 261–274.

Barthelemy, L., Mabin, U., Belaud, A., Peykaud, L. (1977): Encephalic electrical activity in eels submitted to hypoxia or hypercapnia. J. Physiol. (Paris) **73,** 1035–1044.

Bast, H.-D., Klinkhardt, M. B. (1988): Fang eines Silberaales (*Anguilla anguilla* [L. 1758]) im Iberische Becken Nordostatlantik (Teleostei: Anguillidae). Zool. Anz. **221,** 386–398.

Bast, H. D., Strehlow, B. (1990): Length composition and abundance of eel larvae, *Anguilla (Anguilliformes*: Anguillidae) in the Iberian Basin (northeastern Atlantic) during July–September 1984. Helgoländer Meeresunters. **44,** 353–361.

Bastrop, R., Strehlow, B., Jürss, K., Sturmbauer, C. (2000): A new molecular phylogenetic hypothesis for the evolution of freshwater eels. Mol. Phyl. Evol. **14,** 250–258.

Batty, J., Pain, D., Caurant, F. (1996): Metal concentrations in eels, *Anguilla anguilla*, from the Camargue region of France. Biol. Conserv. **76,** 17–23.

Bauch, G. (1954): Die einheimischen Süßwasserfische. Neumann-Verlag, Radebeul & Berlin.

Baumgärtner, R. (1968): Aal in Lang-Saison rund um den Chiemsee. Fisch u. Fang **9,** 109–110.

Beatty, D. D. (1975): Visual pigments of the American eel *Anguilla rostrata*. Vision Res. **15,** 771–776.

Bellini, A. (1907): Expérience sur l'élevage de l'anguille en stabulation à Comacchio. Bull. Soc. Cont. Aquic. Pêche **19,** 81–93, 141–146, 186–196, 215–224.

Bennett, M. B., Rankin, J. C. (1986): The effects of neurohypophyseal hormones on the vascular resistance of the isolated perfused gill of the European eel, *Anguilla anguilla* L. Gen. Comp. Endocrinol. **64,** 60–66.

Bentley, P. J. (1962): Permeability of the skin of the cyclostome *Lampetra fluviatilis* to water and electrolytes. Comp. Biochem. Physiol. **6,** 95–97.

Berg, L. S. (1958): System der rezenten und fossilen Fischartigen und Fische. 310pp., Dt. Verl. Wissensch. Berlin.

Berg, L. S. (1965): Freshwater fishes of the U.S.S.R. and adjacent countries (Translated from the Russian edition of 1949). Smithsonian Institution, Washington.

Berg, R. (1980): Erfahrungen mit spiegelnetzunterstützter Elektrofischerei in Seen. Prot. zur Fischereitechn. **14,** 220–229.

Berg, R. (1987a): Fischschäden durch Turbinen. In: Vorträge der Aalkomm. 1986. Arb. Dt. Fischereiverband 1987, 41–47.

Berg, R. (1987b): Freilanduntersuchungen zur Biologie und Bewirtschaftung des Aales (*Anguilla anguilla* L.) im Bodensee. Dissertation der Naturw. Fakultät Univ. Freiburg/Br.

Berg, R. (1996a): Gedanken zur Problematik "Turbinenschäden an Aalen". In: Vorträge der Aalkomm. 1995. Arb. Dt. Fischereiverbandes **64,** 1996, 21–39.

Berg, R. (1996b): Ableitung von Aalen an Wasserkraftwerken, Stand der Technik, Aussichten. Wertermittlungsforum 3/1996.

Berg, T., Steen, J. B. (1965): Physiological mechanisms for aerial respiration in the eel. Comp. Biochem. Physiol. **15,** 469–484.

Berg, T., Steen, J. B. (1966): Regulation of ventilation in eels exposed to air. Comp. Biochem. Physiol. **18,** 511–516.

Bergmann, A. (1962): Ergebnisse und Erfahrungen mit Aalen. Allg. FischZtg. **87,** 337–341.

Bergmann, A. (1970): Aale gehen an Land. Allg. FischZtg. **95,** 408–410, 440–443, 510–511, 538–541, 566–569.

Berndt, O. (1938): Morphologie und Histologie des Rumpfdarmes von *Anguilla fluviatilis* und die Veränderungen desselben im Individualzyklus. Zool. Jb. (Anat.) **64,** 437–482.

Bernhardt, W. (1963): Aalsterben in der Werra bei extrem niedrigen Temperaturen. Dt. FischZtg. Radebeul, **10,** 327–329.

Berntsson, K. E., Erichsen, L., Öhlund, S. O. (1974): Blankålmärkningar vid Kråkelund 1968–1972. Medd. Havsfiskelabor. Lysekil 159.

Bertin, L. (1956): Eels, a biological study. Cleaver-Hume Press, London.

Bertin, L. (1958): Sexualité et fécondation. In Grassé, P.P. (ed.): Traité de Zoologie **13**(2), 1584–1652. Masson et Cie, Paris.

Betge, E., Rohde, H., Kulow, H. (1965): Untersuchungen über die Reaktion der Fische auf unterschiedliche Farben und Stärken elektrischen Lichtes. Dt. FishZtg., Radebeul, **12,** 286–291.

Beullens, K., Eding, E.H., Ollevier, F., Komen, J., Richter, C.J.J. (1997): Sex differentiation, changes in length, weight and eye size of European eel (*Anguilla anguilla*) maintained in captivity. Aquaculture **153,** 151–152.

Beumer, J. P. (1979): Feeding and movement of *Anguilla australis* and *A. reinhardtii* in Macleods Morass, Victoria, Australia. J. Fish. Biol. **13,** 573–592.

Beumer, J.P. (1985): The eel resources of Fiji. Study Tour Rep. Queensland Dept. of Primary Industries, QS 85010, 1–33.

Beumer, J.P., Pearson, R.G., Pendridge, L.K. (1981): Pacific short-finned eel, *Anguilla obscura* Günther, 1871 in Australia: recent records of its distribution and maximum size. Proc. R. Soc. Qd. **92,** 85–90.

Bezdzenyezhnykh, V. A. (1973): Resorption of oocytes in the European eel. (Beloruss) Vyesti Akad. Nauk, BSSR, Ser. Byal. Nauk **3**(134), 96–101.

Bezdzenyezhnykh, V. A. (1974): State of oocytes of silver eels. (Beloruss) Vyesti Akad. Nauk, BSSR, Ser. Byal. Nauk **3**(142), 99–102.

Bezdzenyezhnykh, V.A., Petukhov, V.B., Petrikov, A.M. (1983a): *Problems of propagation of the eel* (in Russian). Minsk Nauka i Tekhnika, 144pp.

Bezdzenyezhnykh, V.A., Prochorchik, A.M., Petrikov, A.M., Petukhov, V.B., Plyuta, M.V. (1983b): Obtaining larvae of *Anguilla anguilla* L. (Pisces, Anguillidae) under experimental conditions (in Russian, Engl. translation). Dokl. Akad. Nauk **268**(5), 1264–1266.

Bickel, A. (1897): Beiträge zur Rückenmarksphysiologie des Aales. Pflügers Arch. ges. Physiol. **68,** 110–119.

Bielig, W. (1931): Untersuchungen über die Herztätigkeit der Fische II. Z. vergl. Physiol. **15,** 488–513.

Bienarz, K., Eppler, P. 1975: Investigations on accelerating sexual maturity of eelo (*Anguilla anguilla*) males. Counc. Expl. Sea C. M. 1975/M2, Anadr. Catadr. Fish Committee, 1–7.

Bieniarz, K., Eppler, P. (1977): Investigations on inducing sexual maturity in the male eel (*Anguilla anguilla*). J. Fish. Biol. **10,** 555–559.

Bigelow, H. B., Schroeder, W. C. (1953): Fishes of the Gulf of Maine. Fishery Bull. Fish Wildl. Serv., U.S. **53,** 1–577.

Billard, R., Ginsburg, S. (1973): La spermiogénèse et le spermatozoide d'*Anguilla anguilla* L., étude ultrastructurale. Ann. Biol. Anim. Biochim. Biophys., **13,** 523–534.

Biro, P. (1974): Observations on the food of eel (*Anguilla anguilla* L.) in Lake Balaton. Annals Inst. biol. Tihany **41,** 133–152.

Biro, P., László, E. (1970): A Balaton halászata és az Utóbbi évek ichthyológiai problémái. (The Balaton fishery and ichthyological problems of recent years.) Allatt. Közl. **57,** 39–49.

Blancheteau, M. (1972): L'ondulation natatoire chez l'anguille (Poissons, Ostéichthyens, Apodes, Anguillidés, *Anguilla anguilla* L.) spinale. Bull. Mus. Hist. nat. Paris, (Zool.) **59**(80), 917–924.

Blancheteau, M. (1973): L'ondulation natatoire chez l'anguille spinale. C. R. 96. Congr. national Soc. savant., Toulouse 1971, Science, **3,** 541–550.

Blaxter, J. H. S. (1969): Swimming speeds of fish. FAO Fish. Rep. Rome **62**(2), 69–100.

Blin, P. C., Balea, T. (1955): La myologie des poisons: son incidence practique dans la diagnose des poissons courants. Bull. sci. Roumain, Paris **3,** 60–89.

Bloom, S., Polak, J. M. (1981): Gut Hormones. London, Churchill Livingstone.

Bobzin, W. (1965): Neue Perspektiven durch den Einsatz pfahlloser Reusen in der Küstenfischerei. Fischereiforschung **3**(3), 43–59.

Bobzin, W., Finnern, D. (1978): Fangtechnik. Transpress Verlag Berlin.

Boëtius, J. (1967): Experimental indication of lunar activity in European silver eels, *Anguilla anguilla* (L.). Meddr Danm. Fisk. og Havunders. **6,** 1–6.

Boëtius, J. (1976): Elvers, *Anguilla anguilla* and *Anguilla rostrata* from two Danish localities. Size, body weight, developmental stage and number of vertebrae related to time of ascent. Meddr. Danm. Fisk. Havunders., N. S. **7,** 199–120.

Boëtius, J. (1980): Atlantic *Anguilla.* A presentation of old and new data of total numbers of vertebrae with special reference to the occurrence of *Anguilla rostrata* in Europe. Dana **1,** 93–112.

Boëtius, J., Boëtius, I. (1967a): Studies in the European eel, *Anguilla anguilla* (L.). Experimental induction of the male sexual cycle, its relation to the temperature and other factors. Meddr Danm. Fisk. og Havunders. **4,** 339–405.

Boëtius, J., Boëtius, I. (1967b): Eels, *Anguilla rostrata* Le Sueur, in Bermuda. Vidensk. Meddr dansk Naturh. Foren. **130,** 63–84.

Boëtius, J., Boëtius, I. (1980): Experimental maturation of female silver eels, *Anguilla anguilla.* Estimates of fecundity and energy reserves for migration and spawning. Dana **1,** 1–28.

Boëtius, J., Harding, E. F. (1985): A re-examination of Johannes Schmidt's Atlantic eel investigations. Dana **4,** 129–162.

Bogdan, E., Waluga, D. (1980): The effect of transport on the quality of eel stocking material. Aquaculture 20, 139–146.

Bogenschutz, R. P., Clemens, H. P. (1967): Changes in the pituitary gland of goldfish, *Carassius auratus*, during diet-controlled gonadal regression. Copeia 1967, 827–835.

Bohl, M. (1968): Vorläufiger Bericht über Aalfütterungsversuche in Wielenbach. Z.Wasser Abwasserforsch. **1,** 235–239.

Böhlke, J. E. (1966): Order Lyomeri. In: Cohen, D. M. (ed.): Fishes of the Western North Atlantic. Memoirs of the Sears Foundation for Marine Research, Yale University, New Haven, No. **1,** Part **5,** 603–628.

Bohun, S., Winn, H. E. (1966): Locomotor activity of the American eel (*Anguilla rostrata*). Chesapeake Sci. **7,** 235–239.

Boonstra, G. P., Deelder, C. L. (1975): A new method for eel sampling with use of electronically controlled current. EIFAC Paper 23 (supplement 1), 186–193.

Bowman, A. (1913): The distribution of the larvae of the eel in Scottish waters. Scient. Invest. Fishery Bd Scotl. **2,** 1–11.

Bozeman, E. L., Helfman, G. S., Richardson, T. (1985): Population size and home range of American eels in a Georgia tidal creek. Trans. Amer. Fish. Soc. **114,** 821–825.

Brandt, A. v. (1957): Aalfang in Seen. Allg. FischZtg. **82**(16), 301–305.

Brandt, A. v. (1959): Fanggeräte der Kutter- und Küstenfischerei. Land- u. Hauswirtsch. Inform. Dienst. A.I.D. **113,** 1–109, Frankfurt (Main).

Brandt, A. v. (1961): Erfahrungen mit monofilen Aalreusen. Fischereiwelt **11,** 75–78.

Brandt, A. v. (1984): Fish catching methods of the world. Fishing News (Books) Ltd. London, 3rd edition.

Brandt, A. v., Rutkowski, F. (1956): Fischereigeräte der Binnenfischerei. Arbeiten Dt. Fisch. Verb. **6,** 1–84.

Brandt, K. (1905): Über Petersens neue Beiträge zur Aalfrage und seine neue Methode, den Fangertrag an Wanderaalen erheblich zu vermehren. Mitt. Dt. Seefisch.-Verein 333–338.

Branover, G. G., Vasiliev, A. S., Gleiser, S. I., Tsinober, A. B. (1971): A study of the behaviour of eel in artificial and natural magnetic field and the analysis of their mechanism of reception. Vopr. Ikhtiol. **11,** 720–727.

Bräutigam, R. (1961a): Über Versuche zur Intensivierung des Blankaalfanges durch die Kombination von Lichtsperren und Großreusen und ihre grundsätzlichen Bedingungen. Fischereiforschung **4**(1/2), 19–25.

Bräutigam, R. (1961b): Anwendungsmöglichkeiten von Lichterketten zur Intensivierung des Blankaalfanges. Z. Fisch. **10,** 643–651.

Bräutigam, R. (1962): Intensivierung des Blankaalfanges mit Hilfe von Lichtsperrketten und Bedingungen der gewerblichen Anwendung. Fischereiforschung **5**(6), 8–15.

Bray, R. A., Cribb, T. H., Barker, S. C. (1993): Hemiuridae (Digenea) from marine fishes of the Great Barrier Reef, Queensland, Australia. Systematic Parasitology **25,** 37–62.

Breder, C. M., Rasquin, P. (1950): A preliminary report on the role of the pineal organ in the control of pigment in eels and light reactions in recent teleost fishes. Science, **111,** 10–12.

Breitenstein, G. (1954): Geräucherter Aal. Allg. FischZtg. **79,** 329.

Breitenstein, G. (1956): Betrachrungen über Reusen der Binnenfischerei unter besonderer Berücksichtigung der Benennung der Anfertigungsmethoden und der Normierung. Z. Fisch. **5,** 443–476.

Breitenstein, G. (1960): Die Geräte der Seen und Flußfischerei. 1. Das Material für die Anfertigung der Fischereigeräte. In Wundsch, H.H. (ed.): Fischereikunde, 195–206. Neumann, Radebeul & Berlin.

Bremer, H. (1974): Einige histologische und histochemische Untersuchungen epidermaler Proliferationen bei postlarvalen Aalstadien. Int. Revue ges. Hydrobiol. Hydrogr. **59,** 569–572.

Bremer, H., Bayard, J., Klopffleisch, M. (1975): Einige Untersuchungen zu den histologischen, histochemischen und histophysiologischen Veränderungen des exokrinen Pankreas und des Magenfundus bei *Anguilla anguilla* L. im Verlauf der postlarvalen Entwicklung. Gegenbauers morph. Jahrb. Leipzig **121,** 109–120.

Bridge, T. W., Boulenger, G. A. (1958): Fishes. In Harmer, S. F., Shipley, A. E. (eds): Cambridge Natural History. Macmillan, London.

Broadhead, C. L., O'Sullivan, U. T., Deacon, C. F., Henderson, I. W. (1992): Atrial natriuretic peptide in the eel, *Anguilla anguilla* L., its cardiac distribution, receptors and actions on isolated branchial cells. J. Mol. Endocrinol. **9,** 103–114.

Brown, E.E. (1969): The freshwater cultured fish industry of Japan. Univ. Georgia Coll. Agricult. Experim. Sta. Res. Rep. **41,** 1–57.

Brown, J. A., Henderson, I. W, Jackson, B. A., Oliver, J. A. (1980): Angiotensin and single nephron function in the trout, *Salmo gairdneri*. American J. of Physiology **239,** R509–R514.

Brümmerstädt, M. (1955): Der Aalkorb – ein gering beachtetes Fanggerät. Dt. FischZtg. Radebeul **2,** 343–344.

Bruun, A. F. (1937): Contributions to the life histories of the deep sea eels: Synaphobranchidae. Dana Report **9,** 1–39.

Bruun, A. F. (1963): The breeding of the North Atlantic freshwater eels. Adv. mar. Biol., **1,** 137–169.

Buchholz, F. (1948): Die Aalschnur und ihre praktische Anwendung. Allg. FischZtg Radebeul **73,** 153–156.

Buchin, H.E., Müller, H. (1967): Die Aalwirtschaft von West-Berlin. Arch. FischWiss. 18 (Suppl. 1), 447–451.

Buchmann, K., Bjerregaard, J. (1990): Comparative efficacies of benzimidazoles against gill parasites of eels. VII. Congr. Int. Parasitol. Paris, **8,** Suppl. **2,** 1077.

Buckow, F. (1956): Geht der Aal über Land? Dt. FischZtg Radebeul **3,** 92.

Budimawan (1997): The early life history of the tropical eel *Anguilla marmorata* (Quoi & Gaimard, 1894) from four Pacific estuaries, as revealed from otolith microstructural analysis. J. Applied Ichthyology **13,** 57–62.

Buhse, G. (1967): Die Aalwirtschaft im Regierungsbezirk Kassel. Arch. FischWiss., 18 (Suppl. 1), 307–331.

Bukhanevich, I.B. (1969): Novye dannye o skorostyakh dvizheniya ryb. (New data on the swimming speed of fishes). Rybnoe Khozyaistvo Moscow **45**(10), 9–10.

Bullard, E. (1969): The Origin of the oceans. Scient. Am. **221**(3), 66–75.

Burnet, A. M. R. (1952a): Studies on the ecology of the New Zealand freshwater eels. 1. The design and use of an electric fishing machine. Aust. J. Mar. Freshw. Res. **3,** 111–125.

Burnet, A. M. R. (1952b): Studies on the ecology of the longfinned eel, *Anguilla dieffenbachii* Gray. Aust. J. Mar. Freshw. Res. **3,** 32–63.

Burnet, A. M. R. (1959): Some observations on natural fluctuations of trout population numbers. N.Z. J. Sci. **2,** 410–21.

Burnet, A. M. R. (1968): A study of the relationship between brown trout and eels in a New Zealand stream. N. Z. Mar. Dep. Fish. Tech. Rep. No. **26,** 1–30.

Burnet, A. M. R. (1969a): Migrating eels in a Canterbury River, New Zealand. N. Z. J. Mar. Freshw. Res. **3,** 230–244.

Burnet, A. M. R. (1969b): The growth of New Zealand freshwater eels in three Canterbury streams. N.Z. J. Mar. Freshw. Res. **3,** 376–384.

Bursey, C.R. (1982): *Eustrongylides tubifex* (Nitzsch) encystment in an American eel *Anguilla rostrata* (Le Sueur). J. Fish Biol. **21,** 443–442.

Burzawa-Gerard, E., Dumas-Vidal, A. (1991): Effects of 17–estradiol and carp gonadotropin on vitellogenesis in normal and hypophysectomized European silver female eel (*Anguilla anguilla* L.) employing a homologous radioimmunoassay for vitellogenin. Gen. Comp. Endocrinol, **84,** 264–276.

Butler, D. G. (1966): Effect of hypophysectomy on osmoregulation in the European eel (*Anguilla anguilla* L.). Comp. Biochem. Physiol. **18,** 773–781.

Butler, D. G. (1969): Corpuscles of Stannius and renal physiology in the eel *Anguilla rostrata*. J. Fish. Res. Bd Can **26,** 639–654.

Butler, D. G., Oudit, G. Y. (1994): Dorsal aortic and organ blood flow decrease following stanniectomy in freshwater North American eels (*Anguilla restrata* le Sueur). Comp. Biochem. Physiol. **108A**, 229–238.

Butler, D. G., Oudit, G. Y. (1995): Angiotensin I and III mediated cardiovascular responses in the freshwater North American eel, *Anguilla rostrata*: effect of Phe deletion. Gen. Comp. Endocrinol. **97,** 259–269.

Butschek, V., Hofbauer, J. (1956): Versuche über die Schädigung von Aalen durch Kaplanturbinen. Arch. FischWiss. **7,** 178–184.

Byczkowska-Smyk, W. (1958): The respiratory surface of the gills in teleosts. Part II. The respiratory surface of the gills in the eel (*Anguilla anguilla* L.), the loach, (*Misgurnus fossilis* L.), and the perch pike (*Lucioperca lucioperca* L.). Acta biol. cracov., Ser. Zool. **1,** 83–97.

Cahour, A., Chèze, G. (1973): Dynamique pigmentaire rétinienne chez l'Anguille (*Anguilla anguilla*). Soc. Biol. Lyon, Seance Juin 1973, 1253–1257.

Cairns, D. (1942a): Life-history of the two species of New Zealand freshwater eel. I. Taxonomy, age and growth, migration and distribution. N. Z. J. Sci. Technol. Sect. B. **23,** 53–72.

Cairns, D. (1942b): Life-history of the two species of freshwater eel in New Zealand. II. Food, and interrelationship with trout. N. Z. J. Sci. Technol. Sect. B., **23,** 132–148.

Callamand, D. (1943): L'Anguille Européenne. Les bases physiologique de sa migration. Annls. Inst. Océanogr. **21,** 361–440.

Campana, S. E., Neilson, J. D. (1985): Microstructure of fish otoliths. Can. J. Aquat. Sci. **42,** 1014–1032.

Campana-Rouget, Y., Therezien, Y. (1965): Un nouveau *Procamallanus* (Nematoda, Camallanidae), parasite de poissons de mer de Nossi-Be. Ann. Parasitol. hum. comp. **40,** 171–177.

Cano, T. M., Perry, S. F., Fenwick, J. C. (1994): Cholinergic control of stanniocalcin release in the rainbow trout, *Oncorhynchus mykiss* and the American eel, *Anguilla rostrata*. Gen. Comp. Endocrinol. **94,** 1–10.

Cantrelle, I. (1981): Étude de la migration et de la pêche des civelles (*Anguilla anguilla* L. 1758) dans l'estuaire de la Gironde. Thèse Universitè de Paris VI, Ocèanographie Biologique.

Carlisle, D. B., Denton, E. J. (1959): On the metamorphosis of the visual pigments of *Anguilla anguilla* (L.). J. mar. biol. Ass. UK **38,** 97–102.

Carss, D. N., Nelson, K., Kruuk, H. (1998): Unexploited yellow eel stocks in two shallow lakes: spatial and temporal trends (first draft). EIFAC (FAO) Working Party on Eel 1996 in IJmuiden, Holland.

Caruso, J.H., Suttkus, R.D., Gunning, G.E. (1988): Abnormal expression of secondary sex characters in a population of *Anguilla rostrata* (Pisces: Anguillidae) from a dark coloured Florida stream. Copeia **4,** 1077–1097.

Casamajor, M.-L., Prouzet, P., Lazure, P. (2000): Identification des flux de civelles (*Anguilla anguilla*) à partir des relations d'allométrie en fonction hydrodynamiques de l'estuaire de l'Adour. Aquat. Living Resour. **13,** 411–420.

Caspers, H. (1951): Die bulgarische Fischerei im Schwarzen Meer. Ihre biologischen und geographischen Grundlagen, Organisation und Erträge. Abh. Fisch. Hilfswiss. Lief. **4,** 719–786.

Castan, H. (1967): Alles über den Aalfang. Fischwaid **8,** 14–21.

Castle, P. H. J., (1963): *Anguilla* leptocephali in the Southwest Pacific. Zoology Publs Vict. Univ. Wellington, No. **33,** 1–14.

Castle, P. H. J., Williamson, G. R. (1974): On the validity of the freshwater eel species *Anguilla ancestralis* Ege, from Celebes. Copeia 1974, 569–570.

Castonguay, M., Hodson, P. V., Couillard, C. M., Eckersley, M. L., Dutil, J. D., Verreault, G. (1994a): Why is recruitment of the American eel, *Anguilla rostrata*, declining in the St. Lawrence River and Gulf? Can. J. Fish. Aquat. Sc. **51,** 479–488.

Castonguay, M., Hodson, P. V., Moriarty, C., Drinkwater, K. F., Jessop, B. M. (1994b): Is there a role of ocean environment in American and European eel decline. Fish. Oceanogr. **3,** 197–203.

Castonguay, M., McCleave, J. D. (1987): Vertical distributions, diel and ontogenetic vertical migration and net avoidance of leptocephali of *Anguilla* and of other common species in the Sargasso Sea. J. Plankton Res. **9,** 195–214.

Cau, A., Coluccia, E., Deiana, A.M., Pichiri, G., Rossino, R., Salvadori, S., Mezzanotte, R. (1992): Chromosomes and DNA of *Anguilla*: a study with restriction endonucleases. Genome **35,** 838–843.

Cépède, C. (1906): *Myxidium giardi* Cépède et la pretendue immunité des Anguilles a l'égard des infections myxosporidiennes. C. r. Séanc. Soc. Biol. **6,** 170–173.

Chan, D. K. O. (1971): The urophysis and the caudal circulation of teleost fish. Mem. Soc. Endocr. **19,** 391–412.

Chan, D. K. O., Chester Jones, I., Mosley, W. (1968): Pituitary and adrenocortical factors in the control of the water and electrolyte composition of the freshwater European eel (*Anguilla anguilla* L.). J. Endocr. **42,** 91–98.

Chan, D. K. O., Chow, P.H. (1976): The effects of acetylcholine, biogenic amines and other vasoactive agents on the cardiovascular functions of the eel, *Anguilla japonica*. J. exp. Biol. **196,** 13–26.

Chester Jones, I., Chan, D. K. O., Henderson, I. W., Ball, J. N. (1966): The adrenocortical steroids, adrenocorticotropin and the corpuscles of Stannius. In Hoar, W.S., Randall, D.J. (eds): Fish Physiology **2,** 321–376. Academic Press, New York.

Chester Jones, I., Chan, D. K. O., Rankin, J. C. (1969b): Renal function in the European eel: experimental alteration of water and electrolyte composition of the eel (*Anguilla anguilla* L.). J. Endocr. **37,** 297–317.

Chester Jones, I., Chan, D. K. O., Rankin, J. C. (1969a): Renal function in the European eel (*Anguilla anguilla* L.): Changes in blood pressure and renal function of the freshwater eel transferred to sea-water. J. Endocr. **43,** 9–19.

Chester Jones, I., Henderson, I.W., Mosley, W. (1964): Methods for the adrenalectomy of the European eel (*Anguilla anguilla* L.). J. Endocrinol. **30,** 155–156.

Cheung, P. J, Nigrelli, R. F., Ruggieri, G. D. (1979): Studies on cryptocaryoniasis in marine fish: effect of temperature and salinity in the reproductive cycle of *Cryptocaryon irritans* Brown 1951. J. Fish Dis. **2,** 93–97.

Chiarelli, B., Ferrantelli, O., Cucchi, C. (1969): The caryotype of some Teleostei fish obtained by tissue culture in vitro. Experientia, **25,** 426–427.

Chiba, K. (1983): Water quality and eel production in Japanese greenhouse ponds using heated water. Symp. Fr.-Japan. Aquacult., Montpellier 16–12–1983, **1,** 33–63.

Chisnall, B. L. (1989): Age, growth, and condition of freshwater eels (*Anguilla* sp.) in backwaters of the lower Waikato River, New Zealand. N. Z. J. Mar. Freshw. Res. **23,** 459–465.

Chisnall, B. L., Hayes, J. W. (1991): Age and growth of shortfinned eels (*Anguilla australis*) in the lower Waikato basin, North Island, New Zealand. N. Z. J. Mar. Freshw. Res. **25,** 71–80.

Chisnall, B. L., Hicks, B. J. (1993): Age and growth of longfinned eels (*Anguilla dieffenbachii*) in pastoral and forested streams in the Waikato River basin, and in two hydro-electric lakes in the North Island, New Zealand. N. Z. J. Mar. Freshw. Res. **27,** 317–332.

Chisnall, B.L, Kalish, J.M. (1993): Age validation and movement of freshwater eels (*Anguilla dieffenbachii* and *A. australis*) in a New Zealand pastoral stream. N.Z. J. Marine Freshwater Res. **27,** 333–338.

Chow, P. H., Chan, D. K. O. (1975): The cardiac cycle and the effect of neurohumors on myocardial contractility in the Asiatic eel *Anguilla japonica*, Temm. and Schle. Comp. Biochem. Physiol. C. Comp. Pharmacol. **52,** 41–45.

Christiansen, M., Jensen, A. G. C. (1950): On a recent and frequently occuring tumour disease of eel. Rep. Dan. biol. Stn. St. 29–31.

Christomanos, A. A. (1968): Übersichtsreferat zur Struktur der Haemoglobine der Fische der Amphibien und Reptilien 1. Fischhaemoglobine. Fol. biochem. biol. Graeca **5,** 21–84.

Chugunova, N. I. (1963): Age and growth studies in fish. National Science Foundation, Washington.

Churchward, A. S. (1996): Overview of the Severn fishery, England. EIFAC (FAO) Working Party on Eel, Sept. 1996 in IJmuiden, Holland.

Ciccotti, E. (1997): Italy. In Moriarty, C., Dekker, W. (eds): Management of the European eel. Fish. Bull. (Dublin) **15,** 87–96.

Ciccotti, E., Ricci, T., Scardi, M., Fresi, E., Cataudella, S. (1995): Intraseasonal characterization of glass eel migration in the River Tiber: space and time dynamics. J. Fish. Biol. **47,** 248–255.

Clarke, D. A., McCarthy, T. K., Cowx, I. (1993): Foraging activity patterns and diets of yellow eels *Anguilla anguilla* (L.) in Lough Derg. EIFAC (FAO) 1993 in Olsztyn, Poland.

Cohen, D. M., Dean, D. (1970): Sexual maturity and migratory behaviour of the tropical eel, *Ahlia egmontis*. Nature **227,** 189–190.

Colak, A., Yamamoto, K. (1974): Ultrastructure of the Japanese eel spermatozoon. Annotnes. zool. jpn. **47,** 48–54.

Colombo, G., Grandi, G. (1996): Histological study of the development and sex differentiation of the gonad in the European eel. J. Fish Biol. **48,** 493–512.

Colombo, G., Grandi, G., Rossi, R. (1984): Gonad differentiation and body growth in *Anguilla anguilla* (L.). J. Fish Biol. **24,** 215–228.

Comparini, A., Rizzotti, M., Rodino, E. (1977): Genetic control and variability of phosphoglucose isomerase (PGI) in eels from the Atlantic Ocean and the Mediterranean Sea. Mar. Biol. **43,** 109–116.

Comparini, A., Rodino, E. (1980): Electrophoretic evidence for two species of *Anguilla* leptocephali in the Sargasso Sea. Nature **287,** 435–437.

Comparini, A., Schoth, M. (1982): Comparison of electrophoretic and meristic characters of 0–group eel larvae from the Sargasso Sea. Helgoländer Meeresunters. **35,** 289–299.

Cone, D. K., Marcogliese, D. J., Watt, W. D. (1993): Metazoan parasite communities of yellow eels (*Anguilla rostrata*) in acidic and limed rivers of Nova Scotia. Can. J. Zool. **71,** 177–184.

Conlon, J. M., Andrews, P.C., Thim, L., Moon, T. W. (1991): The primary structure of glucagon-like peptide but not insulin has been conserved between the American eel, *Anguilla rostrata* and the European eel, *Anguilla anguilla.* Gen. Comp. Endocrinol. **82,** 23–32.

Conroy, D. A. (1984): Agents Bacteria. In: Kinne, O. (Ed.): Diseases of marine animals. Vol. IV, Part **1,** 48–88.

Conteaux, R., Laurent., P. (1957): Etude au microscope electronique du cœur de l'anguille: observation sur la structure du tissu musculaire de l'oreillette et son innervation. C. r. hebd. Séanc. Acad. Sci. Paris **245,** 2097–2100.

Creutzberg, F. (1961): On the orientation of migrating elvers (*Anguilla vulgaris* Turt.) in a tidal area. Neth. J. Sea Res. **1,** 257–338.

D'Ancona, U. (1927): Richerche sull'ingrandimento d'ell'occhio dell'anguilla in rapporto alla maturità sessuale e considerazione sul suo significato biologica. Atti. Accad. naz. Lincei Rc., Ser. **6, 5,** 36.

D'Ancona, U. (1929): Sui fenomeni di circulazione che accompagnano l'ingrandimento dell'occhio nella maturità sessuale dell'anguilla. Boll. Soc. Eustach. **27,** 1–8.

D'Ancona, U. (1950): Détermination et differenciation du sexe chez les poissons. Arch. Anat. micr. Morph. exp. **39,** 274–294.

D'Ancona, U. (1951a): Tentativi di deviazione sessuale nell'anguilla. Bull. Zool. **18,** 97–101.

D'Ancona, U. (1951b): Intorno alle transformazioni dell'anguilla nella fase conclusiva del suo siclo vitale. Riv. Biol. **43,** 171–186.

D'Ancona, U. (1958): Comparative biology of eels in the Adriatic and in the Baltic. Verh. Int. Verein. theor. angew. Limnol. **13,** 731–735.

D'Ancona, U. (1959a): Old and new solutions to the eel problem. Nature **183,** 1405.

D'Ancona, U. (1959b): Distribution of the sexes and environmental influence in the European eel. Arch. Anat. micr. Morph. exp. **48,** 61–76.

D'Ancona, U. (1960): Une invasion d'*Argulus* dans la Lagune de Venice. Schweiz. Z. Hydrol. **22,** 40–44.

D'Ancona, U. (1961): Biologische und wirtschaftliche Probleme des Aales in Italien. Z. Fisch. 10 N. F., 589–598.

Dabrowski, T., Stodolnik, L., Noch, G. (1970): The investigations on cholesterol content in muscular tissue of some freshwater fishes. Med. Wet. **24,** 299–301.

Dahl, J. (1973): A note on lipid in yellow eels (*Anguilla anguilla* L.). ICES CM 1973/M 9.

Dahl, J. (1991): Eelpasses in Denmark. EIFAC(FAO) Working Party on Eel, Dublin, May 1991.
Dahm, E. (1973): Knüppelhamen mit Rücklaufreuse – eine bemerkenswerte neue Entwicklung in der Hamenfischerei. Fischerblatt (21) **11,** 256–258.
Daniel, W. (1968): Über die Ernährung des Flußaales (*Anguilla vulgaris* Turt.) in einem Marschensielzug. Faun.-Ökol. Mitt. **3,** 229–240.
Day, L.R. (1948): Fishing eels – a profitable side line. Fish. Res. Bd Can. Biol. Stat. St.Andrews, N.B., Gen. Circ. 13.
De Boer, J., Stronck, C. J. N., Traag, W. A., Van Der Meer, J. (1993): Non-ortho and mono-ortho substituted chlorobiphenyls and chlorinated dibenzo-p-dioxins and dibenzofurans in marine and freshwater fish and shellfish from the Netherlands. Chemosphere **26,** 1823–1842.
De Leo, G. A., Gatto, M. (1996): Trends in vital rates of the European eel: evidence for density dependence? Ecological Application **6,** 1281–1294.
De Luze, A. (1987): Function thyroïdienne et sa régulation chez un poisson téléostéen, l'anguille (*Anguilla anguilla* L.). Thèse de Doctorate d'état ès Sciences, Université Pierre et Marie Curie, Paris.
De Luze, A., Leloup, J., Olivereau, M. (1984): Effects of ovine prolactin on iodide metabolism and thyroid activity in the eel. J. Comp. Physiol., 154B, 199–205.
De Nie, H. W. (1982): A note on the significance of larger bivalve molluscs (*Anodonta* spp. and *Dreissena* sp.) in the food of the eel (*Anguilla anguilla*) in the Tjeukemeer. Hydrobiologia **94,** 307–310.
De Nie, H. W. (1987): Food, feeding periodicity and consumption of the eel *Anguilla anguilla* (L.) in the shallow eutrophic Tjeukemeer (The Netherlands). Arch. Hydrobiol. **109,** 421–443.
Dean, B. (1912): Changes in the behavior of the eel during transformation. Ann. N. Y. Acad. Sci. **22,** 321–326.
Deelder, C. L. (1971): A new eel trap developed in the Netherlands. EIFAC Techn. Paper **14,** 97–99.
Deelder, C.L. (1981): *Some data on sex ratio K-factor, fat content and average weight of eel stocks in Dutch inland waters.* EIFAC Working Party Eel, Ferrara 1981, 12pp.
Deelder, C. L., De Veen, J. F. (1958): A calculation on the fishing intensity of the eel trawl on the IJsselmeer. Archs neerl. Zool. **13,** 461–471.
Deelder, C. L., Tesch, F.-W. (1970): Heimfindevermögen von Aalen (*Anguilla anguilla*), die über große Entfernungen verpflanzt wurden. Mar. Biol. Berlin, **6,** 81–92.
Deelder, D. L. (1952): On the migration of the elver (*Anguilla vulguris* Turt.) at sea. J. Cons. perm. int. Explor. Mer **18,** 187–218.
Deelder, D. L. (1954): Factors affecting the migration of the silver eel in Dutch inland waters. J. Cons. perm. int. Explor. Mer **20,** 177–185.
Deelder, D. L. (1957a): Een merkwaardige eigenschap van Schieraal. Visserij Nieuws, **9,** 158–159.
Deelder, D. L. (1957b): On the growth of eels in the IJsselmeer. J. Cons. perm. int. Explor. Mer **23,** 83–88.
Deelder, D. L. (1958): On the behaviour of elvers (*Anguilla vulgaris* Turt.) migrating from the sea into fresh water. J. Cons. perm. int. Explor. Mer **24,** 135–146.
Deelder, D. L. (1960a): The Atlantic eel problem. Nature **185,** 589–591.
Deelder, D. L. (1960b): Ergebnisse der holländischen Untersuchungen über den Glasaalzug. Arch. FischWiss. **11,** 1–10.
Deelder, D. L. (1965): The influence of increased fishing intensity on an eel stock. Proc. 2nd British Coarse Fish Conference, Liverpool, 57–64.
Deelder, D. L. (1970): Synopsis of biological data on the eel *Anguilla anguilla* (Linnaeus) 1758. FAO, EIFAC 70//Gen-6, 22.4.1970.
Deelder, D. L. (1978): A short note on the intensive culture of eel (*Anguilla anguilla* L.). Aquaculture **13,** 289–290.
Degani, G., Gallagher, M. L. (1995): Growth and nutrition of eels. Laser Pages Publishing Ltd. Jerusalem, Israel.
Degani, G., Kushnirov, D. (1992): Effects of 17B-Estradiol and grouping on sex determination of European eel. Prog. Fish Cult. **54,** 88–91.
Dekker, W. (2001): Status of the European eel stock and fisheries. *Proc. Intern. Symp. Advances in Eel Biol.* Univ Tokyo 2001, 50–55.
Delsman, H.L. (1929): The distribution of freshwater eels on the Greater Sunda Islands. Treubia **11,** 287–292.

Dembinski, W., Chmielewski, A. (1971): The application of electricity to the fishing of eels by means of trawls. EIFAC Techn. Paper **14,** 159–161.

Denecke, K. (1935): Nematodenlarven der Gattung *Goezia* als Ursache eines Aalsterbens. Z. Parasitenk. **7,** 702–708.

Denissenko, G. (1882): Einiges über den Bau der Netzhaut des Aales. Arch. mikrosk. Anat. EntwMech. **21,** 1–20.

Deufel, J., Strubbelt, T. (1977): Zur Aalwirtschaft im Bodensee. Oesterr. Fisch. **29,** 189–195.

Diesselhorst, G. (1938): Hörversuche an Fischen ohne Weberschen Apparat. Z. vergl. Physiol. **25,** 748–783.

Dietrich, G., Kalle, K., Krauss, W., Siedler, G. (1975): Allgemeine Meereskunde. 3rd edn. Berlin.

Dijkstra, L.H., Jellyman, D.J. (1999): Is the sub-species classification of the freshwater eels *Anguilla australis australis* Richardson and *Anguilla australis schmidtii* Phillipps still valid? Marine and Freshwater Research **50,** 261–263.

Ditrich, H., Splechtna, H. (1993): The vascularisation of the kidney of the eel *Anguilla anguilla*, Osteichthyes, in the freshwater habitat. Scanning Microscopy **7,** 279–285.

Doby, J. M., Jarecka, L. (1966): Complément a la connaissance de la morphologie et de la biologie de *Proteocephalus macrocephalus* (Creplin, 1825), cestode parasite de l'anguille. Annales Parasit. hum. comp. **41,** 429–442.

Doering, P., Ludwig, J. (1990): Shape analysis of otoliths, a tool of indirect aging of eel, *Anguilla anguilla* (L.)? Int. Revue ges. Hydrobiol **75,** 737–743.

Dogiel, V. A. (1936). Age related changes of the parasite fauna of the eel in connection with its migrations. Uch. zap. LGU, ser. Biol. **3,** 114–122.

Dorfner, K. (1966): Die Aalwirtschaft des Landes Bayern. Regierungsbezirk Oberpfalz. Arch. FischWiss. 16 (Suppl. 1), 50–56.

Dorn, E. (1960): Neue Untersuchungen über den Feinbau der Teleosteerschwimmblase. Verh. dt. Zool. Ges., Suppl. **24,** 223–233.

Dorn, E. (1961): Über den Feinbau der Schwimmblase von *Anguilla vulgaris* L. Licht- und elektronenmikroskopische Untersuchungen. Z. Zellforsch. mikrosk. Anat. **55,** 849–12.

Drapkin, E. I. (1964): On the distribution of the river eel *Anguilla anguilla* (L.) in the Black Sea. Byull. Mosk. Obshch. Isput. Prir. **69,** 140–141.

Dreist, H. H. (1968): Aalbeobachtungen, Fisch u. Fang **9,** 99.

Drilhon, A., Fine, J.M. (1968): Importance de l'étude des transferrins sériques dans la différenciation des espèces: resultats acquis dans genre *Anguilla*. Experientia **24,** 555–557.

Drilhon, A., Fine, J. M. (1971): Les groupes de transferrines dans le genre *Anguilla* L. Cons. Internat. Explor. Mer., Reunion Proc. Verb. L'identification biochimique et sérologique des stocks de poisson. **161,** 122–125.

Drilhon, A., Fine, J. M., Amouche, P., Boffa, G. A. (1967): Les groupes de transferrines chez *Anguilla anguilla*: étude de deux populations méditerranéennes d'orgine géographique différente. C. r. hebd. Séanc. Acad. Sci. Paris **265,** 1096–1098.

Drilhon, A.,Fine, J. M., Boffa, G. A., Amouch, P. (1966): Les groupes de transferrines chez l'anguille. Differences phenotypiques entre les anguilles de l'Atlantique et les anguilles méditerranéenes. C. r. hebd. Séanc. Acad. Sci. Paris, **262,** 1315–1318.

Drimmelen, D. E. Van (1951): Vangst van Schieraal mit behulp van licht. Visserij-Nieuws **3,** 137–140.

Drooglever, F. van Leyden (1917): Over een aal met het linker oog in onder kaak (On an eel having its left eye in the lower jaw). Versl. gewone Vergad. wis-en natuurk. Afd. K. Akad. Wet, Amst. **25,** 1168–1170.

Dröscher, W. (1897): Beiträge zur Kenntnis der Nahrung der Fische. Allg. FischZtg. **22,** 383–386.

Duan, C., Hirano, T. (1992): Effect of insulin-like growth factor-I and insulin on the in vitro uptake of sulphate by eel branchial cartilage. Evidence for the presence of independent hepatic and pancreatic sulphation factors. J. Endocrinol. **133,** 211–219.

Duan, C., Inui, Y. (1990a): Effects of recombinant eel growth hormone on the uptake of [35S] sulfate by ceratobranchial cartilages of the Japanese eel, *Anguilla japonica*. Gen. Comp. Endocrinol. **79,** 320–325.

Duan, C., Inui, Y. (1990b): Evidence for the presence of a somatomedin-like plasma growth factor in the Japanese eel, *Anguilla japonica.* Gen. Comp. Endocrinol. **79,** 326–331.

Dubovskiy, N. A. (1963): Ustroistvo dlya lova migriruyushchevo ugrya. Rybn. Khoz. Moskva **39**(10), 46–48.

Dufour, S. (1994): Neuroendocrinologie de la reproduction de l'anguille: de la recherche fondamentale aux problèmes appliqués. Bull. Fr. Pêche Piscic. **335,** 187–211.

Dufour, S., Lopez, E., Le Menn, F., Le Belle, N., Baloche, S., Fontaine, Y. A. (1988): Stimulation of gonadotropin release and of ovarian development by the administration of a gonadoliberin agonist and of dopamine antagonists, in female silver eel pretreated with estradiol. Gen. Comp. Endocrinol. **70,** 20–30.

Dunajewski, A. (1930): Les vaisseaux lympatiques du tronc de l'anguille (*Anguilla anguilla*). C. r. Acad. Cracovie1930, 7. Die Lymphgefäße im Rumpf des Aale (*Anguilla anguilla* L.). Bull. int. Acad. Sci. Lett. Cracovie B **2,** 467–477.

Durairaj, G. (1970): Thermal acclimation and blood coagulability in the eel, *Anguilla anguilla*. Amer. Zool. **10,** 519–520.

Dutil, J.D., Giroux, A. Kemp, A., Lavoie, G., Dallaire, J. P. (1988): Tidal influence on movements and on daily cycle of activity of American eels. Trans. Amer. Fish Soc. **117,** 488–494.

Duursma, D.K., Nieuwenhuize, J., van Liere, J.M., de Rooy, C.M., Witte, J.I.J., van der Meer, J. (1991): Possible loss of polychlorinated biphenyls from migrating European silver eels: a 3 month simulation experiment. Marine Chemistry **36,** 215–232.

Eales. J. G. (1968): The eel fisheries of eastern Canada. Bull. Fish. Res. Bd. Can. **166,** 1–79.

Edel, R. K. (1975a): The induction of maturation of female American eels through hormone injections. Helgoländer wiss. Meeresunters. **27,** 131–138.

Edel, R. K. (1975b): The effect of shelter availability on the activity of male silver eels. Helgoländer wiss. Meeresunters. **27,** 167–174.

Edelstam, C. G. (1965): Long range navigation in animals. Air Force Office Sci. Res. Grant, AF EOAR **63,** 44pp.

Ege, W. (1939): A revision of the genus *Anguilla* Shaw, a systematic, phylogenetic and geographical study. Dana Rep. **16,** 1–256.

Eggers, J. (1996): Fisch und Fang **6,** 23.

Egginton, S. (1986): Metamorphosis of the American eel, *Anguilla rostrata* LeSueur: 1., Changes in metabolism of skeletal muscle. J. Experimental Zool. **237,** 173–184.

Egusa, S. (1972): Pigmentation in the elvers of the Japanese eel. Bull. Jap. Soc. scient. Fish. **38,** 1203.

Egusa, S. (1978): Anguillicolosis of eel (*Anguilla anguilla* L.) in Japanese eel. In: Infectious diseases of fishes. Koseisha, Koseikako, Tokyo 501–506.

Egusa, S. (1979): Notes on the culture of the European eel (*Anguilla anguilla* L.) in Japanese eel-farming ponds. Rapp. p.-v. Réun. Cons. int. Explor. Mer. **174,** 51–58.

Egusa, S., Hirose, H. (1973): Further notes on sex and growth of European eels in culture ponds. Bull. Jap. Soc. Sci. Fish. **39**(6), 611–616.

Eh (1929): Dr. Harald Nordquist über den Besatz schwedischer Gewässer mit englischer Aalbrut. Fischerbote **21,** 80–81.

Ehrenbaum, E. (1930): Der Flußaal (*Anguilla vulgaris* Turt.) Handb. Binnenfisch. Mitteleur. 3 A, 159–217.

Ehrenbaum, E., Marukawa, H. (1914): Über Altersbestimmungen und Wachstum beim Aal. Z. Fisch. **14,** 89–127.

Eichelbaum, E. (1924): Das Auftreten der Aalbrut an der deutschen Küste und ihre erste Nahrungsaufnahme. Wiss. Meeresunters. (Helgol.) **15,** 1–26.

Eichler, H. (1935): Ovarialcyste bei einem Aal. Z. Fisch., **33,** 401–404.

Eigenmann, C. H., Kennedy, C. H. (1902): The leptocephalus of the American Eel and other leptocephali. Bull. U.S. Fish Comm. **21,** 81–92.

Einsele, W. (1961): Das Wachstum des Aales in oesterreichischen Gewässern. Öst. FischZtg. **14,** 136–138.

Einszporn-Orecka, T. (1979): Fragellates (*Spironuleus anguillae* new species): Parasites of eel *Anguilla anguilla.* Acta Protozool. **18,** 237–242.

Ekman, S. (1932): Prinzipielles über die Wanderungen und die tiergeographische Stellung des Europäischen Aales, *Anguilla anguilla* (L.). Zoogeografica **1,** 85–106.

Ekström, P., Van Veen, T. (1984): Pineal connections with the brain in two teleosts, the crucian carp (*Carassius carassius*) and the European eel (*Anguilla anguilla*). J. Pineal Res. **1,** 245–262.

El Naffar, M.K., Saoud, M.F.A. (1974): *Rhabdochona aegyptiacus* n.sp. (Nematoda: Rhabdochonidae) from some freshwater fishes of the River Nile at Astiut, Egypt. Zool. Soc. Egypt. Bull. **26,** 45–49.

Eldred, B. (1971): 1st records of *Anguilla rostrata* larvae in the Gulf of Mexico and Yucatan Straits. Fla. Dept. Nat. Resourc. Div. Mar. Resourc. Leafl. Ser., pt. I(19), 1–3.

Elie, P. (1979): Contribution a l'étude des montées de civelles d'*Anguilla anguilla* Linné (Poisson, Téléostéen, Anguilliforme), dans l'estuaire de la Loire: pêche, écologie, écophysiologie et élevage. Thése Doctorat Biologie Animale, Université Rennes I, 381pp.

Engelbrecht, H. (1958): Untersuchungen über den Parasitenbefall der Nutzfische im Greifswalder Bodden und Kleinen Haff. Z. Fisch. **7,** 481–511.

Enger, P. S., Kristensen, L., Sand, O. (1976): The perception of weak electric D. C. currents by the European eel (*Anguilla anguilla*). Comp. Biochem. Physiol. 54 A., 101–103.

Engstrom-Heg, R., Loeb, H. A. (1971): Recolonization of a rotenone-treated segment of the Ten Mile River., N. Y. Fish Game. J. **18,** 135–138.

Enomoto, Y. (1969a): Some notes on the fluctuations of the immature erythrocyte number in the blood of cultured fish. Bull. Tokai reg. Fish. Res. Lab. **58,** 173–188.

Enomoto, Y. (1969b): Some notes on the fluctuations of the immature leucocyte number in the blood of cultured fish. Bull. Tokai reg. Fish. Res. Lab. **58,** 191–201.

Epple, A. (1969): The endocrine pancreas. In: Hoar, W.S., Randall, D.J. (eds): Fish Physiology **2,** 275–319. Academic Press, New York and London.

Epple, A., Brinn, J. E. (1986): Pancreatic islets. In Pang, P.K.T., Schreibmann, M.P. (eds): Vertebrate Endocrinology: Fundamentals and Biomedical Implications: Morphological considerations **1,** 279–317. Academic Press, London.

Epple, A., Lewis, T. L. (1975): The effect of pancreatectomy on the survival of *Anguilla rostrata* in different salinities. J. exp. Zool. **193,** 457.

Epple, A., Nibbio, B. (1985): Catecholaminotropic effects of catecholamines in a teleost fish, *Anguilla rostrata.* J. Comp. Physiol., 155B, 285–290.

Eppler, P., Bieniarz, K. (1978): In vitro ovulation of European eel (*Anguilla anguilla* L.) oocytes following in vivo stimulation of sexual maturation. Ann. Biol. Anim. Biochem. Biophys. **18,** 991–995.

Ernst, P. (1975): Catch of an eel (*A. anguilla*) northeast of the Faroe Island. Ann. Biologiques Cons. perm. Explor. Mer. **32,** 175.

Facey, D. E., Labar, G. W. (1981): Biology of American eels in Lake Champlain, Vermont. Trans. Amer. Fish. Soc. **110,** 396–402.

Fagerlund, U. H. M., McBride, J. R., Smith, M., Tomlinson, N. (1963): Olfactory perception in migrating salmon. III. Stimulants for adult sockeye salmon (*Oncorhynchus nerka*) in home stream waters. J. Fish. Res. Bd Can. **20,** 1457–1463.

Falk, K., Lauterbach, R. (1965): Die Süßwasserfischanlandungen aus den Küstengewässern. Dt. FischZtg Radebeul **12,** 113–119.

Fänge, R., Östlund, E. (1954): The effects of adrenalin, noradrenalin, tyramine and other drugs on the isolated heart from marine vertebrates and a cephalopod (*Eledone cirrosa*). Acta zool., Stockholm **35,** 289–305.

FAO (1997a): Aquaculture production statistics 1986–1995. Fisheries Circular No. 815, Rev. 9, Rome, FAO.

FAO (1997b): FAO Fisheries statistics, catches and landings 1995, vol. 80. Rome, FAO.

Feddersen, A. (1893): Alhonnar i Sötvatten. Svensk Fisk. Tidskr., Stockholm, 156–165.

Fent, K., Bucheli, T. D. (1994): Inhibition of hepatic microsomal monooxygenase system by organotins in vitro freshwater fish. Aquat. Toxicol. **28,** 107–126.

Fenwick, J. C., Brasseur, J. G. (1991): Effects of stanniectomy and experimental hypercalcemia on plasma calcium levels and calcium influx in American eels, *Anguilla rostrata*. Gen. Comp. Endocrinol. **82,** 459–465.

Fenwick, J. C., Flik, G., Verbost, P. M. (1995): A passive immunization technique against the teleost hypocalcemic hormone, stanniocalcin provides evidence for the cholinergic control of stanniocalcin release and the conserved nature of the molecule. Gen. Comp. Endocrinol. **98,** 202–210.

Fenwick, J. C., Smith, K., Smith, J., Flik, G. (1984): Efffect of various vitamin D analogs on plasma calcium and phosphorus and intestinal calcium absorption in fed and unfed American eel, *Anguilla rostrata*. Gen. Comp. Endocrinol. **55,** 398–404.

Fenwick, J. C., Vermette, M. G. (1989): Vitamin D_3 and the renal handling of phosphate in American eels. Fish Physiol. Biochem. **7,** 351–358.
Fernandez-Delgado, C., Hernando, J.A., Herrera, M., Bellido, M. (1989): Age and growth of yellow eels, *Anguilla anguilla*, in the estuary of the Guadalquivir River (South west Spain). J. Fish Biol. **34,** 561–570.
Ferrando, M. D., Andreu, E., Fernandez-Casalderrey, A. (1991): Metabolic levels in eel blood after lindane treatment. In: Lahaniatis, E. S., Parlar, H., Lay, J. P., Pfister, G., Bergheim, W., Kotzias, D. (eds): Environmental pollution and its impact on life in the Mediterranean region. Toxicol. Environ. Chem. 31–32, 329–333.
Ferrando, M. D., Andreu-Moliner, E. (1991a): Effect of lindane on the blood of a freshwater fish. Bull. Environ. Contam. Toxicol. **47,** 465–470.
Ferrando, M. D., Andreu-Moliner, E. (1991b): Effects of lindane on fish carbohydrate metabolism. Ecotoxicol. Environ. Saf. **22,** 17–23.
Ferrando, M. D., Gamon, M., Andreu, E. (1992): Accumulation and distribution of pesticides in *Anguilla anguilla* from Albufera Lake (Spain). J. Environ. Biol. **13,** 75–82.
Fiedler, K. (1991): Fische. In Starck, D. (ed.): Lehrbuch der speziellen Zoologie. Vol. 2: Vertebrates. Pt. 2. Fischer, Jena: 498pp.
Fine, J. M., Boffa, G. H., Drilhon, A. (1963): The proteins of eel serum: starch gel electrophoresis and immunological studies. Proc. Soc. exp. Biol. Med. **114,** 651–654.
Fine, J. M., Drilhon, A. (1961): Existence de globulins de faible motilité électrophorétique dans le serum de l'anguille sommise à des injections répétées de sérum humain. C. r. hebd. Séanc. Acad. Sci. Paris, **252,** 3891–3893.
Fine, J. M., Drilhon, A., Boffa, G. A., Amouch, P. (1964): Les types de transferrines chez certains poissons migrateurs. In Peters, H. (ed.): Protides of the Biological Fluids, 165–168. Amsterdam, Elsevier.
Fine, J. M., Drilhon, A., Ridgeway, G. J., Amouche, P., Boffa, G. (1967): Les groupes de transferrines dans le genre *Anguilla.* Différences dans les fréquences phenotypiques de transferrines chez *Anguilla anguilla* et *Anguilla rostrata.* C. r. hebd. Séanc. Acad. Sci. Paris **265,** 58–59.
Firly, S., Fontaine, M. (1932): Sur la teneur en protéines du sérum d'anguille et ses variations au cours des changements de salinité. C. r. hebd. Séanc. Acad. Sci. Paris **194,** 1854–1856.
Fischer, F., Lübbert, H. (1908): Die Organisation des Bezuges von Aalbrut aus England für deutsche Gewässer. Z. Fisch **15,** 17–60.
Fish, M. P. (1927): Contributions to the embryology of American eel *(Anguilla rostrata* Le Sueur). Zoologica, N. Y. **8,** 289–324.
Flechtner, G. L., Anderson, T. (1972): A preliminary survey of the distribution of the American eel (*Anguilla rostrata*) in Newfoundland, MSRL Technical Rep. **7,** 1–47.
Flemming, H. (1954): Schwimmblasen-Entzündung eines Aales *Anguilla anguilla* (L.). Dt. FischZtg., Radebeul, **1,** 158–159.
Florkowski, W. (1930): Die Verteilung der Lymphgefäße im Kopfe des Aales (*Anguilla anguilla*). Bull. int. Acad. Sci. Lett. Cracovie B **2,** 457–464.
Fontaine, M. (1961): L'Anguille européenne succombe-t-elle sans se reproduire? C. r. hebd. Acad. Séanc. Acad. Sci. Paris **252,** 1258–1260.
Fontaine, M., Bertrand, E., Lopez, E., Callamand, O. (1964): Sur la maturation des organs génitaux de l'anguille femelle (*Anguilla anguilla*) et l'émission spontanée des œufs en aquarium. C. r. hebd. Séanc. Acad. Sci. Paris **12, 259,** 2907–2910.
Fontaine, M., Boucher-Firly, S. (1933): Influence de salinité sur la reserve alcaline du sang des poissons. C. r. Séanc. Soc. Biol. **113,** 306–308.
Fontaine, M., Busnel, R. G. (1939): Nouvelle recherches sur la répartition des flavines et de quelques autre pigments fluorescents dans la peau et les yeux des téléostéens. Bull. Inst. océanogr., Monaco, 782.
Fontaine, M., Delattre, S., Callamand, O. (1945): Influence des variations de salinité sur la teneur en hématies de deux téléostéens (*Anguilla anguilla* L. et *Cyprinus carpio* L.). Bull. Inst. oceanogr. Monaco **42**(886), 1–11.
Fontaine, Y.-A., Belle, N.L.E., Lopez, E., Querat, B., Vidal, B., Barthélemy, L., Sebert, P., Alinat, J., Petter, A.J. (1990): Infestation des populations français d'anguilles (*Anguilla anguilla* L.) par les

nematodes (*Anguillicola crassus*): essais therapeutiques evaluation de risques potentiels liés a l'e-cophysiologie de l'hôte. Ann. Parasitol. Hum. Comp. **65,** 64–68.

Fontaine, Y.-A., Dufour, A. M., Fontaine, M. (1985): Un vieux problème: la reproduction des anguilles. Comptes Rendus Acad. Sci., Sér. Générale **2**(1), 1–10.

Fontaine, Y.-A., Dufour, S. (1991): The eels: from life cycle to reproductive endocrinology. Bull. Inst. Zool. Academia Sinica, Monograph **16,** 237–248.

Fontenelle, G., Feunteun, E., Brian, C. (1997): France. Eds: Moriarty, C., Dekker: Management of the European eel. Fish. Bull.(Dublin) **15,** 74–79.

Ford, T. E., Mercer, E. (1986): Density, size distribution and home range of American eel, *Anguilla rostrata*, in a Massachusetts salt marsh. Environm. Biol. Fish. **17,** 309–314.

Forney, J. L. (1977): Evidence of interspecific and intraspecific competition as factors regulating walleye *Stizostedion vitreum vitreum* biomass in Oneida Lake New York. J. Fish. Res. Bd Can., **34,** 1810–1820.

Forrest, D. M. (1974): The cultivation and marketing of eels. Fish Farming International **3,** 50–57.

Forrest, J. N., Cohen, A. D., Schon, D. A., Epstein, F. H. (1973): Na transport and Na-K-ATPase in gills during adaptation to sea water: effects of sea water. Am. J. Physiol. **224,** 709–713.

Forster, M. E. (1976): Effects of catecholamines on the heart and on branchial and peripheral resistance of the eel, *Anguilla anguilla* (L.). Comp. Biochem. Physiol. C. Comp. Pharmacol. **55,** 27–32.

Forster, R. E., Steen, J. B. (1969): The rate of the 'Root Shift' in eel red cells and eel haemoglobin solutions. J. Physiol., London **204,** 259–282.

Franklin, C.E., Davie, P.S. (1991): The pericardium facilitates pressure work in the eel heart. J. Fish Biol. **39,** 559–564.

Fremberg, M., Olivereau, M. (1974): Melanophore responses and intermediate-lobe activity in the eel *Anguilla anguilla* after injection of 6–OH-dopamine. Acta Zool. (Stockholm) **54,** 231–239.

Freytag, G., Horn, W. (1970): Elektrifizierte Aal-Schleppnetze in Binnenseen. Infm. Fischw. **17**(6), 220–224.

Freytag, G., Horn, W. (1972): Winterfischerei mit einem elektrifizierten Aalschleppnetz im Selenter See. Infn. Fischw. **19**(1), 18–20.

Freytag, G., Horn, W. (1975): A description of an electric trawl fishing for eel in freshwater. EIFAC Techn. Pap. 23 (Suppl. **1,** 1), 179–185.

Freytag, G., Horn, W., Steinberg, R. (1971): Erfolgreiche Weiterentwicklung binnenfischereilicher Schleppnetze durch Elektrifizierung. Prot. Z. Fischereitechn. **12,** 417– 426.

Freytag, G., Mohr, H. (1973): Untersuchungen über die Pulsfischerei auf Aale. Protokolle Fischereitechn. **13**(50), 82–106.

Fricke, H., Käse, R. (1995): Tracking of artificially matured eels (*Anguilla anguilla*) in the Sargasso Sea and the problem of the eel's spawning site. Naturwiss. **82,** 32–36.

Fricke, H., Tsukamoto, K. (1998): Seamounts and the mystery of eel spawning. Naturwiss. **85,** 290–291.

Fries, G., Tesch, F.-W. (1965a): Der Einfluss des Massenvorkommens von *Gammarus tigrinus* Sexton auf Fische und niedere Tierwelt in der Weser. Arch. FichWiss. **16,** 133–150

Fries, G., Tesch, F.-W. (1965b): Aufenthalt der Fische im Bereich von Stauwehren. Schweiz. Z. Hydrol. **27,** 257–272.

Fries, L. T., Williams, D. L., Johnson, S. K. (1996): Occurrence of *Anguillicola crassus*, an exotic parasitic swimbladder nematode of eels, in the southeastern United States. Trans. Amer. Fish Soc. **125,** 794–797.

Fritzsche, S. (1970): Aufzucht und Absatz des Aales in Japan. Dt. FischZtg. Radebeul **17,** 315–321.

Frost, W. E. (1945): The age and growth of eels (*Anguilla anguilla*) from the Windermere catchment area. J. anim. Ecol. **14,** 26–36, 106–124.

Frost, W. E. (1946): Observations on the food of eels *Anguilla anguilla* from the Windermere catchment area. J. anim. Eco1.15, 43–53.

Frost, W. E. (1950): The eel fisheries of the River Bann, Northern Ireland and observations on the age of the silver eels. J. Cons. perm. int. Explor. Mer **16,** 358–383.

Frost, W. E. (1952): Predators on the eggs of char in Windermere. Salm. Trout Mag **136,** 192–196.

Frost, W. E. (1954): Observations on the biology of eels (*Anguilla* spp.) of Kenya Colony, East Africa. Colonial Off. Fish. Pub., 6.

Frost, W. E. (1957a): First record of the elver of the African eel *Anguilla nebulosa labiata*. Nature **179,** 594.

Frost, W. E. (1957b): A note on eels, *Anguilla* sp. Piscator, **38,** 104–106.

Frost, W. E. (1961): Einige Beobachtungen über Aale aus dem Windermere Fanggebiet, England. Z. Fisch. 10 (N.F.), 599–607.

Gabriel, O. (1997): Untersuchungen zur Langleinenfischerei auf Dorsch und Aal in der Ostsee. Inf. Fischwirtsch. **44**(2), 69–72.

Gabriel, O. (1998): Mechanisierte Aalfallenfischerei in Holland. Inform. Fischwirtsch. **45**(3), 119–121.

Gabriel, O., Thiele, W. (1997): Untersuchungen zu selektionsverbessernden technischen Maßnahmen in der Ostsee-Aalfischerei. Inf. Fischwirtsch. **44**(4), 171–175.

Gadeau De Kerville, M. (1918): Recherches experimentales sur la durée de la vie de l'anguille commune (*Anguilla anguilla* L.) dans un milieu sec. Bull. Soc. Zool. Fr. **42,** 115–117.

Gaevskaya, A. V. et al. (1975): Key to the parasites of vertebrates of the Black and Azov Seas. Kiev.

Gaitskell, R. E., Chester Jones, I. (1970): Effects of adrenalectomy and cortisol injection on the in vitro movement of water by the intestine of the freshwater European eel (*Anguilla anguilla* L.). Gen. comp. Endocr. **15,** 491–493.

Gaitskell, R. E., Chester Jones, I. (1971): Drinking and urine production in the European eel (*Anguilla anguilla* L.). Gen. comp. Endocr.16, 478–483.

Gandolfi-Hornyold, A. (1918a): Un cas de formation d'une pseudo-nageoire caudale chez la civelle. Revue Pathologie comp., Paris, 162–153.

Gandolfi-Hornyold, A. (1918b): Quelques observations sur la montée a Lisbonne. Mems Soc. port. Sci. nat. (Zool.) **3,** 1–55.

Gandolfi-Hornyold, A. (1921): Determinacion de la edad en algunas anguilas plateadas (maresas) de la Albufera de Valencia. An. Inst. Gen. Valencia, **11,** 27pp.

Gandolfi-Hornyold, A. (1925): Observations sur le sexe des anguilles. C. r. Ass. fr. Avanc. Sci. **49,** 422–424.

Gandolfi-Hornyold, A. (1929): Les parasites intestinaux de l'anguille en Espagne. Int. Congr. Zool. (Budapest) **10,** 995–997.

Gandolfi-Hornyold, A. (1930a): Le sexe et la croissance de la petite anguille de la Roubine du pont de Rousty (Camargue). Bull. Soc. cent. Aquic. Pêche **37,** 11pp.

Gandolfi-Hornyold, A. (1930b): Recherches sur l'age, la croissance er le sexe de la petite anguille argentée du Lac de Tunis. Bull. Stn. oceanogr. Salammbo. **17,** 1–50.

Gandolfi-Hornyold, A. (1932a): Die Aale des Caumasees. Schweiz. FischZtg. 9 & 10.

Gandolfi-Hornyold, A. (1932b): Le sexe de la petite anguille de repeuplement du Marais de la Grande Brière après un séjour de trois et quatre ans dans un aquarium du museum. Bull. Mus. Hist. nat. Paris, **4,** 64–70.

Gandolfi-Hornyold, A. (1934): Observation sur la taille et le sexe de deux cents petites anguilles argentées des Valli de Comacchio. Bull. Soc. oceanogr. Paris, **80,** 1398–1401.

Gandolfi-Hornyold, A. (1935): Le sexe et la croissance de la petite anguille jaune du marais de la Grand Mar (Camargue). Bull. Soc. cent. Aquic. Pêche **42,** 73–84.

Gandolfi-Hornyold, A. (1936): Civelles pigmentées de la Loire de mars et mai 1935. Bull. Soc. cent. Aquic. Pêche, Oct./Dec. 1936.

Gannon, B. J. (1971): A study of the dual innervation of teleost heart by a field stimulation technique.Comp. Gen. Pharmacol. **2**(6), 175–183.

Garcia-Romeu, F., Motais, R. (1966): Mise en evidence d'échanges Na+/NH+ chez l'anguille en eau douce. Comp. Biochem. Physiol. **17,** 1201–1204.

Gascuel, D. (1986): Flow-carried and active swimming migration of the glass eel (*Anguilla anguilla*) in the tidal area of a small estuary on the French Atlantic coast. Helgoländer Meeresunt. **40,** 321–326.

Gascuel, D. (1987): La civelle d'Anguille dans l'estuaire de la Sèvre Niortaise: biologie, écologie exploitation. Publ. Dép. d'halieutique de 17/1.

Gaygalas, K.S. (1966): Sostoyanie zapasov ugrya v Kurskom zalive. Rybn. Khoz. Moskva **42**(3), 12.

Gaygalas, K.S. (1969): Biologicheskie obosnovaniya ratsional'nogo ispol'zovaniya zapasov ugrya *Anguilla anguilla* (L.) v zalive kurskiu mares v svyazi intensifikatsieyi ego dobytsi. Vopr. Ikhtiol. **9**(1), 68–79.

Gedymin, A., Gottwald, S. (1968): Wyniki doswiadczen z przetrzymywanim narybku wegorza. Gospodarka rybna **20**(1), 10–15.

Gehring, F., Wünsche, G. (1965); Versuche über die Haltbarkeit von geräuchertem Aal in Klarsicht-beuteln. Infn. FischW. Auslds. **12,** 88–91.

Gemballa, S. (1995): Vergleichend-anatomische Untersuchungen am Lokomotionsapparat der Actinopterygii. Phylogenetische Rekonstruktion und funktionelle Hypothesen. Dissertation, Univ. Tübingen.

Gemzøe, K. J. (1906): Age and rate of growth of the eel. Rep. Dan. biol. Stn **14,** 10–39.

Gerard, P. (1958): L'appareil excréteur. In Grassé, P.P. (ed.): Traité de Zoologie **13,** 1545–1564. Paris, Masson Cie.

Ghittino, P. (1972): mass mortality by bacterial gill disease (BGD) in elvers of the European eel (*Anguilla anguilla*). Riv. Ital. Piscic. Ittiopatol. **7,** 83–85.

Ghittino, P., Smith, F.-G., Glenn, J. S. (1974): A case report of Myxosporidia (*Myxidium girardi*) in the dermis of an American eel *(Anguilla rostrata*). Riv. ital. Piscic. Ittiopatol. **9,** 13–18.

Gibson, D. J. (1996): Guide to the parasites of fishes of Canada. (Margolis, L., Kabata, Z. eds) Part IV: Trematoda. Can. Spec. Publ. Fish. and Aquat. Sci. 124.

Gibson, W., Boubée, J. A.T. (1992): Matahina dam, elver pass development, installation and fisheries aspects. Rep to ECNZ (New Zealand).

Gilham, I. D., Baker, B.I. (1984): Evidence for the participation of a melanin concentrating hormone in physiological colour change in the eel, *Anguilla anguilla.* J. Endocrinol. **102,** 237–244.

Gill, T. S., Epple, A. (1993): Stress-related changes in the hematological profile of the American eel (*Anguilla rostrata*). Ecotoxicol. Environ. Saf. **25,** 227–235.

Gilloteaux, J. (1969): Note sur l'innervation des branchies chez *Anguilla anguilla*. Experientia **25,** 270–271.

Gilson, G. (1908): L'anguille, sa reproduction, ses migrations et son intérêt économique en Belgique. Annls. Soc. r. zool. malacol. Belg. **42,** 7–58.

Gimeno, L., Ferrando, M. D., Sanchez, S., Andreu, E. (1994): Endosulfan effects on liver and blood of the eel, *Anguilla anguilla*. Comp. Biochem. Physiol. 108C, 343–348.

Gimeno, L., Ferrando, M. D., Sanchez, S., Gimeno, L. O., Andreu, E. (1995): Pesticide effects on eel metabolism. Ecotoxicol. Environ. Saf. **31,** 153–157.

Ginsburg, A., Billard, R. (1972): The ultrastructure of the eel spermatozoid. J. Microsc. (Paris) **14,** 50A–51A.

Gitter, A. (1933): Untersuchungen über die Herztätigkeit der Fische. V. Z. vergl. Physiol **18,** 654–666.

Gleiser, S. I., Khodorkovskii, V. A. (1971): Experimental determination of geomagnetic reception in *Anguilla anguilla*. Dokl. Akad. Nauk SSSR **201,** 964–967.

Godfrey, H. (1957): Feeding of eels in four New Brunswick Salmon streams. Prog. Rep. Atlant. Cst. Stns. **67,** 19–22.

Gollub, H. (1955): Probleme der Aalwirtschaft bei der qualitätsmäßigen Ertragssteigerung in der Binnenfischerei. Dt. FischZtg Radebeul **2,** 36–42.

Gollub, H. (1959): Über die im Bezirk Schwerin durchgeführten Aalförderungsmaßnahmen. Dt. FischZtg Radebeul **6,** 86–89.

Gollub, H. (1963): Erträge und Ertragsziele der Aalwirtschaft im Bereich Schwerin unter besonderer Berücksichtigung der Aalbesatzförderung. Lehrbeiträge staatl. gepr. Fischwirte, Zentralstelle Fachausbildung (Manuskriptdruck), Brieselang, 1–34.

Golovin, P. P. (1977): Monogeneans of eel during its culture using heated water. Investig. Monog. USSR, Zool. Inst. Acad. Sci. 144–150.

Golvan, Y. J. (1969): Systematiques des Acanthocephales. Prem. Part.: L´ordre des Palaeacanthocephala Meyer 1931. Mem. Mus. Nat. Hist.Nat. Ser.A, **57,** 1.

Gordon, J., Shapley, R. W., Kaplan, E. (1978): The eel retina, receptor classes and spectral mechanisms. J. Gen. Physiol. **71,** 123–138.

Gorovaya, S. L. (1969): K stroeniya organa obonyaniya nekotorykh ryb. Vyestsi Akad. Nauk Byelorus. SSR Syer. Byal. Nauk **5,** 117–119.

Gorovaya, S. L. (1973): Morphology of the olfactory analyser of the European eel (*Anguilla anguilla* L.). Vyestsi Akad. Nauk. SSR **5,** 104–108, 138.

Gousset, B. (1990): European eel (*Anguilla anguilla* L.) farming technologies in Europe and in Japan: Application of a comparative analysis. Aquaculture 87 (3/4), 209–235.

Gousset, B. (1992): Eel culture in Japan. Bulletin de l'institute Océanographique de Monaco, no. spécial 10.

Grafflin, A. L. (1937): Observations upon the structure of the nephron in the common eel. Am. J. Anat. **61,** 21–62.

Grassi, B. (1896): The reproduction and metamorphosis of the common eel, *Anguilla vulgaris*. P. R. Soc. London **60,** 260–271.

Grassi, B. (1913): Metamorfosi die murenoidi. Richerche sistematiche ed ecologiche. Regio Comitato Talassografico Itaiano, Monografia **1,** 1–211.

Grassi, B., Calandruccio, S. (1897): Fortpflanzung und Metamorphose des Aales. Allg. FischZtg. **22,** 402–408.

Gray, J. (1936): Studies in animal locomotion IV. The neuromusculatur mechanism of swimming in the eel. J. exp. Biol. **13,** 170–180.

Gray, K. W., Andrews, C. W. (1971): Age and growth of the American eel (*Anguilla rostrata* [Le Sueur]) in Newfoundland waters. Can. J. Zool. **49,** 121–128.

Gregory, W. K. (1933): Fish skulls: a study of the evolution of natural mechanisms. Trans. Amer. Phil Soc. **23,** 75–481.

Gritzke, H. (1980): Experimentelle Untersuchungen über Transporteffekte auf das Verhalten von Glasaalen. Diplomarbeit Fachbereich Biologie, Univ. Hamburg.

Grodzinski, Z. (1954): Contractions of the isolated heart of the European glass eel, *Anguilla anguilla* L. Bull. Acad. polon. Sci. Cl. II Biol., **2**(1), 19–22.

Grunert, F. G. (1969): Aalpöddern auf der Unterweser. Fisch u. Fang, **10,** 118–121.

Guglielmone, R. (1966): Aspetti morfo-funzionali del sistema istiocitario nell'organo renale di *Anguilla vulgaris* (Linn). Atti Accad. Sci. Torino **100,** 463–470.

Guglielmone, R. (1967a): Correlazioni isto-functionali nel convoluto prossimale del nefrone di *Anguilla vulgaris* (L.). Atti Accad Sci. Torino **101,** 383–390.

Guglielmone, R. (1967b): Dotazione stiocitaria dell'organo renale die *Anguilla vulgaris* (L.). Archo zool. ital. **52,** 193–216.

Guibé, J. (1958): Les poissons toxiphores. In Grassé, P.P. (ed.): Traite de Zoologie, **8,** 1935–1943, Paris, Masson & Cie.

Gullestadt, N. (1972): Papillomatosis in the eel, *Anguilla anguilla*. First record from Norway. Fauna. Oslo **25,** 275–276.

Gunning, G. E., Shoop, C. R. (1962): Restricted movements of the American eel, *Anguilla rostrata* (Le Sueur), in freshwater streams, with comments on growth rate. Tulane Stud. Zool. **9,** 265–272.

Gyurko, S. L. (1961): L'anguille (*Angilla anguilla* L.) dans la rivière Crisul Negru. Bul. Inst. Cercet. Pisc. **20,** 44–47.

Haack, H. (1879): Die Acclimatisierung des Aales im Donaugebiet. Bayr. FischZtg. München **4,** 24–25.

Haack, H. (1881): Bericht über die erste größere Aussetzung von Aalen in das Gebiet der Donau. Circ. Dt. FischVer., 123–125.

Hadderingh, R. (1983). Fish Impingement at Power Stations situated along the River Rhine and Meuse in the Netherlands. Hydrobiological Bull **17**(2), 129–141.

Haempel, O., Neresheimer, E. (1914): Über Altersbestimmung und Wachstum des Aales. Z. Fisch. **14,** 265–285.

Haenen, O.L., Banning, P.A. van (1990): Detection of larvae of *Anguillicola crassus* (an eel swimbladder nematode) in freshwater fish species. Aquaculture **87,** 103–109.

Hagen, F. V. (1936): Die wichtigsten Endocrinen des Flußaales. Zool. Jb. (Anat.) **61,** 467–538.

Hagström, O., Wickström, H. (1990): Immigration of young eels to the Skagerrak – Kattegat area 1900–1989. Int. Revue ges. Hydrobiol. **75,** 707–716.

Hahlbeck, E. (1993): Zur Aalschleppnetzfischerei in der Ostsee. Arb. Dt. Fischereiverband, **59,** 1–12.

Hahlbeck, E., Kuhlmann, H. (1997): Germany. In: Moriarty, C., Dekker, W. (eds): Management of the European eel. Fish. Bull. **15,** 57–59.

Hain, J. H. W. (1975): The behaviour of migratory eels, *Anguilla*, in response to current, salinity and lunar period. Helgoländer wiss. Meeresunters. **27,** 211–233.

Haller, B. (1908): Zur Phylogenese des Nierenorgans (Holonephros) der Knochenfische. Jena. Z. Naturw. **43,** 729–801.

Halsband, E. (1971): Fishing for eels by means of electricity. EIFAC Techn. Pap.14, 57–67.

Halsband, E. (1980): Hinweise zur Anwendung besonderer Kathodenarten für die Elektrofischerei. Fisch und Fang **21**(7), 622–623.

Halsband, E. (1989): Nur durch elektrischen Impulsstrom ist das Scheuchen und Leiten von Fischen wirksam möglich. Der Fischwirt **10,** 1989, 1–4.

Halsband, E., Halsband, I. (1971): Physiologische Studien zur Ermittlung des Schädlichkeitsgrades von "Rotschlamm" in Seewasser. Wasser, Luft-Betrieb. **15**(7), 268–273.

Hanek, G., Molnar, K. (1974): Parasites of freshwater and anadromous fishes from Matamek River system, Quebec. J. Fish. Res. Bd Can. **31,** 1135–1139.

Hanson, A. (1966): Differences de teneur en n-acetyl-histidine chez les téleostéens d'eau douce et d'eau de mer. C.r. Séanc. Soc. Biol. **160,** 265–268.

Hanssen, R. G. J. M., Lafeber, F. P. J. G., Flik, G., Wendelaar Bonga, S. E. (1989): Ionic and total calcium levels in the blood of the European eel *(Anguilla anguilla)*: effects of stanniectomy and hypocalcin replacement therapy. J. exp. Biol. **141,** 177–186.

Harden Jones, F. R. (1968): Fish Migration, E. Arnold Ltd. London.

Harder, W. (1964): Anatomie der Fische. Handb. Binnenfisch. Mitteleur. **6,** 2nd,ed., Stuttgart.

Hardt, A. (1981): Anrüchige Würmer mag der Aal nicht. Fisch u. Fang **22,** 527.

Harsanyi, A., Köthke, C. (1988): Einsatz des Aalfanggroßgerätes an der Niederbayerischen Donau. Arb. Dt. Fischereiverband, **47,** 25–33.

Hartmann, F. (1987): Schwimmblasenwürmer beim Aal. Fischer + Teichwirt **38,** 2–3.

Hartmann, F., Nellen, W. (1998). The status of the European eel, after the invasion of the introduced swim bladder parasite, *Anguillicola crassus*, using the eel population in the lower Elbe as an example. J. marine Sci.

Hashimoto, Y., Konosu, S.,Fusetani, N., Nose, T. (1968): Attractants for eels in the extracts of short-necked clam. 1. Survey of constituents eliciting feeding behaviour by the omission test. Bull. Jap. Soc. scient. Fish. **34,** 78–83.

Hasler, A. D. (1956): Perception of pathways by fishes in migration. Q. Rev. Biol. **31,** 20–209.

Hasler, A. D. (1966): Underwater guideposts, 155pp. Madison, Univ. Wisconsin Press.

Hasler, A. D., Wisby, W. J. (1951): Discrimination of stream odours by fishes and its relation to parent stream behaviour. Amer. Nat. **85,** 223–238.

Haslewood, G. A. D. (1969): In Schiff, L., Carey, J. B., Dietschy, J.(eds): Bile Salt Metabolism, 151–159. Thomas, Springfleld, Illinois.

Hass, G. (1964): Aalwirtschaft im hessischen Rheinstromgebiet. Fischwirt **14,** 1–12.

Hata, M., Hata, M. (1967): Fatty acid composition of esters in fish liver. I. Separation of fatty acid of vitamin A 1 ester. II. Eel (*Anguilla anguilla*) liver and *Channa argus* (Cantor) liver. Tohoku J. agric. Res. **18,** 225–231, 267–273.

Hatai, K., Horose, H., Hioki, M., Miyakawa, M., Egusa, S. (1979): *Dermocystidium anguillae* found on the gills of the European eel, *Anguilla anguilla* cultures in Japan. Fish Pathol. **13,** 205–210.

Hatakeyama, T., Ukamara, T. (1928): Über die Kenntnis von der Fischgalle. III. Die Galle von *Cyprinus carpio*, *Carassius auratus*, *Anguilla japonica* und *Anago anago*. Z. Biochem. **9,** 333–335.

Hathaway, C. B., Epple, A. (1989): The sources of plasma catecholamines in the American eel, *Anguilla rostrata*. Gen. Comp. Endocrinol. **74,** 459–465.

Hathaway, C. B., Epple, A., 1990: Catecholamines, opioid peptides and true opiates in the chromaffin cells of the eel: immunohistochemical evidence. Gen. Comp. Endocrinol. **79,** 393–405.

Hattop, H.-W. (1979): Der Stand der Entwicklung von Elektrozugnetzen zur Abfischung von Seen und Teichen. Z. Binnenfischerei DDR **26**(7), 214–218.

Hattop, H. W., Predel, G. (1969): Die Anwendung elektrischer Schleppnetze in den Binnengewässsern der Deutschen Demokratischen Republik. Z. Fisch. **17,** 215–226.

Hebrank, M. R. (1980): Mechanical properties and locomotor functions of eel skin. Biol. Bull. Mas. Biol. Lab., Woods Hole **158**(1), 58–68.

Heermans, W., Willigen, J. A. van (1979): Investigations on the condition-factor and the sex-ratio of brown eels in fresh and brackish waters in the Netherlands. Abstract of report ZA 79–01 Netherlands Institute of Fisheries Investigations, 1–4.

Heincke, F. (1894): Die Fische Helgolands. Wiss. Meeresunters. (Helgol.), **1,** 97–120.

Heinsbroek, L. T. N. (1991): A review of eel culture in Japan and Europe. Aquaculture and Fisheries Management **22,** 57–72.

Heinsbroek, L.T.N., Kamstra, A. (1990): Design and performance of water recirculation systems for eel culture. Aquaculture Engineering **9,** 187–207.

Heinsbroek, L. T. N., Kamstra, A. (1995): The River Eels. In Nash, C.E., Novotny, A.J. (eds): Production of aquatic animals, 109–131. World Animal Science. Elsevier, Amsterdam.

Heitz, F. A. (1918): *Salmo salar* L., seine Parasitenfauna und seine Ernährung in Meer und Süßwasser. Eine parasitologisch-biologische Studie. Arch. Hydrobiol. **12,** 311–372.

Heldt, H., Heldt, H. (1929a): Études sur les civelles de Sidi-Daoud. Bull. Stn. oceanogr. Salammbo, **16,** 1–23.

Heldt, H., Heldt, H. (1929b): Les civelles du lac de Tunis. Considération sur les époques de présence, la taille et le poid. Bull. Stn oceanogr. Salammbo, **14,** 1–40.

Helfman, G. S., Bozeman, E. L., Brothers, E. B. (1984): Size, age, and sex of American eels in a Georgia river. Trans. Amer. Fish. Soc **113,** 132–141.

Hellier, R. H. (1967): The fishes of the Santa Fé river system. Bull. Fla. St. Mus. biol. Sci. **2,** 1–46.

Henck, B. (1965): Bundgarnfischerei an der Ostküste Schleswig-Holsteins. Fischerblatt **13,** 1–9.

Henderson, I. W. (1997): Endocrinology of the Vertebrates. In Dantzler, W.H. (ed.): Comparative Physiology. Vol **1,** Handbook of Physiology Section **13,** Oxford University Press/American Physiological Society, New York and Oxford, 623–749.

Henderson, I. W., Chester Jones, I. (1967): Endocrine influences on the net extrarenal fluxes of sodium and potassium in the European eel, *Anguilla anguilla* (L.). J. Endocr. **37,** 319–325.

Henderson, I. W., Deacon, C. F. (1993): Phylogeny and comparative physiology of the renin-angiotensin system. In Robertson, J.I.S., Nicholls, M.G. (eds): The renin-angiotensin system. Vol **1,** 2.1.-2.8, Gower, London.

Henderson, I. W., Garland, H. O. (1980): The interrenal gland of Pisces: Physiology. In Chester Jones, I., Henderson, I.W. (eds): General, Comparative and Clinical Endocrinology of the Adrenal Cortex. Volume **3,** Academic Press, London, New York, San Francisco, 471–523.

Henderson, I. W., Hazon, N., Hughes, K. (1985): Hormones, ionic regulation and kidney function in fishes. In Laverack, M.S. (ed.): Physiological adaptations of marine animals. Company of Biologists, Cambridge, Symposia of the Society for Experimental Biology **29,** 245–265.

Hendriks, A. J., Pieters, H. (1993): Monitoring concentrations of microcontaminants in aquatic organisms in the Rhine delta: a comparison with reference values. Chemosphere **26,** 817–836.

Henking, H. (1908): Der Aalfang in den Lagunen von Comacchio und Venedig. In: Mitt. Dt. Seef.-Verein, 366–393.

Henrikson, R. C., Matoltsy, A. G. (1968a): The fine structure of teleost epidermis. I. Introduction and filament containing cell. J. Ultrastruct. Res. **21,** 194–212.

Henrikson, R. C., Matoltsy, A. G. (1968b): The fine structure of teleost epidermis. II. Mucous cells. J. Ultrastruct. Res. **21,** 213–221.

Henrikson, R. C., Matoltsy, A. G. (1968c): The fine structure of teleost epidermis. III. Club cells and other cell types. J. Ultrastruct. Res. **21,** 222–232.

Herold, W. (1933): Über einen Zug von Jungaalen in der Swine. Zool. Anz. **101,** 161–164.

Herrmann, G. (1957): Bemerkenswerter Aalbrutaufstieg. Fischwirt **7,** 171–172.

Herrmann, G. (1967): Die Aalwirtschaft im Lande Schleswig-Holstein. Arch. FischWiss. **18** (Suppl. 1), 405–424.

Herrmann, G., Marre, G. (1961): Aalfang und Aalwirtschaft in Schleswig-Holstein. Z. Fisch. **10,** 611–634.

Heusden, G. P. H. van (1943): De trek van den glasaal naar het IJsselmeer. Doctor. Thesis, Utrecht.

Hilge, V., Klinger, H. (1978): Changes in the hemogram of the male European eel (*Anguilla anguilla* L.) during induced maturation. ICES CM 1978/M4.

Hill, B. J., Williams, R. F., Smale, C. J., Underwood, B. O., Brown, F. (1980): Physiochemical and serological characterization of two rhabdoviruses isolated from eels. Intervirology **14,** 208–212.

Hill, L. G. (1969): Reactions of the American eel to dissolved oxygen tensions. Tex. J. Sci. **20,** 305–313.

Hine, P. M. (1980): Distribution of helminths in the digestive tracts of New Zealand freshwater eels 1. Distribution of digeneans. N. Z. J. Mar.Freshw. Res. **14,** 329–338.

Hine, P. M., Boustead, N. C. (1974): A guide to disease in eel farms. New Zealand Ministr. Agricult. Fisheries, Fish. Res. Div., Occasional Publ. No. **6,** 1–28.

Hine, P.M., Kennedy, C.R. (1974): Observations on the distribution, specificity and pathogenicity of the acanthocephalan *Pomphorhynchus laevis* (Müller). J. Fish Biol. **6,** 521–535.

Hirano, T. (1967): Effects of hypophysectomy on water transport in isolated intestine of the eel, *Anguilla japonica*. Proc. Japan Acad. **43,** 152–155.

Hirano, T. (1974): Some factors regulating water intake by the eel, *Anguilla japonica.* J. exp. Biol. **61,** 737–748.

Hirano, T. (1992): Endocrine control of osmoregulation in migratory fishes. Marine Biol. **2,** 3–14.

Hirano, T., Hasegawa, S., Yamauchi, H., Orimo, H. (1981): Further studies on the absence of hypocalcemic effects of eel calcitonin in the eel, *Anguilla japonica*. Gen. Comp. Endocrinol., **43,** 42–50.

Hirano, T., Satou, M., Utida, S. (1972): Central nervous control of osmoregulation in the eel, *Anguilla japonica*. Comp. Biochem. Physiol. 43A, 537–544.

Hirofugi, S. (1966): Stero-bile acids and bile alcohols. LXXXIII. Metabolism of deoxycholic acid and 5beta-cholestane-3a, 12a-diol in the eel. J. Biochem., Tokyo **60,** 514–518.

Hoa, L. V., Khue, P. N. (1968): Relation entre les Seuratoidea et les Cucullanidae par l'intermédiaire d'un nouveau nématode *Campanarougetia campana rougetae* n. g. n.sp., parasite des anguilles du Sud Viet Nam. Bull. Soc. Path. exot., **60,** 393–398.

Hoar, W. S. (1955): Phototactic and pigmentary responses of sockeye salmon smolts following injury to the pineal organ. J. Fish. Res. Bd Can. **12,** 178–185.

Hobbs, D. I. (1949): Migrating eels in Lake Ellesmere. Trans. R. Soc. N. Z., **77,** 228–232.

Hofer, B. (1897): Welche Aussicht hat der Versuch zur Einbürgerung des Aales im Donaugebiet? Allg. FischZtg **22,** 445–449.

Hoffmann, E., Hansen, H., Christensen, P. M. (1979): Changes in catch per unit effort and catch composition of silver eels in south-eastern Denmark 1949–1977. ICES CM 1979/M 28.

Hoffmeister, H. (1970): Die kleine Hochsee- und Küstenfischerei Schleswig-Holsteins i. J. 1969. Fischerbl. **18,** 80–88.

Hofmann, J. (1960): Haben Aalfänge noch eine Berechtigung und wie lassen sich neue einrichten. Allg. FischZtg.85, 301–303.

Hohl, R. (1987): Wandernde Kontinente. Urania-Verlag, Leipzig. 128pp.

Hölke, H. (1964): Aalfang mit Hilfe von elektrischem Licht in der Flußfischerei. Dt. Fisch Ztg., Radebeul, **11,** 82–88.

Holmberg, B., Jensen, S., Larsson, A., Lewander, K., Olsson, M. (1972): Metabolic effects of technical Pentachlorphenol (PCP) on the eel *Anguilla anguilla* L. Comp. Biochem. Physiol. 43 B, 171–183.

Holmes, W. N., Donaldson, E. M. (1969): The body compartments and the distribution of electrolytes. In Hoar, W.S., Randall, D.J. (eds): Fish Physiology **1,** 1–90. Academic Press, New York, London.

Holmgren, K. (1996a): Otolith growth scaling of eel, *Anguilla anguilla* (L.) and back-calculation errors revealed from alizarin labelled otoliths. Nordic J. Freshw. Res. **72,** 71–79.

Holmgren, K. (1996b): Effect of water temperature and growth variation on the sex ratio of experimentally reared eels. Ecol. Freshw. Fish. **5,** 203–212.

Holmgren, K., Mosegaard, H. (1996a): Implications of individual growth status on the future sex of the European eel. J. Fish Biol. **49,** 910–925.

Holmgren, K., Mosegaard, H. (1996b): Plasticity in growth of indoor reared European eel. J. Freshw. Res. **72,** 63–70.

Holmgren, K., Wickström, H. (1993): Sex dimorphism in cultured eels (*Anguilla anguilla* L.). Nordic J. Freshw. Res. **68,** 80–90.

Holmgren, K., Wickström, H. (1996): Note on otolith growth in elvers, *Anguilla anguilla* L., and the relative otolith size during somatic growth. Arch. Ryb. Pol. **4,** 223–234.

Holmgren, K., Wickström, H., Clevestam, P. (1997): Sex related growth of European eel, *Anguilla anguilla*, with focus on median silver eel age. Can. J. Fish. Aquat. Sci. 54.

Honma, Y. (1966): Notes on the catadromous eels obtained from off the coast of Niigata, the Sea of Japan, with special reference to the histology of some of the organs. La Mer **4,** 241–260.

Hopkins, C. L. (1971): Production of fish in two small streams in the North Island of New Zealand. N. Z. Jl. Mar. Freshw. Res. **5,** 280–290.

Hoshina, T. (1952): Notes on some myxosporidian parasites on fishes of Japan. J. Tokyo Univ. Fish. **39,** 69–89.

Hoshina, T., Ookubo, M. (1956): On a fungi-disease of eel. J. Tokyo Univ. Fish. **42,** 1–13.

House, C. R., Maetz, J. (1974): On the electrical gradient across the gill of the seawater-adapted eel. Comp. Biochem. Physiol. 47A, 917–924.

Huang, Y. S., Schmitz, M., Le Belle, N., Chang, C.F., Querat, B., Dufour, S. (1997): Androgens stimulate gonadotropin-II-subunit in eel pituitary cells in vitro. Mol. Cell. Endocrinol. **131,** 157–166.

Huggins, A. K., Colley, L. (1971): The changes in the non-protein nitrogenous constituents of muscle during the adaptation of the eel *Anguilla anguilla* L. from freshwater to seawater. Comp. Biochem. Physiol. 38B, 537–541.

Hulbert, W. C., Moon, T. W. (1978): A histochemical, light and electron microscopic examination of eel, *Anguilla rostrata* red and white muscle. J. Fish. Biol. **13,** 527–534.

Hulet, W. H. (1978): Structure and functional development of the eel leptocephalus *Ariosoma balearicum* (de la Roche, 1809). Philos. Trans. R. Soc. London 282 (987), 107–138.

Hurley, D. A. (1972): The American eel (*Anguilla rostrata*) in eastern Lake Ontario. J. Fish. Res. Bd Can. **29,** 535–543.

Hurley, D. A. (1973): The commercial fishery for American eel, *Anguilla rostrata* (Le Sueur), in Lake Ontario. Trans. Amer. Fish. Soc. **102,** 169–377.

Hussein, S. A. (1981): The population density, growth and food of eel *Anguilla anguilla* L. in some tributaries of the River Tweed. Proc. 2nd British Freshwater Fish. Conf., Univ. of Liverpool, 120–128.

Hvidsten, N. R. (1985): Yield of silver eel and factors affecting downstream migration in the stream Imsa, Norway. Rep. Inst. Freshwater Res., Drottningholm **62,** 71–74.

Hwang, D. F., Lin, J.F., Jeng, S. S. (1996): Comparative toxicity of copper and zinc to isolated eel hepatocytes. Chem. Ecol. **12,** 109–114.

Ibrahim, K. H. (1961): Up-stream migration of elvers of *Anguilla nebulosa (= bengalensis)* over first anicut of the River Godavari. J. Bombay nat. Hist, Soc. **58**(3), 810–811.

Inui, Y. (1969): Hepactectomy in eels. Its operation technique and effects on blood glucose. Bull. Jap. Soc. scient. Fish. **35,** 975–978.

Inui, Y., Egusa, S. (1967): Histological changes observed in glass eel liver during starvation. Bull. Jap. Soc. Scient. Fish. **33,** 181–189.

Inui, Y., Ishioka, H. (1983a): Effects of insulin and glucagon on the incorporation of ^{14}C glycine into the protein of the liver and opercular muscle of the eel in vitro. Gen. Comp. Endocrinol. **51,** 208–212.

Inui, Y., Ishioka, H. (1983b): Effects of insulin and glucagon on amino acid transport into the liver and opercular muscle of the eel in vitro. Gen. Comp. Endocrinol. **51,** 213–218.

Inui, Y., Ishioka, H. (1985): In vivo and in vitro effects of growth hormone on the incorporation of ^{14}C leucine into protein of liver and muscle of the eel. Gen. Comp. Endocrinol. **59,** 208–212.

Ishida, M. (1958): On the neurobiological relationship of sensory, motor and reticular nuclei of the medulla oblongata in teleosts, especially in *Anguilla japonica.* Hiroshima J. med. Sci. **6,** 203–215.

Ishikawa, S. (1916): Über im Auge des Japanischen Aales schmarotzende *Filaria anguillae* n. sp. (orig. jap.). Doobutsugaku Zasshi **28,** 332.

Itazawa, Y. (1970): Characteristics of respiration of fish considered from the arteriovenous difference of oxygen content. Bull. Jap. Soc. scient. Fish. **36,** 571–577.

Itida, N., Turitzin, S., Steen, J. B. (1970): Root-shift in eel hemoglobin. Respiration Physiol. **8**(3), 276–279.

Ito, H. (1974): Fine structure of the torus semicircularis of some teleosts. J. Morph. **142,** 137–152.

Ivashkin, V. M., Sobolev, A. A., Khromova, L. A. (1977): Camallanata of animals and man and diseases caused by them. Moskva 1971, translat.: Israel Prog–Sci.Transl., Jerusalem.

Jackson, P. B. N. (1959): New records and little known species of fish. Occ. Pap. natn. Mus. Sth Rhodesia **3,** 295–305.

Jacob, R. (1970): Die Wollhandkrabbe als Angelköder. Fisch u. Fang **11,** 242–244.

Jacobsen, J. P., Johansen, A. C. (1922): On the causes of the fluctuations in the yield of some of our fisheries. II. The eel fisheries. Meddr Kommn Havunders. **6**(9), 1–32.

Jaeger, R. (1975): Zur Wirkung von Adrenalin und Noradrenalin auf Ruhe und Aktionspotential bei Fischen *Salmo gairdneri* Richardson and *Anguilla vulgaris* L. Revue Suisse Zool. **82,** 229–235.

Jäger, K. (1960): Herstellung und Anwendung der wichtigsten Fanggeräte sowie deren Behandlung. In Wundsch, H.H. (ed.): Fischereikunde. 207–260 Radebeul, Berlin.

Jahn, L. A. (1969): Movements and homing of cutthroat trout (*Salmo clarki*) from open-water areas of Yellowstone Lake. J. Fish. Res. Bd Can. **26,** 1243–1261.

Jakubowski, M. (1960a): The structure and vascularization of the skin of the eel (*Anguilla anguilla* L.) and the viviparous blenny (*Zoarces viviparus* L.). Acta biol. cracov., Ser. Zool. **3,** 1–22.

Jakubowski, M. (1960b): The structure and vascularization of the skin of the leather carp (*Cyprinus carpio var. nuda*) and flounder (*Pleuronectes flesus lascus* Pall.). Acta biol. cracov., Ser. Zool. **3,** 139–162.

Jamieson, A., Turner, R. J. (1980): Muscle protein differences in two eels *Anguilla anguilla* L. and *Anguilla rostrata* (LeSueur). Biol. J. Linn. Soc. **13,** 41–45.

Jampol, L. M., Epstein, F. H. (1970): Sodium-potassium activated adenosine triphosphatase and osmotic regulation by fishes. Am. J. Physiol. **218,** 607–611.

Janicki, R. H., Kinter, W. B. (1971): DDT Disrupted osmoregulatory events in the intestine of the eel *Anguilla rostrata* adapted to seawater. Science, N. Y. **173,** 1144–1148.

Janicki, R.H., Lingus, J. (1970): Mechanisms of ammonia production from aspartate in teleost liver Comp. Biochem. Physiol. **37,** 101–105.

Jankowski, H.-J. (1968): Versuche zur Adaption der Fische in normalen Temperaturbereich. Helgoländer wiss. Meeresunters. **18,** 317–362.

Järvi, T. H. (1909): Über das Vorkommen des Aales im Wassersystem des Kymi-Flusses (Finnland). Meddn Soc. Fauna Flora fenn. **35,** 222–226.

Jellyman, D. J. (1977): Invasion of a New Zealand freshwater stream by glass-eels of two *Anguilla* spp. N. Z. J. Mar. Freshw. Res. **11,** 193–209.

Jellyman, D. J. (1979a): Upstream migration of glass eels (*Anguilla* spp.) in the Waikato River. N. Z. J. Mar. Freshw. Res. **13,** 13–22.

Jellyman, D. J. (1979b): Scale development and age determination in New Zealand freshwater eels (*Anguilla* spp.). N. Z. J. Mar. Freshw. Res. **13,** 23–30.

Jellyman, D. J. (1987): Review of the marine life history of Australasian temperate species of *Anguilla*. Amer. Fish. Soc. Symp. **1,** 276–285.

Jellyman, D. J. (1993): A review of the fishery for freshwater eels in New Zealand. N. Z. Freshw. Res. Rep, **10,** 1–51.

Jellyman, D.J. (2001): The influence of growth rate on the size of migrating female eels in Lake Ellesmere, New Zealand. J. Fish Biol. **58,** 725–736.

Jellyman, D. J., Chisnall, B. L., Dijkstra, L. H., Boubee, J. A. T. (1996): First record of the Australian longfinned eel, *Anguilla reinhardtii*, in New Zealand. Mar. Freshw. Res. **47,** 1037–1040.

Jellyman, D. J., Coates, G. D. (1976): The farming of freshwater eels in New Zealand. IPFC (FAO) 76/Sym/33, 1–5.

Jellyman, D. J., Glova, G. J. (1996): Movements of shortfinned eels, *Anguilla australis*, in Lake Ellesmere, New Zealand: results from mark-recapture studies and sonic tracking. N. Z. J. Mar. Freshw. Res. **30,** 371–381.

Jellyman, D., Todd, P. (1980): New South Island eel fishery. Catch **7**(5), June.

Jellyman, D.J., Glova, G.J., Todd, P.R. (1996): Movements of short-fined eels, *Anguilla australis*, in Lake Ellesmere, New Zealand: results from mark-recapture studies and sonic tracking. N.Z. J. Mar. Freshw. Res. **30,** 371–381.

Jens, G. (1953): Über den lunaren Rhythmus der Blankaalwanderung. Arch. FischWiss. **4,** 94–110.

Jens, G. (1967): Die Aalwirtschaft des Landes Rheinland-Pfalz. Die Mosel. Arch. FischWiss. 17 (Suppl.1), 264–275.

Jens, G. (1969): Von Aalleitern und Fischschleusen. Allg. FischZtg. **94,** 46–48.

Jens, G., (1982): Der Bau von Fischwegen, Fischtreppen, Aalleitern und Fischschleusen. Verlag Paul Parey Hamburg. 93pp.

Jensen, A. J. C. (1961): Die Schwankungen des Aalbestandes und der Aalfischerei in den dänischen Küstengewässern. Z. Fisch. **10,** 635–641.

Jensen, A. S. (1937): Remarks on the Greenland eel, its occurrence and reference to *Anguilla rostrata*. Meddr. Grönland **118, 9,** 8.

Jerkoe, H., Turunen-Rise, I., Enger, P.S., Sand, O. (1989): Hearing in the eel (*Anguilla anguilla*). J. Comp. Physiol. A **165**(4), 455–459.

Jespersen, P. (1942): Indo-Pacific leptocephalids of the genus *Anguilla*. Dana Rep. **22,** 1–128.

Johansen, A. C. (1929): Mortality among porpoises, fish and the larger crustaceans in the waters around Denmark in severe winters. Rep. Dan. biol. Stn. **35,** 59–91.

Jollie, M. (1986): A primer of bone names for the understanding of the actinopterygian head and pectoral girdle skeleton. Can. J. Zool. **64,** 365–379.

Jones, J. W. (1959): Eel migration. Nature **184,** 1281.

Jones, S., Pantulu, V. R. (1954): A remarkable case of albinism in the freshwater eel, *Anguilla bengalensis* Gray. J. Bombay nat. Hist. Soc. **51,** 285–287.

Jubb, R. A. (1961): The freshwater eels (*Anguilla* spp.) of Southern Africa. An introduction to their identification and biology. Ann. Cap. prov. Mus. **1,** 15–48.

Jubb, R. A. (1964): The eels of South African rivers and observation on their ecology. Monographia biol. **14,** 186–205.

Kagawa, H. (2002). Artificial induction of oocyte maturation and ovulation. In: Advances in Eel Biology (Proceedings of an International Symposium, Tokyo 2002).

Kaiya, H., Takei, Y. (1996): Changes in plasma atrial and ventricular natriuretic peptide concentration after transfer of eels from fresh water to sea water or vice versa. Gen. Comp. Endocrinol. **104,** 337–343.

Kaiya, H., Takei, Y. (1997): Interaction of osmotic and volemic mechanisms in secretion of atrial and ventricular natriuretic peptides in eels. Gen. Comp. Endocrinol. **107,** 322–326.

Kakaceva-Avramova, D. (1983): Helminths of freshwater fishes in Bulgaria. Publ. House Bulg. Acad. Sci. Sofia.

Kaluza, J. (1959): The blood vessels of the olfactory organ of the eel. Acta biol. cracov. Ser. Zool. **2,** 89–96.

Kamiya, M., Utida, S. (1968): Changes in activity of sodium-potassium activated adenosinetriphosphate in cells during adaptation of the Japanese eel to seawater. Comp. Biochem. Physiol. **26,** 675–685.

Kamstra, A., Davidse, W. P. (1991): Evaluation of biological, technical and economical aspects of eel farming in intensive recirculation systems. Report for the European Community.

Kanai, K., Wakabayashi, H., Egusa, S. (1977): Comparison of intestinal microflora between healthy and diseased pond-cultured eels. Fish. Pathol. **12,** 199–204.

Kaneko, T., Hasegawa, S., Sasai, S. (2001): Chloride cells in Japanese eel (*Anguilla japonica*) during their early life stages and downstream migration. *Proc. Intern. Symp. Adv. in Eel Biol.* Univ. Tokyo, 2001, 95–97.

Karlsson, L. (1984): Migration of European silver eels, *Anguilla anguilla*. Acta Univ. Uppsaliensis, Abstracts of Uppsala Dissertations from the Faculty of Science 754.

Karlsson, L., Bartel, R., Tesch, F.-W. (1988): Migration and orientation of tagged silver eels released in the south-eastern Baltic. ICES CM 1988/M17.

Käse, R.H., Zenk, W. (1996): Structure of the Mediterranean water and Meddy. Characteristics in the North-Eastern Atlantic. In Kraus, W. (ed.): The warm water sphere in the North Atlantic ocean. 366–395. Gebr. Bornträger, Berlin, Stuttgart.

Katz, E. J. (1969): Further study of a front in the Sargasso Sea. Tellus **21,** 259–269.

Kaup, J. J. (1856): Catalogue of the Apodal Fish, in the Collection of the British Museum. British Museum, London.

Kawamoto, N. (1929): Physiological studies on the eel. II. The influence of temperature and of the relative volumes of the red corpuscles and plasma upon the haemoglobin dissociation curve. Sci. Rep. Tôhoku Univ., Ser. **4, 4,** 635–659.

Kenyon, A. J. (1967): The role of the liver in the maintenance of plasma proteins and amino acids in the eel, *Anguilla anguilla* L., with reference to amino acid deamination. Comp. Biochem. Physiol. **22,** 169–175.

Kerr, I. (1942): A comparative study of some teleost pituitaries. Proc. zool. Soc. London **112,** 137–156.

Kerstan, S. L. (1992): Der Befall von Fischen aus dem Wattenmeer und dem Nordatlantik 1988–1990 mit Nematodenlarven und eine Bibliographie über parasitische Nematoden in Fischen und Seesäugern. Ber. Inst. Meereskunde Univ. Kiel, 219.

Keys, A. B. (1931): Chloride and water secretion and absorption by the gills of the eel. Z. vergl. Physiol. **15,** 364–388.

Keys, A. B. (1933): The mechanism of adaption to varying salinity in the common eel and the general problem of osmotic regulation in fishes. Proc. R. Soc., B, **112,** 184–199.

Keys, A. B., Willmer, E. N. (1932): Chloride secreting cells in the gills of fishes, with special reference to the common eel. J. Physiol. **76,** 368–378.

Kiekhäfer, H., Quoss, H., Wiegner, H. (1974): Die Entwicklung und der Einsatz eines neuen Reusentypes für tiefe natürliche Gewässer und Baggerseen. Fischwirt **24,** 69–71.

Kienast, E. (1961): Das Aalstechen im Bereich der FGS Wismar. Dt. FischZtg., Radebeul, **8,** 359–378.

Kiener, A. (1965): Contribution à l'étude écologique et biologique des eaux saumâtres malagaches. Vie Milieu 16 (2C), 1013–1149.

Kim, In-Bae, Jo, Yae-Yoon (1977): Rearing of the early stage of the eel *Anguilla japonica*. Bull. Korean Fish. Soc. **7,** 179–186.

Kimura, I., Miyake, T., Kubota, S., Kamata, A., Morikawa, S., Ito, Y. (1976): Denomatous polyps in the stomachs of hatchery-grown salmonids and other types of fishes. Prog. exp. Tumor Res. **20,** 181–194.

Kinne, O. (Ed.) (1980–1990): Diseases of marine animals. Vol. I–IV. Hamburg. Vol. IV, Part **1,** Pisces, 1984.

Kirsch, R., Laurent, P. (1975): L'oesophage, organe effecteur de l'osmorégulation chez un Téléostéen euryhalin, l'anguille (*Anguilla anguilla* L.). C. R. Hebd. Séances Acad. Sci., Paris (D). **280,** 2013–2015.

Kirsch, R., Nonnotte, G. (1977): Cutaneous respiration in three freshwater teleosts. Respir. Physiol. **29,** 339–354.

Kirsche, W. (1966): Biometrische Untersuchungen am Nucleus nervi oculomotorii von Süßwasser-Knochenfischen in Beziehung zur Verhaltensweise. Biol. Zbl. **85,** 579–595.

Kirschner, L. B. (1969a): Ventral aortic pressure and sodium fluxes in perfused eel gills. Am. J. Physio1. **217,** 596–604.

Kirschner, L. B. (1969b): ATPase activity in gills of euryhaline fish. Comp. Biochem. Physiol. **29,** 871–874.

Kishida, M., Hirano, T., Kubota, J., Hasegawa, S., Kawauchi, H., Yamaguchi, K., Shirahata, K., 1987: Hypocalcamemic activity of trout hypocalcin and bovine parathyroid hormone in stanniectomized eels. J. exp. Biol. **140,** 199–208.

Kleckner, R. C. (1980a): Swimbladder volume maintenance related to initial oceanic migratory depth in silver-phase *Anguilla rostrata*. Science **208,** 1480–1482.

Kleckner, R. C. (1980b): Swimbladder wall guanine enhancement related to migratory depth in silver phase *Anguilla rostrata*. Comp. Biochem. Physiol. A **65,** 351–354.

Kleckner, R.C., McCleave, J.D. (1980): Spatial and temporal distribution of *Anguilla rostrata* and *Anguilla anguilla* leptocephali found in North American ichthyoplankton collections. ICES CM 1980/M21.

Kleckner, R.C., McCleave, J.D. (1988): The northern limit of spawning by Atlantic eels (*Anguilla* spp.) in the Sargasso Sea in relation to the thermal fronts and surface water masses. J. Mar. Res. **46,** 647–667.

Kleckner, R. C., McCleave, J. D., Wippelhauser, G. S. (1983): Spawning of American eel, *Anguilla rostrata*, relative to thermal fronts in the Sargasso Sea. Environmental Biol. Fish. **9,** 289–293.

Klust, G. (1965): FAO Catalogue of fishing gear designs. Data Sheet No. 505.

Klust, G. (1971): Eel stownets in German rivers. EIFAC Techn. Pap. **14,** 39–56.

Klust, G., Schärfe, J. (1955): Vorbereitende Untersuchungen zur Einführung des Scherbretthamens in die Hamenfischerei der Unterweser. Protok. Fisch. Techn. **3,** 237–248.

Klyszejko, B., Muzykiewicz, J., Wegrzynowicz, B. (1977): Effect of deterlon (DBS) on the electrocardiogram of the eel (*Anguilla anguilla*). Acta physiol. Pol. **28**(2), 179–186.

Knight, B. (1982): Body dimensions of farmed eels (*Anguilla anguilla* L.) in relation to condition factors, grading, sex and feeding. Aquacult. Eng. **1**(4), 297–310.

Knopf, K., Naser, K., Van der Heijden, M.H.T., Taraschewski, H. (1999): Die Immunantwort des Europäischen Aals auf den Schwimmblasennematoden *Anguillicola crassus*. Arb. Dtsch. Fischereiverb. **14,** 63–76.

Knopf, K., Würtz, J., Sures, B., Taraschewski, H. (1998): Impact of low temperature on the development of *Anguillicola crassus* in the final host *Anguilla anguilla*. Dis. Aquat. Org. **33,** 143–149.

Koehn, R.K., Williams, G.C. (1978): Genetic differentiation without isolation in the American eel, *Anguilla rostrata*. 2: Temporal stability of geographic patterns. Evolution **32,** 624–637.

Koie, M. (1988): Parasites in European eel *Anguilla anguilla* (L.) from Danish freshwater, brackish and marine localities. Ophelia **29,** 93–118.

Koie, M. (1988): The swimbladder nematodes *Anguillicola* spp. and the gill mongeneans *Pseudodactylogyrus* spp. in European eels. ICES Min-Symposium, Bergen, Norway. No.3.

Koie, M. (1991): Swimbladder nematodes (*Anguillicola* spp.) and gill monogeneans (*Pseudodactylogyrus* spp.) parasitic on the European eel (*Anguilla anguilla*). J. Cons. int. Explor. Mer 47. 391–398.

Kokhnenko, S.W. (1958): Biologiya I rasprostranenie ugrya. Izv. Akad. Nauk SSR Belorus, 132pp.

Kokhnenko, S.W. (1959): On the broadness and narrowness of the head of the European eel (*Anguilla anguilla* L.). Vopr. Ikhtiol. **12,** 19–27.

Kokhnenko, S.W. (1967): Zaselit vodoemy ugrem. Rybovodstvo Rybolovstvo Moskva **10**(2), 4–5.

Kokhnenko, S.W. (1975): American eels in the southern regions of the USSR. Rybovodstvo Rybolovstvo Moskva **2,** 15.

Kokhnenko, S. W., Bezdenyezhnykh, V. A. (1973a): Development of the ovocytes of the eel as a function of its size (Beloruss.). Vyestsi Akad. Nauk BSSR Ser. Byal. Nauk **2,** (141), 90–94.

Kokhnenko, S. W., Bezdenyezhnykh, V. A. (1973b): Development of the testes in the European eel in the freshwater period (Beloruss.). Vyestsi Akad. Nauk BSSR. Ser. Byal. Nauk, **1**(142), 95–99.

Kokhnenko, S. W., Borovik, E. A. (1958): Results obtained after two years' investigations on growrh and development of eel fry in the water reservoirs of Belorussia. Bull. Inst. Biol. Minsk **3,** 269–272.

Kokhnenko, S. W., Gorovaya, S. L. (1965): Nekotorye osobennosty formirovaniya pola u evropeitskovo ugrya. Trudy 12. Naukh. Konf. po lzursch. Vodoem. Pribaltiki, Vilnius, 22.–24.9.1995, 116–119.

Koops, H. (1962): Die Wanderung des Aales (*Anguilla anguilla* L.) in Abhängigkeit vom Sommerhochwasser. Kurze Mitt. Inst. Fisch. Biol. Univers, Hamburg. **12,** 19–26.

Koops, H. (1965): Fütterung von Aalen in Teichen. Arch. FischWiss. **16,** 33–38.

Koops, H. (1966a): Ein Fütterungsversuch mit Satzaalen in der Flussteichwirtschaft Müden an der Mosel. Arch. FischWiss. **17,** 36–44.

Koops, H. (1966b): Die Aalproduktion in Japan. Arch. FischWiss. **17,** 44–50.

Koops, H. (1967a): Ergebnisse der Aalwirtschaft im Ausland und Folgerungen für unsere einheimische Aalwirtschaft. Fischwirt **17,** 29–40.

Koops, H. (1967b): Die Aalwirtschaft im anderen Teil Deutschlands. Arch. FischWiss. **18,** (Suppl. 1), 453–465.

Koops, H. (1968): Glasaalaussatz im zeitigen Frühjahr. Fischwirt **18,** 1–2.

Koops, H. (1971): Ergebnisse einer Fütterung von Teichaalen mit Frischfisch und mit Trockenfutter. Fischwirt **21,** 73–77.

Koops, H., Kuhlmann, H. (1980a): Eel farming in the thermal effluents of a conventional power station in the harbour of Emden. EIFAC (FAO) 80 Symp.: E/4, 1–15.

Koops, H., Kuhlmann, H. (1980b): Annual variation in feeding and growth rate of eels, farmed in thermal effluents of a conventional power station. EIFAC (FAO)/80/Symp., E/3, 1–7.

Koops, H., Mann, H. (1966): The cauliflower disease of eels in Germany. Bull. Off. int. Epiz. **65,** 991–998.

Koops, H., Mann, H. (1969): Die Blumenkohlkrankheit der Aale, Vorkommen und Verbreitung der Krankheit. Arch. FischWiss. 20 (Suppl. 1), 5–15.

Kostyuchenko, A. A., Prishchepov, G. P. (1972): Commercial return of the eel (*Anguilla anguilla* L.) from Byelorussian lakes and the determining factors. J. Ichthyol. **12,** 968–975.
Köthke, M. (1964a): Bericht über die Einführung der Aalreuse aus Kunststoff. Fischwirt **14,** 29–33.
Köthke, M. (1964b): Bericht über Aalpaßversuche. Allg. FischZtg. **89,** 268–269 u. Fischwirt **14,** 68–70.
Köthke, M. (1968): Ist der Aal in unseren Gewässern ein gefährlicher Räuber? Fischwirt **18**(7), 169–172.
Köthke, M., Klust, G. (1956): Der Scherbretthamen. Arch. FischWiss., **7,** 93–119.
Kracht, R. (1982): On the geographic distribution and migration of I- and II-group eel larvae as studied during the 1979 Sargasso Sea expedition. Helgoländer Meeresunt. **35,** 321–327.
Kracht, R., Tesch, F.-W. (1981): Progress report on the eel expedition of R. V. "Anton Dohrn" and R. V."Friedrich Heincke" to the Sargasso Sea 1979. Env. Biol. Fish. **6,** 371–375.
Krause, P. H. (1961): Aale legen Röhren an. Fisch u. Fang, **2,** 42.
Kreutzmann, H.-L. (1973): Veränderungen des Haemoglobingehaltes, des Haematokrits und der Erythrocytenzahl von *Anguilla anguilla* nach Erkrankung an Salzwasserseuche. Wiss. Z. Univ. Rostock **22,** (6–7, Part 2), 721–725.
Kreutzmann, H.-L. (1976): Studies on the morphology of the blood of the European eel (*Anguilla anguilla*): I. The erythrocytes and their development stages. Folia Haematol. (Leipzig) **103,** 226–235.
Kreutzmann, H.-L. (1977): The effects of chloramphenicol and oxytetracyline on haematopoesis in the European eel (*Anguilla anguilla*). Aquaculture **10,** 323–334.
Krishnamurthy, V. G., Bern, H. A. (1969): Correlative histology of the corpuscles of Stannius and the juxtaglomerular cells of teleost fishes. Gen. Comp. Endocr. **13,** 313–335.
Kroezus, E. (1954): Massensterben unserer Blankaale, Fischwirt **4,** 12.
Krogh, A. (1924): Anatomie und Physiologie der Kapillaren. Springer Verlag, Berlin.
Kubota, S., Funahashi, N., Kimura, I. (1974): Histology of adenomatous polyps in the stomach in fishes. Proc. Jap. Cancer Ass. 33.
Kühl, H. (1955): Über die Wanderungen der Glasaale. Landw. Angew. Wiss. Landw. Verlag Hiltrup i. W., 59–63.
Kühl, H. (1965): Aale und Quappenwurm vor der Elbmündung. Allg. FischZtg. **90,** 696.
Kuhlmann, H. (1974): Experimentelle Untersuchungen über den Einfluß von Wassertemperatur und qualitativem Futterangebot auf das Wachstum und die Geschlechtsdifferenzierung des europäischen Aales. Dissertation Fachbereich Biologie, Hamburg.
Kuhlmann, H. (1975): Der Einfluß von Temperatur, Futter, Größe und Herkunft auf die sexuelle Differenzierung von Glasaalen (*Anguilla anguilla*). Helgoländer wiss. Meeresunters. **27,** 139–155.
Kuhlmann, H. (1998): Zur Bestandssituation des Europäischen Aales. Arbeiten Dtsch. Fisch. Verb. **69**, 47–61.
Kukla, M. (1954): The arrangement of blood vessels on the surface and in the wall of the alimentary canal of the eel (*Anguilla anguilla* L.). Folia morph. Warszawa **5**(13), 27–38.
Kukla, M. (1958): The strucure and vascularization of the pancreas of the eel (*Anguilla anguilla* L.), Acta biol. cracov. Ser. Zool. **1,** 99–111.
Kuo, S.-C., Kuo, G.-H. (1978): *Pseudomonas anguilliseptica* isolated from red spot disease of pond cultured eel, *Anguilla japonica*. Rep. Inst. Fish. Biol. Minist. Econ. AFF. Natl. Taiwan Univ. **3**(3), 19–23.
Kurokawa, T., Kagawa, H., Ohta, H., Tanaka, H., Okuzawa, K., Hirose, K. (1995): Development of digestive organs and feeding ability in larvae of Japanese eel (*Anguilla japonica*). Can J. Fish. Aquat. Sci. **52,** 1030–1036.
Kusaka, T. (1973): A study on facial bones of fish, especially urohyal bone. Mer (Tokyo), **11,** 181–183.
Kusaka, T. (1975): The feature of urohyal in the suborder Anguilloidei. Mer (Tokyo) **13,** 144–149.
Kvasova, I. P. (1966): 0 belkovykh fraktsiakh syvorotki krovi sudaka, ugrya, nalima. (Observations on blood serum protein fractions of pike-perch, eel and burbot.) In: Tezisy dokladov. Vsesoyuznye soveshchaniya po ekologii i fiziologii ryb. Moscow, 104–105.
Kwon, H.C., Mugiya, Y. (1994): Involvement of growth hormone and prolactin in the induction of vitellogenin synthesis in primary hepatocyte culture in the eel, *Anguilla japonica*. Gen. Comp. Endocrinol. **93,** 51–60.

L'Hermite, A., Ferrand, R., Dubois, M. P., Andersen, A. C. (1985): Detection of endocrine cells by immunofluorescence method in the gastrenteropancreatic system of the adult eel, glass-eel and leptocephalic larva (*Anguilla anguilla* L.) Gen. Comp. Endocrinol. **58,** 347–359.

LaBar, G. W., Casal, J. A., Delgado, C. F. (1987): Local movements and population size of European eels (*Anguilla anguilla*) in a small lake in southwestern Spain. Environm. Biol. Fish. **19,** 111–117.

LaBar, G. W., Facey, D. E. (1983): Local movements and inshore population sizes of American eels in Lake Champlain, Vermont. Trans. Amer. Fish. Soc. **112,** 111–116.

Lacey, S.M., Williams, I.C., Carpenter, A.C. (1982): A note on the occurrence of the digenetic trematode *Sphaerostoma bramae* (Müller) in the intestine of the European eel, *Anguilla anguilla* (L.). J. Fish. Biol. **20,** 593–596.

Ladiges, W. (1936): Untersuchung über den Aalbestand im Hamburgischen Hafengebiet. Z. Fisch. **34,** 21–34.

Lafeber, F.P.G.J., Hanssen, R.G.M., Choy, Y.M., Flik, G., Herrmann-Erlee, M.P.M., Pang, P.K.T., Wendelaar Bonga, S.E. (1988a): Identification of hypocalcin (teleocalcin) isolated from trout stannius corpuscles. Gen. Comp. Endocrin. **69,** 19–30.

Lafeber, F.P.G.J., Hanssen, R.G.M., Wendelaar Bonga, S.E. (1988b): Hypocalcaemic activity of trout hypocalcin and bovine parathyroid hormone in stanniectomised eels. J. exp. Biol. **140,** 199–208.

Laibach, E. (1937): Das Geruchsorgan des Aals (*Anguilla vulgaris*) in seinen verschiedenen Entwicklungsstadien. Zool. Jb. (Anat.) **63,** 37–72.

Lamarque, P. (1967): Electrophysiology of fish subjected to the action of an electric field. Fish News.

Lammens, E. H. R. R., Visser, J. T. (1989): Variability of mouth width in European eel, *Anguilla anguilla*, in relation to varying feeding conditions in three Dutch lakes. Environm. Biol. Fish. **26,** 63–75.

Lamp, F. (1980): Lohnt sich eine herbstliche Schleppnetzfischerei auf Aal in der westlichen Ostsee? Inform. Fischwirtsch. **27,** 241–242.

Lang, H. J. (1972): Korrelation und Kausalität bei lunaren Periodizitätserscheinungen. Nachr. Akad. Wiss. Göttingen, **8,** 30–34.

Lantz, A. W. (1966): Special products from freshwater fish. Bull. Fish. Res. Bd Can. **151,** 1–45.

Larinier, M., Dartiguelongue, J. (1989): La Circulation des Poissons Migrateurs: Le Transit à Travers les Turbines des Installations Hydroelectriques. Conseil Superieur de la Peche, Boves, France.

Larouche, M., Beaulien, G., Bergeron, J. (1974): Quelques données sur la croissance de l'anguille d'Amerique (*Anguilla rostrata*) de l'estuaire du Saint-Laurent. Rapp. ann. 1973. Dir. gen. Pêches marit., Dir. Rech., 109–116.

Larsen, K. (1955): Fishpopulation analyses in some small Danish trout streams by means of D. C. electrofishing. Meddr Danm. Fisk. og Havunders. 1 (N. S.), 1–69.

Larsen, K. (1961a): The fish population of a peat pit as determined by Rotenone poisoning. Meddr Danm. Fisk. Og Havunders. 3 (N. S.), 117–132.

Larsen, K. (1961b): Fish populations in small Danish streams. Verh. int. Verein. theor. angew. Limnol. **14,** 769–772.

Larsen, K. (1972): Studies on the biology of Danish stream fishes. III. On seasonal fluctuations in the stock density of yellow eel in shallow stream biotopes and their cause. Meddr Danm. Fisk. og Havunders. (N. S.) **7,** 23–46.

Larsson, A., Fänge, K. (1969): Chemical differences in the blood of yellow and silver phases of the European eel *Anguilla anguilla* L. Archs int. Physiol. Biochem. **77,** 701–709.

Larsson, P., Hamrin, S., Okla, L. (1990): Fat content as a factor inducing migratory behavior in the eel (*Anguilla anguilla* L.) to the Sargasso Sea. Naturwiss. **77,** 488–490.

Lassleben, P. (1966): Die Aalwirtschaft des Landes Bayern. Regierungsbezirk Niederbayern. Arch. FischWiss. **16,** (Suppl. 1), 42–50.

Lauder, G. V., Liem, K. F. (1983): The evolution and interrelationships of the actinopterygian fishes. Bull. Mus. Comp. Zool. **150,** 95–197.

Lauterbach, R. (1967): Möglichkeiten einer rationellen Aalproduktion im Bereich unserer Küsten. Fischereiforschung **5**(2), 43–46.

Le Bras, Y. M. (1979): Circadian rhythm in brain catecholamine concentration in the teleost: *Anguilla anguilla*. Comp. Biochem. Physiol. C. Comp. Pharmacol. **62,** 115–118.

Leatherland, J. F., Dodd, J. M. (1968): Studies on the structure, ultrastructure and function of the subcommissural organ - Reissner's fibre complex of the European eel (*Anguilla anguilla* L.) Z. Zellforsch. mikrosk. Anat. **89,** 533–549.

Lecomte-Finiger, R. (1994): The early life history of the European eel. Nature **370,** 424.

Lecomte Finiger, R., Prodon, R. (1979): Étude experimentale du comportement fouisseur de la civelle (*Anguilla anguilla* L.): le choix d'un substrat. C.r. Acad. Sc. Paris **289,** 741–743.

Lee, Tae-Won (1979): Dynamique des populations d'Anguille *Anguilla anguilla* (L.) des Lagunes du Bassin d'Arcachon. Dissert. Univ. Langedoc 1.218.

Lee, Tae-Won, Lee, Kwan-Soon (1989): Daily growth increments and lunar pattern in otolith of the eel, *Anguilla japonica*, in the freshwater. Bull. Korean Fish. Soc. **22**(1), 36–40.

Lee, W. J. (1979): Studies on *Vibrio anguillarum* of the eel (*Anguilla japonica*). Bull. Tongeong Fish. Jr. Coll.14, 9–11.

Legault, A. (1991a): Eel ladders in France. EIFAC(FAO) Working Party on Eel, Dublin, May 1991.

Legault, A. (1991b): Study of some selectivity factors in eel ladders. EIFAC(FAO) Working Party on Eel, Dublin, May 1991.

Leiby, M. M. (1979: Morphological development of the eel *Myrophis punctatus* (Ophichthidae) from hatching to metamorphosis with emphasis on the developing head skeleton. Bull. Mar. Sci. **29,** 509–521.

Leiby, M. M. (1981): Larval morphology of the eels *Bascanichthys bascanium*, *B. scuticaris*, *Ophichthus melanoporus* and *O. ophis* (Ophichthidae), with a discussion of larval identification methods. Bull. Mar. Sci. **31,** 46–71.

Leich, H. (1929): Der Aalaufstieg am Emswehr Herbrum vom 6. bis 8. Mai 1928. Fischerboote **21,** 103–105.

Leicht, R. (1969): Der Sauerstoffverbrauch von *Anguilla vulgaris* in seiner Abhängigkeit von Temperaturdifferenzen zwischen Kopfbereich und Rumpf. Mar. Biol. Berlin **3,** 28–45.

Lemaire-Gony, S. (1993): Stude experimentale de l'action du cadmium sur les juveniles (civelles et anguillettes) de l'anguille européenne *Anguilla anguilla*. Final reports on research projects dealing with the effects of pollutants on marine communities and organisms. UNEP Mediterranean Action Plan, Athens, Greece. MAP-Tech. Rep. Ser. **73,** 1–34.

Lemaire-Gony, S., Lemaire, P. (1992): Interactive effects of cadmium and benzo(a)pyrene on cellular structure and biotransformation enzymes of the liver of the European eel *Anguilla anguilla*. Aquat. Toxicol. **22,** 145–160.

Lemaire-Gony, S., Mayer-Gostan, N. (1994): In vitro dose-response study of the effect of cadmium on eel (*Anguilla anguilla*) gill Na+/K+-ATPase activities. Ecotoxicol. Environ. Saf. **28,** 43–52.

Lemoine, A.-M., Olivereau, M. (1971): Présence d'acide N-acétyleuraminique dans la peau d'*Anguilla anguilla* L. Interet de son dosage dans l'étude de l'osmorégulation. Z. vergl. PhysioI. **73,** 22–33.

Leopold, M. (1976): Efficiency and prospectives of eel management/summary, some economic approaches. ICES/EIFAC Symposium on eel Research and Management, Helsinki **17,** 1–17.

Leopold, M. (1986): Problems of analyzing effectiveness of eel stocking in interconnected lakes. Vie Milieu **36,** 291–293.

Levander, K. M. (1909): Beobachtungen über die Nahrung und die Parasiten der Fische des Finnischen Meerbusens. Finnl. hydrogr.-biol. Unters. 5.

Lewis, T. L., Epple, A. (1972): Pancreatectomy in the eel: effects on serum glucose and cholesterol. Science 178 (4067), 1286–1288.

Lewis, T. L., Epple, A. (1984): Gonadal histology and plasma steroid profiles in wild New Zealand fresh water eels (*Anguilla dieffenbachii* and *A. australis*) before and at the onset of the naturalspawning migration. Males. Fish Physiol. Biochem.

Lewis, T. L., Wesley, W. P., Epple, A. (1977): Pancreatectomy in a teleost fish, *Anguilla rostrata* (American eel). Lab. Quim. Sci. **27,** 102–109.

Lhotte, R. (1945): Influence des variations de salinité sur le rhythme cardiaque de la jeune anguille (civelle). C.r. Séanc. Soc. Biol. **139,** 27–28.

Liao, I.C., Chang, S.L. (2001): Induced spawning and larval rearing of Japanese eel. J. Taiwan Fish. Res. **8**(1) 97–108.

Lick, R. R. (1991). Untersuchungen zu Lebenszyklus (Krebse – Fische – marine Säuger) und Gefrierresistenz anisakider Nematoden in Nord- und Ostsee. Ber. Inst. Meereskunde Univ. Kiel, 218.

Lieder, U. (1963): Über vermutliche Gonosomen bei *Perca*, *Acerina* und *Anguilla* (Vertebrata. Pisces). Biol. Zbl. **82,** 297–302.
Liew, P. K. L. (1974): Age determination of American eels based on the structure of their otoliths. In: Bagenal, T. B. (Ed.): Symposium on the aging of fish. Unwin Bros., Old Woking, UK 234pp., 124–136.
Liewes, E. W., Schamine-Mann, S. (1987): Onderzoek aad parasiet vordest. Aquacultuur **2,** 5–17.
Ligny, W., Pantelouris, E. M. (1973): Origin of the European eel. Nature **246,** 518–519.
Lillelund, K. (1967): Messungen der Aktivität von Fischen mit Hilfe magnetischer Marken. Ber. dt. wiss. Kommn Meeresforsch. **19,** 1–12.
Lin, Y.-S., Poh, Y.-P., Tzeng, G.S. (2001): A phylogeny of freshwater eels inferred from mitochondrial genes. Mol. Phyl. Evol. **20,** 252–261.
Linde, A. R., Arribas, P., Sanchez-Galan, S., Garcia-Vazquez, E. (1996): Eel (*Anguilla anguilla*) and brown trout (*Salmo trutta*) target species to assess the biological impact of trace metal pollution in freshwater ecosystems. Arch. Environ. Contam. Toxicol. **31,** 297–302.
Lindquist, A. (1972): Eel larvae in the Skagerrak. Meddn Havsfiskelab., **127,** 1–3.
Lindroth, A. (1953): Internal tagging of salmon smolts. Ann. Rep. Drottningholm **34,** 49–57.
Lindroth, A. (1979): Eel catch and lunar cycle on the Swedish east coast. Rapp.p.-v. Reun. Cons. int. Explor. Mer **174,** 124–126.
Lindsey (1978): Form, function and locomotory habits in fish. In: Hoar, W. S., Randall, D. J. (Eds.): Fish Physiology. 1–100. Academic Press, London, New York.
Lintas, C., Hirano, J., Archer, S. (1998): Genetic variation of the European eel (*Anguilla anguilla*). Mol. Mar. Biol. Biotech. **7,** 263–269.
Linton, E. (1897): Notes on larval cestode parasites. Proc. U.S. Nat. Mus. **19,** 787–824.
Lissner, H. (1923): Das Gehirn der Knochenfische. Wiss. Meeresunters. Helgol. **14,** 125–188.
Lobon-Cervia, L., Utrilla, C. G., Rincon, P. A. (1995): Variations in the population dynamics of the European eel *Anguilla anguilla* (L.) along the course of a Cantabrian river. Ecol. Freshwater Fish **4,** 17–27.
Loebell, R. (1965): So fängt man Aale. Verlag P. Parey, Hamburg u. Berlin.
Loebell, R. (1966): Wichtiges vom Angeln im Meer. Fisch u. Fang, **7,** 40–41.
Lohmann, K. J., Lohmann, C. M. F. (1996): Orientation and open-sea navigation in sea turtles. J. exp. Biol. **199,** 73–81.
Lokman, P. M., Vermeulen, G. J., Lambert, J.G. D., Young, G. (1998): Gonadal histology and plasma steroid profiles in wild New Zealand fresh water eels (*Anguilla dieffenbachii* and *A. australis*) before and at the onset of the natural spawning migration. Males. Fish Physiol. Biochem.
Lokman, P. M., Young, G. (1995): Plasma sex steroids in female New Zealand freshwater eels (*Anguilla* spp.) before and after onset of the spawning migration. In Goetz, F.W., Thomas, P. (eds): Proc. Fifth Int. Symp. on the Reproductive Physiology of Fish. 221–223. Univ. Texas, Austin.
Lom, J. (1970): Observations on trichodinid ciliates from freshwater fishes. Arch. Protistenk. 112.
Lom, J., Molnar, K., Dykova, E. (1986): *Hoferellus gilsoni* Debaseius 1925 comb. n. (Myxozoa, Myxosporea): redescription and mode of attachment to the epithelium of the urinary bladder of its host, the European eel. Protistologica **22,** 405–413.
Lopez, E. (1970): L'os cellulaire d'un poisson téléostéen *Anguilla anguilla* L., I. Etude histocytologique et histophysiologique. Z. Zellforsch. mikrosk., Anat. **109,** 552–565.
Lopez, E., Fontaine, M. (1967): Reponse des corpuscules de Stannius de l'anguille (*Anguilla anguilla* L.) à des blessures experimentales. C.r. Séanc. Soc. Biol. **161,** 36–39.
Lopez, E., Lee, H. S., Baud, C. A. (1970): Etude histophysique de l'os d'un téléostéen *Anguilla anguilla* L. au cours d'une hypercalcemie provoquée par la maturation expérimentale. C. r. hebd. Séanc. Acad. Sci. Paris **270,** 2015–2017.
Lopez, E., Peigneux-Deville, J., Lallier, F., Martelly-Bagot, E., Milet, C. (1976): Effect of ultimobranchialectomy (UEX) upon the bone catabolism and anabolism and on some aspects of the calcium metabolism regulation in *Anguilla anguilla* L. Calcif. Tiss. Res. 21 (Suppl.), 88–95.
Lopez, E., Peignoux-Deville, J., Lallier, F., Colston, K. W., Macintyre, I. (1977): Responses of bone metabolism in the eel (*Anguilla anguilla*) to injections of **1,** 25–$(OH)_2$–D_3. Calcif. Tiss. Res. 22 (Suppl), 19–23.

Lopez, E., Tisserand-Jochem, E. M., Eyquem, A., Milet, C., Hillyard, C., Lallier, F., Vidal, B., Macintyre. I. (1984): Immunocytochemical detection in eel corpuscles of Stannius of a mammalian parathyroid-like hormone. Gen. Comp. Endocrinol. **53,** 28–36.

Loretz, C. A., Pollina, C., Kaiya, H., Sakaguchi, H., Takei, Y. (1997): Local synthesis of natriuretic peptides in the eel intestine. Biochem. Biophys. Res. Commun. **238,** 817–822.

Löwe, E. (1930): Zwei Beobachtungen von Glasaalzügen in der Weser. Fischerboote **22,** 281–282.

Lowe, R. H. (1951): Factors influencing the runs of elvers of the River Bann, Northern Ireland, J. Cons. perm. int. Explor. Mer **17,**(3), 299–315.

Lowe, R. H. (1952): The influence of light and other factors on the seaward migration of the silver eel (*Anguilla anguilla* L.).J. anim. Ecol, **21,** 275–309.

Löwenberg, U. (1982): Die Aalschleppnetzfischerei in der Deutschen Bucht. Arbeiten Dtsch. Fisch. Verb. **36,** 7–17.

Loyakano, H., Crane, S. (1977): Control of *Ichthyophthirius* on the American Eel. Prog. Fish. Cult. **39**(1), 39–40.

Lübben, G. (1968): Wie fange ich den größten Aal? Fisch u. Fang, **9,** 124–127.

Lübben, G., Tesch, F.-W. (1966): Sommer und Winteraufenthalt der Aale. Fisch u. Fang **9,** 56–57.

Lübbert, H. (1908): Der Aalfang in der Lagune von Comacchio. Mitt. dt. Seefisch. Ver. **8,** 1–11.

Lübbert, H. (1910): Die Besetzung deutscher Gewässer mit Aalbrut im Jahr 1909. Z. Fisch. **15,** 182–233.

Lübbert, H. (1930): Beobachtungen über den Aufstieg und die Größenverhältnisse von Glasaalen in Ägypten. Fischerbote **22,** 377–383.

Lühmann, M., Mann, H. (1956): Beobachtungen über die Blumenkohlkrankheit der Aale. Arch. FischWiss. **7,** 229–239.

Lühmann, M., Mann, H. (1958): Wiederfänge markierter Elbaale von der Küste Dänemarks. Arch. FischWiss. **9,** 200–202.

Lühmann, M., Mann, H. (1961): Untersuchungen über den Aalfang in der Elbe. Fischwirt **11,** 165–176.

Lundbeck, J. (1927): Untersuchungen über die Beschädigungen von Fischen, besonders Aalen, in den Turbinen des Kraftwerkes Friedland (Ostpreussen) Z. Fisch. **25,** 439–465.

Määr, A. (1947): Über die Aalwanderung im Baltischen Meer auf Grund der Wanderaalmarkierungsversuche im Finnischen und Livischen Meerbusen in den Jahren 1937–1939. Meddn. St. Unders. o. ForsAnst Sötvattfisk Drottningholm **27,** 1–56.

Maetz, J. (1971): Fish gills: mechanisms of salt transfer in fresh water and sea water. Phil. Trans. R. Soc. (B)262, 209–249.

Maetz, J., Nibelle, J., Bornancin, M., Motais, R. (1969): Action sur l'osmoregulation de l'anguille de divers antibiotiques inhibiteurs de la synthèse des protéeines ou de renouvellement cellulaire. Comp. Biochem. Physiol. **30,** 1125–1151.

Maetz, J., Rankin, J. C. (1969): Quelques aspects du role biologique des hormones neurohypophysaires chez les poissons. In: La specificité zoologique des hormones hypophysaires et de leur activité. Colloques int. Cent. natn. Res. scient. **177,** 45–55.

Maetz, J., Skadhauge, E. (1968): Drinking rates and gill ionic turnover in relation to external salinities in the eel (*Anguilla anguilla*). Nature **217**(5126), 371–373.

Maniukas, J. (1961): Die Fischfauna, der Zustand der Fischbestände und die Ertragsverhältnisse im Kurischen Haff. Verh. int. Verein. theor. angew. Limnol. **14,** 737–739.

Mann, H., 1960: Untersuchungen über die Hälterung von Aalen. Arch. FischWiss. **11,** 145–151.

Mann, H., 1961: Der Aalaufstieg in der Aalleiter an der Staustufe Geesthacht. Fischwirt **11,** 69–74.

Mann, H., 1962: Beobachtungen über die Krankheiten und Parasiten an Elbfischen. Fischwirt **12,** 300–309.

Mann, H., 1963: Beobachtungen über den Aalaufstieg in der Aalleiter an der Staustufe Geesthacht im Jahre 1961. Fischwirt **13,** 182–186.

Mann, H., 1964: Die Elbe als Fischgewässer. Fischwirt **14,** 47–48.

Mann, H., 1965: Über das Rückkehrvermögen verpflanzter Flußaale. Arch. FischWiss. **15,** 177–185.

Mann, H., 1968: Der Einfluß der Staustufe Geesthacht auf den Aalfang in der Elbe. Allg. FischZtg. **93,** Nr. 19.

Mann, H., Klust, G. (1952): Aale in Dänemark. Fischwirt **2,** 376–377.

Mann, R. H. K., Blackburn, J.H. (1991): The biology of the eel *Anguilla anguilla* (L.) in an English chalk stream and interactions with juvenile trout *Salmo trutta* L. and salmon *Salmo salar* L. Hydrobiologia **218,** 65–76.

Mann, R. H. K., Penczak, T. (1994): The efficiency of a new electro-fishing technique in determining fish numbers in a large river in central Poland. J. Fish Biol. **24,** 173–185.

Manter, H. W. (1955): The zoogeography of trematodes of marine fishes. Expl. Parasit., **4,** 62–86.

Marchelidon, J., Schmitz, M., Houdebine, L. M., Vidal, B., Le Belle, N., Dufour, S. (1996): Development of a radioimmunoassay for European eel growth hormone and application to the study of silvering and experimental fasting. Gen. Comp. Endocrinol. **102,** 360–369.

Marcus, K. (1919): Über Alter und Wachstum des Aales. Mitt. Zool. St. Inst. Hamb. **36,** 1–70.

Margolis, L., Arthur, J. R. (1979): Synopsis of the parasites of fishes of Canada. Bull. Fish. Res. Board Canada, 199.

Margolis, L., Kabata, Z. (Eds.) (1988, 1989, 1996): Guide to the parasites of fishes of Canada. Part I. General Introduction. Beverly-Burton, M.: Monogenea + Turbellaria. Part II. Kabata, Z., Rafi, F., Bousfield, E.-L. (1988): Crustacea. Part III. Arai, H. P. (1989): Acanthocephala. 1–90. Arai. M. N. (1989): Cnidaria. 91–95. Part IV. Gibson, D. I. (1996): Trematoda.

Markiewicz, F., Migala, K. (1980): Trichodinid invasion on young eels (*Anguilla anguilla* L.) grown in aquaria. Acta Hydrobiol. **22,** 229–236.

Marquet, G. (1992): L'Etude de recrutement de la physiologie des anguilles de la polynesie française permet-elle de cerner leure aire de ponte? Bull. Inst Océanogr. Monaco Special no. **10,** 129–147.

Marsigliante, S., Muscella, A., Vilella, S., Nicolardi, G., Ingrosso, L., Ciardo, V., Zonno, V., Vinson, G.P., Ho, M.M., Storelli, C. (1996): A monoclonal antibody to mammalian angiotensin II AT-1 receptor recognizes one of the angiotensin II receptor isoforms expressed by the eel (*Anguilla anguilla*). J. Molecular Endocrin. **16**(1), 45–56.

Marsigliante, S., Muscella, A., Vinson, G.P., Storelli, C. (1997): Angiotensin II receptors in the gill of sea water- and freshwater-adapted eel. J. Molecular Endocrin. **18**(1), 67–76.

Martin, E. (1974): *Paracardiocolides yamagutii* gen. et. spec. n. from an Australian eel (Trematoda: Sanguinicolidae). Proc. helminthol. Soc. Wash. **41,** 22–25.

Martinköwitz, H. (1960): Erste Ergebnisse von Blankaalmarkierungen an der ostrügenschen Küste im Herbst 1959. Fischereiforschung 3 (5/6), 9–13.

Martinköwitz, H. (1961): Ergebnisse von Blankaalmarkierungen an der ostrügenschen Küste und Möglichkeit ihrer Nutzung für die Fangsteigerung durch neuartige Reusenkonstruktionen. Z. Fisch. **10,** 653–663.

Masoni, A., Isaia, J. (1973): Influence du mannitol et de la salinité extrême sur l'équilibre hydrique et l'aspect morphologique de la branchie d'anguille adaptée à l'eau de mer. Arch. Anat. micr. Morph. exp. **62,** 293–306.

Mathiesen, C. (1984): Structure and innervation of inner ear sensory epithelia in the European eel (*Anguilla anguilla* L.). Acta Zool. Stockh. **65, 4,** 189–207.

Matschenis, A. S. (1965): 0 polovoi differenziazii ugreyi v vodoyemakh litovskoi SSR. Trudy 12. Nauch. Konf. po izuch. Vnutr. Vodoyem. Pribaltiki, Vilnius, 22.–24.9., 120–124.

Matsui, I. (1952): Studies on the morphology, ecology and pondculture of the Japanese Eel (*Anguilla japonica* Temminck u. Schlegel). J. Shimonoseki Coll. Fish. **2**(2), 1–245.

Matsui, I. (1957): On the records of a leptocephalus and catadromous eels of *Anguilla japonica* in the waters around Japan with a presumption of their spawning places. J. Shimonoseki Coll. Fish **7,** 151–167.

Matsui, I. (1971): Leptocephalae of the eel *Anguilla japonica* found in the waters of Ryukyu Deep. J. Shimonoseki Univ. Fish. **20,** 13–18.

Matsui, I. (1980): Theory and practice of eel culture. Verl. A. A. Balkava, Rotterdam, 133pp.

Matsui, I. (1986): Theory and Practice of Eel Culture. Balkema Rotterdam.

Matsui, I., Takai, T. (1959): Osteology of the Japanese eel. J. Shimonoseki Coll. Fish. **8,** 53–62.

Matsui, I., Takai, T., Kataoka, A. (1968): Anguillid leptocephalus found in the Japan current and its adjacent waters. J. Shimonoseki Univ. Fish. **17,** 17–23.

Matsui, I., Takai, T., Kataoka, A. (1970): Leptocephalae of the eel *Anguilla obscura* found in the stomachs of skipjack tuna *Katsuwonus pelamis* J. Shimonoseki Univ. Fish. **19,** 25–28.

Matsumoto, A., Arai, Y. (1992): Hypothalamus. In Matsumoto, A., Ishii, S. (eds): Atlas of Endocrine Organs. 25–38. Springer-Verlag: Berlin, Heidelberg, New York.

Mattern, O. (1971): Über Winterverluste in unseren Seen. Dt. FischZtg, Radebeul **18,** 6–8.

Mattheis, T. (1964a): Das Vorkommen von *Vibrio anguillarum* in Ostseefischen. Z. Fischerei **12,** 250–263.

Mattheis, T. (1964b): Fälle von Zystenbildung und Geschwülsten bei Aalen (*Anguilla vulgaris*). Z. Fisch. **12,** 709–715.

May, D., Todd, C. M., Rand-Weaver, M. (1997). cDNA cloning of eel (*Anguilla anguilla*) somatolactin. Gene **188,** 63–67.

Mayer, N., Nibelle, J. (1970): Kinetics of the mineral balance in the eel *Anguilla anguilla* in relation to external salinity changes and intravascular saline infusions. Comp. Biochem. Physiol. **35,** 553–566.

Mayer-Gostan, N., Flik, G., Pang, P. K. T. (1992): An enzyme linked immunosorbent assay for stannioclacin, a major hypocalemic hormone in teleosts. Gen. Comp. Endocrinol. **86,** 10–19.

Mayer-Gostan, N., Hirano, T. (1976): The effects of transecting the IXth and Xth cranial nerves on hydromineral balance in the eel *Anguilla anguilla*. J. exp. Biol. **64,** 461–475.

McBean, R.L., Neppel, M.J., Goldstein, L. (1966): Glutamate dehydrogenase and ammonia production in the eel (*Anguilla rostrata*). Comp. Biochem. Physiol. **18,** 909–920.

McCleave, J. D. (1967): Homing and orientation of cutthroat trout (*Salmo clarki*) in Yellowstone Lake, with special reference to olfaction and vision. J. Fish. Res. Bd Can. **24,** 2011–2044.

McCleave, J. D., Kleckner, R. C. (1982): Selective tidal stream trasport in the estuarine migration of glass eels of the American eel (*Anguilla rostrata*). J. Cons. int. Explor. Mer. **38,** 262–271.

McCleave, J. D., Rommel, S. A., Cathcart, C. L. (1971): Weak electric and magnetic fields in fish orientation. In animal orientation: sensory basis. Ann. N. Y. Acad. Sci. **188,** 270–282.

McCosker, J.E. (1977): The osteology, classification, and relationships of the eel family Ophichthidae. Proc. Calif. Acad. Sci., Ser. **4, 41,** 1–123.

McDowall, R. M., Jellyman, D. J., Dijkstra, L. H. (1998): Arrival of an Australian anguillid eel in New Zealand: an example of transoceanic dispersal. Envir. Biol. Fish. **51,** 1–6.

McGrath, C. J., O'Leary, D. P., Sharkey, P. J., Murphy, D. G. (1976): Experimental electrical guidance system for eel at Killaloe eel weir on the River Shannon. Joint ICES/EIFAC Symp. Helsinki, Doc. Nr. 66.

McInerney, J. E. (1964): Salinity preference, an orientation mechanism in salmon migration. J. Fish. Res. Bd Can. **21,** 995–1018.

Medcof, J. C. (1966): Incidental records on behaviour of eels in Lake Ainslie, Nova Scotia. J. Fish. Res. Bd Can. **23,** 1101–1105.

Medcof, J. C. (1969): Fishermen's reports of freshwater and saltwater migrations of Nova Scotia eels (*Anguilla rostrata*). Can. Fld Nat. **83,** 132–138.

Meike, H. (1970): Vom Räuchern. Fisch u. Fang **11,** 83–85.

Meister, E. (1970): Jungaale überwinden auf ihrer Wanderung den Rheinfall. Schweiz. FischZtg. **78**(2), 10–11.

Meizies, A., Reichwald, J. (1973): Die Lipide im Fleisch und im Rogen frischer und geräucherter Fische. Z. Ernähr. Wiss. **12,** 248–251.

Mellergaard, S. (1988): Alens svomme blaereorm *Anguillicola* en ny parasit i den europaeiske alebestand. Nordisk Aqvakultur **4,** 50–54.

Menzies, W. J. M. (1936): The run of elvers in the River Bann, Northern Ireland. J. Cons. perm. int. Explor. Mer **11,** 249–259.

Mercier, L., Poisson, R. (1927): A propos d'un cas de macrophthalmie chez une anguille. C.r. hebd. Séanc. Acad. Sci. Paris **184,** 123–125.

Meske, C., (1968): Warmwasseraufzucht von Glasaalen. Fischwirt **18,** 12.

Meske, C., (1969): Aufzucht von Aalbrut in Aquarien. Arch. FischWiss. **20,** 26–32.

Meske, C., (1980): Nassfutter oder Trockenfutter für Glasaale? Inform. Fischwirtsch. **27,** 192–194.

Meske, C., Cellarius, O. (1973): Experimentell erzeugte Geschlechtsreife und Beobachtung zur Sexualdifferenzierung bei Aalen (*Anguilla anguilla*) in Warmwasserhaltung. Arch. FischWiss. **24,** 191–197.

Meyer, P. F. (1938): Die Beeinflussung des Blankaalfanges an der Rügenschen Küste durch meteorologische und kosmische Faktoren. Z. Fisch. **36,** 643–680.

Meyer, P. F. (1943): Qualitätsuntersuchungen an Nutzfischen der Ostsee. Z. Fisch., **41,** 47–67.

Meyer, P. F. (1951): Die Aalbrutfangstation von Herbrum in Oldenburg. Fischwirt **1,** 207–212.

Meyer, P. F., Kühl, H. (1952/53): Welche Milieufaktoren spielen beim Glasaalaufstieg eine Rolle? Arch. FischWiss. **4,** 87–94.

Meyer-Rochow, V. B. (1978): Skin papillae as possible electroreceptors in the deep-sea eel *Cyema atrum* (Cyemidae: Anguilloidei). Marine Biol. **46,** 277–282.

Meyer-Waarden, P. F. (1964): Untersuchungen von Gewässern in Bayern und Baden-Württemberg auf ihre Eignung als Aalgewässer. Veröffentl. Inst. Küsten- und Binnenfisch., **19,** 1–55.

Meyer-Waarden, P. F. (1965): Die wundersame Lebensgeschichte des Aales. In: Keune, J.A., Struck, H. (eds): Der Aal. 7–47. H. A. Keune Verlag, Hamburg.

Meyer-Waarden, P. F., Halsband, E. (1975): Einführung in die Elektrofischerei. H. Heenemann GmbH, Berlin.

Meyer-Waarden, P. F., Koops, H. (1968): Qualitätsbeurteilung von gefütterten Teichaalen. Allg. FischZtg **93,** 54–55.

Micheler, C. (1967): Biologische Untersuchungen an den Aalen des Chiemsees, Dissertation Vet. med. Fakult. München, 1–107.

Miles, S. G. (1968a): Laboratory experiments on the orientation of the adult American eel, *Anguilla rostrata*. J. Fish. Res. Bd Can. **25,** 2143–2155.

Miles, S. G. (1968b): Rheotaxis of elvers of the American eel (*Anguilla rostrata*) in the laboratory to water from different streams in Nova Scotia. J. Fish. Res. Bd Can. **25,** 1591–1602.

Milet, C., Martelly, E., Lopez, E. (1989): Partial purification of parathyrin from the corpuscles of Stannius of the eel (*Anguilla anguilla* L.) Gen. Comp. Endocrinol. **76,** 83–94.

Milne, K. P., Ball, J. N., Chester Jones, I. (1971): Effects of salinity, hypophysectomy and corticotrophin on branchial Na and K-activated ATPase in the eel, *Anguilla anguilla* L. J. Endocr. **49,** 177–178.

Mishina, S., Takei, Y. (1997): Characterisation of natriuretic peptide receptors in eel gill. J. Endocrinol. **154,** 415–422.

Mishra, T. N., Chubb, J. C. (1969): The parasite fauna of the fish of the Shropshire Union Canal, Cheshire. J. Zool., Lond. **157,** 213–224.

Mislin, M. (1960): Zur Funktionsanalyse des lymphatischen Kaudalherzens beim Aal. (*Anguilla anguilla* L.). Revue suisse Zool. **67,** 262–269.

Miura, T., Yamauchi, K., Takahashi, H., Nagahama, Y. (1991): Human chorionic gonadotrophin induces all spermatogenesis in the male Japanese eel (*Anguilla japonica*). Developmental Biol. **146,** 258–262.

Miyazaki, T., Egusa, S. (1977): Histopathological studies of red spot disease of the Japanese eel (*Anguilla japonica*): I. Natural infection. Fish. Pathol. **12,** 39–59.

Mizuhira, V., Aiviakawa, T., Yasmashina, S., Shirai, N., Utida, S. (1964): Electronmicroscopic studies on the localization of sodium ions and sodium potassium activated adenosinetriphosphatase in chloride cells of eel gills. Exptl Cell Res. **59,** 346–348.

Mohr, H. (1963a): Versuche mit Aalkörben aus synthetischem Material. Protokolle Fisch. Techn. **8**(38), 292–303.

Mohr, H. (1963b): Aalkörbe aus Plastik. Fischwirt **13,** 112–114.

Mohr, H. (1966): Versuche über die Scheuchwirkung von teerhaltigen Netzkonservierungsmitteln. Infn Fischw. Auslds **13,** 237–238.

Mohr, H. (1969): Über Reaktionen des Aales auf chemische Reize. Fischwirt **19,** 62–65.

Mohr, H. (1970): Aktivitätsperiodik bei Aalen. Infn. Fischw. Auslds **17,** 105–107.

Mohr, H. (1971): Untersuchungen über die auslösenden Faktoren der Schwimmaktivität bei Gelbaalen. Protokolle Fisch. Techn. **12**(58), 425–436.

Möller, H. (1975): Bestimmungstafeln für die Fischparasiten der Kieler Bucht. Ber. Inst. Meereskunde **10,** 1–32.

Möller, H. (1975): Der Einfluß von Temperatur und Salzgehalt auf Entwicklung und Verbreitung von Fischparasiten. Diss. Kiel 1975.

Möller, H., Anders, K. (1986): Diseases and Parasites of marine fishes. Kiel.

Möller, H., Holst, S., Lüchtenberg, H., Petersen, F. (1991): Infection of eel (*Anguilla anguilla*) from the River Elbe estuary with two nematodes *Anguillicola crassus* and *Pseudoterranova decipiens*. Dis. Aquat. Org. **11,** 193–199.

Möller-Christensen, J. (1964): Burning of otoliths, a technique for age determination of soles and other fish. J. Cons. perm. int. Explor. Mer **29,** 73–81.

Molnar, K. (1984): Occurrence of new monogeneans of far-east origin on the gills of fishes in Hungary. Acta vet. Hung. **32,** 153–157.

Monten, E. (1985): Fish and Turbines. Vattenfall Verlag, Stockholm, 1–111.

Montero, M., Le-Belle, N., King, J. A., Millar, R. P., Dufour, S. (1995): Differential regulation of the two forms of gonadotropin-releasing hormone (mnRH and CGnRH-II) by sex steroids in the European female silver eel (*Anguilla anguilla*). Neuroendocrinology **61,** 525–535.

Moravec, F. (1966): The finding of the nematode *Paraquimperia tenerrima* (v. Linstow, 1878) from eels *Anguilla anguilla* (L.) in Czechoslovakia. Folia parasit. **13**(2), 182–186.

Moravec, F. (1977): 1st record of the nematode *Spinitectus inermis*, new record from eels of Czecholovakia. Folia Parasitol. (Prague) **24,** 85.

Moravec, F., Koie, M. (1987): *Daniconema anguillae* gen. et sp. n., a new nematode of a new family Daniconematidae fam. n., parasitic in European eels. Folia parasit. **34,** 335–340.

Moravec, F., Taraschewski, H. (1988): Revision of the genus *Anguillicola* Yamaguti 1935 (Nematoda: Anguillicolidae) of the swimbladder of eels, including descriptions of two new species, *A. novaezelandiae* sp. n. and *A. papernai* sp. n. Folia Parasitol. **35,** 125–146.

Moriarty, C. (1969): Biological studies of Irish eels. J. Cons. perm. int. Explor. Mer **8,** 2pp.

Moriarty, C. (1972): Studies of the eel *Anguilla anguilla* in Ireland. 1. In the lakes of the Corrib System. Ir. Fish. Invest. (A), **10,** 1–30.

Moriarty, C. (1973a): A technique for examining eel otoliths. J. Fish. Biol. **5,** 183–184.

Moriarty, C. (1973b): Studies of the eel *Anguilla anguilla* in Ireland. 2. In Lough Conn, Lough Gill and North Cavan Lakes. Ir. Fish. Invest.(A) **13,** 1–13.

Moriarty, C. (1974): Diet and development of eels in Ireland. ICES CM 1973/M21, 1–2.

Moriarty, C. (1978): Eels. David and Charles, Newton Abbot.

Moriarty, C. (1990): European catches of elvers of 1928–1988. Int. Revue ges. Hydrobiol. **75,** 701–706.

Moriarty, C. (1996): The European eel fishery in 1993 and 1994. Fish. Bull.No. 14.

Moriarty, C., Bninska, M., Leopold, M. (1990): Eel *Anguilla anguilla* (L.), stock and yield in Polish lakes. Aquac. Fish. Mgmt. **21,** 347–356.

Moriarty, C., Dekker, W. (eds) (1997): Management of the European eel. Fish. Bull. No. 15.

Moriarty, C., Nixon (1990): Eel biomass in a small lowland stream. Int. Revue ges. Hydrobiol. **75,** 781–784.

Moriarty, C., Reynolds, J. (1997): Ireland. In: Moriarty, C., Dekker, W.: Management of the European eel. Fish. Bull. No. **15,** 61–65.

Moriarty, C., Tesch, F.-W. (1996): Possible increase in catch of Atlantic elver *Anguilla anguilla* in 1993 and 1994. Ecol. Freshwater Fish **5,** 213–215.

Morovic, D. (1970): Quelques observation sur l'anguille, *Anguilla anguilla* L., de la côte orientale de l'Adriatique. Bilj Inst. Oceanogr. Ribarst, Notes No.27.

Motais, R. (1970): Effect of actinomycin D on the branchial Na-K dependent ATPase activity in relation to sodium balance of the eel. Comp. Biochem. Physiol. **34**, 497–501.

Motais, R., Garcia-Romeu, F. (1972): Transport mechanisms in the teleostean gill and amphibian skin. A. Rev. Physiol. **34,** 141–176.

Motais, R., Isaia, J., Rankin, J. C., Maetz, J. (1969): Adaptive changes of the water permeability of the teleostean gill epithelium in relation to external salinity. J. exp. Biol. **51,** 529–546.

Mott, J. C. (1950): Radiological observations on the cardiovascular system in *Anguilla anguilla.* J. exp. Biol. **27,** 324–333.

Mott, J. C. (1951): Some factors affecting the blood circulation in the common eel (*Anguilla anguilla*). J. Physiol. **114,** 387–398, London.

Motte, I. de la (1963): Untersuchungen zur vergleichenden Physiologie der Lichtempfindlichkeit geblendeter Fische. Naturwissenschaften **50,** 363.

Motte, I. de la (1964): Untersuchungen zur vergleichenden Physiologie der Lichtempfindlichkeit geblendeter Fische. Z. vergl. Physiol. **49,** 58–90.

Mourad, M. H. (1990): Effects of lindane on the electrocardiogram of eel, *Anguilla anguilla* L. Acta Ichthyol. Pisc. **20,** 77–84.

Mourad, M. H. (1991): Cardiovascular and respiratory changes in eel, *Anguilla anguilla* L., during exposure to lethal doses of lindane at different water temperatures. Acta Ichthyol. Pisc. **21,** 73–79.

Mourad, M. H., Kiyszejko, B., Ciereszko, K. (1992): Bioaccumulation of PCB's (Aroclor 1242) and its effects on the electrocardiogram of the eel, *Anguilla anguilla* L. Bull. Natl. Inst. Oceanogr. Fish. Egypt **18,** 141–147.

Müller, H. (1952): Die produktionsbiologischen Verhältnisse märkischer Seen in der Umgebung Storkows. Z. Fisch.1, 95–246.

Müller, H. (1967): Über den Stand der Aalintensivhaltung. Dt. FischZtg Radebeul, **14,** 1–7.

Müller, K. (1962): Versuche über Aalverluste in Turbinen. Fischwirt, **12,** 40–42.

Murai, E. (1971): Tapeworms (Cestodes) parasitizing eels introduced into Lake Balaton. Parasit. hungar. **4,** 145–155.

Murygin, I.L., Anokhina, Yu. I. (1967): Vyshivanie nekotoryi vidov ryb v vozdushnoi srede (Survival of some species of fish in an air environment). Vopr. Ikhtiol. **7**(4), 690–695.

Muus, B. J. (1967): The fauna of Danish estuaries and lagoons. Distribution and ecology of dominatingspecies in the shallow reaches of the mesohaline zone. Meddr Danm. Fisk. og Havunders. N. S. 5 (I), 1–316.

Nagasawa, K., Kim, Y.-G., Hirose, H. (1994): *Anguillicola crassus* and *A. globiceps* (Nematoda: Dracunculoidea) parasitic in the swimbladder of eels (*Anguilla japonica* and *A. anguilla*) in East Asia: a review. Folia Parasitol. **41,** 127–137.

Nagel (1907): Blumenkohlkrankheit der Aale auch in den deutschen Binnengewässern beobachtet. Dt. FischZtg. Radebeul **4,** 160.

Nagieç, M., Bahsawy, M. H. (1990): Age and growth of female eels, *Anguilla anguilla* L., in a Polish lake, Jeziorak Lake, Mazurian Lake district, Poland. Aquacult. Fish. Managem. **21,** 459–470.

Nair, R. V., Dorairaj, K. (1975): Observations on the scales of the shortfinned eel, *Anguilla bicolor* McClelland and their utility in age determination. Curr. Sci. **44**, 53–55.

Nakao, T. (1976): Hydrography south of the Okinawa Islands where leptocephali were collected. J. Fac. Mar. Sci. Technol. Tokai Univ. **9,** 33–46.

Narva, M. E., Engelhardt, F. R. (1980): Compartmentalization of ingested petroleum in tissue and bile of the American eel (*Anguilla rostrata*). Bull. Environ. Contam. Toxicol. **24**, 879–885.

Nedelec, C., Prado, J. (1990): Definition and classification of fishing gear categories. FAO Fish. Techn. Paper 222 Rev. 1.

Neill, R.M. (1940). On the existence of two types of chromatic in teleostean fishes. J.exp. Biol. **17,** 74–95.

Nellen, W., Hempel, G. (1970): Beobachtungen am Ichthyoneuston der Nordsee. Ber. Dt. Wiss. Komm. Meeresforsch. **21,** 311–348.

Nelson, G. J. (1966): Gill arches of teleostean fishes of the order Anguilliformes. Pacif. Sci. **20**(4), 391–408.

Nelson, G. J. (1967): Branchial muscles in representatives of five eel families. Pacific Sci. **21,** 348–363.

Nelson, G. J. (1969): Gill arches and the phylogeny of fishes, with notes on the classification of vertebrates. Bull. Am. Mus. Nat. Hist. **141,** 475–552.

Nelson, G. J. (1973): Relationships of clupeomorphs, with remarks on the structure of the lower jaw in fishes. In: Greenwood, P. H., Miles, R. S., Patterson, C. (Eds.): Interrelationships of Fishes. J. Linn Soc. London (Zool.), Suppl. **1,** Academic Press, London & New York, 333–349.

Neubaur, R. (1924): Abnorm gefärbter Aal. FischZtg., Neudamm **27,** 486.

Neubaur, R. (1933): Änderungen der Umwelt beim Aufstieg der Jungaale. Z. Fisch. **31,** 491–513.

Neuhaus, E. (1967): Die Aalwirtschaft des Landes Schleswig-Holstein. Ostseegebiet. Arch. FischWiss. **18** (Suppl. 1), 431–440.

Neuhaus, E. (1968/69): Die Schleswig-Holsteinische kleine Hochsee- und Küstenfischerei. Fischerblatt 15/1968, 139–143, 16/1968, 99–104, 17/1969, 87–95.

Neuhaus, W., Appelbaum, S. (1980): Quantitative determination of olfactory adaptation in *Phoxinus laevis* (L.) and the importance of adaptation for the olfactory orientation in migratory fishes. Zool. Jb. Physiol. **84,** 458–471.

Neveu, A. (1981): Densité et microrépartition des différentes espèces de poissons dans la basse Nivelle, petit fleuve côtier des Pyrenées atlantiques. Bull. Français Piscicult. **280,** 86–103.

Nicol, J. A. (1963): Some aspects of photoreception and vision in fishes. Adv. mar. Biol., I. Acad. Pr., London & New York.

Nielsen, J. G., Smith, D. G. (1978): The eel family Nemichthyidae (Pisces, Anguilliformes). Dana Rep. **88,** 1–73.

Nikanorov, J., Smotryaev, M. (1962): Eels in the Seligersee. Rybovodstv i rybolovstvo, Moskva **5**(3), 12–14.

Nilsson, S. (1860): Aufenthalt, Lebensweise, Nahrung und Fortpflanzung des Süßwasseraales (*Muraena anguilla* Linn.). Z. ges. naturw. Halle 15 (7/8), 15–31.

Nishi, G, Imai, S. (1969): On the juvenile of Anguilla marmorata Auoy & Gaimard in Yakushima (Yaku Island). Its ecology and morphology. Mem. Fac. Fish. Kagoshima Univ. **18,** 65–76.

Nishi, G., Sadahiko, T., Imai, S. (1969): On the juvenile of *Anguilla marmorata* Quoy & Gaimard in Yakushima (Yaku Island). Its ecology and morphology. Mem. Fac. Fish. Kagoshima Univ. **18,** 65–76.

Nishimura, H., Ogawa, M. (1973): The renin-angiotensin system in fishes. Amer. Zool. **13,** 823–838.

Nishiwaki, M. (Ed.) (1974): Preliminary report of the Hakuho Maru Cruise KH–73–2, February 20 March **27,** 1973 Western North Pacific waters, adjacent to Ryukyu and Taiwan Islands. Ocean Res. Inst. Univ. Tokyo.

Niwa, H. (1979): Effect of environmental salinity on the L-response of the S-potential as an index of spectral sensitivity. Comp. Bioch. Physiol. 64 A, 81–83.

Nolte, K. (1975): Teilaufzucht von Glasaalen in Teichen. Z. Binnenfisch.22, 12–16.

Nolte, W. (1938): Die Blankaalfischerei mit großen Reusen an den Küsten Rügens. Z. Fisch, **36,** 681–725.

Nolte, W. (1967): Die Aalwirtschaft der Länder Niedersachsen und Bremen. Arch. FischWiss. 18 (Suppl. 1), 398–404.

Norberger, B., Björnsson, B. T., Brown, C. L., Wichardt, U. P., Deftos, L. J., Haux, C. (1989): Changes in plasma vitellogenin, sex steroids, calcitonin and thryoid hormones related to sexual maturation in female brown trout (*Salmo trutta*). Gen. Comp. Endocrinol. **75,** 316–326.

Nordeng, H. (1971): Is the local orientation of anadromous fishes determined by pheromones? Nature **233,** 411–413.

Nordquist, O. F. (1904): Alfiskeförsök och alundersökningar i södra Finland. Fisk.-Tidskr. Finl.13.

Nordquist, O. F., Alm, G. (1920): Undersökningar om alens alder, storlek och tillväxthastighet i Sverige. Svenska hydrogr. Biol. Kommn. Fyrskeppunders **7,** 1–16.

Norman, J. R. (1926): The development of the chondrocranium of the eel (*Anguilla vulgaris*), with observations on the comparative morphology of the chondrocranium in bony fishes. Philos. Trans. R. Soc. London, Ser. B **214,** 369–464.

Noseda, V., Chiesa, F., Marchetti, R. (1963): Vector-cardiogram of *Anguilla anguilla* L. and *Pleurodeles waltii* Mich. Nature 197 (4869), 816–818.

Nozaki, M., Fujita, I., Saito, N., Tsukahara, T., Kobayashi, H., Ueda, K., Oshima, K. (1985): Distribution of LHRH-like immunoreactivity in the brain of the Japanese eel (*Anguilla japonica*) with special reference to the nervus terminalis. Zool. Sci. (Tokyo) **2,** 537–548.

Nybelin, O. (1935): Untersuchungen über den bei den Fischen krankheitserregenden Spaltpilz *Vibrio anguillarum.* Meddn St. Unders. O. Fors. Ant. Sötvattfisk. **8,** 55–62.

Ochiai, A., Ogawa, M., Umeda, S., Taniguchi, N. (1975): Changes in blood properties of maturing Japanese eel, *Anguilla japonica,* with hormone injection. Bull. Jpn. Soc. Sci. Fish. **41,** 609–614.

O'Connor, J. F., Power, G. (1971): Trout production and eels in Bill Lake, Saguenay County, Quebec. J. Fish Res. Bd Can.30, 1398–1401.

Odiorne, J. M. (1957): Color changes. In Brown, M.E. (ed.): The physiology of fishes **2,** 387–401. Academic Press, New York.

Oets, J. (1950): Electrocardiograms of fishes. Physiol. comp. Oecol. **2,** 181–186.

Ogata, M., Miyake, Y. (1979): Identification of organic sulfur compounds transferred to fish from petroleum suspension by mass chromatography. Water Res. **13,** 1179–1185.

Ogawa, K., Egusa, S. (1978): Seven species of *Gyrodactylus* (Monogenea: Gyrodactylidae) from *Plecoglossus altivelis* (Plecoglossidae), *Cyprinus carpio* (Cyprinidae) and *Anguilla* spec. (Anguillidae). Bull. Jap. Soc. Sci. Fish **44,** 613–618.

Ogawa, M. (1962): Chloride cells in Japanese common eel, *Anguilla japonica.* Scient. Rep. Saitama Univ., Ser. B., Biol. Earth. Sci.4, 131–137 (1961).

Ogawa, M. (1992): Corpuscles of Stannius. In Matsumoto, A., Ishii, S. (eds): Atlas of Endocrine Organs 181–183. Springer-Verlag Berlin, Heidelberg, New York.

Ogden, J. C. (1970): Relative abundance, food habits, and age of the American eel, *Anguilla rostrata* (Le Sueur), in certain New Jersey streams. Trans. Amer. Fish. Soc. **99,** 54–59.

Ohno, S., Christian, L., Romero, M., Dofuku, R., Ivey, C. (1973): On the question of American eels, *Anguilla rostrata* versus European eels, *Anguilla anguilla*. Experientia **29,** 891.

Oide, M., Utida, S. (1967a): Sur la chromoproteine verte du serum de l'anguille japonaise (*Anguilla japonica*). C.r. Séanc. Soc. Biol. **161,** 1205.

Oide, M., Utida, S. (1967b): Changes in water and ion transport in isolated intestines of the eel during salt adaption and migration. Mar. biol. Wash.1, 102–106.

Oide, M., Utida, S. (1968): Changes in intestinal absorption and renal excretion of water during adaptation to sea-water in the Japanese eel. Mar. biol. Wash.1, 172–177.

Okamura, A., Zhang, H., Yamada, Y., Tanaka, S. Horie, N., Mikawa, N., Utoh, T., Oka, H.P. (2000): Re-examination of the spermatozoal ultrastructure of eels: observations of the external morphology of spermatozoa in three species. J. Fish. Biol. **57,** 161–169.

Okuhara, H. (1958): Illustration of Japanese Fishing Boat and Fishing Gear. Japan Association for Agriculture and Forestry, Tokyo.

Olivereau, M. (1967): Observations sur l'hypophyse de l'anguille femelle, en particulier lors de la maturation sexuelle. Z. Zellforsch. mikrosk. Anat. **80,** 286–306.

Olivereau, M. (1971): Structure histologique du rein et électrolytes plasmatiques chez l'anguille après autotransplantation de l'hypophyse. Z. vergl. Physiol. **71,** 350–364.

Olivereau, M., Lemoine, A.-M. (1971): Action de la prolactine chez l'anguille intacte et hypophysectomisée. VII. Effect sur la teneur en acide sialique (N-acétyleneuramine) de la peau. Z. vergl. Physiol. **73,** 34–43.

Olivereau, M., Olivereau, J. (1977): Effects of transfer to sea water and back to fresh water on the histological structure of the eel kidney. J. comp. Physiol. **115,** 223–239.

Olivereau, M., Olivereau, J. (1985): Effect of 17 alpha methyltestosterone on the skin and gonads of freshwater male silver eels. Gen. Comp. Endocrin. **57,** 64–71.

Oordt, G. J. van, Bretschneider, L. H. (1942): Über den Einfluß gonadotroper Stoffe auf die Entwicklung der Hoden des Aales (*Anguilla anguilla* L.). Arch. Entw. Mech. Org. **141,** 45–57.

Opuszynski, K. (1965): Obserwacje nad wiekiem biologia mlodego Wegorza (*Anguilla anguilla* L.). (Observation on young eel, *Anguilla anguilla* L. age and biology). Rocz. Nauk. Roln. 86–B-2, 393–411.

Opuszynski, K., Leszczynski. (1967): The food of young eels (*Anguilla anguilla* L.) from several lakes and a river in northern Poland. Ekol. pol., Ser. A. **15,** 410–424.

Oray, I. K. (1968): Über die Alterung von Aalreusen aus Polyamiddrähten und ihr Einfluß auf die Fängigkeit. Fischwirt **18,** 29–43.

Orimo, H., Fujita, T., Yoshikawa, M., Watanabe, S., Otini, M., Abe, J. (1972): Ultimobranchial calcitonin of the eel, *Anguilla japonica*. Endocrinol. Japon. **19,** 299–302.

Orlov, Yu. I. O. (1966): On the naturalization of eels in USSR waters. Ryb. Khoz. Moskva **42,**(8), 21.

Ossenkopp, K.-P., Barbeito, R. (1979): Bird orientation and the geomagnetic field. A review. Neurosci. Biobehav. Rev. **2,** 255–270.

Ostermeier, U. (1999): Vitamine in Fischen. Inf. Fischwirtsch. **46**(3), 42–50.

Otani, M., Yamauchi, H., Meguro, T., Kitazawa, S., Watanabe, S., Orimo, H. (1976): Isolation and charcterization of calcitonin from pericardium and esophagus of eel. J. Biochem. **79,** 345–352.

Oudit, G. Y., Butler, D. G. (1995): Cardiovascular effects of arginine vasotocin, atrial natriuretic peptide, and epinephrine in freshwater eels. Amer. J. Physiol. **265,** R1273–R1280.

Palayer, P. (1962): Influence de l'alimentation et de la thyroxine sur la morphologie du pancreas de l'anguille (*Anguilla anguilla* L.), C.r. Séanc. Soc. Biol. **156,** 768–790.

Palayer, P. (1967): Le pancreas endocrine de l'anguille, étude quantitative et morphologique, variations naturelles. Bull. Cent. Etud. Rech. scient. Biarritz **6,** 667–718.

Pang, P. K. T. (1971): Calcitonin and ultimobranchial glands in fishes. J. exp. Zool. **178,** 89–99.

Pang, P. K. T., Yang, M., Oguro, C., Phillips, J. G., Yee, J. A. (1980): Hypotensive actions of parathyroid hormone preparations in vertebrates. Gen. Comp. Endocrinol. **41,** 135–138, 423–431.

Pankhurst, N.W. (1982): relation of visual changes to the onset of sexual maturation in the European eel *Anguilla anguilla* (L.). J.Fish Biol. **21,** 127–140.

Pankhurst, N.W., Lythgoe, J.N. (1983): Changes in vision and olfaction during sexual maturation in the European eel *Anguilla anguilla* (L.). J. Fish Biol. **23,** 229–240.

Pankhurst, N.W., Sorensen, P.W. (1984): Degeneration of the alimentary tract in sexually maturing European *Anguilla anguilla* (L.) and American eels *Anguilla rostrata* (Le Sueur). Can. J. Zool. **62,** 1143–1149.

Pantelouris, E. M., Arnason, A., Tesch, F.-W. (1970): Genetic variation in the eel: 2. Transferrins, haemoglobins and esterases in the eastern North Atlantic. Possible interpretations of phenotypic frequency differences. Genet. Res. **16,** 277–284.

Pantelouris, E. M., Arnason, A., Tesch, F.-W. (1971): Genetic variation in the eel: 3. Comparisions of Rhode Island and Icelandic populations. Implications for the Atlantic eel problem. Mar. Biol. Wash.9, 242–249.

Pantelouris, E. M., Payne, R. H. (1968): Genetic variation in the eel. I. The detection of haemoglobin and esterase polymorphisms. Genet. Res., **11,** 319–325.

Pantulu, V. R. (1956): On the biology of the Indian fresh-water eel, *Anguilla bengalensis* Gray. Proc. natn. Inst. Sci. India **22,** 259–280.

Pantulu, V. R., Singh, V. D. (1962): On the use of otoliths for the determination of age and growth of *Anguilla nebulosa* McClelland. Proc. Indian Acad. Sci.55, (B), 263–275.

Pape, A. (1952): Untersuchungen über die Erträge der Fischerei der Mittelelbe und Auswirkungen ihres Ertragsniederganges. Z. Fisch. **1,** 45–72.

Pape, A. (1970): Die kleine Hochsee- und Küstenfischerei Schleswig-Holsteins im Jahre 1969. Fischerblatt **18,** 110–117.

Pape, R. (1966): Auf Weseraal. Fisch und Fang **7,** 107.

Park, E. H., Kang, Y. S. (1979): Karyological confirmation of conspiuous ZW sex chromosomes in two species of Pacific anguilloid fishes (Teleostomi, Anguilliformes). C. Cytogen. Cell. Genet. **23,** 33–38.

Parker, S. J. (1995): Homing ability and home range of yellow-phase American eels in a tidally dominated estuary. J. mar. biol. Ass. U.K **75,** 127–140.

Parker, S. J., McCleave, J. D. (1997): Selective tidal stream transport by American eels during homing movements and estuarine migration. J. Mar. Biol. Ass. U.K. **77,** 707–725.

Parry, G. (1966): Osmotic adaptation in fishes. Biol. Rev. **41,** 392–444.

Parson, J., Vickers, K. U., Warden, Y. (1977): Relationship between elver recruitment and changes in the sex ratio of silver eels *Anguilla anguilla* L. migrating from Lough Neagh, Northern Ireland. J. Fish. Biol. **10,** 211–229.

Passakas, T. (1976): Further investigations on the chromosomes of *Anguilla anguilla.* Folia Biol. (Krakow) **24,** 239–244.

Passakas, T. (1981): Comparative studies on the chromosomes of the European eel (*Anguilla anguilla* L.) and the American eel (*Anguilla rostrata* Le Sueur). Folia Biol. (Krakow) **29**(1), 42–57.

Passakas, T., Klekowski, R. Z. (1973): Chromosomes of European eel (*Anguilla anguilla*) as related to in vivo sex determination. Polskie. Archwm. Hydrobiol. **20**, 517–519.

Passakas, T., Tesch, F.-W. (1980): Karyological and gonadal sex of eels (*Anguilla anguilla*) from the German Bight and the lower River Elbe. Helgoländer Meeresunters. **34,** 159–164.

Patterson, C. (1975): The braincase of pholidophorid and leptolepid fishes, with a review of the actinopterygian braincase. Phil. Trans. Roy. Soc. London, Ser. B **269,** 275–579.

Patterson, C. (1977): Cartilage bones, dermal bones and membrane bones, or the exoskeleton versus the endoskeleton. Linn. Soc. Symp. Ser. **4,** 77–121.

Patterson, C., Johnson, G. D. (1995): The intermuscular bones and ligaments of teleostean fishes. Smithsonian Contrib. Zool. **559,** 1–83.

Pavesi, P. (1894): Curioso metacromatismo in *Anguilla.* R. c. Inst. lomb. Sci. Lett. **27,** 688–692.

Pedersen, M. (1997): Denmark. In Morairty, C., Dekker, W. (eds): Management of the European eel. Fish. Bull. **15,** 53–56.

Pelster, B., Scheid, P. (1991): Activities of enzmes for glucose metabolism in the swimbladder of the European eel *Anguilla anguilla*. J. exp. Biol. **156,** 207–214.

Pelster, B., Scheid, P. (1992): The influence of gas gland metabolism and blood flow on gas deposition into the swimbladder of the European eel *Anguilla anguilla*. J. exp. Biol. **173,** 205–216.

Peñaz, M., Tesch, F.-W. (1970): Geschlechtsverhältnis und Wachstum beim Aal (*Anguilla anguilla*) an verschiedenen Lokalitäten von Nordsee und Elbe. Ber. dt. wiss. Kommn Meeresforsch. **21,** 290–310.

Pequignot, J., Labat, R., Serfaty, A. (1968): Vagotonicité et resistance à la salinité chez quelques téléostéens d'eau douce. Hydrobiologia **32,** 570–576.

Pequignot, J., Serfaty, A. (1965): Influence de la salinité sur la respiration tissulaire chez les téléostéens. Experientia **21,** 227–228.

Perlmutter, A. (1951): An aquarium experiment on the American eel as a predator on larval lampreys. Copeia 1951, 173–174.

Pernkopf, E. (1930): Beiträge zur vergleichenden Anatomie des Vertebratenmagens. Z. Anat. Entwgesch. **31,** 329–390.

Pesaro, M., Balsamo, M., Gandolfi, G., Togiorgo, P. (1981): Discrimination among different kinds of water in juvenile eels, *Anguilla anguilla* (L.). Monitore Zool. Ital. (N.S.) **15,** 183–193.

Peters, G. (1977): The papillomatosis of the European eel (*Anguilla anguilla* L.): Analysis of seasonal fluctuations in the tumor incidence. Arch. Fischwiss. **27,** 251–263.

Peters, G., Delventhal, H., Klinger, H. (1980): Physiological and morphological effects of social stress in the eel (*Anguilla anguilla* L.). Arch. Fischwiss. **30,** 157–180.

Peters, N. (1935): Angeln. In: Willer, H. (ed.): Handb. Seefisch. Nordeuropas, Stuttgart, **4**(4), 1–48.

Peters, N., Peters, G. (1970): Tumorgenese, ein Energieproblem der Zelle? Untersuchungen an Papillomen des Europäischen Aals, *Anguilla anguilla* (L.). Arch. FischWiss.21, 238–257.

Petersen, C. G. J. (1906): The infuence of light on the migrations of the eel. Rep. Dan. biol. Stn. **14,** 3–9.

Pevzner, R. A. (1978): Elektronnomikroskopicheskoe issledovanie vkusovykh hukovits evropeiskogo ugryo. (Electron miscrospopic study of the taste buds of the eel, *Anguilla anguilla*). Tsitologiya **20,** 1112–1118.

Pfeiffer, W. (1960): Über die Schreckreaktion bei Fischen und die Herkunft des Schreckstoffes. Z. vergl. Physiol. **43,** 578–614.

Pfitzner, J. (1969): Zur Aetiologie der Blumenkohlkrankheit der Aale. Arch. FischWiss., 20 (Suppl. 1) 24–35.

Piatek, M. (1970): Age, growth and changes of weight proportions and of chemical composition of eel during its life in Polish waters. Acta Ichtyol Pisc. **1,** 73–96.

Plehn, M. (1924): Praktikum der Fischkrankheiten. In: Handbuch Binnenfisch. Mitteleur. **1,** 301–470.

Poluhovich, J. J., Park, R. P. (1972): A comparative study of blood chemestry and respiration in marine and freshwater eels. Prog. Fish. Cult. **34,**(1), 33–38.

Poole, W. R. (1994): A population study of the European eel (*Anguilla anguilla* L.) in the Burrishoole system, Ireland, with special reference to growth and movement. Thesis Dublin Univ. Degree Doct. Philosophy.

Poole, W. R., Reynolds, J. D.,Moriarty, C. (1990): Observations on the silver eel migration of the Burrishoole River system, Ireland, 1959 to 1988. Int. Revue ges Hydrobiol. **75,** 807–815.

Portier, P. (1938): Physiologie des animaux marines. Flammarion, Paris.

Portier, P., Duval, M. (1922): Pression osmotique du sang de l'anguille essuyée en modifications de salinité du milieu extrieur. C.r. hebd. Séanc. Acad. Sci. Paris, **175,** 1105–1107.

Post, A., Tesch, F.-W. (1982): Midwater trawl catches of adolescent and adult anguilliform fishes during the Sargasso Sea eel expedition 1979. Helgoländer Meeresunters. **35,** 341–356.

Pradet-Balade, B., Schmitz, M., Salmon, C., Dufour, S., Querat, B. (1997): Down regulation of TSH subunit mRNA levels by thyroid hormones in the European eel. Gen. Comp. Endocrinol. **108,** 191–198.

Predel, G., Gollub, H. (1973): Neue Erkenntnisse beim Einsatz elektrifizierter Schleppnetze in der Binnenfischerei. Z. Binnenf. DDR (12), 372–374.

Prokhorchik, G.A., Petukhov, V.B., Petrikov, A.M. (1987): Embryonic development of European eel, *Anguilla anguilla*, under experimental conditions. J. Ichthyol. **27**(3), 37–43.

Puqin, W., Xiumin, L. (1976): Some nematodes of the suborder Camallanata from Fujian Province, with notes on their life histories. Acta Zool. Sin. **21,** 350–358.

Querat, B., Cardinaud, B., Hardy, A., Vidal, B., D'Angelo, G. (1994): Sequence and regulation of European eel prolactin mRNA. Mol. Cell. Endocrinol. **102,** 151–160.

Quinn, T. P. (1980): Evidence for celestial and magnetic compass orientation in lake migrating sockeye salmon fry. J. Comp. Physiol. **137,** 243–248.

Quinn, T. P., Merrill, R. R., Brannon, E. (1981): Magnetic field detection in sockeye salmon. J. exp. Zool. **217,** 137–142.

Quoss, H. (1976): Fangversuche mit einer randparallel montierten Aalkette im Bodensee. Fischwirt **26,** 17–19.

Raben, K. v. (1955): Kaplanturbinen und Fische. Wasserwirtschaft, Stuttgart, **45,** 196–200.

Raben, K. v. (1958): Zur Beurteilung der Schädlichkeit der Turbinen für Fische. Wasserwirtschaft, Stuttgart, **48,** 60–63.

Radford, D. S. (1972): The first record of *Anguilla rostrata* from Alberta. J. Fish. Res. Bd Can. **29,** 1084.

Radke, R.J., Eckmann, R. (1996): Piscivorous eels in Lake Constance: can they influence year class strength of perch? Ann. Zool. Fennici **33,** 489–494.

Radulescu, N., Angelescu, N. (1972): A papilloma case in a European eel (*Anguilla anguilla* L.) captured in the Black Sea. Bul. Cerc. pisc. **31,** 133–136.

Rahn, J. (1955a): Untersuchungen über Alter und Wachstum des Aalbestandes im Sakrower See. Z. Fisch. **4,** 235–256.

Rahn, J. (1955b): Ein Beitrag zur Satzaalfrage. Dt. FischZtg Radebeul **2,** 130–132.

Rahn, J. (1957a): Satzaale oder Aalbrut? Dt. FischZtg Radebeul **4,** 70–72.

Rahn, J. (1957b): Die Fischerei auf dem Sakrower See in den Jahren 1949 bis 1955. Z. Fisch. 6 (N. S.), 561–588.

Rahn, J. (1957c): Beitrag zur Altersbestimmung an Aalen. Z. Fisch. **6,** 393–396.

Rahn, J. (1963): Beobachtungen an Seen während des Winters 1962/63. Dt. FischZtg Radebeul, **10,** 313–323.

Rameil, H. (1967): Die Aalwirtschaft des Landes Hessen. Regierungsbezirke Darmstadt und Wiesbaden. Arch. FischWiss.18 (Suppl. 1), 291–305.

Randall, D. J. (1968): Functional morphology of the heart of fishes. Amer. Zool. **8,** 179–189.

Randall, D. J. (1970): The circulatory system. In Hoar, W.S., Randall, D.J. (eds): Fish Physiology, **4,** 133–172. New York and London, Academic Press.

Rand-Weaver, M. A., Kawauchi, H. (1993): Growth hormone, prolactin and somatolactin: a structural overview. In Hochachka, P., Mommsen, T.P. (eds): Biochemistry and Molecular Biology of Fishes. 39–56. Elsevier Amsterdam.

Rankin, J. C., Maetz, J. (1971): A perfused teleostean gill preparation: vascular actions of neurohypophysial hormones and catecholamines. J. Endocr. **51,** 621–635.

Rankovic, N. J. (1957): Jugoslawische Versuche mit Aalschnüren. Fischwirt **7,** 15–16.

Rasmussen, C. J. (1951): Two Danish finds of female eel (*Anguilla anguilla*) in spawning or partial spawning dress. Rep. Dan. biol. Stn. **53,** 35–39.

Rasmussen, C. J. (1952): Size and age of the silver eel (*Anguilla anguilla* L.) in Esrum Lake. Rep. Dan. biol. Stn. **54,** 3–36.

Rathcke, P.-C. (1994): Untersuchung über Schädigungen von Fischen durch Turbine und Rechen im Wasserkraftwerk Dringenauer Mühle. In: Vorträge der Aalkomm. 1993. Arb. Dt. Fischereiverband 1994, 37–74.

Rathcke, P.-C. (1995): Maßnahmen zur Vermeidung fischereilicher Schäden durch Turbinen. In: Schriftenr. Verb. Dt. Fischereiverwaltungsb. u. Fischereiwiss. **9,** Kleinwasserkraftanlagen und Gewässerökologie, 34–41.

Rehbronn, E., Rutkowski, F. (1997): Das Räuchern von Fischen. 7. Edn., Parey Buchverlag, Berlin.

Reichenbach-Klincke, H. H. 1952: Ein Aal mit doppelter Leber. Nachr.-naturwiss. Mus. Aschaffenburg **37,** 51–54.

Reichenbach-Klinke, H. H. (1956): Ein Aal (*Anguilla vulgaris* Turton) mit einem Stein in der Leibeshöhle. Nachr. naturw. Mus. Aschaffenb. **53,** 23–28.

Reichenbach-Klinke, H. H. (1966): Ein Aal (*Anguilla vulgaris* Turton) mit doppelter Leber. Nachr. naturw. Mus. Aschaffenb. **37,** 51–54.

Reichenbach-Klinke, H.H. (1975): Bestimmungsschlüssel zur Diagnose von Fisch-krankheiten. Gustav Fischer Verlag, Stuttgart. 98pp.

Reichenbach-Klinke, H. H. (1980): Krankheiten und Schädigungen der Fische. 2. Edn., Stuttgart.

Reichwald, J., Meizies, A. (1973): Die Fettsäuren der Lipide im Fleisch von Süßwasserfischen und Seefischen. Z. ErnährWiss.12, 86–91.

Reid, S. G., Fritsche, R., Jonsson, A. C. (1995): Immunohistochemical localization of bioactive peptides and amines associated with chromaffin tissue of five species of fish. Cell Tiss. Res. **280,** 499–512.

Reid, S. G., Perry, S. F. (1995): Cholinergic-mediated control of catecholamine release from chromaffin cells in the American eel, *Anguilla rostrata*. J. Comp. Physiol. 165B, 464–470.

Reimer, L. (1966): Digene Trematoden und Cestoden der Ostseefische, ihre Entwicklung, Ökologie und ihre Verwendung als natürliche Fischmarken. Habil.-Schrift Math-Nat. Fakult. Univ. Greifswald.

Reimer, L. W. (1970): Digene Trematoden und Cestoden der Ostseefische als natürliche Fischmarken. Parasitol. Schriftenr. 20.

Reimer, L. W. (1987): Bewegungen der Aale der Ostsee vor der Laichwanderung auf Grund ihrer Parasitierung. Wiss. Z. P. H Güstrow, Math.-Nat. Fak. 157–166.

Reimer, L. W. (1995): Parasites especially of piscean hosts as indicators of the eutrophication in the Baltic Sea. Appl. Parasitol. **36,** 124–135.

Reimer, L.W. (2000): Parasiten von Weserfischen. Naturschutz Landschaftspflege, Sonderheft**10**, 35pp.

Reimer, L.M., Walter, U. (in press): Veränderungen im Parasitenbefall der Aale der südlichen Ostseeküste in den 90er Jahren. Mitt. Landesforschungsanst. Landwirtsch. Fischerei Mechlenb.-Vorpommern.

Reimer, L.W., Hildebrand, A., Scharberth, D., Walter, U. (1994): *Anguillicola crassus* in the Baltic sea: field data supporting transmission in brackish water. Dis. Aquat. Org. **18,** 77–79.

Reinsch, H. H. (1968): Fund von Fluß-Aalen, *Anguilla anguilla* (L.), im Nordatlantik. Archiv. FischWiss. **19,** 62–63.

Remane, A., Schlieper, C. (1958): Die Biologie des Brackwassers. In Thienemann (ed.): Die Binnengewässer, 22.

Reynier, B., Foulquier, L., Angelier, E., Grauby, A. (1970): Premières donnés expérimentales sur la dynamique du césium-137 chez *Anguilla anguilla* (L.). C.r. hebd. Séanc. Acad. Sci. Paris **270,** 862–865.

Richard, A. (1956): Phénomènes inhibiteurs, d'origine sensorielle sur la locomotion de l'anguille. J. Physiol., **48,** 700–702.

Richard, H. (1938): Aalfang mit Aalpuppen. Fischztg., Neudamm, **41,** 172–173.

Riggs, A. (1952): The oxygen equilibrium of the hemoglobin of the eel, *Anguilla rostrata*. J. gen. Physiol. **35,** 41–44.

Robak, S. (1994): Stocking of young eels and spring downstream migration of small eel *Anguilla anguilla* (W <300 g) in the River Sapina basin (Great Mazurian lake district – Poland). Bull. Sea Fish. Inst. **2**(132), 3–11.

Robins, C. R., Cohen, D. M., Robins, C. H. (1979): The eels, *Anguilla* and *Histiobranchus,* photographed on the floor of the deep Atlantic in the Bahamas. Bull. Mar. Sci.29, 401–405.

Rocca, E., Chiretti, F. (1964): A toxic protein from eel serums *(Anguilla anguilla).* Toxicon **2,** 79–80.

Rodino, E., Comparini, A. (1978a): Biochemical polymorphism in teleosts. The eel problem. Boll. Zool., 45 (Suppl. 2), 47–61.

Rodino, E., Comparini, A. (1978b): Genetic variability in the European eel, *Anguilla anguilla* L. In: Battaglia, B., Beardmore J. A., eels: Marine organism, genetics, ecology and evolution. Nato Conf., Ser. IV(2), Plenum Pr. New York, London, 389–424.

Romanello, M.G., Scapolo, P.A., Luprano, S., Mascarello, F. (1989): Post-larval growth in the lateral white muscle of the eel *Anguilla anguilla*. J. Fish Biol. **30,** 161–172.

Rosengarten, J. (1954): Der Aufstieg der Fische im Moselfischpaß Koblenz im Frühjahr 1952 und 1953. Z. Fisch. **3,** 489–552.

Rossi, R. (1979): An estimate of the production of eel population in the valli of Comacchio (Po Delta) during 1974–1976. Bollitino di Zool. **46,** 217–223.

Rossi, R. A. M., Carrieri, A., Franzoi, G., Cavallini, G., Gnes, A. (1987, 1988): A study of eel (*Anguilla anguilla* L.) population dynamics in the Comacchio lagoons (Italy) by mark-recapture method. Oebalia **14,** 1–14.

Rossi, R., Bianchini, M., Carrieri, A., Franzoi, P. (1987): Observations on movements of yellow eels, *Anguilla anguilla* L., after displacement from coastal waters to sea. J. Fish. Biol. **31,** 155–164.

Rossi, R., Colombo, G. (1976): Sex ratio, age and growth of silver eel in two brackish water lagoons in the northern Adriatic Valli of Commacchio and Valli nuova. Arch. Oceanogr. and Limnol. 18 (Suppl. 3), 327–341.

Rossi, R., Colombo, G. (1979): Some observations on age, sex, and growth of silver eels (*Anguilla anguilla* L.) in North Adriatic lagoon. Rapp. P. V. Reun. Cons. int. Explor. Mer. **174,** 64–69.

Roth, D. (1997): Kurze Schatten – lange Aale. Fisch u. Fang **6,** 36.

Rowchai, S., Chiba, K., Hirano, R. (1986): Influence of dissolved oxygen on growth of young eel. Bull. Jpn. Soc. Sci. Fish. **52**(4): 597–607.

Royce, W. F., Smith, L. S., Hartt, A. C. (1968): Models of oceanic migration of Pacific Salmon and comments on guidance mechanisms. Fishery Bull. Fish Wildl. Serv. U.S. **66,** 441–462.

Rudsinke, W. (1960): Aale haben keinen "Wassersinn"! Fisch u. Fang **1,** 153–154.

Rumphorst, H. (1929): Zwei Fälle von Xanthorismus beim Aal. Mitt. dt. Seefisch. Ver., 45, 123–124.

Rumphorst, H. (1930): Die Blankaalfischerei im Kreise Rügen. Z. Fisch.28, 403–432.

Runge, K. (1965): Mein Krebsbach. Fisch u. Fang **6,** 169.

Rünger, F., Meyer, R. (1941): Die Aalfangversuche mittels Aalhamen in der schlesischen Oderstrecke bei Neusalz. Z. Fisch **39,** 259–283.

Rupp, R. S., Deroche, S. E. (1965): Standing crops of fishes in three small lakes compared with C14 estimates of net primary productivity. Trans. Amer. Fish. Soc. **94,** 9–25.

Sadler, K. (1981): The toxicity of ammonia to the European eel (*Anguilla anguilla* L.). Aquaculture **26,** 173–181.

Saglio, P., Escaffre, A.M., Blanc, J.M. (1988): Structural characteristics of the epidermal mucosa in yellow and silver European eel *Anguilla anguilla* (L.). J. Fish Biol. **32,** 505–514.

Saglio, P., Fauconneau, B., (1988): Free amino-acid content in the skin mucus of yellow and silver eel, *Anguilla anguilla* (L.). Comp. Biochem. Physiol. 91A, 101–104.

Saint-Paul, U. (1977): Einige Aspekte der deutschen Satzaalfischerei. Arch. FischWiss. **28,** 123–133.

Sancho, E., Ferrando, M. D., Andreu, E. (1996): Physiological stress responses of *Anguilla anguilla* to fenitrothion. J. Environ. Sci. Health, Part B, Pestic., Food Contam., Agric. Wastes B31, 87–98.

Sancho, E., Ferrando, M. D., Andreu, E., Gamon, M. (1993): Bioconcentration and excretion of diazinon by eel. Bull. Environ. Contam. Toxicol. **50,** 578–585.

Sang, T. K., Chang, H. Y., Chen, C. T., Hui, C. F. (1994): Population structure of the Japanese eel, *Anguilla japonica*. Mol. Biol. Evol., **11,** 250–260.

Sano, T. (1976): Viral diseases of cultured fishes in Japan. Fish Path. **10,** 221–226.

Sano, T., Okamoto, N., Dnishimura, T. (1981): A new viral epizootic of *Anguilla japonica* Temminck and Schlegel. J. Fish Diseases **4,** 127–139.

Santos, M. A., Hall, A. (1990a): Influence of inorganic lead on the biochemical blood composition of the eel, *Anguilla anguilla*. Ecotoxicol. Environ. Saf. **20,** 7–9.

Santos, M. A., Hall, A. (1990b): Influence of inorganic mercury, under different temperatures, on the biochemical blood composition in the eel *Anguilla anguilla* L. Rev. Biol. Univ. Aveiro **3,** 61–70.

Santos, M. A., Weber, M. (1992): Growth studies on monthly captured glass eels from the Rio Minho in two recirculation systems. Irish Fish. Invest. Ser. A., No. **36,** 35–40.

Sarava, A., Chubb, J.C (1989): Preliminary observations on the parasites of *Anguilla anguilla* (L.) from Portugal. Bull. Eur. Ass. Fish Pathology **9**(4), 88–89.

Saroglia, M., Fantin, P., Arlati, G. (1985): Eel production in Italy – Problems and perspectives. EIFAC Working Party on Eel, Perpignan, September 17–18.

Saroglia, M., Iandoli, C., Ingle, E., Angle, G. (1991): Recent market trends and economic remarks on the eel sector in Italy. EIFAC Working Party on Eel, Dublin May 20–25.

Sato, N., Aida, K. (2000): Induction of ovarian maturation by two distinct gonadotropins in Japanese eel. J. Taiwan Fish. Res. **8**(1), 93–5.

Satoh, H., Nakamura, N., Hibiya, T. (1962): Studies on the sexual maturation of the eel.-I. On the sex differentiation and the maturing process of the gonads. Bull. Japan. Soc. scient. Fish. **28,** 579–584.

Schalensky, G. (1975): Aalpöddern in Frankreich. Fisch u. Fang **16,** 385.

Schäperclaus, W. (1927): Die Rotseuche des Aales im Bezirk von Rügen und Stralsund. Z. Fisch **25,** 99–128.

Schäperclaus, W. (1949): Bewirtschaftung von Binnenseen mit Karpfen. Neumann, Radebeul u. Berlin.

Schäperclaus, W. (1953): Die Blumenkohlkrankheit der Aale und anderer Ostseefische. Z. Fisch.2, 105–124.

Schäperclaus, W. (1956): Bekämpfung der infektiösen Bauchwassersucht des Karpfens durch Antibiotika. Z. Fisch. **5,** 3–59.

Schäperclaus, W. (1967, 1998): Lehrbuch der Teichwirtschaft. 3rd. Edn. 1967, 4th Edn. 1998, Parey Buchverlag Berlin.

Schäperclaus, W. (1979): Fischkrankheiten. 4th Edn. Berlin.

Scheer, D. (1934): Ein Aal mit Nierengeschwulst. Z. Fisch.32, 585–586.

Scheer, D. (1964): Lehren aus den Aalverlusten im Winter 1962/63. Dt. FischZtg. Radebeul **11,** 131–134.

Scheuring, L. (1927): Ein wellenförmig gekrümmter Aal. Allg. FischZtg. **52,** 349–350.

Schicker, D. (1981): Aalangeln mit allen Schikanen. Paul Parey, Hamburg u. Berlin.

Schiemenz, F. (1930): Über Aalbesatz und Aalaufstieg in den Gewässern Hannovers. Allg. FischZtg **55,** 87.

Schiemenz, F. (1940): Abänderungen und neue Formen der Aaltreppe. FischZtg., Neudamm, **43,** 147–147, 152–153.

Schiemenz, F. (1950): Wie soll das Unterende der Fischtreppe in das Hauptgewässer münden? Versuche mit Glasaalen. Wasserwirtschaft, Stuttgart **40,** 130–135.

Schiemenz, F. (1952): Versuche mit Glasaalen. Mitt. Hannover Versuchsanst. Grund-Wasserbau, Techn. Hochschule, Hannover, No. **2,** 24–33.

Schiemenz, F. (1958): Die Enge des Abstandes der Stäbe von Schutzrechen vor Turbinen und Sielpumpen in ihrer Bedeutung für die Fischerei wie für die Kraftwirtschaft. Fischereiwelt **8,** 285–289.

Schiemenz, F. (1960a): Die Talabwanderung der Aale, die Turbinen und die Schutzrechen. Müllerei, Stuttgart, **13,** 15.

Schiemenz, F. (1960b): Unterschied der Wanderungen der Fische, insbesondere der Aale verschiedenen Reifegrades, im unkanalisierten Strom und im kanalisierten Strom und die Bedeutung der Fischtreppen. Z. Fisch.9, 133–154.

Schiemenz, F. (1962): Das Fischereirecht in Mitteleuropa. In: Handbuch Binnenfisch. Mitteleur. **4,** 417–510.

Schiemenz, F., Kühne, M. (1964): Eine weitere Beobachtung zum Aufstieg von Jungaalen in Aalschlängeltreppen. Fischwirt **6,** 333–339.

Schiemenz, P. (1910): Was frißt der Aal? Fischerbote, **2,** 197–200.

Schlicher, J. (1927): Vergleichend-physiologische Untersuchungen der Blutkörperchenzahlen bei Knochenfischen. Zool. Jb. (Allg. Zool.) **43,** 121–200.

Schlieker, E. (1957): Probleme des Baues und der Anwendung von Großreusen. Z. Fisch. **6,** 531–558.

Schlieker, E. (1964a): Berechnung, Bau und Einsatz von Scherbretthamen. Z. Fisch. **12,** 355–372.

Schlieker, E. (1964b): Hinweise zur Fischerei mit Scherbretthamen. Dt. FischZtg. Radebeul **11,** 151–164.

Schlieper, C. (1933): Über die osmoregulatorische Funktion der Aalkiemen. Z. vergl. Physiol.18.

Schmalhausen, J. J. (1913): Bau und Phylogenese der unpaaren Flossen und insbesondere der Schwanzflosse der Fische. Z. wiss. Zool. **104,** 1–80.

Schmeidler, E. (1957): Entwicklung des Glasaalaufstieges in der Ems, seine zeitliche Begrenzung und die auf ihn einwirkende Wassertemperatur. Z. Fisch. **6,** 141–144.

Schmeidler, E. (1963): Beobachtungen über die mengenmäßige Entwicklung und den zeitlichen Ablauf des Glasaalaufstieges in der Ems bei Hebrum von 1950 bis 1962. Dt. Gewässerkundl. Mitt. **7,** 84–86.

Schmid, J. (1962): Aalfischerei in Oberbayern. Fischwirt **12,** 245–246.

Schmid, J. (1964): Vom Riegsee. Allg. FischZtg. **8,** 417.

Schmid, J. (1966): Die Aalwirtschaft des Landes Bayern. Regierungsbezirk Oberbayern. Arch. FischWiss. 16 (Suppl. 1), 29–42.

Schmid, O. J. (1969): Beitrag zur Histologie und Ätiologie der Blumenkohlkrankheit der Aale. Arch. FischWiss. 20 (Suppl. 1), 16–23.

Schmidt, J. (1906): Contributions to the life-history of the eel (*Anguilla vulgaris,* Flem.). Rapp P.-V. Reun. Cons. perm. int. Explor. Mer **5,** 137–264.

Schmidt, J. (1909a): Remarks on the metamorphosis and distribution of the larvae of the eel (*Anguilla vulgaris* Turt.). Meddr Kommn Havunders., Ser. Fisk.3(3), 1–17.

Schmidt, J. (1909b): On the distribution of the freshwater eels (*Anguilla*) throughout the world. 1. Atlantic Ocean and adjacent region. Meddr Kommn Havunders., Ser. Fisk.3, 1–45.

Schmidt, J. (1912): Danish researches in the Atlantic and Mediterranean on the life-history of the freshwater eel (*Anguilla vulgaris* Turt). Int. Revue ges. Hydrobiol. **5,** 317–342.

Schmidt, J. (1913): First report on eel investigations 1913. Rapp. p.-v. Reun. Cons. perm. int. Explor. Mer **18,** 1–30.

Schmidt, J. (1916): On the early larval stages of the freshwater eels (*Anguilla*) and some other North Atlantic Muraenoides. Meddr. Kommn. Havunders. Rer. Fisk. **5**(4), 1–20.

Schmidt, J. (1923): The breeding places of the eel. Phil. Trans. R. Soc. 211, 179–208. (Int. Revue ges. Hydrobiol. **11, 140,** 1923).

Schmidt, J. (1925a): The breeding places of the eel. Ann. Rep. Smithsonian Inst. 1924, 279–316.

Schmidt, J. (1925b): On the distribution of the freshwater eels (*Anguilla*) throughout the world. II. Indopacific region. Skr. Kongel. Dansk. Selsk. Nat. Afd. Kobh.10, 329–382.

Schmidt, J. (1929): Introduction to the oceanographical reports. Oceanogrl Rep. „Dana„ Exped. 1920–22, Nr. **1,** 1–86.

Schnakenbeck, W. (1932): Blankaale im Meer. Fischerboote **24,** 200.

Schnakenbeck, W. (1934): Veränderungen im Verdauungstraktus bei Blankaalen. Zool. Anz., **108,** 85–91.

Schnakenbeck, W. (1940): Stehende Geräte. In: Handbuch Seefisch. Nordeuropas **4**(3), 1–48.

Schnakenbeck, W. (1942): Schleppnetze, Waden. In Handbuch Seefisch. Nordeuropas 4 (1/2), 1–52.

Schnakenbeck, W. (1944): Untersuchungen über Fragen um den Aal. Z. Fisch. **42,** 365–389.

Schnakenbeck, W. (1953a): Die deutsche Seefischerei. Krögers Buchbinderei Blankenese.

Schnakenbeck, W. (1953b): Entwicklung des "Laichhabitus" bei Aalen. Zool. Anz. **150,** 268–275.

Schnakenbeck, W. (1954): Zur Satzaalfrage. Dt. FischZtg. Radebeul **1,** 181–183.

Schnakenbeck, W. (1955): Pisces. Handbuch Zool., Berlin, **6,** 1. II, 551–1115.

Schnare, K. (1960): Merkwürdige Quartiere der Aale. Fisch u. Fang **1,** 105.

Schneider, G. (1904): Beiträge zur Kenntnis der Helminthenfauna des Finnischen Meerbusens. Acta Soc. Fauna Flora fenn. 26.

Schoeniger, E. (1951): Untersuchungen über die Wirkung von Lockstoffen auf Aalbrut. Protokolle Fisch. Techn. **4**(8), 1–28.

Schooten, F.J. van, Maas, L.M., Moonen, E.J.C., Kleinjans, J.C.S., Oost, R. van der (1995): DNA-dosimetry in biological indicator species living on PAH-contaminated soils and sediments. Ecotoxicol. Environ. Saf. **30,** 171–179.

Schoth, M. (1982): Taxonomic studies on the 0-Group eel larvae (*Anguilla* spec.) caught in the Sargasso Sea in 1979. Helgoländer Meeresunt. **35,** 279–287.

Schoth, M., Tesch, F.-W. (1982): Spatial distribution of 0–group eel larvae (*Anguilla* spec.) in the Sargasso Sea. Helgoländer Meeresunters. **35,** 309–320.

Schoth, M., Tesch, F.-W. (1984): The vertical distribution of small 0-Group *Anguilla* larvae in the Sargasso Sea with reference to other anguilliform leptocephali. Meeresforsch. **30,** 188–195.

Schrieven, H. (1935): Aktionspotentiale des Nervus lateralis bei Reizung der Seitenorgane von Fischen. Pflügers Arch. ges. Physiol. **235,** 771–784.

Schubert, G. (1969): Elektromikroskopische Untersuchungen an der Haut mit Blumenkohlkrankheit behafteter Aale. Arch. FischWiss. 20 (Suppl. 1), 36–49.

Schubert, H.-J. (1999): Die neue Fischaufstiegsanlage am Elbewehr Geesthacht. Arb. Dtsch. Fischereiverb. **14,** 47–62.

Schultze, E. (1965): Beiträge zur Temperaturadaptation des Aales. Z. wiss. Zool. 172, 104–133.

Schultze, E. (1972): Untersuchungen an der Regio olfactorio des Anles *Anguilla anguilla* L. : Reinstrukter des Reichepithels. Z. Zellforsch. mitkroskop. Anatom. **125,** 210–228.

Schulz, D. (1975): Salinitätspräferenzen bei Glasaalen und jungen Gelbaalen (*Anguilla anguilla*). Helgoländer wiss. Meeresunters. 27, 199–210.

Schulze (1980): Goldgelber Aal, gefangen im Sommer 1978 im Gülper See der PwF Brandenburg (Cover picture). Binnenfisch DDR **27**(1).

Schwanz-Pfitzner, I. (1973): Untersuchungen zur Klärung der Ätiologie der "Blumenkohlkrankheit der Aale" und zum Einfluß dieses Tumors auf Ernährung und Wachstum. Verh. int. Verein. theor. angew. Limnol. **18,** 1666–1673.

Sedlar, J., Dobrota, P. (1966): A contribution to recognition of the age of the eel (*Anguilla anguilla*) in the rivers of Slovakia. Acta zootech., Nitra **13,** 153–158.

Sedlar, J., Krcmárik, J. (1967): Contribution to the knowledge of the composition of the main food components of the eel (*Anguilla anguilla* L.) in the Slovak rivers. Pol'nohospodarstva **13,** 58–62.

Seliger, H. H. (1963): Direct action of light in naturally pigmented muscle fibres. J. gen. Physiol. **46,** 333–342.

Seligo, A. (1926): Die Fischerei in den Fliessen, Seen und Strandgewässern Mitteleuropas. Handbuch Binnenfisch. Mitteleuropa, **5,** 1–422.

Sella, M. (1929): Estese migrazioni dell'anguilla in acque sotterranee. Monografie R. Com. Talassogr. ital. **158,** 1–17.

Seutter, E., Trijbels, J. M. F., Sutorius, A. H. M., Urselmann, E. J. M. (1970): The sweat gland an amucous gland. Dermatologica **141,** 397–408.

Sharratt, B.M., Bellamy, D., Chester Jones, I. (1964a): Adaptation of the silver eel (*Anguilla anguilla* L.) to sea water and to artificial media together with observations on the role of the gut. Comp. Biochem. Physiol. **11,** 19–30.

Sharratt, B. M., Chester Jones I., Bellamy, D. (1964b): Water and electrolyte composition of the body and renal function of the eel (*Anguilla anguilla* L.). Comp. Biochem. Physiol. **11,** 9–18.

Sheldon, W. W. (1974): Elvers in Maine – Techniques of locating, catching and holding. Sea Grant Publication 4266.3030.

Shirai, N., Utida, S. (1970): Development and degeneration of the chloride cell during seawater and freshwater adaptation of the Japanese eel, *Anguilla japonica.* Z. Zellforsch. mikrosk. Anat. **103,** 247–264.

Shulman, S. S. (1966): Mikrosporidii. Fauny SSSR. Moskva, Leningrad.

Shulman, S.S. (1984): Parasitische Einzeller. Vol.I. In Bauer, O. N.: Keys to the parasites of freshwater fishes of the USSR. Leningrad.

Sick, K. M., Bahn, B., Frydenberg, O., Nielsen, J. T., Wettstein, D. von (1967): Haemoglobin polymorphism of the American fresh water eel *Anguilla*. Nature **214,** 1141–1142.

Sick, K., Westergaard, M., Frydenberg, O. (1962): Haemoglobin pattern and chromosome number of American, European and Japanese eels (*Anguilla*). Nature, London, **193,** 1001–1002.

Simmoneaux, V., Humbert, W., Kirsch, R. (1988): Structure and osmoregulatory functions of the intestinal folds in the seawater eel, *Anguilla anguilla*. J. comp. Physiol. **158,** 45–55.

Simon, B., Sébert, P., Cann-Moisan, C., Barthélémy, L. (1992): Muscle energetics in yellow freshwater eels (*Anguilla anguilla* L.) exposed to high hydrostatic pressure (101 ATA) for 30 days. Comp. Biochem. Physiol. **102B**, 205–208.

Sinha, V. R. P. (1969): A note on the feeding of larger eels *Anguilla anguilla* (L.). J. Fish. Biol., **1,** 279–283.

Sinha, V. R. P., Jones, J.W. (1966): On the sex and distribution of the freshwater eel (*Anguilla anguilla*). J. Zool., Lond. **150,** 371–385.

Sinha, V. R. P., Jones, J.W. (1967a): The Atlantic eel problem. Proc. 3rd. British Coarse Fish Conference, Liverpool, 70–73.

Sinha, V. R. P., Jones, J.W. (1967b): On the food of the freshwater eels and their feeding relationship with the salmonids. J. Zool., London, **153,** 119–137.

Sinha, V. R. P., Jones, J.W. (1967c): On the age and growth of the freshwater eel (*Anguilla anguilla*). J. Zool., London, **153,** 99–117.

Sitowski, L. (1937): O masowym wyste powaniu pijawki *Cystobranchus respirens* Troschel w Dunajcu i jego doplywach. Przegl. ryb. **10,** 185–187.

Sivertsen, E., (1938): Undersøkelser over forholdet mellem spisog bredhodet al og deres naering. Fisk. Dir. Skr. **5**(8), 1–22.

Skadhauge, E. (1969): The mechanism of salt and water absorption in the intestine of the eel (*Anguilla anguilla*) adapted to various salinites. J. Physiol. London, **204,** 139–158.

Skadhauge, E. (1974): Coupling of transmural flows of NaCl and water in the intestine of the eel (*Anguilla anguilla*). J. expl. Biol. **60,** 535–546.

Smith, D. G. (1989a): Order Anguilliformes. Family Anguillidae. In: Böhlke, E. B. (Ed.): Fishes of the Western North Atlantic. Memoirs of the Sears Foundation for Marine Research, Yale University, New Haven, No. **1,** Part **9,** Vol. **1,** Orders Anguilliformes and Saccopharyngiformes, 25–47.

Smith, D. G. (1989b): Family Congridae. In: Böhlke, E. B. (Ed.): Fishes of the Western North Atlantic. Memoirs of the Sears Foundation for Marine Research, Yale University, New Haven, No. **1,** Part **9,** Vol. **1,** Orders Anguilliformes and Saccopharyngiformes, 460–567.

Smith, D.G., Castle, P. H. J. (1972): The eel genus *Neoconger* Girard: systematics, osteology, and life history. Bull. mar. Sci. **22,** 196–249.

Smith, H. W. (1930): The absorption and secretion of water and salts by marine teleosts. Am. J. Physiol. **93,** 480–505.

Smith, M. W. (1941): Treatment of Potter's Lake, New Brunswick, with rotenone. Trans. Amer. Fish. Soc. **70,** 347–355.

Smith, M. W. (1955): Control of eels in a lake by preventing the entrance of the young. Can. Fish Cult. **17,** 13–17.

Smith, M. W. (1966): Amount of organic matter lost to a lake by migration of eels. J. Fish. Res. Bd. Can. **23,** 1799–1801.

Smith, M. W., Saunders, J. W. (1955): The American eel in certain freshwaters of the maritime provinces of Canada. J. Fish. Res. Bd Can. **12,** 238–269.

Smith, S. W. (1968): Otolith age reading by means of surface structure examination. J. Cons. perm. int. Explor. Mer **32,** 270–277.

Sokabe, H., Mizogami, S., Murase, T., Sukai, F. (1966): Renin and euryhalinity in the Japanese eel, *Anguilla japonica.* Nature **212,** 952–953.

Sokabe, H., Ogawa, M. (1974): Comparative studies of the juxtaglomerular apparatus. Intl. Rev. Cytol. **37,** 271–327.

Sokabe, H., Oide, H., Ogawa, M., Utida, S. (1973): Plasma renin activity in Japanese eels (*Anguilla japonica*) adapted to sea water or in dehydration. Gen. Comp. Endocr. **21,** 160–167.

Sola, L., Camerini, B., Cataudella, S. (1984): Cytogenetics of Atlantic eels: C- and G-banding, nucleolus organizer regions, and DNA content. Cytogenet. Cell Genetics **38,** 206–210.

Someren, V. D. van, Whitehead, P. J. P. (1959): Records of young eels in Kenya rivers. Nature **183, 950,** 971.

Sörensen, I. (1951): An investigation of some factors affecting the upstream migration of the eel. Rep. Inst. Freshwater Res. Drottningholm, **32,** 126–172.

Sorokin, V. P., Konstantinov, K. G. (1960): The eel *Anguilla anguilla* in the Kola Bay. Zool. Zh. **39,** 621–622.

Souci, S. W., Fachmann, W., Kraut, H. (1994): Die Zusammensetzung der Lebensmittel, Nährwerttabellen. 5. Aufl. Wissenschaftlich Verlagsgesellschaft, Stuttgart. 378–496.

Souza, J. J., Poluhovich, J. J., Guerra, J. G. (1988): Orientation responses of American eels, *Anguilla rostrata*, to varying magnetic fields. Comp. Biochem. Physiol. 90A(1), 57–61.

Spangenberg, R. (1975): Eine Kiemenkrankheit beim Aal, verursacht durch *Dermocystidium anguillae* n. sp. Z. Binnenfischerei **22,** 363–367.

Spärck, R. (1930): Über den Sauerstoffverbrauch bei Aallarven und einigen anderen pelagischen Tieren. Arch. zool. ital. Torino **16,** 900–904.

Sparrenberg, C. (1934): Die Herstellung von Aal in Aspik. Fischwaren- und Feinkostindustrie, Hamburg, 7.

Spazier, E., Storch, V., Braunbeck, T. (1992): Cytopathology of spleen in eel *Anguilla anguilla* exposed to a chemical spill in the Rhine River. Dis. Aquat. Org. **14,** 1–22.

Sprengel, G., Lüchtenberg, H. (1991): Infection by endoparasites reduces maximum swimming speed of European smelt (*Osmerus eperlanus*) and European eel (*Anguilla anguilla*). Dis. aquat. Org. **11,** 31–35.

Stafford, J. (1904): Trematodes from Canadian fishes. Zool. Anz. **27,** 481–495.

Stafford, J. (1907): Preliminary report on the trematodes of Canadian marine fishes. Contrib. Can. Biol. 1902–1905, 91–94.

Starrett, W. C., Fritz, A. W. (1965): A biological investigation of the fishes of lake Chautanqua, Illinois. Bull. Ill. nat. Hist. Surv. **29,** 1–104.

Stasko, A. B., Rommel, S. A. (1974): Swimming depth of adult American eels (*Anguilla rostrata*) in a saltwater bay as determined by ultrasonic tracking. J. Fish. Res. Bd Can. **31,** 1148–1150.

Steen, J. B. (1963a): The physiology of the swimbladder of the eel *Anguilla vulgaris.* I. The solubility of gases and the buffer capacity of the blood. Acta physiol. scand., **58,** 124–137.

Steen, J. B. (1963b): The physiology of the swimbladder of the eel *Anguilla vulgaris*. II. The reabsorption of gases. Acta physiol. scand., **58,** 138–149.

Steen, J. B. (1963c): The physiology of the swimbladder in the eel. *Anguilla vulgaris.* III. The mechanism of gas secretion. Acta physiol. scand., **59,** 221–241.

Steen, J. B., Berg, T. (1966): The gills of two species of haemoglobin – free fishes compared to those of other teleosts – with a note on severe anaemia in an eel. Comp. Biochem. Physiol. **18,** 517–526.

Steen, J. B., Kruysse, A. (1964): The respiratory function of teleostean gills. Comp. Biochem, Physiol. **12,** 127–142.

Steen, J. B., Sund, T. (1977): Gas deposition by counter-current multiplication in the eel swim-bladder: Experimental verification of a mathematical model. J. Physiol., London, **267,** 697–702.

Steinberg, R. (1964): Versuchsfischerei mit Tuckzeesen in norddeutschen Binnenseen. Fischwirt 14.

Stépánek, M. (1968): Beiträge zur Bioakustik der Binnengewässer. 1. Die Reaktionen der Süßwasserorganismen auf die von der Wasseroberfläche kommenden Geräusche. Arch. Hydrobiol. (Suppl.) **33,** 397–422.

Stinton, F. C. (1975): Fish otoliths from the English Eocene: Part I. Palaeontogr. Soc. Monogr., London, 129 (544), 1–56.

Stolck, A. (1956): Plecospondylic spinal column in the eel *Anguilla anguilla*. Beaufortia **5,** 143–148.

Stramke, D. (1972): Veränderungen am Auge des europäischen Aales (*Anguilla anguilla* L.) während der Gelb- und Blankaalphase. Arch. FischWiss. **23,** 101–117.

Stray-Pedersen, S., Nicolaysen, A. (1975): Qualitative and quantititive studies of the capillary structure in the rete mirabile of the eel, *Anguilla vulgaris* L. Acta physiol. scand., **94,** 339–357.

Strehlow, B. (1996): Biometric studies on premetamorphic eel larvae of *Anguilla anguilla* (Anguilliformes: Anguillidae) in comparison with younger developmental stages of the species. Helgoländer Meeresunters. **50,** 415–424.

Strophal, E. E. (1930): Der Aal in der deutschen Fischwirtschaft. Samml. fischereil. Zeitfragen, Neudamm, **18,** 1–20.

Strubberg, A. (1913): The metamorphosis of elvers as influenced by outward conditions. Meddr Kommn Havunders., Ser. Fisk. **4**(3), 1–11.

Struck, H. (1965): Die industrielle Verarbeitung des Aales. In: Keune, J. A., Struck, H., (eds): Der Aal. 72–81. H. A. Keune-Verlag, Hamburg.

Studnitz, G. von (1933): Beiträge zur Adaptation der Teleostier. Z. vergl. Physiol., **18,** 307–338.

Subklew, H. J. (1955): Der Greifswalder Bodden, fischereibiologisch und fischereiwirtschaftlich betrachtet. Z. Fisch. **4,** 545–588.

Sugimoto, Y., Takeuchi, Y. (1976): Induced maturation of female Japanese eels (*Anguilla japonica*) by administration of salmon pituitaries with notes on changes of oil droplets in eggs of matured eels. Bull. Fac. Fish. Hokkaido Univ. **27,** 107–120.

Summer, J.F., Hopkirk, G. (1976): Lipid composition of New Zealand eels. J. Sci. Fd. Agr. **27,** 933–938.

Summer, J.F., Beumer, J.P., Chambers, A., Mobley, M.C. (1984): The lipid content and fatty acid composition of wild Australian eels, *Anguilla australis australis* Richardson. Food Technol. In Australia **36,** 287–289.

Susuki, S. (1992): Thyroid gland. In Matsumoto, A., Ishii, S. (eds): Atlas of Endocrine Organs. 25–38. Springer-Verlag: Berlin, Heidelberg, New York.

Suzuki, R., Yasuda, A., Kondon, J., Kawauchi, H., Hirano, T. (1991): Isolation and characterization of Japanese eel prolactins. Gen. Comp. Endocrinol. **81,** 391–402.

Suzuki, Y., Tsutui, S., Nishikawa, J., Nishida, S. (2000): Immunological detection of gelatinous plankton as natural food of eel leptocephali. J. Taiwan Fish. Res. **8**(1).

Svärdson, G. (1969): Alen som kaftpredator (Eels prey on crayfish.). Zool. Revy **30**(2), 61–72.

Svärdson, G. (1972): The predatory impact of eel (*Anguilla anguilla* L.) on a population of crayfish (*Astacus astacus* L.). Rep. Inst. Freshw. Res. Drottningholm **52,** 149–191.

Svärdson, G. (1976): The decline of the Baltic eel population. Rep. Inst. Freshw. Res. Drottningholm **55,** 136–143.

Svedäng, H, Neuman, E, Wickström, H. (1996): Maturation patterns in female eel: age and size at the silver eel stage. J. Fish Biol. **48,** 342–351.

Sverdrup, H. U., Johnson, M. W., Fleming, R. M. (1942): The oceans. Prentice-Hall Inc., New York.

Swarts, W. (1969): Blood studies of some marine teleosts. Trans. Amer. Fish. Soc., **98,** 328–331.

Swierzowska, L., Dembinski, W. (1978): A trial at estimating effectiviness of catches by means of electric assemblies and eel pair nets. Roczniki Nauk Rolniczych 98, Ser. H, Z.4. 131–146.

Swierzowski, A. (1964): Zastosowanie przegród swietlnych do polowow negorza splywajacego. Rybna **16** (8), 11–14.

Swierzowski, A., Dembinski, W. (1978): Selectivity of an eel trap net and size structure of eel run from Lake Sasek Wielki. Roczniki Nauk Rolniczych 98, Ser. H, Z.4. 148–157.

Syrski, S. (1876): Lecture on the organs of reproduction and the fecundation of fishes and especially of eels. Rep. U. S. Commnr. Fish **3,** 719–734.

Tabeta, O., Nosaka, S., Sakurai, N. (1979): Occurrence of elvers of the European eel, *Anguilla anguilla* Linnaeus in Shizuoka and Kanagawa Prefectures, Japan, in 1977. Bull. Jap. Soc. Sci. Fish. **45,** 437–441.

Tabeta, O., Takai, T. (1973): An elver of *Anguilla japonica* found in the Northern part of the South China Sea. J.Shimonoseki Univ. Fish. **22,** 49–53.

Tabeta, O., Takai, T. (1975): Leptocephali of *Anguilla japonica* found in the waters South of the Okinawa Islands. Bull. Jap. Soc. Sci. Fish.41, 137–145.

Tabeta, O., Takai, T., Matsui, I. (1976): The sectional counts of vertebrae in the anguillid elvers. Jap. J. Ichthyol. **22,** 195–200.

Tabeta, O., Takai, T., Matsui, I. (1977): Record of short finned eel from Nagata River, Shimonoseki, Japan. Bull. Jpn. Soc. Sci. Fish.42, 1333–1338.

Tagliavini, J., Gandolfi, B., Cau, A. (1996a): Mitchondrial DNA variability in *Anguilla anguilla* (L.). Boll. Zool.

Tagliavini, J., Gandolfi, G., Deiana, A. M., Salvadori, S. (1996b): Phylogenetic relationships among two Atlantic and three Indo-Pacific *Anguilla* species (Osteichthyes, Anguillidae). Ital. J. Zool., **63,** 271–276.

Takahashi, H., Sugimoto, Y. (1978): A spontaneous hermaphrodite of the Japanese eel, *Anguilla japonica,* and its artificial maturation. Japan. J. Ichthyol. **24,** 239–245.

Takai, T. (1959): Studies on the morphology, ecology and culture of the important apodal fishes, *Muraenesox cinereus* (Forskål) and *Conger myriaster* (Brevoort). J. Shimonoseki Univ. Fish **8,** 209–555.

Takashima, Y., Takano, K., Hara, A. (1979): Changes in female specific serum proteins during the course of induced maturation of female Japanese eels *(Anguilla japonica*). Bull. Fac. Fish. Hokkaido Univ. **30,** 23–33.

Takei, Y., Takahashi, A., Watanabe, T. X., Nakajima, K., Ando, A. (1994): Eel ventricular natriuretic peptide: isolation of a low molecular size form and charaterization of plasma form by homologous radioimmunoassay. J. Endocrinol. **141,** 81–89.

Tammar, A. R. (1974): Bile salts in Fishes. In Florkin, M., Scheer, B.T. (eds): Chemical Zoology **8,** 595–612. Academic Press, New York & London.

Tanaka, S. (1975): Collection of leptocephali of the Japanese eel in waters South of the Okinawa Islands. Bull. Jap. Soc. Sci. Fish. **41,** 129–136.

Tanaka, H., Kagawa, H., Ohta, H., Okuzawa, K., Hirose, K. (1995): The first report of eel larvae ingesting rotifers. Fisheries Sci. **61**(1), 171–172.

Taniguchi N., Morita, T. (1979): Identification of European, American and Japanese eels by lactate dehydrogenase and malate dehydrogenase isozyme patterns. Bull. Jap. Soc. Sci. Fish. **45,** 37–42.

Taniguchi, N., Numachi, K. (1978): Genetic variation of 6–phosphogluconate dehydrogenase, isocitrate dehydrogenase, and glutamic-oxaloacetic transaminase in the liver of Japanese eel. Bull. Jap. Soc. Sci. Fish., **44,** 1351–1355.

Taniguchi, N., Takahashi, J., Konishi, Y. (1972): Studies on a biochemical method for identification of the European and Japanese freshwater eel. Bull. Jap. Soc. Scient. Fish. **38,** 627–631.

Tåning, A. V. (1938): Deep-sea fishes of the Bermuda Oceanographic Expedition. Family Anguillidae. Zoologica, N. Y. **23,** 313–318.

Taraschewski, H., Moravec, F., Lamah, T., Anders, K. (1987): Distribution and morphology of two helminths recently introduced into European eel populations: *Anguillicola crassus* (Nematoda, Dracunculoidea) and *Paratenuisentis ambiguus* (Acanthocephala, Tenuisentidae). Dis. aquat. Org. **3,** 167–176.

Teichmann, H. (1954): Vergleichende Untersuchungen an der Nase der Fische. Z. Morph. Ökol. Tiere **43,** 171–212.

Teichmann, H. (1957): Das Riechvermögen des Aales (*Anguilla anguilla* L.). Naturwissenschaften **44,** 242.

Teichmann, H. (1959): Über die Leistung des Geruchssinnes beim Aal (*Anguilla anguilla* L.). Z. vergl. Physiol. 42.

Teklenberg, H. (1972): Product highly vulnerable to rough handling. Comm. Fish. Auckland **11**(5), 12–13.

Telegut, M. (1955): Un peste rar in Bega. Bull. Inst. Cerc. pisc., **14,** 89–90.

Tesch, F.-W. (1956): Kritische Betrachtungen über allgemeine und spezielle Resistenz von wechselwarmen Wassertieren. Biol. Zbl. **75,** 397–407.

Tesch, F.-W. (1958): Weiterer Fund eines Steines in der Leibeshöhle eines Aales (*Anguilla vulgaris* Turton). Nachr. naturw. Mus. Aschaffenb. **60,** 61–62.

Tesch, F.-W. (1964): Die Ursache eines Aal- und Barbensterbens in der oberen Weser. Fischwirt **14,** 39–40.

Tesch, F.-W. (1965): Verhalten der Glasaale (*Anguilla anguilla*) bei ihrer Wanderung in den Ästuarien deutscher Nordseeflüsse. Helgoländer wiss. Meeresunters. **12,** 404–419.

Tesch, F.-W. (1966): Der Einfluß der Weserstauwehre auf die Jungaalwanderung. Fischwirt **16,** 29–37.

Tesch, F.-W. (1967a): Homing of eels (*Anguilla anguilla*) in the southern North Sea. Mar. Biol., Berlin **1,** 1–9.

Tesch, F.-W. (1967b): Die Aalwirtschaft der Länder Niedersachsen und Bremen. Arch. FischWiss. 18 (Suppl. 1), 361–404.

Tesch, F.-W. (1967c): Glasaal- und Aalfischerei in Nordirland. Fischwirt **17,** 165–178.

Tesch, F.-W. (1967d): Aktivität und Verhalten wandernder *Lampetra fluviatilis, Lota lota* und *Anguilla anguilla* im Tidegebiet der Elbe. Helgoländer wiss. Meeresunters. **16,** 92–111.

Tesch, F.-W. (1968a): Salinitätstoleran des parasitischen Ciliaten Ichthyophthirius multifiliis. Helgoländer wiss. Meeresunters. **18,** 130–135.

Tesch, F.-W. (1968b): Heimfindevermögen von Aalen in der Nordsee. Umschau Wiss. Techn. **68**(8), 247–248.

Tesch, F.-W. (1970): Heimfindevermögen von Aalen *Anguilla anguilla* nach Beeinträchtigung des Geruchssinnes, nach Adaptation oder nach Verpflanzungen in ein Nachbar-Ästuar. Mar. Biol., Berlin, **6,** 148–157.

Tesch, F.-W. (1971): Aufenthalt der Glasaale (*Anguilla anguilla*) an der südlichen Nordseeküste vor dem Eindringen in das Süßwasser. Vie Milieu Suppl. **22,** 381–392.

Tesch, F.-W. (1972): Versuche zur telemetrischen Verfolgung der Laichwanderung von Aalen (*Anguilla anguilla*) in der Nordsee. Helgoländer wiss. Meeresunters. **23,** 165–183.

Tesch, F.-W. (1973): Der Aal. Ist Edn. P.Parey, Hamburg. 306pp.

Tesch, F.-W. (1974a): Speed and direction of silver and yellow eels, *Anguilla anguilla,* released and tracked in the open North Sea. Ber. dt. wiss. Komm. Meeresforsch. **23,** 181–197.

Tesch, F.-W. (1974b): Influence of geomagnetism and salinity on the directional choice of eels. Helgoländer wiss. Meeresunters. **26,** 382–395.

Tesch, F.-W. (1974c): Aale mit Ultraschallsendern. Umschau Wiss. Techn., Frankfurt, **74,** 554–555.

Tesch, F.-W. (1975a): Migratory behaviour of displaced homing yellow eels (*Anguilla anguilla*) in the North Sea. Helgoländer wiss. Meeresunters. **27,** 190–198.

Tesch, F.-W. (1975b): Fishes: Orientation in space: animals. In Kinne. O. (ed.): Marine Ecology **2**(2), 657–707. Wiley, London.

Tesch, F.-W. (1977): *The eel.* Chapman & Hall, London 434pp.

Tesch, F.-W. (1978a): Horizontal and vertical swimming of eels (*Anguilla anguilla*) during the spawning migration at the edge of the continental shelf. In: Schmidt-Koenig, K., Keeton, W. T. (eds): Animal migration, navigation and homing. 378–393. Springer-Verlag, Berlin, Heidelberg.

Tesch, F.-W. (1978b): Telemetric observations on the spawning migration of the eel *Anguilla anguilla* west of the European continental shelf. Env. Biol. Fish. **3,** 203–209.

Tesch, F.-W. (1978c): Erste Ergebnisse zur Wanderung der Blankaale in der Weser. Arbeiten Dt. Fischereiverband. **26,** 114–115.

Tesch, F.-W. (1979a): Tracking of silver eels (*Anguilla anguilla* L.) in different shelf areas of the Northeast Atlantic. Rapp. P.-v. Reun. Cons. int. Explor. Mer **174,** 104–114.

Tesch, F.-W. (1979b): Bericht über die Reisen der Forschungsschiffe "Anton Dohrn" und "Friedrich Heincke" 1979 im Sargassomeer. Arbeiten Dt. Fischereiverband. **29,** 54–66.

Tesch, F.-W. (1980): Occurrence of eel *Anguilla anguilla* larvae west of the European continental shelf, 1971–1977. Envir. Biol. Fish **5,** 185–190.

Tesch, F.-W. (1982): The Sargasso Sea eel expedition 1979. Helgoländer Meeresunters. **35,** 263–277.

Tesch, F.-W. (1983): Der Aal. 2nd edition. P. Parey, Hamburg. 340pp.

Tesch, F.-W. (1986): Der Aal als Konkurrent von anderen Fischarten und von Krebsen. Österreichs Fischerei **39,** 5–20.

Tesch, F.-W. (1989): Changes in swimming depth and direction of silver eels (*Anguilla anguilla* L.) from the continental shelf to the deep sea. Aquat. Living Resour. **2,** 9–20.

Tesch, F.-W. (1991): *Anguilla anguilla* (Linnaeus, 1758) In: The freshwater fishes of Europe. Clupeidae, Anguillidae Vol. **2,** 389–437. Aula-Verlag, Wiesbaden.

Tesch, F.-W. (1992): Insignificance of tidal currents for silver eel migration as studied by eel trackings and current measurements. Irish Fish. Invest. Ser A, No. **36,** 105–109.

Tesch, F.-W. (1994): Verfolgung von Blankaalen in Weser und Elbe. Fischökologie **7,** 47–59.

Tesch, F.-W. (1995): Vertical movements of migrating silver eels (*Anguilla anguilla*) in the sea. Bull. Sea Fish. Inst. **2**(135), 24–30.

Tesch, F.-W. (1998): Age and growth rates of North Atlantic eel (*Anguilla* spp.) based on published length data. Helgoländer Meeresunters. **52**(1).

Tesch, F.-W. (2001): Ways to investigate the problems of the sustainability of eel stocks. J. Taiwan Fish. Res. **8**(1), 259–267.

Tesch, F.-W., Fries, G. (1963): Die Auswirkungen des eingebürgerten Flohkrebses *Gammarus triginus* auf Fischbestand und Fischerei in der Weser. Fischwirt **13,** 319–326.

Tesch, F.-W., Kracht, R., Schoth, M., Smith, D. G., Wegner, G. (1979): Report on the eel expedition of FRV (Anton Dohrn) and R. K. "Friedrich Heincke" to the Sargasso Sea 1979. ICES CM 1979/M6, 1–29.

Tesch, F.-W., Lelek, A. (1973a): Directional behaviour of transplanted stationary and migratory forms of the eel (*Anguilla anguilla*) in a circular tank. Neth. J. Sea Res., **7,** 46–52.

Tesch, F.-W., Lelek, A. (1973b): An evaluation of the directional choice in the eel, in captivity. Arch. FischWiss **24,** 237–251.

Tesch, F.-W., Niermann, U. (1992): Stock density of eel larvae *Anguilla anguilla* (L.) on the European continental slope, based on collections made between 1985 and 1989. Irish Fish Invest. Ser. A, **36,** 110–113.

Tesch, F.-W., Niermann, U.,, Plaga, A. (1986): Differences in development stage and stock density of larval *Anguilla anguilla* off the west coast of Europe. Vie Milieu **36,** 255–260.

Tesch, F.-W., Wegner, G. (1990): The distribution of small larvae of *Anguilla* sp. related to hydrographc conditions 1981 between Bermuda and Puerto Rico. Int. Revue ges. Hydrobiol. **75,** 845–858.

Tesch, F.-W., Wendt, T., Karlsson, L. (1992): Influence of geomagnetism on the activity and orientation of the eel, *Anguilla anguilla* (L.), as evident from laboratory experiments. Ecol. Freshw. Fish **1,** 52–60.

Tesch, F.-W., Westerberg, H., Karlsson, L. (1991): Tracking studies on migrating silver eels in the Central Baltic. Meeresforsch **35,** 193–196.

Tesch, J. J. (1928): On sex and growth investigations of the freshwater eel in Dutch waters. J. Cons. perm. int. Explor. Mer. **3,** 52–69.

Theret, C., Palayer, P. (1967): Etude comparative des ultrastructures de la cellule beta du pancreas endocrine de quelques téléostéens. C.r. hebd. Séanc. Acad. Sci., Paris, **265,** 915–918.

Thomas, J. D. (1962): The food and growth of brown trout (*Salmo trutta* L.) and its feeding relationship with the salmon parr (*Salmo salar* L.) and the eels (*Anguilla anguilla* L.) in the River Teify, West Wales. J. Anim. Ecol. **31,** 175–205.

Thomas, M. L. H. (1968): Overwintering of American lobsters (*Homarus americanus*) in burrows in Bideford River, Prince Edward Island. J. Fish. Res. Bd Can. **25,** 2725–2727.

Thornton, V. F., Howe, C. (1974): The effect of change of background colour on the ultrastructure of the pars intermedia of the pituitary of the eel (*Anguilla anguilla*). Cell Tiss. Res. **151,** 103–115.

Thulin, J., Höglund, J., Lindesjöö, E. (1989): Fisksjukdomar i Kustvatten. Naturvårds verket, Stockholm.

Thumann, M. E. (1953): Ein goldgelber Aal aus der Havel. Z. Fisch., **2,** 321–324.

Thurow, F. (1957): Beiträge zur Biologie des Flußaales *Anguilla vulgaris* Turt. Phil. Diss. Kiel.

Thurow, F. (1958): Untersuchungen über die spitz- und breitköpfigen Varianten des Flußaales. Arch. FischWiss., **9,** 79 ff.

Thurow, F. (1959): Über Fangerträge und Wachstum des Aales in der westlichen Ostsee. Z. Fisch., **8,** 597–626.

Tierney, M. L., Luke, G., Cramb, G., Hazon, N. (1995): The role of the renin-angiotensin system in the control of blood pressure and drinking in the European eel, *Anguilla anguilla*. Gen. Comp. Endocrinol. **100,** 39–48, 210.

Timmermann (1939): Aalsterben durch Massenbefall mit *Ichthyophthirius*. FischZtg., Neudamm, **42,** 344.

Tisserand-Jochem, E. M., Lopez, C., Milet, C., Vidal, B., Magnac, C., Eyquem, A., Cohn, D. V. (1987): Co-localization and secretion of parhyrin of Stannius corpuscles (immunoreactive parathyroid hormone) and of secretory glycoproteins including secretory protein-I in the European eel (*Anguilla anguilla* L.). Bone Miner **2,** 163–175.

Todd, P. R. (1973): First record of the freshwater eel *Anguilla dieffenbachi* Gray to be caught at sea. J. Fish. Biol. **5,** 231–232.

Todd, P. R. (1977): Lake Ellesmere – home for most NZ eels. Catch(2), March.

Todd, P. R. (1979): The hormone – induced maturation of New Zealand freshwater eels. Rapp. p., v. Reunion Cons. int. Explor. Mer **174,** 91–97.

Todd, P. R. (1980): Size and age of migrating New Zealand freshwater eels (*Anguilla* spp.). N. Z. J. Marine Freshwater Res. **14,** 283–293.

Todd, P. R. (1981a): Timing and periodicity of migrating New Zealand freshwater eels (*Anguilla* spp.). N. Z. J. Mar. Freshw. Res. **15,** 225–235.

Todd, P.R., (1981b): Hormone-induced maturation in male New Zealand freshwater eels (*Anguilla* spp.). N.Z. J. Mar. Freshw. Res. **15**(3), 237–246.

Todd, P. R. (1982): Morphometric changes, gonad histology and fecundity estimates in migrating New Zealand freshwater eels. N. Z. J. Mar. Freshw. Res. **16,** 155–170.

Topp, R., Paulerson, R. (1973): Elver investigations in the Southeast. Marine Fish. Rev. **35,** 45–47.

Törlitz, M. (1922): Anatomische und entwicklungsgeschichtliche Beiträge zur Artfrage unseres Flußaales. Z. Fisch. **21,** 1–48.

Tosi, L., Spampanaio, A., Sola, C., Tongiorgi, P. (1990): Relation of water odor, salinity and temperature to ascent of glass eels, *Anguilla anguilla* (L.): a laboratory study. J. Fish Biol. **36,** 327–348.

Tright, H. v. (1913): Das Elektrogramm des kaudalen Aalherzens. Z. Biol. **62,** 217–242.

Troemer, B. (1932): Untersuchungen über die Herztätigkeit der Fische. IV. Z. Vergl. Physiol. **16,** 463–470.

Trybom, F., Schneider, G. (1908): Markierungsversuche mit Aalen und die Wanderungen gekennzeichneter Aale in der Ostsee. Rapp. p.-v. Reun., Cons. perm. int. Explor. Mer. **9,** 51–59.

Tsukamoto, K. (1992): Discovery of the spawning area for Japanese eel. Nature. **356,** 789–791.

Tsukamoto, K., Nakai, I., Tesch, F.-W. (1998): Do all freshwater eels migrate? Nature, **396** (6712), 635–636.

Tucker, D. W. (1959): A new solution to the Atlantic eel problem. Nature **183,** 495–501.

Tulonen, J., Vuorinen, P. J. (1996): Concentrations of PCBs and other organochlorine compounds in eels (*Anguilla anguilla,* L.) of the Vanajavesi watercourse in southern Finland, 1990–1993. Sci. Total Environ. **187,** 11–18.

Tzeng, W. N. (1984): Dipersal and upstream migration of marked Anguillid eel, *Anguilla japonica*, elvers in the estuary of the Shuang River, Taiwan. Bull. Jap. Soc. Fish. Oceanogr.

Tzeng, W. N. (1985): Immigration timing and activity rhythms of the eel, *Anguilla japonica*, elvers in the estuary of Northern Taiwan, with emphasis on environmental influences. Bull. Jap. Soc. Fish. Oceanogr. No. 47/48, 11–28.

Tzeng, W. N. (1990): Relationship between growth rate and age at recruitment of *Anguilla japonica* elvers in a Taiwan estuary as inferred from otolith growth increments. Marine Biol. **107,** 75–81.

Tzeng, W.N., Severin, K.P., Wickström, H. (1997): Use of otolith microchemistry to investigate the environmental history of European eel *Anguilla anguilla*. Mar. ecol. Prog. Ser. **149,** 73–81.

Tzeng,W.N., Lin, H.R, Wang, C.H., Xu, S.N. (2000): Differences in size and growth rates of male and female migrating Japanese eels in Pearl River, China. J. Fish Biol. **57,** 1245–1253.

Ubisch, L. V. v. (1924): Stimmen die Ergebnisse der Aalforschung mit Wegners Theorie der Kontinentalverschiebung überein? Naturwissenschaften 24.

Uesaka, T., Yano, K., Yamasaki, M., Nagashima, K., Ando, M. (1994): Somatostatin-related peptides isolated from the eel gut: effects on ion and water absorption across the intestine of the seawater eel. J. Exp. Biol. **188,** 205–215.

Usui, A. (1974): Eel culture. Fishing News (Books), West Byfleet & London.

Utida, S., Hirano, T. (1973): Effects of changes in environmental salinity on salt and water movement in the intestine and gills of the eel, *Anguilla japonica.* In Chavin, W. (ed.): Responses of fish to environmental changes. 240–269. Proceedings of a US-Japan seminar held in Tokyo (Japan), 23–26 Nov. 1970. C. C. Thomas (Publisher), Springfield, Illinois (USA).

Utida, S., Hirano, T., Kamiya, M. (1969): Seasonal variations in the adjustive responses to seawater in the intestine and gills of the Japanese cultured eel, *Anguilla japonica.* Proc. Japan Acad. **45,** 293–297.

Utida, S., Hirano, T., Oide, H., Ando, M., Johnson, D. W., Bern, H. A. (1972): Hormonal control of the intestine and urinary bladder in teleost osmoregulation. Gen. Comp. Endocr. Suppl. **3,** 317–327.

Utida, S., Isono, N. (1967): Alkaline phosphatase activity in intestinal mucosa of the eel adapted to fresh or sea water. Proc. Japan Acad. **43,** 789–792.

Utida, S., Isono, N., Hirano, T. (1967): Water movement in isolated intestine of the eel adapted to fresh water or sea water. Zool. Mag. Tokyo, **6,** 128–129.

Utida, S., Kamiya, M., Shirai, N. (1971): Relationship between the activity of Na+K+-activated adenosinetriphosphatase and the number of chloride cells in eel gills with special reference to sea water adaptation. Comp. Biochem. Physiol. 38A, 443–447.

Utrecht, W. L. van, Holleboom, M.A. (1985): Notes on eel larvae (*Anguilla anguilla* Linnaeus 1758) from the central and eastern North Atlantic and on glass eels from the European continental shelf. Bijdragen tot de Dierkunde **55**(2), 249–262.

Vaillant, L. (1896): Sur la présence de l'anguille commune en haute mer. C. r. hebd. Séanc. Acad. Sci., Paris **126,** 1429–1430.

Van der Oost, R., Heida, H., Opperhuizen, A., Vermeulen, N. P. E. (1991): Bioaccumulation of organic micropollutants in different aquatic organisms: sublethal toxic effects on fish. In: Mayes, M. A., Barron, M. G. (eds): Aquatic toxicology and risk assessment. **14,** 166–180. American Society for Testing and Materials, Philadelphia,

Van der Oost, R., Opperhuizen, A., Satumalay, K., Heida, H., Vermeulen, N. P. E. (1996): Biomonitoring aquatic pollution with feral eel (*Anguilla anguilla*). 1. Bioaccumulation: biota-sediment ratios of PCBs, OCPs, PCDDs and PCDFs. Aquat. Toxicol. **35,** 21–46.

Van Rijsingen, J. (1995): European Aquaculture Production Trends & Outlook: Eel. Eurofish Report trade Conference 15–16 November 1995, Brussels.

Veen, T. van, Andersson, M. (1982): Threshold for synchronization of locomotor activity to visible radiation in the eel *Anguilla anguilla*. Oikos **38,** 21–26.

Veen, T. van, Hartwig, H. G., Müller, K. (1976): Light dependent motor activity and photonegative behavior in the eel (*Anguilla anguilla* L.), evidence for extraretinal and extrapineal photoreception. J. comp. Physiol. **111,** 209–219.

Vigh-Teichmann, I., Korf, H.W., Nurnberger, F., Oksche, A., Olsson, R. (1983): Opsin-immunoreactive outer segments in the pineal and parapineal organs of the lamprey (*Lampetra fluviatilis*), the eel (*Anguilla anguilla*) and the rainbow trout (*Salmo gairdneri*). Cell Tiss.Res. **230,** 289–307.

Villani, P., Lumare, F. (1975): Nota sull'accrescimento ovarico indotta in *Anguilla anguilla* (L.). Investigacion pesq. **39,** 187–197.

Vilter, V. (1942): Régulation sympathico-hypophysaire de migration pigmentaires dans le rétine de l'anguille (*Anguilla vulgaris* L.). Arch. Physiol. biol.16 (suppl.), 6–8.

Vilter, V. (1943): Métamorphose des larves d'anguille dans ses rapports avec l'activité hypophysaire. C. r. Séanc Soc. Biol. **137,** 534–535.

Vilter, V. (1945a): Développment du tractus gastro-intestinal au cours de la métamorphose de larve d'anguille (civelles). C.r. Séanc. Soc. Biol. **139,** 223–224.

Vilter, V. (1945b): Inhibition experimentale de la croissance stomacale au cours de la métamorphose larvaire de l'anguille. C. r. Séanc. Soc. Biol. **139,** 817–818.

Vilter, V. (1946): Determinisme, hypophysaire du comportement "halophobe" de civelles immigrantes. C. r. Séanc. Soc. Biol. **140,** 483–484.

Vilter, V. (1951): Intervention probable de la lumière chez la naissance des structures rétiniennes révelée par l'étude comparée de la rétine chez les anguilles normales et cavernicoles. C. r. Séanc. Soc. Biol. 145.

Virchov, H. (1882): Über die Glaskörper und Netzhautgefäße des Aales. Morph. Jb. **7,** 573–590.

Visurgis (1954): Geräucherter Aal. Allg. FischZtg **79,** 384–385.

Vivien, L. (1958): Les glandes endocrines. In Grassé, P.P. (ed.): Traitée de Zoologie 13/2, 1470–1544. Paris, Masson & Cie.

Vladykov, V. D. (1964): Quest for the true breeding area of the American eel (*Anguilla rostrata* Le Sueur). J. Fish. Res. Bd Can. **21,** 1523–1530.

Vladykov, V. D. (1966): Remarks on the American eel (*Anguilla rostrata* Le Sueur). Sizes of elvers entering streams, the relative abundance of adult males and females, and present economic importance of eels in North America. Verh. Int. Verein. theor. angew. Limnol. **16,** 1007–1017.

Vladykov, V. D. (1971): Homing of the American eel *Anguilla rostrata*, as evidenced by returns of transplanted tagged eels in New Brunswick. Can. Fd Nat. **85**(3), 241–248.

Vladykov, V. D. (1973): Macrophthalmia in the American eel (*Anguilla rostrata*). J. Fish. Res. Bd Can. **30,** 689–693.

Volf, F., Smíšek, J. (1955): Aale in tschechoslowakischen Gewässern. Sb. C.S. Akad. zemed. Ved. **28,** 395–410.

Vøllestad, L. A. (1992): Geographic variation in age and length at metamorphosis of maturing European eel: environmental effects and phenotypic plasticity. J. anim. Ecol. **61,** 41–48.

Vøllestad, L. A., Jonsson, B., Hvidsten, N. A., Næsje, T. F., Haraldstad, Ø., Ruud-Hansen, J. (1986): Environmental factors regulating the seaward migration of European silver eels (*Anguilla anguilla*). Can. J. Fish. Aquat. Sci. **43,** 1909–1916.

Vøllestad, L. A., Lecomte-Finiger, R., Steinmetz, B. (1988): Age determination of *Anguilla anguilla* (L.) and related species. EIFAC(FAO) Occasional Paper No. 21.

Wald, G. (1958): The significance of vertebrate metamorphosis. Science, **128,** 1481–1490.

Walter, E. (1910): Der Flußaal, eine biologische und fischwirtschaftliche Monographie. Verlag J. Neumann, Neudamm. 346pp.

Watson, R. A. (1984): The life cycle and morphology of *Tetracerasta blepta* gen. et sp. nov., and *Stegodexamene callista* sp. nov. (Trematoda: Lepocreadiidae) from the long-finned eel, *Anguilla reinhardti* Steindachner. Aust. J. Zool. **32,** 177–204.

Watts, S. L., Watts, D. C. (1974): Nitrogen metabolism in fishes. In Florkin, M., Scheer, B. T. (eds): Chemical Zoology **8,** 369–446, Academic Press, New York, London.

Weber, M. (1986): Fishing method and seasonal occurrence of glass eels (*Anguilla anguilla* L.) in the Rio Minho, West coast of the Iberian Peninsula. Vie Milieu **36,** 243–350.

Wegner, G. (1982): Main hydrographic features of the Sargasso sea in spring 1979. Helgoländer Meeresunters. **35,** 385–400.

Wegrzynowicz, R., Zbanyszek, R., Klyszejko, B., Glebocka, G. (1979): Effect of crude oil on tissue respiration of eel, *Anguilla anguilla* (L.) kept in water at 18 °C and 24 °C. Acta Ichthyol. Piscatoria. **9,** 31–35.

Wehrmann, L. (1968): Messungen der Bewegungsaktivität des Aales (*Anguilla anguilla* L.) unter Verwendung einer Markierung mit Magnetmarken. Diplomarbeit Math.-Nat.-Fakultät Univ. Hamburg.

Weiss, A. (1995): Aalangeln, fein fängt besser. Fisch und Fang **8,** 32–53.

Wenner, C. A. (1973): Occurrence of American eels, *Anguilla rostrata,* in waters overlying the eastern North American continental shelf. J. Fish. Res. Bd Can. **30,** 1752–1755.

Wenner, C. A., Musick, J. A. (1974): Fecundity and gonad observations of the American eel, *Anguilla rostrata,* migrating from Chesapeake Bay, Virginia. J. Fish. Res. Bd Can. **31,** 1387–1391.

Wenner, C. A., Musick, J. A. (1975): Food habits and seasonal abundance of the American eel, *Anguilla rostrata,* from the lower Chesapeake Bay. Chesapeake Sci. **16,** 62–66.

Westenburg, J. (1952): Acoustical aspects of some Indonesian fishes. J. Cons. perm. int. Explor. Mer **18,** 311–325.

Westerberg, H. (1975): Counter current orientation in the migration of the European eel (*Anguilla anguilla* L.) Göteborgs Univ. Oceanogr. Inst. Rep., **9,** 1–18.

Westerberg, H. (1979): Counter current orientation in the migration of the European eel (*Anguilla anguilla* L.). Rapp. p.-v. Réun. Cons. int Explor. Mer **175,** 134–143.

White, E. M., Knights, B. (1994): Factors affecting the upstream migration of elvers and juvenile eels, *Anguilla anguilla* L., in the Rivers Severn and Avon, England. SEI (ICES) EIFAC (FAO) W. P. on Eel, Oviedo 1994, (Ecol. Freshw. Fish).

Whitefield, R. E., Kolenosky, P. (1978): Prototype eel ladder in the St. Lawrence River. Prog. Fish. Cultur. **40,** 152–154.

Whitehouse, R. H. (1910): The caudal fin of Teleostomi. Proc. Zool. Soc., London **2,** 590–627.

Wiberg, U. (1983): Sex determination in the European eel (*Anguilla anguilla* L.). A hypothesis based on cytogenetic results correlated with the finding of skewed sex ratios in eel culture ponds. Cytogen. Cell Genetics **36,** 589–598

Wickins, J. F. (1987): Effects of size, culling and social history on growth of cultured elvers, *Anguilla anguilla* (L). J. Fish Biol. **31,** 71–82.

Wickström, H., Hamrin, S. (1997): Sweden. In Moriarty, C., Dekker, W. (eds): Management of the European Eel. Fish. Bull. **15,** 46–52.

Wickström, H., Westin, L., Clevestam, P. (1996): The biological and economical yield from a long-term eel stocking experiment. Ecol. Freshw. Fish. **5,** 140–147.

Wiehr, H. (1932): Über die Chemie und Biologie der Fettsubstanzen vom Aal (*Anguilla vulgaris*). *Z.* Fisch. **30,** 169–196.

Wiehr, H. (1967): Beschaffung von Aalbrut und Satzaalen in der Bundesrepublik. Arch. FischWiss. 18 (Suppl. 1), 467–473.

Wierzbicka, J. (1986): *Sphaerospora sphaerocapsularae* sp. n. (Myxospora, Bivalvulida) a parasite of eel, *Anguilla anguilla* (L.). Acta protozool. **25,** 355–358.

Wiesner, E. (1966): Die Aalwirtschaft des Landes Bayern Regierungsbezirk Schwaben. Arch. FischWiss., 16 (Suppl. 1), 114–130.

Wilke, H. (1975): Aalaufstieg – leicht gemacht. Fisch u. Fang **16,** 160–161.

Willemse, J. J. (1975): Some remarks on the structure and function of the musculus lateralis in the European eel, *Anguilla anguilla* (L.) (Pisces, Teleostei). Z. Morph. Tiere **81,** 195–208.

Willemse, J. J. (1979): Guide to the internal morphology of the European eel, *Anguilla anguilla* (L.) (Pisces, Teleostei). Aquaculture **17,** 91–103.

Willemse, J. J., Ruiter, A. de (1979): A quantitative identification of four types of fibres in lateral musculature of immature European eel, *Anguilla anguilla* (L.) (Pisces, Teleostei). Aquaculture **17,** 105–111.

Willemse, J.J., van den Berg, P. G. (1978): Growth of striated muscle fibres in the m. lateralis of the European eel, *Anguilla anguilla* (L.) (Pisces, Teleostei). J. Anat. **125,** 447–460.

Williams, G. C., Koehn, R. K., Mitton, J. B (1973): Genetic differentiation without isolation in the American eel, *Anguilla rostrata.* Evolution **27,** 192–204.

Williamson, G. R., Castle, P. H. I. (1975): Large-eyed specimen of eel *Anguilla bicolor* from a freshwater well. Copeia 1975, 561–564.

Wiltschko, W., Wiltschko, R. (1996): Magnetic orientation in birds. J. exp. Biol. **199,** 129–138.

Winn, H. E., Richkus, W. A., Winn, L. K. (1975): Sexual dimorphism and natural movements of the American eel (*Anguilla rostrata*) in Rhode Island streams and estuaries. Helgoländer wiss. Meeresunters. **27,** 156–166.

Wippelhauser, G. S., McCleave, J. D (1987): Precision of behaviour of migrating juvenile American eels (*Anguilla rostrata*) utilizing selective idal stream transport. J. Cons. Explor. Mer. **44,** 80–89.

Wippelhauser, G.S., McCleave, J.D. (1988): Rhythmic activity of migrating juvenile American eels *Anguilla rostrata.* J. mar. biol Ass. UK **68,** 81–91.

Wirth, T., Bernatchez, L., (2001): Genetic evidence against panmixia in the European eel. Nature **409,** 1037–1041.

Woest, O. (1977): Entwicklung der Aalzeese. Z. Binnenfisch. **24,** 106–111.

Wolf, K. (1984): Agents: Virales. In: Kinne, O. (Ed.): Diseases of marine animals. Vol. IV, Part **1,** 17–47.

Wolff, B. (1912): Über ein Blastom bei einem Aal, nebst Bemerkungen zur vergleichenden Pathologie der Geschwülste. Virchows Arch. pathol. Anat. Physiol. **210,** 369–385.

Woods, C. S. (1964): Fisheries aspects of the Tongariro power project. Fish. tech. Rep. N. Z., **10,** 1–213.

Wunder, W. (1936): Physiologie der Süßwasserfische Mitteleuropas. In: Handbuch Binnenfisch. Mitteleur. 2B.

Wunder, W. (1963, 1964): Körpergestalt und Lebensweise der Fische. Allg. Fischztg. **88,** 22; **89,** 1–4.

Wunder, W. (1967): Die Abschnürung des Magens und seine Umbildung zu einer mit Flüssigkeit gefüllten Blase (Zyste) beim Aal (*Anguilla anguilla*). Allg. FischZtg. **92**(22).

Wunder, W. (1968): Hochrückigkeit beim Aal (*Anguilla anguilla* L.) bedingt durch Wirbelsäulenverkürzung. Biol. Zbl. **87,** 325–331.

Wunder, W. (1969): Wirbelanalyse mit Hilfe von Röntgenstrahlen bei einem Aal (*Anguilla anguilla* L.) mit Wellenkrümmung der Wirbelsäule (Plekospondylie). Biol. Zbl. **88,** 611–627.

Wundsch, H. H. (1916): Neue Beiträge zu der Frage nach dem Alter und Wachstum des Aales. Z. Fisch. **18,** 55–88.

Wundsch, H. H. (1953): Das Vorkommen von Aalen in "vorgeschrittenem Reifezustand" in einem märkischen Binnengewässer. Z. Fisch. **2,** 1–18.

Wundsch, H. H. (1954): Über Alter und Geschlechtsverhältnis bei dem zur Zeit im Handel befindlichen Satzaalmaterial. Dt. FischZtg. Radebeul **1,** 9–15.

Wurtz-Arlet, J. (1961): Verteilung und Ausnutzung der Aalbrut und der Speiseaale in Frankreich. Z. Fisch. **10,** 671–675.

Yamada, K., Yokote, M. (1975): Morphochemical analysis of mucosubstances in some epithelial tissues of the eel (*Anguilla japonica*). Histochem. **43,** 161–172.

Yamada, Y., Hama, K. (1972): Fine structure of the lateral-line organe of the common eel, *Anguilla japonica.* Z. Zellforsch. **124,** 454–464.

Yamada, Y., Zhano, H., Okamura, A., Tanaka, S., Horie, N., Mikawa, N., Utoh, T., Okah, H.P. (2001): Morphological and histological changes in the swim bladder during maturation of the Japanese eel. J. Fish Biol. **58,** 804–814

Yamagata, Y., Niwa, M. (1979): The toxicity of nitrite to eel. The Aquiculture **27**(1): 5–11.

Yamagata, Y., Niwa, M. (1982): Acute and chronic toxicity of ammonia to eel *Anguilla japonica*. Bull. Jap. Soc. Sci. Fish. **48**(2), 171–176.

Yamagishi, I., Nagumo, N., Oshiro, S., Suzuki, S., Takashi, O., Muraki, T. (1969): Electron microscopic observations on the absorptive epithelium of eel intestine. J. Med. Soc. Toho Univ. **16,** 101–108.

Yamagushi, K., Hashimoto, K., Matsuura, F. (1968): Studies on a blue-green serum-pigment of eel. IV. Seasonal variation of concentration of pigment in serum. Bull. Jap. Soc. scient. Fish. **34,** 826–835.

Yamaguti, S. (1952): Studies on the helminth fauna of Japan. Part **49,** Cestodes of fishes II. Acta Med. Okayama, **8,** 1–78.

Yamaguti, S. (1959): Systema helminthum. Vol. II. The cestodes of vertebrates. New York.

Yamaguti, S. (1963a): Systema helminthum. Vol. V. Acanthocephala. New York.

Yamaguti, S. (1963b): Parasitic Copepoda and Branchiura of fishes. New York.

Yamaguti, S. (1971): Synopsis of digenetic trematodes of vertebrates. Vols I & II, Tokyo.

Yamamoto, K., Morioka, T., Hiroi, O., Omori, M. (1974a): Artificial maturation of Japanese eels by the injection of salmonid pituitary. Bull. Jap. Soc. scient. Fish. **40,** 1–7.

Yamamoto, K., Omori, M., Yamauchi, J. (1974b): Oogenesis of the Japanese eel. Bull. Jap. Soc. scient. Fish **40,** 9–15.

Yamamoto, K., Yamauchi, K. (1974): Sexual maturation of Japanese eel and production of eel larvae in the aquarium. Nature **251,** 220–222.

Yamamoto, M., Ueda, K. (1978): Comparative morphology of fish olfactory epithelium: IV. Anguilliformes and Myctophiformes. Bull. Jap. Soc. Sci. Fish. **44,** 1207–1212.

Yang, H. N., Chen, H. C. (1996): The influence of temperature on the acute toxicity and sublethal effects of copper, cadmium and zinc to Japanese eel, *Anguilla japonica*. Acta Zool. Taiwan **7,** 29–38.

Yoshioka, M., Hamanda, K., Okazaki, T., Kajita, A., Shukuya, R. (1968): Hemoglobins from erythrocytes of the eel, *Anguilla japonica*. II. Further studies on the properties of the two hemoglobins. J. Biochem., Tokyo, **63,** 70–76.

Zacchei, A.M., Tavolaro, P. (1988): Lateral line system during the life cycle of *Anguilla anguilla* (L.). Boll. Zool. **3,** 145–153.

Ziepke, S. (1974): Fliegen schürfende Aale. Fisch u. Fang **15,** 323.

Index

Index Note: page numbers in bold denote figures and tables